Applied Condition Monitoring 22

The book series Applied Condition Monitoring publishes the latest research and developments in the field of condition monitoring, with a special focus on industrial applications. It covers both theoretical and experimental approaches, as well as a range of monitoring conditioning techniques and new trends and challenges in the field. Topics of interest include, but are not limited to: vibration measurement and analysis; infrared thermography; oil analysis and tribology; acoustic emissions and ultrasonics; and motor current analysis. Books published in the series deal with root cause analysis, failure and degradation scenarios, proactive and predictive techniques, and many other aspects related to condition monitoring. Applications concern different industrial sectors: automotive engineering, power engineering, civil engineering, geoengineering, bioengineering, etc. The series publishes monographs, edited books, and selected conference proceedings, as well as textbooks for advanced students.

** Indexing: Indexed by SCOPUS, WTI Frankfurt eG, SCImago

Ali Akrout · Moez Abdennadher · Nabih Feki ·
Mohamed Slim Abbes · Fakher Chaari ·
Mohamed Haddar
Editors

Advances in Acoustics and Vibration IV

Proceedings of the Fourth International Conference on Acoustics and Vibration (ICAV2022), December 19–21, 2022, Sousse, Tunisia

Springer

Editors
Ali Akrout
Faculty of Science
National School of Engineers of Sfax
Sfax, Tunisia

Nabih Feki
Higher Institute of Applied Sciences
and Technology
University of Sousse
Sousse, Tunisia

Fakher Chaari
National School of Engineers of Sfax
Sfax, Tunisia

Moez Abdennadher
Preparatory Engineering Institute of Sfax
Sfax, Tunisia

Mohamed Slim Abbes
National School of Engineers of Sfax
Sfax, Tunisia

Mohamed Haddar
National School of Engineers of Sfax
Sfax, Tunisia

ISSN 2363-698X ISSN 2363-6998 (electronic)
Applied Condition Monitoring
ISBN 978-3-031-34189-2 ISBN 978-3-031-34190-8 (eBook)
https://doi.org/10.1007/978-3-031-34190-8

This Springer imprint is published by the registered company Springer Nature Switzerland AG
The registered company address is: Gewerbestrasse 11, 6330 Cham, Switzerland

Preface

The Fourth International Conference on Acoustics and Vibration (ICAV'2022) was organized by the Tunisian Association of Industrial Acoustics and Vibration (ATAVI). The ICAV'2022 was held at Sousse, Tunisia, from 19 to 21 December 2022. The main objectives of this conference were to collect high-level research works in the fields of acoustics and vibrations and to encourage communication and collaboration between scientists from different countries. After three successful editions in 2016, 2018 and 2021 with proceedings published in the Springer Applied Condition Monitoring (ACM) book series, the organizers of this edition were honored by the presence of the following eminent professors who kindly agreed to share their knowledge with a hundred of participants, and held very interesting plenary session:

- Professor **Mabrouk Bentahar**, Roberval Laboratory, University of Technology of Compiegne, France.
- Professor **Pierre-Olivier MATTEI**, Deputy Director of Mechanical and Acoustics Laboratory (LMA), CNRS, Marseille, France.
- Professor **Tarek BELGASEM**, currently working at Honda R&D Americas Inc. as Materials Research Engineer, who is responsible for researching and developing new materials related to automotive components (body, interior, chassis, and/or powertrain).
- Professor **Weidong ZHU**, Department of Mechanical Engineering, University of Maryland, USA.
- Professor **Li CHENG**, Department of Mechanical Engineering, Director of the Hong Kong Polytechnic University.
- Professor **Abdelkhalek ELHAMI**, Mechanical Engineering Department, National Institute of Applied Sciences in Rouen (INSA de Rouen), France.

During the 3 days of the conference, about 100 attendees discussed several topics such as:

- Dynamics and vibration of structures and machinery,
- Nonlinear dynamics,
- Modeling and simulation,
- Numerical techniques,
- Fault diagnosis and prognosis,
- Vibration control of mechatronic systems,
- Fluid–structure interaction and computational vibro-acoustics,
- Vibration field measurements,
- Material behavior in dynamics.

All the 49 selected chapters from the presented papers, rigorously reviewed by referees, are included in this book. We would like to thank all persons who contributed to the success of the ICAV'2022: organizing committee, scientific committee and all participants which are from Tunisia, Algeria, France, USA, Hong Kong and Saudi Arabia. A lot of thanks are also addressed to Springer for continuous support of ICAV editions.

December 2022

Ali Akrout
Moez Abdennadher
Nabih Feki
Mohamed Slim Abbes
Fakher Chaari
Mohamed Haddar

Organization

Organizing Committee

Conference Chairs

Abdennadher Moez	IPEIS Sfax, Tunisia
Akrout Ali	ENIT Tunis, Tunisia

Honorays Chairs

Fakhfakh Taher	ENIS Sfax, Tunisia
Karra Chafik	IPEIS Sfax, Tunisia

Organizing Committee Chairs

Feki Nabih	ISSAT Sousse, Tunisia
Tounsi Dhouha	ISM Sfax, Tunisia
Dhafer Ghribi	Safran Transmission Systems, France

Scientific Committee Chairs

Abbes Mohamed Slim	ENIS Sfax, Tunisia
Taktak Mohamed	FSS Sfax, Tunisia
Bouaziz Slim	ENIS Sfax, Tunisia
Bouguecha Anas	ENIS Sfax, Tunisia

Program Chairs

Walha Lassaad	ENIS Sfax, Tunisia
Hentati Taissir	ENIS Sfax, Tunisia
Hammami Ahmed	ENIG Gabes, Tunisia
Trabelsi Hassen	ISSIG Gabes, Tunisia

Logistics Chairs

Barkallah Maher	ENIS Sfax, Tunisia
Beyaoui Moez	ENIS Sfax, Tunisia

Hentati Hamdi	ESSTHS, Tunisia
Djemal Fathi	ENIG Gabes, Tunisia

Publication Chairs

Chaari Fakher	ENIS Sfax, Tunisia
Jarraya Abdessalem	UJ, Saudi Arabia
Ben Souf Mohamed Amine	ENIS Sfax, Tunisia
Chaari Riadh	ISAMS Sfax, Tunisia

Contents

On the Vibrations of Functionally Gradient Porous Shells

Souhir Zghal[1,2](✉), Najah Joueid[2], Mouldi Chrigui[2], and Fakhreddine Dammak[3]

[1] High Institute of Applied Science and Technology of Gabes (ISSAT-Gabes), University of Gabes, Rue Omar ibn el Khattab - CP, 6072 Gabes, Tunisia
souhirzghal@yahoo.fr

[2] Modeling, Mechanics, Energetic and Materials (M2EM) Unit, National Engineering School of Gabes, University of Gabes, Av. Omar Ibn El Khattab, Zrig Eddakhlania, 6029 Gabes, Tunisia
mouldi.chrigui@enig.rnu.tn

[3] Laboratory of Electrochemistry and Environment (LEE), National Engineering School of Sfax, (ENIS), University of Sfax, Sfax, Tunisia
fakhreddine.dammak@enis.tn

Abstract. In this work, forced vibrations of functionally gradient (FG) porous shell subjected to dynamic load is presented. The mechanical properties of the FG shell are graded smoothly in the thickness direction and they are assumed to be porosity dependent. The governing equations are obtained using a finite element formulation with the consideration of the transverse shear strains. The resolution procedure is carried out via the Newmark's integration technique. The accuracy of the formulation is verified by comparing the present natural frequencies of the first four modes of vibrations with the existing studies in the literature. The effects of material compositions like porosity volume fraction, types of porosity distribution patterns and FG power index on time-deflection response of FG porous shell are also presented. The results show that the increase in porosity volume fraction and power FG index induces an augmentation in temporal deflections of the FG shell due to the reduction in the stiffness of the shell. Furthermore, the type of porosity distributions, namely even and uneven types, has a significant role on the dynamic behaviour of FG porous shell, because the uniform distribution of the pores, in even case, diminishes further the flexural rigidity of the shell compared to randomly distribution of the pores, in the case of uneven porosity distribution.

Keywords: functionally gradient · vibrations · finite element · time · deflections

1 Introduction

The determination of the dynamic responses of structures under impulse load is of great interest in vibration domain. Recently, with the progress in the material sciences and with the discovery of functionally gradient materials (Koizumi 1997), many researchers have been focused on the analysis of vibrational behaviour of such structures. To deal with this issue, (Chen et al. 2016) have presented free and forced vibrations of FG porous beams using Ritz method and Newmark algorithm. As well as, (Wattanasakulpong et al.

A. Akrout et al. (Eds.): ICAV 2022, ACM 22, pp. 1–7, 2023.
https://doi.org/10.1007/978-3-031-34190-8_1

2018) have reported the effect of porosity parameter and its forms of distributions on vibration behaviour of imperfect (FG) porous beams using the Chebyshev collocation approach. In another work, (Wattanasakulpong and Mao. 2015) presented the dynamic responses of (FG) porous beams under classical and non-classical boundary conditions within the Chebyshev collocation method. Free vibration analysis of perfect (FG) beams using a high-order beam model is presented in (Zghal et al. 2020). A separation of the displacement and stresses fields is achieved within a mixed formulation. As well as, post-buckling analysis of perfect (FG) structures is achieved using a modified first-order shear deformation finite element model (Zghal et al. 2022a). For porous structures, (Trinh and Kim, 2019, Trinh et al. 2020) presented various analysis on mechanical behaviour of (FG) porous plates and shells using either a refined shear deformation theory or a semi-analytical approach. The effects of porosity coefficient, types of distributions and gradient index are investigated and explored. Recently, transient vibration response of (FG) porous plates is presented by (Zghal et al. 2022b).

From this brief review, it can be concluded that dynamic analysis of FG porous shells is a new axis and few papers are dealing within. For that, we propose in this paper, a finite element formulation for the study of the dynamic behaviour of FG porous spherical shell subjected to suddenly impulse load, where even and uneven porosity distributions are considered. The effect of power FG index is also examined. The obtained results show that the porosity parameter and the types of distributions patterns play an important role in the prediction of temporal responses of spherical shells. Consequently, the volume fraction of porosity constitutes a crucial parameter for vibration control of such FGM structures.

2 Finite Element Formulation

In this section, the necessary relations for elasto-dynamic analysis of FG shell including porosities are presented. The inner and outer thickness, the middle surface radius and the span angle of the spherical shell are denoted by h, H, R_s and β respectively. The FG shell is ceramic-rich at its top surface and metal-rich at its bottom surface as shown in Fig. 1. The FG shell is made of porous material, and according to a power law function, the elastic Young's modulus can be given, for two types of porosity distributions, as:

- Even porosity pattern

$$E(z, \xi) = (E_m - E_c)\left(\frac{1}{2} + \frac{z}{h}\right)^p + E_c - \frac{\xi}{2}(E_m + E_c) \tag{1}$$

- Uneven porosity pattern

$$E(z, \xi) = (E_m - E_c)\left(\frac{1}{2} + \frac{z}{h}\right)^p + E_c - \frac{\xi}{2}(E_m + E_c)\left(1 - \frac{2|z|}{h}\right) \tag{2}$$

where, p is the power FG index, that determines the variation profile of material properties across the shell thickness. ξ is the porosity volume fraction or coefficient which vary from zero to 1 ($0 \le \xi \prec 1$). E_m and E_c stand with metal and ceramic Young's moduli.

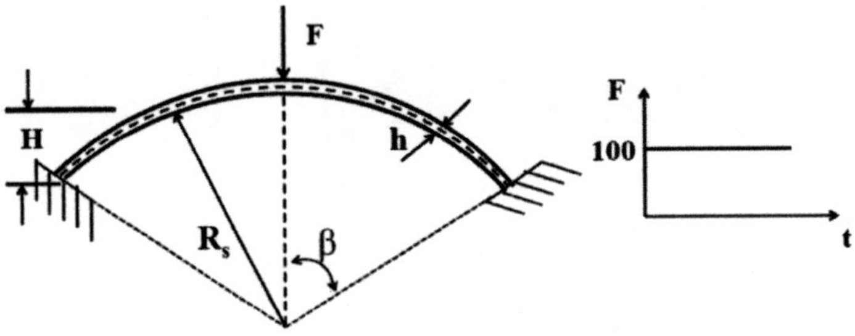

Fig. 1. Geometry of the shell subjected to impulse load

The strain energy of the shell element is computed using the Lagrangian weak form of equilibrium as

$$G = \int_A \delta \Sigma^T . \mathbf{R} dA - G_{ext} = 0 \tag{3}$$

where $\mathbf{\Sigma}$ and $\mathbf{R}$ respresent the generalized resultant of strains and stresses vectors, respectively. G_{ext} is the virtual work of external load.

$$\mathbf{R} = \begin{bmatrix} \mathbf{N} \\ \mathbf{M} \\ \mathbf{T} \end{bmatrix}, \quad \Sigma = \begin{bmatrix} \mathbf{e} \\ \chi \\ \gamma \end{bmatrix} \tag{4}$$

In addition, the strain-stress relationship can be expressed as:

$$\begin{bmatrix} \mathbf{N} \\ \mathbf{M} \\ \mathbf{T} \end{bmatrix} = \begin{bmatrix} \mathbf{H}_{11} & \mathbf{H}_{12} & 0 \\ \mathbf{H}_{12} & \mathbf{H}_{22} & 0 \\ 0 & 0 & \mathbf{H}_{33} \end{bmatrix} \begin{bmatrix} \mathbf{e} \\ \chi \\ \gamma \end{bmatrix} \tag{5}$$

where $\mathbf{H_{ij}}$ is the elasticity matrix and the stiffness components are expressed as:

$$(\mathbf{H}_{11}, \mathbf{H}_{12}, \mathbf{H}_{22}) = \int_{-h/2}^{h/2} \left(1, z, z^2\right) \mathbf{H} dz; \quad \mathbf{H}_{33} = \int_{-h/2}^{h/2} f(z)^2 \, \mathbf{H}_\tau dz \tag{6}$$

$\mathbf{H}$ and $\mathbf{H}_\tau$ are the in-plane and out-of plane elastic sub-matrices of the FG shell. *f(z)* is the shear function that allows a parabolic distribution of the transverse shear stresses, given as:

$$f(z) = 5/4\left(1 - 4(z/h)^2\right) \tag{7}$$

2.1 Displacement Field for the FG Shell

According to the first order shear deformation theory (FSDT), the displacement and strain fields can be written as:

$$\mathbf{X}_q\left(\xi^1, \xi^2, z\right) = \mathbf{X}_p\left(\xi^1, \xi^2\right) + z\mathbf{D}\left(\xi^1, \xi^2\right) \qquad \mathbf{x}_q = \mathbf{x}_p + z\mathbf{d} \tag{8}$$

$$\mathbf{e} = \begin{Bmatrix} e_{11} \\ e_{22} \\ 2e_{12} \end{Bmatrix}, \boldsymbol{\chi} = \begin{Bmatrix} \chi_{11} \\ \chi_{22} \\ 2\chi_{12} \end{Bmatrix}, \boldsymbol{\gamma} = \begin{Bmatrix} \gamma_1 \\ \gamma_2 \end{Bmatrix} \tag{9}$$

where: $z \in [-h/2, h/2]$ is the thickness variable and **d** is the shell director vector. **e**, $\boldsymbol{\chi}$ and $\boldsymbol{\gamma}$ are the membrane, the bending and the shear components.

2.2 Governing Equations and Solution Procedure

To derive the equations of motion, the Variational principle of Hamilton is used and the mass and stiffness matrices are constructed, at nodal level I, as

$$M^I = \left[m_{1I}\ m_{1I}\ m_{1I}\ m_{2I}\ m_{2I} \right]_{diag} \tag{10a}$$

$$m_{kI} = \frac{\int_A \bar{\rho}_k dA \int_A \bar{\rho}_k N^T N^T dA}{\int_A \left(\sum \bar{\rho}_k N^T N^T\right) dA}, k = 1, 2 \tag{10b}$$

where N^T are the standard isoparametric shape functions for a four-node shell finite element and $\overline{\rho}_k$ are the mass densities of the FG shell, defined as:

$$\overline{\rho}_k = \int_A \rho(z)(1, f^2) dA \tag{11}$$

For the stiffness matrix is given as:

$$\mathbf{K} = \int_A \mathbf{B}^T \mathbf{H}_{ij} \mathbf{B} dA \tag{12}$$

With **B** is the strain-displacement matrix and $\mathbf{H}_{ij}$ is the elasticity matrix as mentioned previously. Finally, the governing equations are obtained as:

$$\mathbf{M}\ddot{\mathbf{U}}_n + \mathbf{K}\mathbf{U}_n = \mathbf{F}(\mathbf{t}) \tag{13}$$

In Eq. (13), the pairs (**M**, **K**) and ($\mathbf{U_n}$, **F(t)**) refer to global mass, stiffness matrices and the nodal load vector and the external applied load, respectively. The resolution of this equation is carried out by the Newmark's algorithm (Newmark, 1959; Chang, 2008) where low computational cost, stability and accuracy are the main features of this method and it is an efficient and applicable numerical technique for dynamic problems. To obtain stable and convergent results, the constant average acceleration method is employed, in this paper, for pulse distributed load ($\gamma = 0.5$; $\beta = 0.25$).

3 Results for FG Spherical Shell with Porosities

Results are presented by two evaluations. The first one presents a validation of the formulation in the prediction of natural frequencies of spherical shell and the second evaluation shows the influence of porosity volume fraction, type of distribution patterns and power FG index on time-deflection responses of FG porous shells.

3.1 Model Validation

The dimensionless natural frequency parameter $\tilde{\omega} = \omega R\sqrt{\rho/E}$ of the first four modes of vibration for an isotropic clamped (CC) spherical shell ($R = 1$ m, $\nu = 0.3$) is compared with those existing studies as shown in Table 1. A good agreement is noted with the finite element method (FEM) solutions (Singh and Mirza 1985) and the semi-analytical approach (Du et al. 2019). Consequently, the proposed model can provide accurate natural frequencies of the isotropic spherical shells.

Table 1. Dimensionless natural frequencies of the isotropic spherical shell

Source	Vibration modes			
	1	2	3	4
Current	1.0820	1.2942	1.6015	1.9340
FEM (Singh and Mirza, 1985)	1.0788	1.2930	1.5696	1.9292
Semi-analytic (Du et al. 2019)	1.0790	1.2931	1.5700	1.9300

3.2 Parametric Results

In this section, parametric analysis is performed. The material of the structure are FGMs and they are porosity-dependent. Indeed, the outer surface is ceramic ($Al_2\,O_3$) and the inner surface is metal (Al). The mechanical properties of materials are: Young's modulus ($E_m = 70$ GPa; $E_c = 380$ GPa; density; $\rho^c = 3960$ kg/m^3; $\rho^m = 2702$ kg/m^3; Poisson's ratio $\upsilon^m = \upsilon^c = 0.3$). Figures 2–3 show the time-deflection responses of clamped (CC) FG porous spherical shells for different porosity volume fractions, types of distributions and also with different power FG indexes.

In Fig. 2, it is observed that maximum deflection is obtained with high value of porosity volume fraction ($\xi = 0.4$) and for even type of porosity distribution, whereas the minimum deflection is obtained for uneven porosity pattern with a lower value of porosity volume fraction ($\xi = 0$). This is because, increasing the porosity volume fraction leads to decreasing the flexural rigidity of the shell and so the FG spherical shell becomes more flexible. As well as, it can be seen that the oscillations are well stable and periodic in the time interval [0–0.02] s and the steady state response is well established around the equilibrium position.

On other hand, it should be noted that the type of porosity distributions has a significant impact on temporal responses of FG porous shells. In fact, since the even type of porosity induces a uniform spread of the pores along the thickness direction of the shell, while the uneven type shows a concentration of the pores in the mid-surface and a vanish in the extremes surfaces the stiffness of the shell becomes more weaker with even type in comparison with uneven type and the even FG porous shell presents, hence, more deflections compared to uneven one.

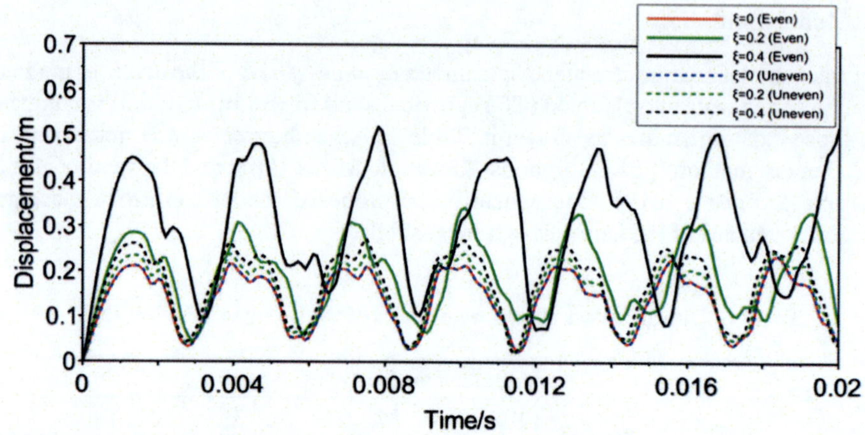

Fig. 2. Transient vibration response of FG porous spherical shell (p = 1) subjected to suddenly impulse load.

Figure 3 shows that the power FG index has a high impact on time-deflection responses of FG porous shells ($\xi = 0.2$) when moving from ceramic (p = 0) to metal (p = 2) material composition. It can be concluded that by increasing the FG power index, the stiffness of the FG porous shell tends to decrease and eventuates to an augmentation in deflection response. Therefore, more the FG shell is reinforced by metal portions, more it deflects easily.

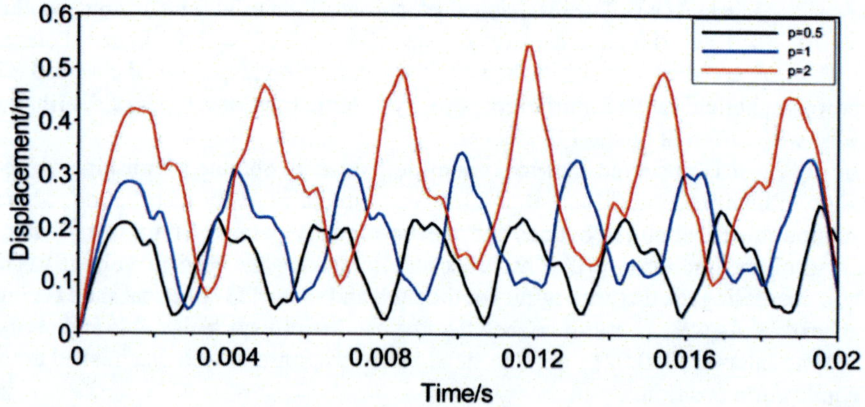

Fig. 3. Transient vibration response of FG porous spherical shell ($\xi = 0.2$, even) with various power FG indexes.

4 Conclusion

In this study, time-deflection responses of FG porous spherical shell subjected to an impulse sudden load was presented using the finite element formulation taking into account the transverse shear strains. The material properties of the FG porous shell are supposed to be porosity dependent and vary continuously along the thickness direction according to a power-law function. Within the variational principle of Hamilton's, the governing equations were obtained and the resolution procedure was performed via the Newmark's algorithm. Accuracy of the formulation was confirmed via a comparison evaluation of the present natural frequencies results with literature ones. The effects of different parameters like porosity volume fraction, types of distributions and FG power index on time-deflection analysis of FG porous shell was carried out. The results were shown that these material properties play an important role on elasto-dynamic analysis of FG porous shell. As perspective of the present study, the damping effect will be introduced in the formulation in the future works to show its impact on dynamic behaviour of FG porous shells.

References

Koizumi, M.: FGM activities in Japan. Compos. B **28**, 1–4 (1997)

Chen, D., Kitipornchai, S., Yang, J.: Free and forced vibrations of shear deformable functionally graded porous beams. Int. J. Mech. Sci. **108–109**, 14–22 (2016)

Wattanasakulpong, N., Chaikittiratana, A., Pornpeerakeat, S.: Chebyshev collocation approach for vibration analysis of functionally graded porous beams based on third-order shear deformation theory. Acta Mech. Sin. **34**, 1124–1135 (2018)

Wattanasakulpong, N., Mao, Q.: Dynamic response of timoshenko functionally graded beams with classical and non-classical boundary conditions using chebyshev collocation method. Compos. Struct. **119**, 346–354 (2015)

Trinh, M.C., Kim, S.E.: A three variable refined shear deformation theory for porous functionally graded doubly curved shell analysis. Aerosp. Sci. Technol. **94**, 105356 (2019)

Trinh, M.C., Mukhopadhyay, T., Kim, S.E.: A semi-analytical stochastic buckling quantification of porous functionally graded plates. Aerosp. Sci. Technol. **105**, 105928 (2020)

Zghal, S., Dammak, F.: Vibrational behavior of beams made of functionally graded materials by using a mixed formulation. Proc. Inst. Mech. Eng. Part C: J. Mech. Eng. Sci. **234**(18), 3650–3666 (2020)

Zghal, S., Trabelsi, S., Dammak, F.: Post-buckling behavior of functionally graded and carbon-nanotubes based structures with different mechanical loadings. Mech. Based Des. Struct. Mach. **50**(9), 2997–3039 (2022)

Zghal, S., Trabelsi, S., Dammak, F.: Transient response of functionally graded porous plate. In: Ben Amar, M., Bouguecha, A., Ghorbel, E., El Mahi, A., Chaari, F., Haddar, M. (eds.) A3M 2021. LNME, pp. 150–155. Springer, Cham (2022). https://doi.org/10.1007/978-3-030-84958-0_16

Newmark, N.M.: A method of computation for structural dynamics. J. Eng. Mech. Div. ASCE **85**, 67–94 (1959)

Chang, Y.S.: Accuracy evaluation of Newmark method referring to theoretical solutions. Earthquake Eng. **12**, 1–16 (2008)

Singh, A.V., Mirza, S.: Asymmetric Modes and Associated Eigenvalues for Spherical Shells. J. Pressure Vessel Technol. **107**(1), 77–82 (1985)

Du, Y., Huo, R., Pang, F., Li, S., Huang, Y., Zhang, H.: Free vibration of spherical cap subjected to various boundary conditions. Adv. Mech. Eng. **11**(9), 1–12 (2019)

Effects of the Mechanical Characteristics on the Dynamic Behavior of a Bolted Structure Under Transient Excitation

Ahmed Ben Saidane, Charfeddine Mrad(✉), and Jamel Chakhari

Laboratory of Applied Mechanics and Engineering (LMAI), National Engineering School of Tunis (ENIT), University of Tunis el Manar (UTM), BP 37, Le Belvedere, 1002 Tunis, Tunisia
charfeddine.mrad@enit.rnu.tn, jamel.chakhari@enit.utm.tn

Abstract. A prismatic structure composed of two mechanical parts assembled by a bolt is studied. The structure is clamped-free and is excited longitudinally by a transient force applied at its free end. A simplified model is used to analyze the dynamic behavior of the structure when varying its mechanical characteristics. The model formulation gives a system of nonlinear equations. The differential equations resolution is conducted using the Euler method, to determine the structure dynamic response. The study focuses firstly on the effect of the clamped part characteristics on the dynamic response of the structure, the clamped part mass, stiffness, and damping are varied and the structure dynamic response is noted. Secondly, the study focuses on the effect of the free part characteristics on the dynamic response of the structure, the free part mass, stiffness, and damping are varied and the structure dynamic response is recorded. The effects of the parts mechanical characteristics on the bolted structure dynamic behavior are identified, which helps to optimize the structure design.

Keywords: Bolted structure · Transient excitation · Dynamic behavior · Nonlinear modeling · Mechanical characteristics

1 Introduction

Most of the mechanical structures in different fields are composed of several parts assembled by mechanical joints. Among the various mechanical joints are bolted joints. Bolted joints have several advantages such as strength, safety, and durability. However, their inherent dynamics are too complex to be easily modeled and analyzed.

Many researchers investigated bolted structures. Lecomete et al. (2013) have developed an analytical model able to account for load transfers in a two-bolt aluminum composite double lap joint. Jalali et al (2007) established a detailed parameter model to characterize the bolted joint contact interface by the force-state mapping method. Liao et al (2016) developed an analytical model of a bolted structure of nonlinear dynamic characteristics under transient excitation. Jamia et al (2021) constructed a detailed model of bolted flanges to capture the frictional behavior in the contact interface.

A. Akrout et al. (Eds.): ICAV 2022, ACM 22, pp. 8–16, 2023.
https://doi.org/10.1007/978-3-031-34190-8_2

In finite element modeling of bolted assemblies, Verwaerde et al (2020) presented a nonlinear finite element model to represent the quasi-static behavior of a bolted connection. Jie et al. (2022), Tong and Yongbo (2022), used the identification connector parameters approach to represent the behavior of the joint at the bolted connection, which allowed to account for the friction between the various assembled elements.

In this context, we propose to study the effects of the mechanical characteristics on the dynamic behavior of a bolted prismatic structure. The structure is clamped-free, and is subjected to an impact force. The nonlinear dynamic model used integrates a cubic stiffness to characterize the contact interface.

2 Structure Presentation

2.1 Physical Model

The bolted structure to study is shown in Fig. 1. The structure is clamped-free and is subjected to a transient excitation applied longitudinally at the free end.

Figure 2 specifies the dimensions of the studded structure. A bolt, in iron alloy, is used to join the two parts, clamped and free, together. The clamped part (1) is in iron alloy while the free part (2) is in aluminum alloy.

Fig. 1. Structure geometry

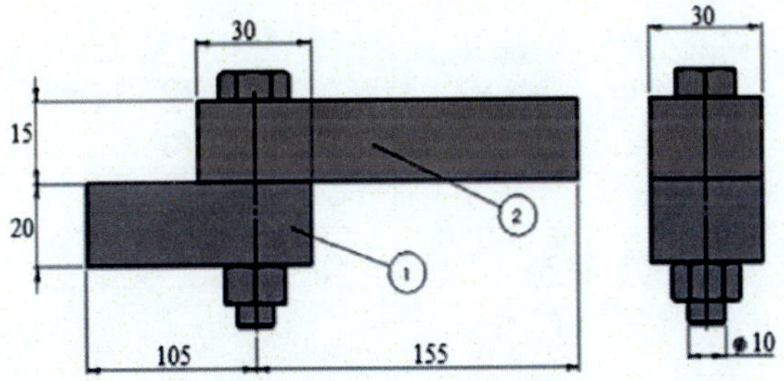

Fig. 2. Structure dimensions

2.2 Mathematical Model

Figure 3 shows the model used to study the structure. It is a discrete model with two degrees of freedom. m_1 and m_2 are respectively the masses of the clamped and the free parts. k_1 and k_2 are respectively the linear rigidities of the clamped and the free parts. c_1 and c_2 are respectively the viscous damping coefficients of the clamped and the free parts. And k_3 is a nonlinear stiffness to account for the contact interface of the two parts.

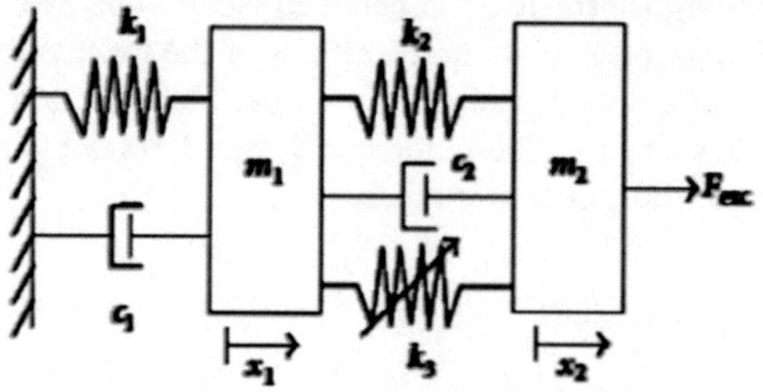

Fig. 3. Structure dynamic model

The equations of motion are as follows:

$$m_1\ddot{x}_1 + c_1\dot{x}_1 + k_1x_1 + c_2(\dot{x}_1 - \dot{x}_2) + k_2(x_1 - x_2) + k_3(x_1 - x_2)^3 = 0 \tag{1}$$

$$m_2\ddot{x}_2 + c_2(\dot{x}_2 - \dot{x}_1) + k_2(x_2 - x_1) + k_3(x_2 - x_1)^3 = F_{exc} \tag{2}$$

To solve the differential equations, the Euler method is applied using MATLAB software. The reference parameters are noted in Table 1. The assembly parameters are according to the physical model, the bolting stress is of 200 MPa.

Table 1. Reference parameters.

Parameters	Units	Values
m_1; m_2	Kg	$m_1 = 0.56$; $m_2 = 0.27$
c_1; c_2	Ns/m	$c_1 = 0.48$; $c_2 = 0.15$
k_1; k_2	N/m	$k_1 = 1.05 \times 10^6$; $k_2 = 0.2 \times 10^6$
k_3	N/m^3	$k_3 = 0.05 \times 10^6$
F_{exc}	N	100
Step time	s	2×10^{-5}

3 Assembled Parts Effects

3.1 Clamped Part Effect

The effect of the mechanical characteristics of the clamped part (mass, rigidity, and damping) on the dynamic behavior of the studied structure is examined. The tested values are around the reference values, inferior and superior to them.

Figure 4 shows the effect of m_1 on the clamped part response, while Fig. 5 illustrates the effect of m_1 on the free part response. Figure 6 shows the effect of k_1 on the clamped part response, while Fig. 7 illustrates the effect of k_1 on the free part response. And Fig. 8 shows the effect of c_1 on the clamped part response, while Fig. 9 illustrates the effect of c_1 on the free part response.

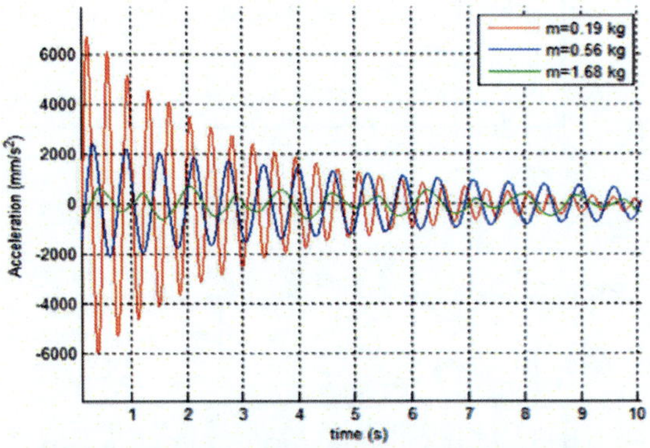

Fig. 4. Effect of m_1 on the clamped part response

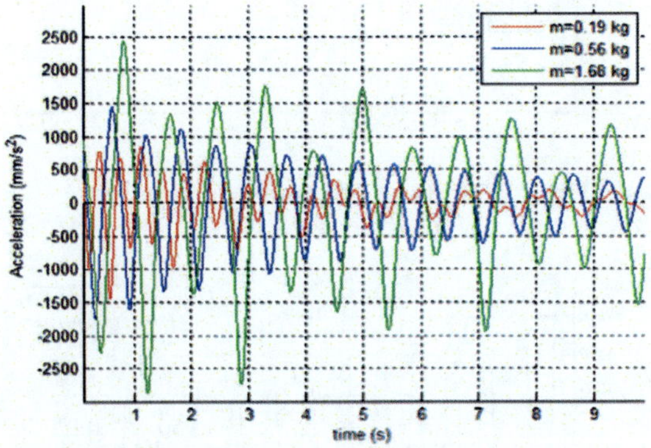

Fig. 5. Effect of m_1 on the free part response

The increase of m_1 decreases the acceleration amplitude of the clamped part, but increases the acceleration amplitude of the free part. The increase of k_1 increases the acceleration amplitude of the clamped part, and increases the acceleration amplitude of the free part. And the increase of c_1 decreases the acceleration amplitude of the clamped part, and decreases the acceleration amplitude of the free part. Besides, the nonlinearity effect is seen in both parts accelerations.

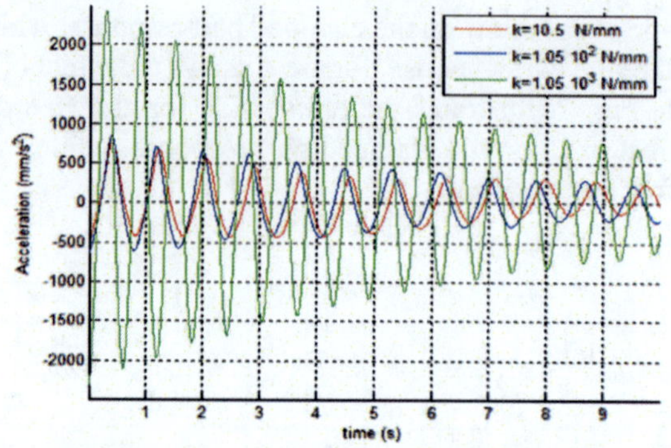

Fig. 6. Effect of k_1 on the clamped part response

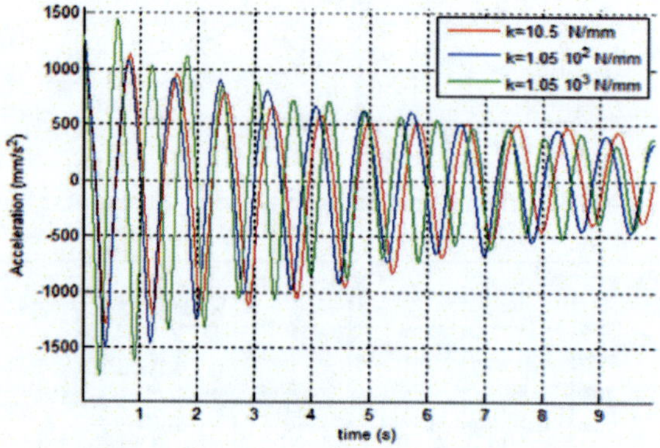

Fig. 7. Effect of k_1 on the free part response

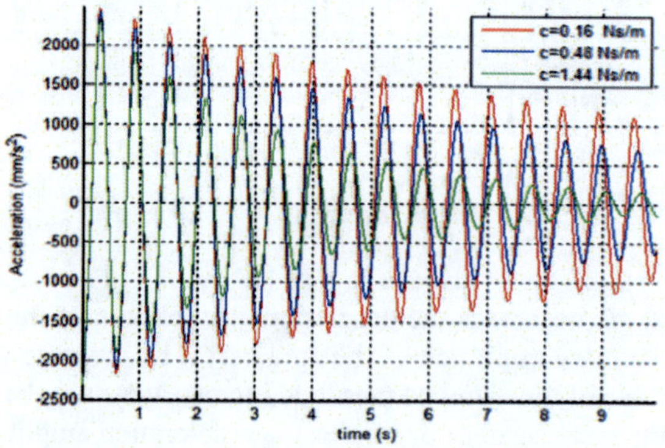

Fig. 8. Effect of c_1 on the clamped part response

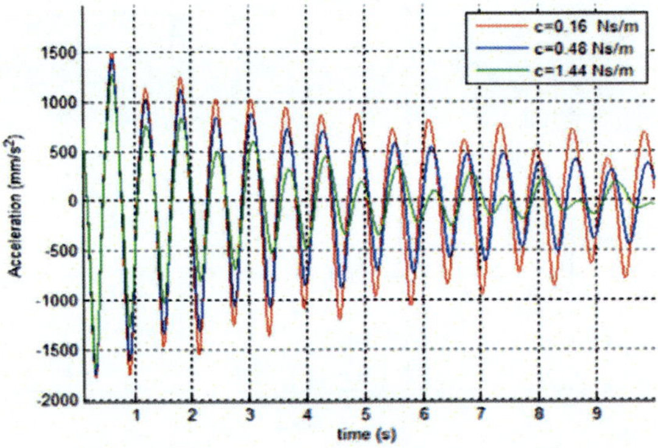

Fig. 9. Effect of c_1 on the free part response

3.2 Free Part Effect

The effect of the mechanical characteristics of the free part (mass, rigidity, and damping) on the dynamic behavior of the studied structure is investigated. The tested values are around the reference values, inferior and superior to them.

Figure 10 shows the effect of m_2 on the clamped part response, while Fig. 11 illustrates the effect of m_2 on the free part response. Figure 12 shows the effect of k_2 on the clamped part response, while Fig. 13 illustrates the effect of k_2 on the free part response. And Fig. 14 shows the effect of c_2 on the clamped part response, while Fig. 15 illustrates the effect of c_2 on the free part response.

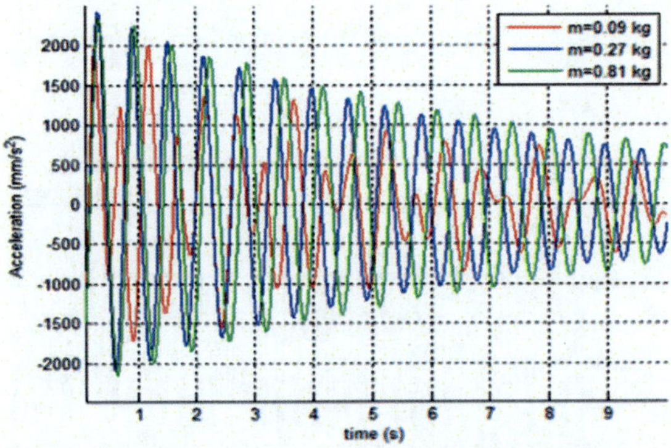

Fig. 10. Effect of m_2 on the clamped part response

The increase of m_2 increases the acceleration amplitude of the clamped part, but decreases the acceleration amplitude of the free part. The increase of k_2 increases the acceleration amplitude of the clamped part, and increases the acceleration amplitude of the free part. And the increase of c_2 decreases the acceleration amplitude of the clamped part, and decreases the acceleration amplitude of the free part. In addition, the nonlinearity effect is seen in both parts accelerations.

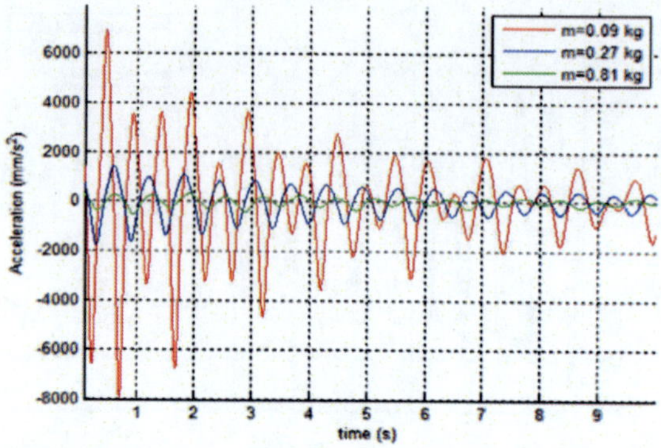

Fig. 11. Effect of m_2 on the free part response

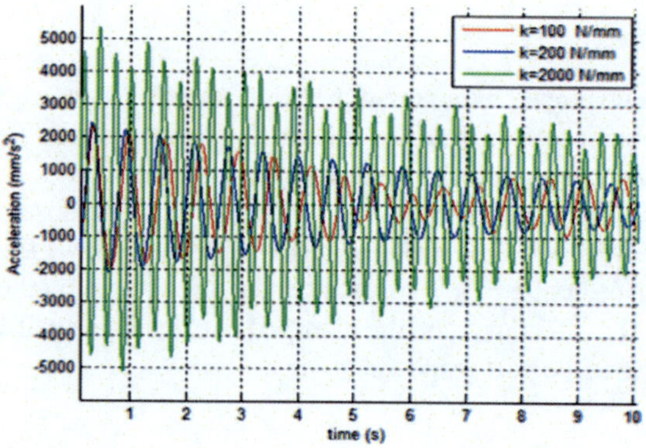

Fig. 12. Effect k_2 on the clamped part response

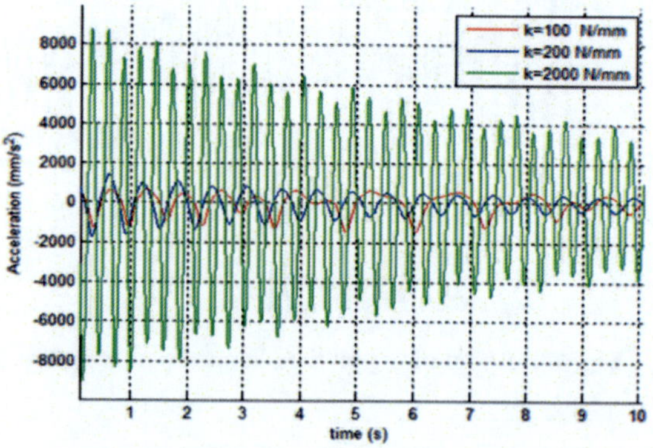

Fig. 13. Effect of k_2 on the free part response

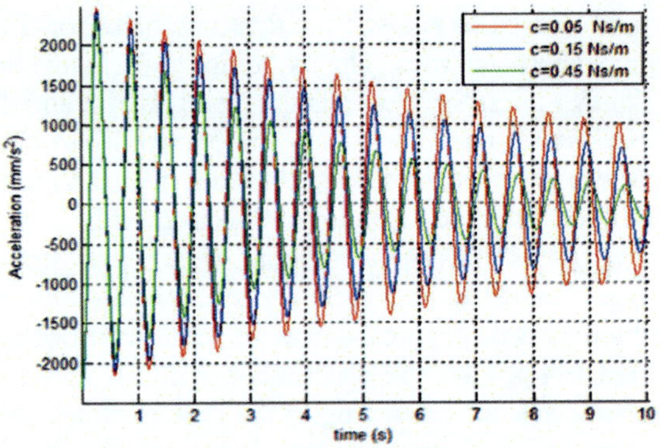

Fig. 14. Effect c_2 on the clamped part response

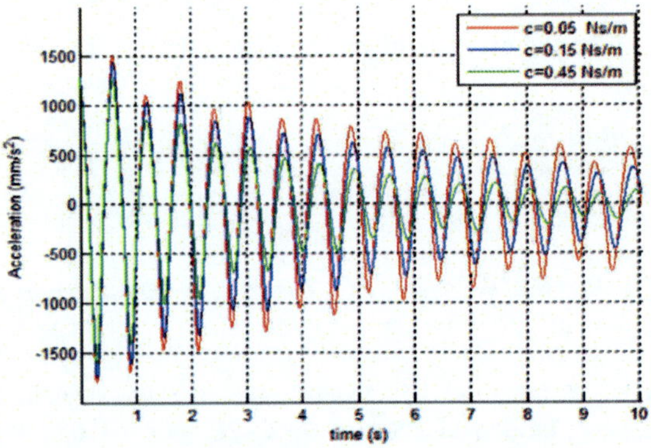

Fig. 15. Effect of c_2 on the free part response

4 Discussion

The mechanical characteristics variation of the clamped and the free parts have similar effects in stiffness and in damping, on the parts accelerations, but have different effect in mass, the clamped and the free parts behave inversely when varying masses. Besides, the nonlinearity effect is seen in both parts responses.

The obtained results show that the structure dynamic response depends on the mechanical characteristics of the clamped and the free parts, and not only on the assembly characteristics (bolting stress and contact interface) and the excitation force.

5 Conclusion

A prismatic structure composed of two mechanical parts assembled by a bolt was studied. The structure is clamped-free and is excited longitudinally by a transient force applied at its free end.

A nonlinear model is used to analyze the dynamic behavior of the structure when varying its mechanical characteristics. The dynamic model integrates a cubic stiffness characterizing the contact interface, between the clamped part and the free part.

The study focused firstly on the effect of the clamped part characteristics on the dynamic response of the structure. Secondly, the study focused on the effect of the free part characteristics on the dynamic response of the structure.

The mechanical characteristics of the structure were found decisive for its dynamic behavior, and the nonlinearity effect is clearly seen in the structure responses. The effects of the assembled parts mechanical characteristics on the bolted structure dynamic behavior are identified, to optimize the structure design.

The assembly characteristics consisting of the bolting stress, the contact interface, and the excitation force, are thus not the only parameters to be taken into account when studying the dynamic behavior of bolted structures, the mechanical characteristics of the assembled parts are also to be considered.

Acknowledgements. The authors gratefully acknowledge the helpful comments and suggestions of the reviewers.

References

Lecomte, J., Bois, C., Wargnie, H., Wahl, J.C., Bautista, A.: Influence of geometric defects on the behavior of bolted metal-composite assemblies. In: 21st French Congress of Mechanics, France (2013)

Jalali, H., Ahmadian, H., Mottershead, J.E.: Identification of nonlinear bolted lap-joint parameters by force-state mapping. Int. J. Solids Struct. **44**(25–26), 8087–8105 (2007)

Liao, X., Jianrun, Z., Xiyan, X.: Analytical model of bolted joint structure and its nonlinear dynamic characteristics in transient excitation. Shock Vibr. **2016**, 1–11 (2016)

Jamia, N., Jalali, H., Taghipour, J., Friswell, M.I., Khodaparast, H.H.: An equivalent model of a nonlinear bolted flange joint. Mech. Syst. Signal Process. **153**, 107507 (2021)

Verwaerde, R., Guidault, P.-A., Boucard, P.-A.: A nonlinear finite element connector for the simulation of bolted assemblies. Comput. Mech. **65**(6), 1531–1548 (2020). https://doi.org/10.1007/s00466-020-01833-1

Jie, Y., Yekai, S., Christoph, S., Loic, S.: Computation of damped nonlinear normal modes for large-scale nonlinear systems in a self-adaptive modal subspace. Mech. Syst. Signal Process. **162**, 108082 (2022)

Tong, Z., Yongbo, P.: Efficient reliability analysis based on deep learning-enhanced surrogate modeling and probability density evolution method. Mech. Syst. Signal Process. **162**, 108064 (2022)

Experimental Investigation on the Dynamics of a Hydraulic Dual Tube Automobile Strut Damper

Amina Ben Abdelwahed[1], Jamel Chakhari[1], Charfeddine Mrad[1](✉), and Lotfi Mezghani[2]

[1] Laboratory of Applied Mechanics and Engineering (LMAI), National Engineering School of Tunis (ENIT), University of Tunis El Manar (UTM), BP 37, Le Belvedere 1002, Tunis, Tunisia
jamel.chakhari@enit.utm.tn, charfeddine.mrad@enit.rnu.tn
[2] Leading Technology in Mechanics (LTM), ZI El Agba, BP 233, Denden, 2011 Tunis, Tunisia

Abstract. The dynamic behavior of a hydraulic dual tube automobile strut damper is studied. The purpose is to explore the piston and cylinder valves softening on the damper force response for different excitation velocities. Indeed, an experimental investigation is conducted on a dual tube hydraulic damper for automobile front axle. A test bench is used to load the strut damper at different reference velocities, and to measure its dynamic response. The dynamic response diagrams are obtained in compression and in extension, for different reference configurations of the piston and cylinder valves. The results depend on the mechanical and the hydraulic parameters of the strut damper, but also on the testing conditions. First, the experimental results are analyzed, the force-displacement responses are examined to conclude on the calibration forces. Second, the calibration forces are checked, a force-velocity diagram is established to conclude on the force evolution with velocity. The effects of the piston and cylinder valves configuration and of the excitation velocity are determined, which helps to tune the damper design to the automobile requirements.

Keywords: Experimental investigation · Strut damper · Dynamic behavior · Excitation velocity · Valves configuration

1 Introduction

The suspension system of an automobile is composed of springs, dampers, and mechanical links to connect the axles to the chassis. Strut and shock dampers play a major role in the safety and the comfort for an automobile, they limit chassis oscillatory motion and reduce wheels bouncing.

Dual tube dampers present the advantage of being shorter than monotube dampers. The required damping characteristics depend on the automobile weight, axles, and springs.

Several works were carried out on hydraulic automobile dampers. Lang (1977) performed a study to obtain an understanding of the frequency-dependent behavior of the

A. Akrout et al. (Eds.): ICAV 2022, ACM 22, pp. 17–23, 2023.
https://doi.org/10.1007/978-3-031-34190-8_3

automotive damper. Skagerstrand (2014) proposed a force-velocity based model of an automobile damper, requiring parameters tuning. Chahine (2011) developed a damper for rally applications, which allows high ground grip together with good chassis control. Gallsjo and Johansson (2012) studied and simulated the behavior of a CES (Continuously controlled Electronic Suspension) valve, to estimate the characteristics of the valves. Ferdek and Luczko (2012) developed a physical-mathematical model for a dual tube hydraulic damper, the effect of the excitation amplitude and frequency as well as the fluid flow parameters were examined, to conclude on the damping force. Konieczny (2016) determined the damping characteristics of an automotive damper entailing the stroke value, to reach force-displacement and force-velocity diagrams.

In this context, we propose to study experimentally the dynamic response of a dual tube hydraulic front axil automobile damper, in compression and in extension, at different excitation velocities, and for various valves configurations.

2 Strut Damper

The strut damper to study is a dual tube hydraulic damper. It consists mainly of a piston, two coaxial tubes, and two valves, Fig. 1.

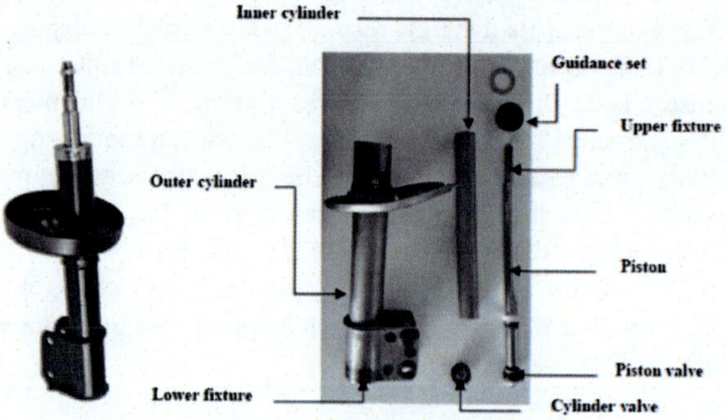

Fig. 1. The strut damper

The lower chamber of the piston is the work chamber, the upper chamber of the piston is the reserve chamber, and the chamber between the two cylinders is the compensation chamber. The compensation chamber is filled with oil to its 2/3, 1/3 of its volume is thus filled with air. The piston and cylinder valves present wholes throttled by spring washers.

3 Operation Phases

The operation phases are of two, compression when the strut damper is loaded, and relaxation when the strut damper is unloaded, Fig. 2.

During relaxation (blue arrows), the fluid flows from the compensation and reserve chambers to the work chamber, while during compression (red arrows), the fluid flows inversely, from the work chamber to the compensation and reserve chambers, Fig. 3

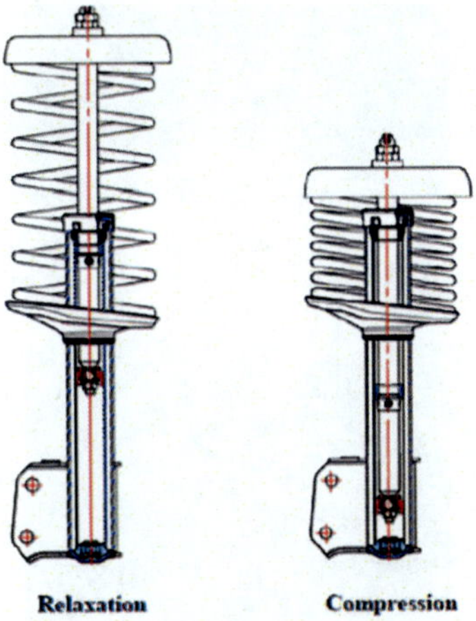

Fig. 2. The operation phases

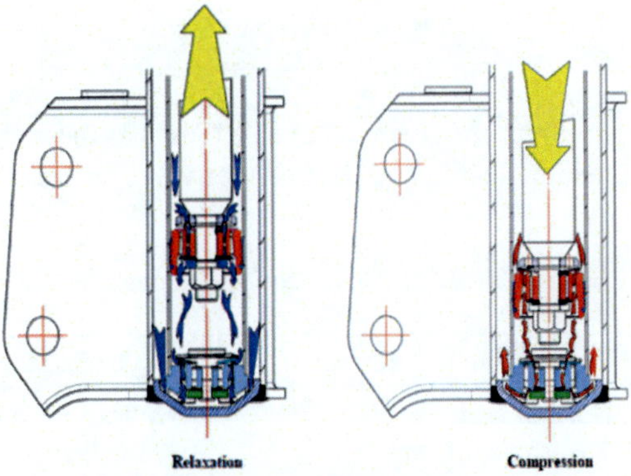

Fig. 3. The fluid flow

4 Test Bench

The test bench is a MTS Damper Test System, it is used to explore the damping force evolution with the excitation velocity and the piston and cylinder valves configuration.

The damper is fixed at one end, and is driven in cyclic displacement at the other end, Fig. 4. A load sensor is installed on the fixed end, and a data acquisition system records the force as a function of the displacement, to give force-displacement diagrams.

5 Results and Discussion

The tests are conducted at three reference excitation velocities, and for three reference piston and cylinder valves configurations. The valves configurations are to explore the valves stiffness softening, while the reference velocities are to investigate the excitation velocity increasing.

Fig. 4. The test bench

The valves configurations are given in Table 1. Test 1 is the reference test, the piston valve stiffness is softened in Test 2, and the cylinder valve stiffness is softened in Test 3.

Table 1. Valves configurations.

Test	Piston valve	Cylinder valve
1	Notches seal 3 Washers 0,1 mm 1 Washer 0,2 mm Strangler Membrane 2 Holes base	5 Washers 0,1 mm Membrane 6 Holes base
2	Notches seal 2 Washers 0,2mm Strangler Membrane 2 Holes base	5 Washers 0,1 mm Membrane 6 Holes base
3	Notches seal 3 Washers 0,1 mm 1 Washer 0,2 mm Strangler Membrane 2 Holes base	1 Washers 0,1 mm 1 Washer 0.2 mm Membrane 6 Holes base

The obtained force-displacement diagrams are shown in Fig. 5 (Test 1), Fig. 6 (Test 2), and Fig. 7 (Test 3). The diagrams upper part corresponds to extension (positive forces), and the diagrams lower part corresponds to compression (negative forces).

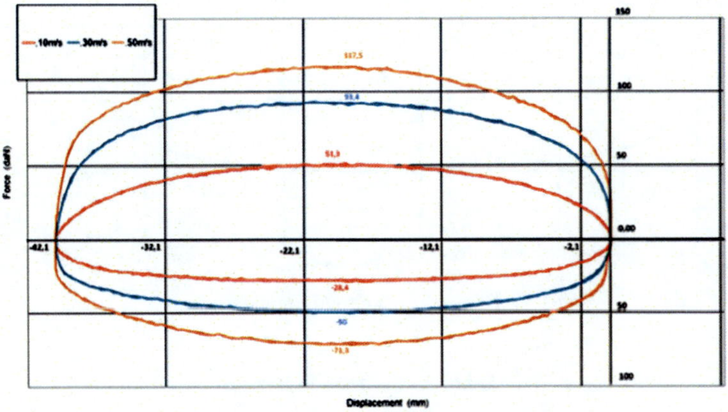

Fig. 5. Force-displacement diagrams: Test 1

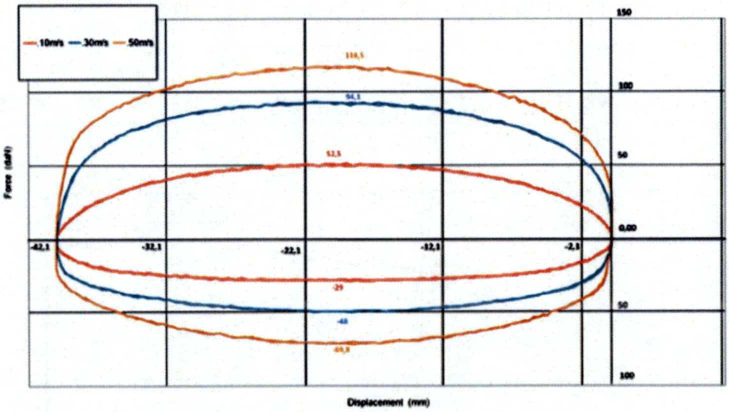

Fig. 6. Force-displacement diagrams: Test 2

The three tests calibration forces (maximal absolute values), in compression and in extension, are noted in Table 2. A force-velocity diagram is hence obtained, Fig. 8.

Relatively to Test1, Test 2 conduced to slight force increase in compression and slight force decrease in extension, while Test 3 produced high force increase in compression and low force decrease in extension.

The piston valve stiffness softening led then to slight compression hardening and extension softening, and the cylinder valve stiffness softening led to considerable compression hardening and slight extension softening.

The difference is thus in compression, the valves softening varies extension forces similarly but modifies compression forces differently, according to a velocity increase.

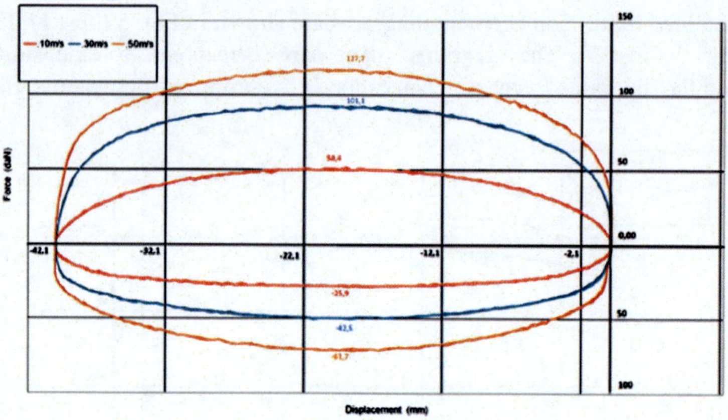

Fig. 7. Force-displacement diagrams: Test 3

Table 2. Calibration forces.

Velocity (m/s)	Compression (daN)			Extension (daN)		
	0,1	0,3	0,5	0,1	0,3	0,5
Test 1	51,3	93,4	117,5	28,4	50.0	71,3
Test 2	52,5	94,1	118,5	29.0	48.0	69,8
Test 3	58,4	101,1	127,7	28,1	49,6	70,5

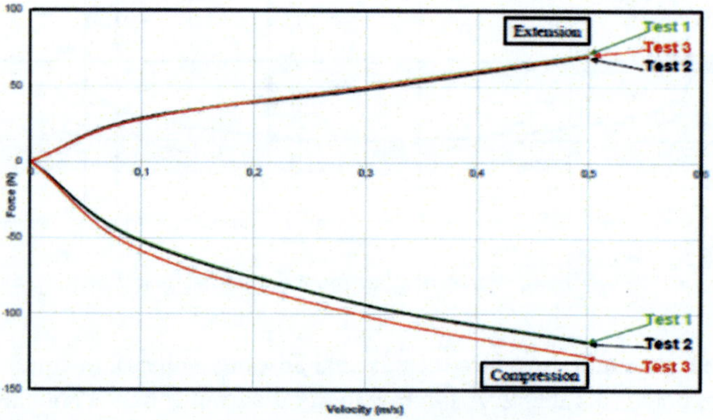

Fig. 8. Calibration force-velocity diagrams

6 Conclusion

The dynamic behavior of a hydraulic dual tube automobile strut damper was studied. Indeed, an experimental investigation was conducted on a dual tube hydraulic damper for automobile front axle.

A test bench is used to excite the strut damper at different velocities, and to measure its dynamic response. The dynamic response diagrams are determined for different piston and cylinder valves configurations, in compression and in extension. The results depend on the mechanical and the hydraulic parameters, but also on the testing conditions.

The experimental results are analyzed, the force-displacement responses are examined to conclude on the calibration forces. Then, the calibration forces are checked, a force-velocity diagram is established to conclude on the force evolution with velocity. The effects of the piston and cylinder valves configuration and of the excitation velocity are determined, to optimize the damper design.

The increase of the excitation velocity increases the calibration forces. The piston and cylinder valves stiffness softening varies extension forces similarly but modifies compression forces differently. The valves stiffness hardening is expected to give opposite effects.

The decomposition of thick spring washers into several thin spring washers is expected to generate a contact effect, to be studied for further results.

Acknowledgements. The authors are thankful to the LTM Company for providing the experiment equipment.

The authors gratefully acknowledge the helpful comments and suggestions of the reviewers.

References

Lang, H.H.: A study of the Characteristics of Automotive Hydraulic Dampers at High Stroking. University of Michigan, USA (1977)

Skagerstrand, H.: Shock Absorber Modelling. Chalmers University of Technology, Sweden (2014)

Chahine, R.: Modeling of a World Rally Championship Car Damper and Experimental Testing of its Components. The Royal Institute of Technology, Sweden (2011)

Gallsjo, A., Johansson, M.: Physical Modelling and automatic Configuration of CES Valve. Linkoping University, Sweden (2012)

Ferdek, U., Luczko, J.: Modeling and analysis of a twin-tube hydraulic shock absorber. J. Theor. Appl. Mech. (Poland) **50**(2), 627–638 (2012)

Konieczny, A.: Analysis of simplifications applied in vibration damping modeling for a passive car shock absorber. Shock Vibr. (2016)

Numerical Study of Machining Vibration Effect on Machined Surface Roughness in Orthogonal Milling

Wael Baklouti, Charfeddine Mrad(✉), and Rachid Nasri

Laboratory of Applied Mechanics and Engineering (LMAI), National Engineering School of Tunis (ENIT), University of Tunis el Manar (UTM), BP 37, Le Belvedere, 1002 Tunis, Tunisia
{charfeddine.mrad,rachid.nasri}@enit.rnu.tn

Abstract. Machine tools may vibrate during machining, especially when the cutting conditions are not optimized. Bad cutting conditions are the most destructive, they generate chatter vibration to cause cutting instability. Cutting instability is unfavourable for the whole cutting system: workpiece, tool, and machine. Cutting instability may be studied analytically, numerically, or experimentally. Commercial softwares opened recently different opportunities to investigate machining problems. This work aims to study numerically the machining vibration effect on the machined surface roughness in orthogonal milling. The arithmetic average roughness is tracked using two methods: mathematical and physical. The mathematical method is based on the workpiece surface nodes while the physical method is based on the cutting tool displacement. The numerical cutting is conducted using the ABAQUS software. The Johnson-Cook laws of strength and fracture are adopted for the workpiece, the Coulomb law is considered for the tool-workpiece contact, and an explicit scheme with a Lagrangian formulation is used. Firstly, the milling process was presented, the numerical model was described, and a numerical stability lobe was established. Secondly, the workpiece surface roughness was explored, basing on the arithmetic average roughness criterion and using the two proposed methods. Both methods confirm that the cutting vibration unfavourableness for the workpiece surface quality, and the physical method is more realistic.

Keywords: Numerical cutting · Machining vibration · Machined Surface · Surface roughness · Orthogonal milling

1 Introduction

Machine tools may vibrate during machining, especially when the cutting conditions are not optimized. Three types of vibration may occur during machining: free vibration, forced vibration, and self-excited vibration. Self-excited vibration, due to bad cutting conditions, is the most destructive for the cutting system: workpiece, tool, and machine.

Several works were carried out on cutting vibration in turning and milling, to investigate the cutting stability or the surface roughness. These works are analytical such as

A. Akrout et al. (Eds.): ICAV 2022, ACM 22, pp. 24–31, 2023.
https://doi.org/10.1007/978-3-031-34190-8_4

those conducted by Tlusty and Polacek (1963), and Faassen et al. (2003), experimental such as those performed by Erol et al. (2012), and Altintas et al. (2008), and numerical such as those elaborated by Mahnama and Movahhedy (2010), Baklouti et al. (2018), or Baklouti et al. (2020).

This work is to estimate numerically the machined surface arithmetic average roughness in orthogonal milling, using two methods, and to analyse the results.

2 Numerical Cutting

2.1 Milling Process

The model used in milling is 2D. The tool is an eight teeth disc cutter. The tool radial depth of cut is ae, and the tool axial depth of cut is ap, fixed at 0.11 mm. The feed direction is perpendicular to the depths of cut.

Figure 1 shows the milling process details.

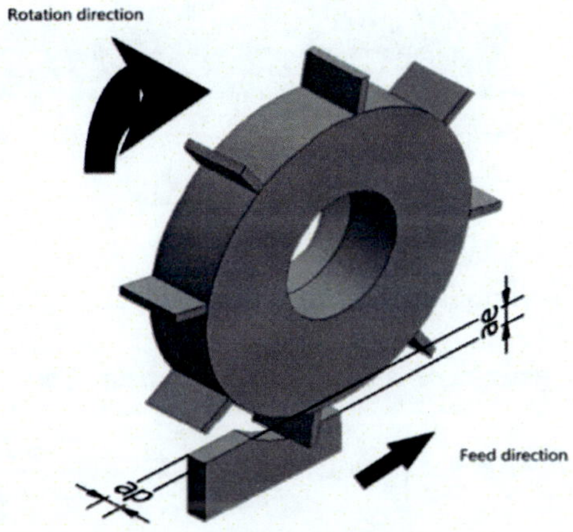

Fig. 1. Orthogonal milling

2.2 Numerical Model

The cutting tool is considered as a linear dynamic system of one degree of freedom along the radial direction. The tool details are shown in Fig. 2.

The tool-workpiece contact is governed by the Coulomb law. The mesh elements are CPE4RT. The workpiece details are shown in Fig. 3.

2.3 Cutting Stability

The numerical cutting is conducted using the ABAQUS software. The Johnson-Cook laws of strength and fracture are adopted for the workpiece, the Coulomb law is

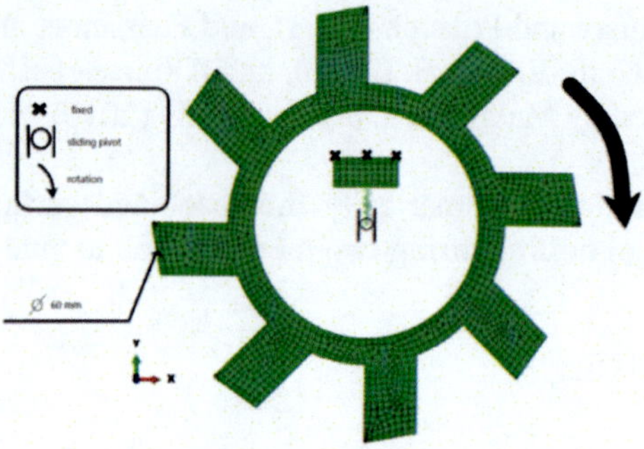

Fig. 2. Tool details

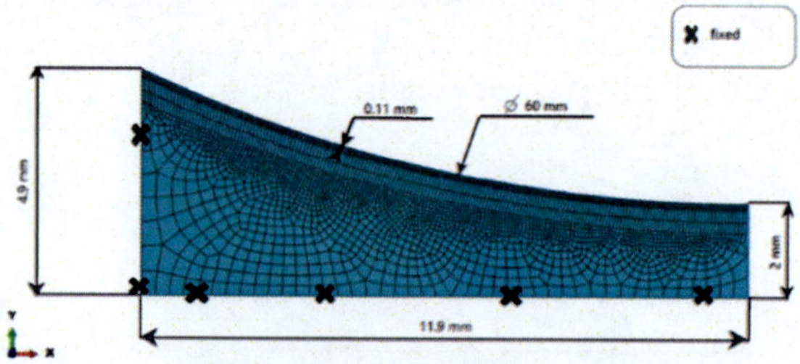

Fig. 3. Workpiece details

considered for the tool-workpiece contact, and an explicit scheme with a Lagrangian formulation is used.

The strength of the workpiece is according to Johnson-Cook, Eq. 1.

$$\sigma = [A + B\varepsilon^n] \times \left[1 + C\ln\frac{\dot{\varepsilon}}{\dot{\varepsilon}_0}\right] \times \left[1 - \left(\frac{T - T_r}{T_m - T_r}\right)^m\right] \tag{1}$$

The strength constants for the used 7075T6 alloy, are: $A = 546$ MPa, the elasticity limit; $B = 676$ MPa, the hardening modulus; and $n = 0.71$, the hardening exponent. $C = 0.024$ is the hardening coefficient, $\dot{\varepsilon}$ is the equivalent plastic strain rate, and $\dot{\varepsilon}_0$ is the reference strain rate. T is the ambient temperature, $T_r = 30$ °C is the reference temperature, and $T_m = 430$ °C is the melting temperature. $m = 1.56$ is the thermal softening exponent.

The Fracture of the workpiece is according to Johnson-Cook law, Eq. 2.

$$\varepsilon_f = [D_1 + D_2 exp(-D_3\eta)] \times \left[1 + D_4\ln\frac{\dot{\varepsilon}}{\dot{\varepsilon}_0}\right] \times \left[1 - D_5\left(\frac{T - T_r}{T_m - T_r}\right)^m\right] \tag{2}$$

The fracture constants for the used 7075T6 alloy are: $D_1 = -0.068$, $D_2 = 0.451$, $D_3 = -0.952$, $D_4 = 0.036$, and $D_5 = 0.697$.

The tool-workpiece contact is according to the Coulomb law, Eq. 3.

$$\tau_{cri} = \mu_0 \times p \tag{3}$$

where $\mu_0 = 0.3$ is the friction coefficient, τ_{cri} is the critical shear stress, and p is the contact normal pressure.

A numerical stability lobe is established, as shows Fig. 4. The lobe separates the stable cut zone and the unstable cut zone.

The green points are in the stable zone, the red points are in the unstable zone, while the orange points are on the border of stability.

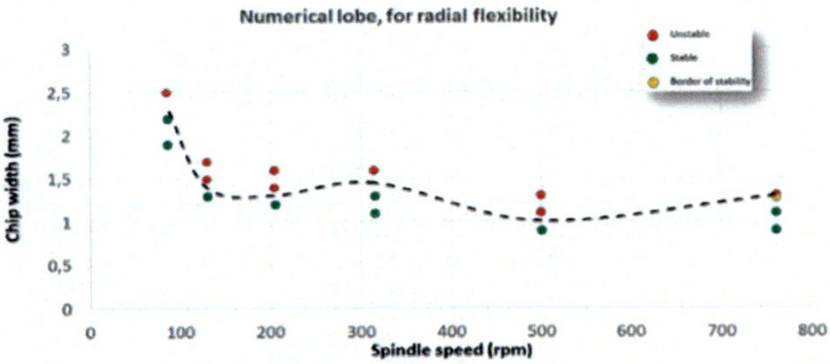

Fig. 4. Stability lobe

3 Surface Roughness

3.1 Roughness Criterion

The machined surface quality is investigated using the arithmetic average roughness criterion, Eq. 4, with z_i are the heights along the basic length, $i = 1 \ldots n$.

$$R_a = \frac{|z_1| + \cdots + |z_n|}{n} \tag{4}$$

The average line is determined using the least square method, it is considered as a straight line, as shows Fig. 5.

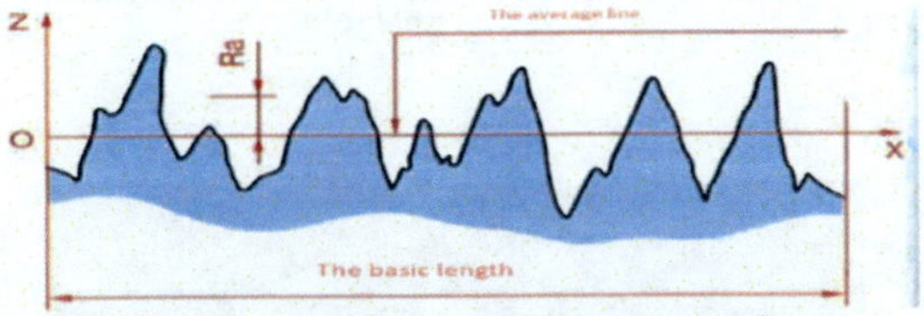

Fig. 5. Arithmetic average roughness

3.2 Mathematical Method

The mathematical method uses the nodes of the machined surface to determine the machined profile, as shows Fig. 6, where the nodes are in red colour.

When cutting, the tool produces a circle arc on the workpiece. The nodes of the machined surface, in the radial direction, are then measured, relatively to the arc center. These measurements constitute the profile of the machined surface, the average line is then determined, and the Ra values are calculated, as noted in Table 1.

Table 1. Ra values-Mathematical method.

N (rpm)	ap (mm)	Stability	Ra (μm)
760	0.9	stable	34.31
	1.1	stable	36.40
	1.3	unstable	49.87
500	0.9	stable	33.00
	1.1	unstable	39.84
	1.3	unstable	56.12
315	1.1	stable	33.61
	1.3	stable	34.94
	1.6	unstable	41.96
205	1.2	stable	34.28
	1.4	unstable	47.37
	1.6	unstable	50.74
130	1.7	stable	34.70
	2.0	stable	37.44
	2.3	unstable	44.82
85	1.9	stable	18.29
	2.2	stable	31.12
	2.5	unstable	53.49

Fig. 6. Machined surface

3.3 Physical Method

The physical method uses the displacement of the tool to determine the machined profile. Figure 7 shows an example of tool displacement.

The measured displacements constitute the profile of the machined surface, the average line is then determined, and the Ra values are calculated, as noted in Table 2.

Table 2. Ra values-Physical method.

N (rpm)	ap (mm)	Stability	Ra (μm)
760	0.9	stable	7.53
	1.1	stable	7.66
	1.3	unstable	8.27
500	0.9	stable	5.74
	1.1	unstable	9.59
	1.3	unstable	9.75
315	1.1	stable	6.70
	1.3	stable	6.88
	1.6	unstable	8.19
205	1.2	stable	6.95
	1.4	unstable	8.42
	1.6	unstable	8.51
130	1.7	stable	5.43
	2.0	stable	7.67
	2.3	unstable	8.19
85	1.9	stable	6.15
	2.2	stable	7.02
	2.5	unstable	8.66

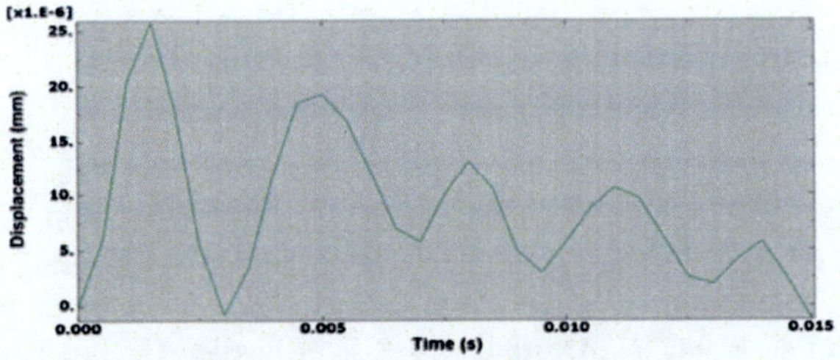

Fig. 7. Tool displacement example

4 Comparison

For the mathematical method, all the values of the arithmetic average roughness are less than 40 μm in the event of stability, they exceed 40 μm in the event of instability. Thus, the reference arithmetic average roughness is of 40 μm.

For the physical method, all the values of the arithmetic average roughness are less than 8 μm in the event of stability, they exceed 8 μm in the event of instability. Thus, the reference arithmetic average roughness is of 8 μm.

The reference arithmetic average roughness of 8 μm is more realistic then the reference arithmetic average roughness of 40 μm, which leads to conclude that the physical method is more suitable to estimate the machined surface quality.

5 Conclusion

This work aimed to study numerically the machining vibration effect on the machined surface roughness in orthogonal milling.

The arithmetic average roughness is determined using two methods: mathematical and physical. The mathematical method is based on the workpiece surface nodes while the physical method is based on the cutting tool displacement.

The numerical cutting is conducted using the ABAQUS software. The Johnson-Cook laws of strength and fracture are retained for the workpiece, the Coulomb law is considered for the tool-workpiece contact, and an explicit scheme with a Lagrangian formulation is adopted for the material removal.

Both methods confirm that the cutting vibration unfavourableness for the workpiece surface quality, but the physical method is more suitable to estimate the machined surface quality.

The mathematical Ra is approximately seven times higher than the physical Ra. Thus, the mathematical method, based on the surface nodes, overestimates Ra, while the physical method, based on the tool displacement, gives common Ra.

The obtained results prove the possibility to predict the machined surface quality and to conclude on the milling process stability, according to the cutting conditions.

Acknowledgements. The authors gratefully acknowledge the helpful comments and suggestions of the reviewers.

References

Tlusty, J., Polacek, M.: The stability of machine tools against self-excited vibrations in machining. In: Proceedings of the International Research in Production Engineering Conference, USA, ASME, pp. 465–474 (1963)

Faassen, R.P.H., Van de Wouw, N., Oosterling, J.A.J., Nijmeijer, H.: Prediction of regenerative chatter by modeling and analysis of high-speed milling. Int. J. Mach. Tools Manuf. **43**(14), 1437–1446 (2003)

Erol, T., Sezan, O., Suleyman, N., Suleyman, Y.: Decomposition of process damping ratios and verification of process damping model for chatter vibration. Measurement **45**, 1380–1386 (2012)

Altintas, Y., Eynian, M., Onozuka, H.: Identification of dynamic cutting force coefficients and chatter stability with process damping. CIRP Ann. Manuf. Technol. **57**(1), 371–374 (2008)

Mahnama, M., Movahhedy, M.R.: Prediction of machining chatter based on FEM simulation of chip formation under dynamic conditions. Int. J. Mach. Tools Manuf. **50**, 611–620 (2010)

Baklouti, W., Mrad, C., Nasri, R.: Numerical study of the chatter phenomenon in orthogonal turning. Int. J. Adv. Manuf. Technol. **99**(1–4), 755–764 (2018). https://doi.org/10.1007/s00170-018-2528-2

Baklouti, W., Mrad, C., Nasri, R.: Numerical determination of cutting stability lobes in orthogonal milling. In: Aifaoui, N., et al. (eds.) CMSM 2019. LNME, pp. 392–398. Springer, Cham (2020). https://doi.org/10.1007/978-3-030-27146-6_42

Vibration Effect on the Electro-Thermal Mechanical Behavior of the BGA Solder Joints

Sinda Ghenam[1,2](✉), Abdelkhalak Elhami[1], Ali Akrout[2], Wajih Gafsi[2], and Mohamed Haddar[2]

[1] Laboratory of Mechanics of Normandy (LMN), National Institute of Applied Sciences of Rouen (INSA – Rouen), Rouen, France
sinda.ghenam@enis.tn, abdelkhalak.elhami@insa-rouen.fr
[2] Laboratory of Mechanics, Modelling and Production (LA2MP), National School of Engineers of Sfax (ENIS), University of Sfax, Sfax, Tunisia
wajih.gafsi@enis.tn, moamed.haddar@enis.rnu.tn

Abstract. The widespread usage of electronic equipments in all vital existing sectors requires that these equipments are functional as long as necessary. In order to improve their lifetime, a reliability study should be carried out. Among the most decisive factors for the lifespan of electronic devices are the applied loads which are basically power, thermal load and mechanical load such as vibrations. Experience has revealed that the most impacted component in these devices are the solder joints. The aim of this paper is to assess the effectiveness and durability of SAC305 in comparison with traditional Sn63Pb37 under electro-thermomechanical loads. The finite element method (FEM) is used to simulate the performance of the brazing joints of the BGA assembly in order to evaluate their response after being exposed to the mentioned process. Stresses and strains are evaluated. Based on the present results, it has been concluded that SAC305 is more effective when subjected to electrothermal loading while Sn63Pb37 is better in vibrational response.

Keywords: life span · solder joints · electrothermal load · vibration · stress-strain output

1 Introduction

The future market leaders in the consumer electronics sector will be distinguished by their success in providing increasingly miniaturized products at lower cost. Electronic equipment consists of a set of several assembled electronic boards, interconnected with each other in order to perform the required functions. In modern electrical devices, the reliability of the entire system is considerably limited by the failure of the semiconductor components (Wang et al. 2013). The in-service operating environment of embedded systems requires the manufacture of electronic components and printed circuits capable of operating under severe conditions of temperature, vibration, humidity.

Surveys conducted by (Wiese et al. 2001), (Nguyen et al. 2010) and (Xiao et al. 2004) indicated that with the transition from Pb-based solder to Pb-free solder due

A. Akrout et al. (Eds.): ICAV 2022, ACM 22, pp. 32–42, 2023.
https://doi.org/10.1007/978-3-031-34190-8_5

to government restrictions on the use of lead (Pb) by the European Union (EU), an alternative solder material equivalent to the Pb-based SnPb eutectic solder alloy that has traditionally been used in electronic equipment assembly has been sought by electronics manufacturers.

Focusing on an enlarged appraisal of different solder compositions, electronics companies appear to be introducing a tin-silver-copper (SAC) solder alloy as an agreed-upon alternative to the lead-free solder alloy. In particular, many investigators have made considerable progress in describing the material behavior and sustainability of these weld alloys. Although the specific composition of SAC solder is an ongoing issue, they have shown that 96.5Sn3.0Ag0.5Cu (SAC305) solder is increasingly recognized as the favored lead-free substitute for surface mount assemblies that are extensively stressed under cyclic thermal conditions.

In previous studies, researchers report an assessment of the service life of SAC and lead-free solder alloys for electronic equipment subjected to thermal cycling and vibration, but they underestimate the action of power. In order to complete their investigation and fill the gap in their reliability study, this paper is conducted.

2 Theoretical Study

Solder interconnections, or solder joints, are fundamental constituents of advanced electronics. Functionally, solder joints ensure the thermal, mechanical and electrical connectivity between package components and retain their continuity throughout the production process and under operating conditions. In order to have a reliable assembly, it is important that these joints that connect the electronic box and the PCB ensure its function as it should without failure. The role of these solders is depicted in Fig. 1 as follows:

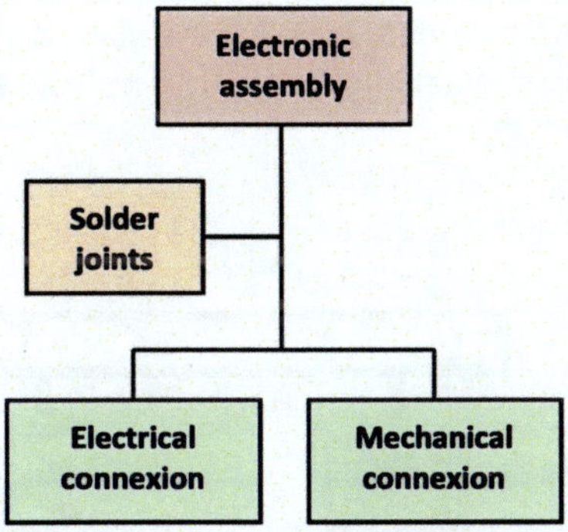

Fig. 1. Plot of solder joints roles

For reliability matters, each of these functions must be accomplished until the end of the lifetime of the assembly.

In order to identify the mathematical equations that define the electro- thermomechanical coupling, it is required to determine the equations linking the electrical, thermal and mechanical fields. For each domain, previous researches have tackled this point like the work of Ghenam et al. (Ghenam et al. 2022a) and (Makhloufi et al. 2016). The constitutive coupled thermoelectric equations are as follow:

$$\{q\} = [\pi].\{J\} - [K].\{\nabla T\} \tag{1}$$

$$\{J\} = [\sigma].\{E\} - [S].\{\nabla T\} \tag{2}$$

With: $[\pi] = T.\ [\alpha]$
[S]: Seebeck coefficients matrix
$[\pi]$: Peltier coefficients matrix
{q}: heat flux vector
{J}: electric current density
[K]: Thermal conductivity matrix
{E}: electric field

The constitutive coupled thermoelastic equations are as follows:

$$\{\sigma\} = [D].\{\varepsilon\} - [\beta].\{\Delta T\} \tag{3}$$

$$Q = T_0.\{\beta\}^T.\{\varepsilon\} + \rho.C_v.\{\Delta T\} \tag{4}$$

$[\beta]$: Vector of thermoelastic coefficients
[D]: elastic stiffness matrix
$\{\varepsilon\}$: Total strain vector

As for the vibratory study, we have to concretize the problem by simplifying it with masses, springs and shock absorbers, then we translate it by equation of movements, hence the construction of a mathematical model. Here's an explanatory diagram of the simplified system: (see Fig. 2).

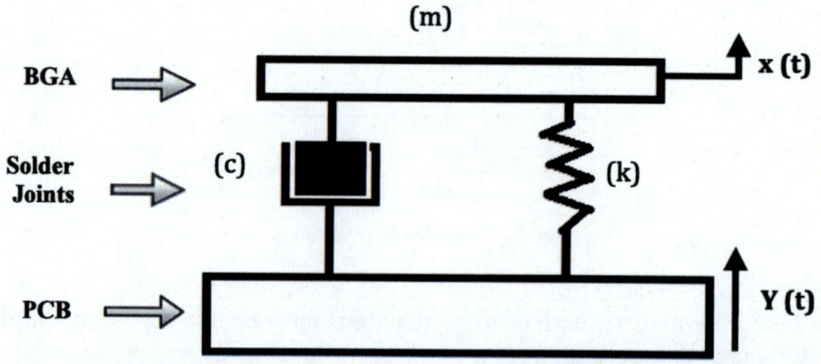

Fig. 2. Explanatory simplified diagram of the BGA assembly

The equation of motion of a vibrating system is the second order differential equation whose coefficients are constant matrices. It is established according to the kinematic quantities (displacement, speed and acceleration) expressing the movement of the system. It represents the mathematical model of the system studied and describes its behavior. The resolution of a differential equation of motion requires looking for the displacement, the speed or the acceleration of the mechanical system studied. Let the equations of motion of our system obtained by Newton's formalism:

$$m\ddot{x} + \dot{c} + kx = ky + c\dot{y} \quad (5)$$

3 Materials and Methods

One of the key causes of the breakdown of electronic devices is electro-thermomechanical stresses. In practice, the devices are exposed to temperature fluctuations resulting from subjecting the package to an imposed cyclic external temperature, as shown in Fig. 3 below, in addition to the Joule effect resulting from exposing the device to an electrical load, as studied in a previous study by (Ghenam et al. 2022b). The combination of thermal stress and the dissimilar material properties of the assembled layers leads to a considerable accumulation of inelastic stresses and strains in the weld joints. This leads to cracking and interconnection breakdowns.

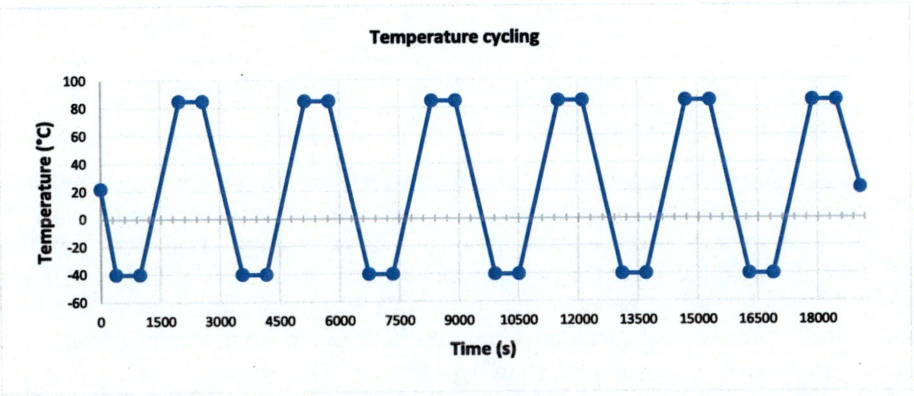

Fig. 3. Temperature cycling

In the interest of a simpler and more accurate model, some hypotheses are made. We consider that the imposed electrical pulses are vertical in the direction of the Z-axis, the starting temperature T (t = 0 s) = 22 °C is assigned to be homogeneous in all parts of the system, the radiation phenomenon is overlooked and $T < T_{Melt} = 217$ °C and no relative displacement of any component occurs under any external excitation.

Behavioral laws are related principally to the stress and strain analysis of assemblies. Two types of behaviour are generally defined for solder alloys: creep and viscoplastic behaviour.

The Anand and the Busso model are the two most important ones. Busso has established a viscoplastic model for the SnPb alloy. Concerning the lead-free solder behaviour (SnAgCu), researchers have developed the Anand equations which integrates viscoplasticity and time dependent plasticity (Assif et al. 2013). Here's below the Anand parameters for the two solder alloys that are going to be studied through this paper in Table 1. [(Depiver et al. 2021)].

Table 1. Anand parameters of the solder joints

Parameters	Solder joints type	
	SAC305	Sn63Pb37
S_0 (MPa)	45.9	3.1522
Q/R	7460	6.526
A	5.87e6	6220
ξ	2	3.33
m	0.0942	0,27
h_0 (MPa)	9350	60.599
S^ (MPa)	58.3	63.86
n	0.015	0.022
a	1.5	1,7811

Due to the time-consuming and costly process of electrothermal cycling testing, finite element modelling is privileged in the analysis of the reliability of the components.

The BGA system is meshed with tetrahedral finite elements. This mesh is further refined at the weld balls and the two contact surfaces. As a result, the FE model was adequately meshed before being input into the ANSYS software where it was simulated and evaluated for its static structural response to induced electrothermal loads. The suggested dimensions are derived from the literature (Ghenam et al. 2022a). Using a 1/4 of the full model, as shown in Fig. 4 below, minimizes the resolution time of the simulation and yet provides precise and consistent results.

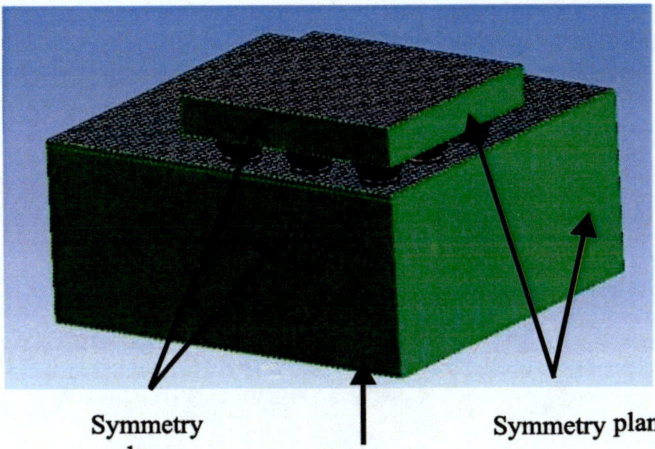

Fig. 4. Quarter model with boundary conditions

Since the impact of the dissimilarity of materials in the electro-thermomechanical simulation of the system is significant, it is crucial to look at the material properties of each part of the assembly. Based on the work of (Halouani 2020) Table 2. below summarizes, for each material, the values of the electrical, thermal and mechanical properties.

Table 2. Material properties

	FR4	Copper	Silicon	Prepreg
λ	0.3	400	130	0.5523
α	18×10^{-6}	17×10^{-6}	$2{,}6 \times 10^{-6}$	$1{,}69 \times 10^{-5}$
Cp	1369	385	700	1069
E	22	110	170	26,4
P	1900	8960	2330	1857
υ	0.28	0.35	0.28	0,1543
ρ	8×10^{11}	$1{,}7 \times 10^{-8}$	4×10^{3}	-------

4 Numerical Results

Following the application of the voltage which is equal to 3V on the upper side of the silicon chip and equal to 0 V on the lower side of the PCB to obtain a DDP we get the following result (see Fig. 5).

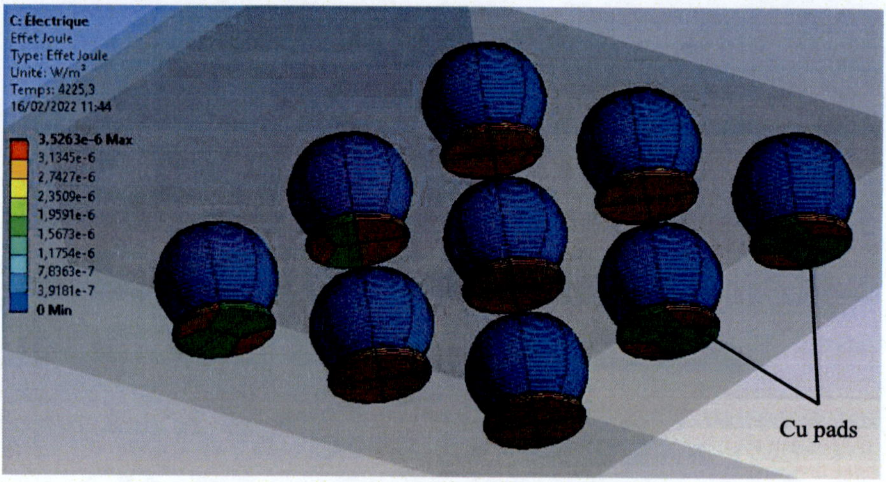

Fig. 5. Joule's effect depicted in the Cu pads

This result has the same distribution as in the current density and in heat flux as shown in Fig. 6:

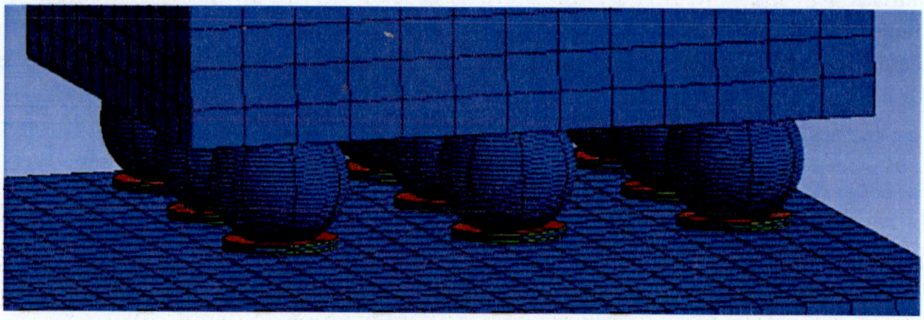

Fig. 6. Heat flow depicted in the Cu pads

This distribution of outputs highlights the weakness of the interconnection of these solder joints (Cu pads) with the other layers that compose the BGA packaging when they are subjected to an electrothermal load. Once these results are obtained, we apply a harmonic mechanical load which is an acceleration of the order of 100 m/s^2. The obtained frequencies are listed in Table 3. and the mechanical results as stresses and strains are listed in Table 4. below:

According to the numerical findings, we observed that the Sn63Pb37 has the lowest deformation while SAC305 has the highest deformation. Our obtained results reveal that the elastic modulus of SAC305 (44.6 GPa) and Sn63Pb37 (56 GPa) showing that the better the elastic modulus and mass density, the more the solder resists deformation when subjected to harmonic vibrations. Contrary to the thermal deformation which is less significant in SAC305 because it contains silver and copper in its formulation. That

Table 3. Natural frequencies of the solder joints

Modes	Frequencies (Hz)
1	218
2	416
3	614
4	812
5	1010
6	1208
7	1406
8	1604
9	1802
10	2000

Table 4. Mechanical outputs of the solder joints

Results	Mechanical outputs	
	Sn63Pb37	SAC305
Current density (A/m^2)	1850.2	2696.7
Heat flux (W/m^2)	9756	11550
Equivalent stress (Pa)	7.5 E−6	1.435 E−7
Thermal strain (m/m)	8.6 E−6	6.12 E−7
Elastic strain (m/m)	1.2 E−5	5.8 E−4

leads to the conclusion that SAC305 is a better performer when it comes to electrothermal loading.

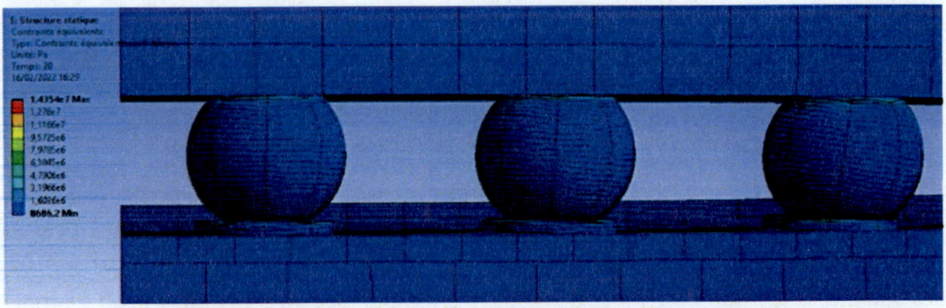

Fig. 7. Equivalent stress accumulated in solder joints

Fig. 8. Elastic strain of the solder joints

As for the stress and strain distribution as depicted in Fig. 7 and Fig. 8, we observe a higher accumulation of these mechanical results in the solder joint corners. This is due to the mismatch of the CTE coefficients in the first place and the weakness of this interconnection. There is a tendency to break between the brazing joints and the copper pads as a result for the imposed loads.

In order to more concentrate on the effect of this loadings on the BGA assembly, we used the fatigue tool implemented in ANSYS Workbench and we obtained the results showed in Fig. 9 and Fig. 10.

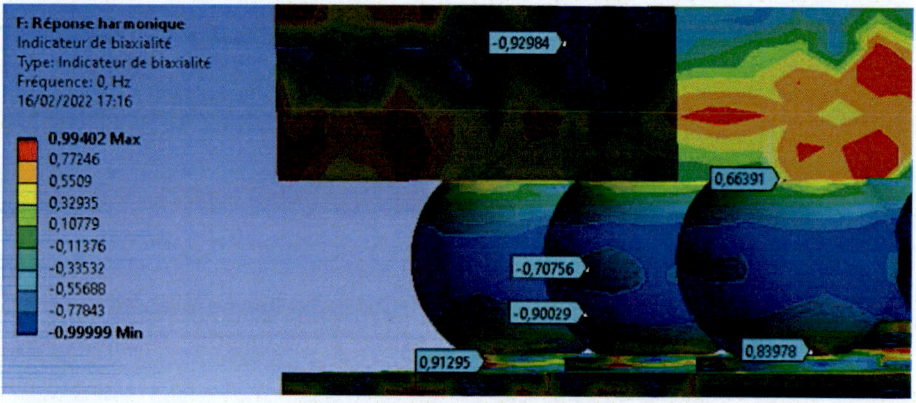

Fig. 9. Probe results for the biaxiality indicator

This result (see Fig. 9) gives an indication of the stress state on the model and how to interpret the results. A biaxiality of 0 indicates uniaxial stress; a value of −1 indicated pure shear and a value of 1 indicates a pure biaxial state. To this end, I used a probe to highlight what each part of the BGA assembly is subjected to.

The highest value of the safety factor is 15 as colored in blue. For fatigue safety factor, values less than 1 indicates premature failure. This is marked with the red color.

Fig. 10. Results for the safety factor

The concentration of this color is accumulated is Cu pads mostly and in the prepreg layer which threatens the failure of the interconnection and therefore the breakdown of the entire system.

5 Conclusion

In this paper, we have studied the impact of coupling the three main domains on the behavior of the BGA assembly. As well as the submission of the system to a vibration load. Although the lead-based solder joint has better responses in vibration tests, it has modest thermal properties compared to SAC305. Given the harmful environmental factor of the lead in SnPb, and the extreme importance of the joint's resistance to temperature in electronic domain, it is recommended to use SAC305 because, as indicated in numerical results sections, thermal deformation is lower compared to SnPb.

Appendix: Notations

$[C^t]$: Elementary specific heat matrix, $[C^{\mathrm{P}}]$: Elementary dielectric permittivity matrix, $[K^t]$: Thermal conductivity matrix, $[K^{\mathrm{P}}]$: Elementary electrical conductivity matrix, $[K^{\mathrm{P}t}]$: Matrix of elementary Seebeck coefficients, [Q]: Elementary heat density vector, $[Q^p]$: Peltier effect heat flux vector, $[C^{tu}]$: Thermo- elastic damping matrix, $[K^{ut}]$: Thermo-elastic stiffness matrix, ε: Strain, σ: Stress,

Φ: heat flux, P: power in Watt, $\mathbf{h_f}$: exchange coefficient in W. m^{-2}. K^{-1}, dt/dx: the temperature gradient in K . m^{-1}, dt: basic time in s, $\boldsymbol{T_0}$: initial temperature of the system, λ: thermal conductivity of the material in W. m^{-1}. K^{-1}, T: temperature in K, Q: dissipated power in J.

References

Halouani, A., Abel, C., Mohamed, H.: Modélisation et Simulation Multi-Physique Pour La Fiabilisation Des Composants Électroniques (2020)

Depiver, J.A., Mallik, S., Amalu, E.H.: Thermal fatigue life of ball grid array (BGA) solder joints made from different alloy compositions. Eng. Fail. Anal. **125**, 10544 (2021)

Ghenam, S., Elhami, A., Akrout, A., Gafsi, W., Haddar, M.: Electro-thermomechanical modelling of a BGA assembly subjected to a damaging displacement and to random vibrations. In: Amar, M.B., Bouguecha, A., Ghorbel, E., El Mahi, A., Chaari, F., Haddar, M. (eds.) A3M 2021. LNME, pp. 353–364. Springer, Cham (2022). https://doi.org/10.1007/978-3-030-84958-0_38

Wang, H., Liserre, M., Blaabjerg, F.: Toward reliable power electronics: challenges, design tools, and opportunities. IEEE Ind. Electron. Maga. **7**(2), 17–26 (2013)

Makhloufi, A., Aoues, Y., el Hami, A.: Reliability based design optimization of wire bonding in power microelectronic devices. Microsyst. Technol. **22**(12), 2737–2748 (2016)

Nguyen, T.T., Lee, D., Kwak, J.B., Park, S.: Effect of glue on reliability of flip chip BGA packages under thermal cycling. Microelectron. Reliabil. **50**(7), 1000–1006 (2010)

Assif, S., Abdelkhalak, E.H., Mohamed, A.: Fiabilité et Optimisation Des Structures Mécaniques À paramètres Incertains : Application Aux Cartes Électroniques (2013). https://tel.archives-ouvertes.fr/tel-00950354

Ghenam, S., et al.: Electro Thermomechanical Coupling of a BGA Assembly Subjected to Combined and Alternating Power and Thermal Cycles (2022)

Wiese, S., et al.: Constitutive behaviour of lead-free solders vs. lead- containing solders-experiments on bulk specimens and flip-chip joints. In: 2001 Proceedings, 51st Electronic Components and Technology Conference (Cat. No.01CH37220), pp. 890–902. IEEE (2001)

Xiao, Q., Bailey, H.J., Armstrong, W.D.: Aging effects on microstructure and tensile property of Sn3.9Ag0.6Cu solder alloy. J. Electron. Packa. Trans. ASME **126**(2), 208–212 (2004)

Large Deflection of Smart Magneto-Electro-Elastic Cylindrical Shell

Hanen Jrad[1,2](✉), Hajer Ellouz[1], Mondher Wali[1,2], and Fakhreddine Dammak[3]

[1] Laboratory of Electromechanical Systems (LASEM), National Engineering School of Sfax, University of Sfax, Route de Soukra km 4, 3038 Sfax, Tunisia
hanen.j@gmail.com, hanen.jrad@essths.u-sousse.tn

[2] Higher School of Sciences and Technology of Hammam Sousse, University of Sousse, Rue Lamine Abassi, 4011 Hammam Sousse, Tunisia

[3] Laboratory of Electrochemistry and Environment (LEE), National Engineering School of Sfax, ENIS, University of Sfax, Sfax, Tunisia
Fakhreddine.dammak@enis.tn

Abstract. The geometrically nonlinear response of Magneto-electro-elastic composite materials (MEE) shells is investigated. In the presented model, based on the First-order shear deformation theory (FSDT), a constant transverse shear is adopted, and a shear correction component is needed. In order to prevent transverse shear locking, the presumed natural strain method (ANS) is assimilated in the FE formulation. A discrete system of geometrically nonlinear equilibrium equations find out by the Newton-Raphson approach. Numerical results of a pinched cylinder with end diaphragms are compared with pioneer works. A good agreement is obtained, and static response of MEE under large deformations and finite rotations is conducted to outline the present model's accuracy.The accurate FE method based on the Mindlin's deformation theory is proposed in this work, to explore the 3D-MEE-shell structures' geometrically nonlinear static response. The new model is validated by comparing the obtained findings to the literature for a pinched cylinder with end diaphragms. The real and reference findings accord well.

Keywords: MEE Composites · FSDT · cylindrical shell

1 Introduction

With the extended innovation of technology in the area of intelligence, the classic single-phase materials are often incapable to fulfil the needs and requirements of production technology. Magneto-electro-elastic composite materials (MEE) have been attracting attention between the many multiphase composite materials, as smart material that combines piezomagnetic phase material and piezoelectric phase. In fact, the MEE materials have magneto-electric coupling effect that carry out the mutual conversion of mechanical energy, electrical energy and magnetic energy, Nan et al. (2008); Carrera and Nali (2010); Kondaiah et al. (2012); Milazzo (2013).

A. Akrout et al. (Eds.): ICAV 2022, ACM 22, pp. 43–51, 2023.
https://doi.org/10.1007/978-3-031-34190-8_6

To investigate MEE materials, several researchers have used various theories and methods. Pan (2001) proposed an exact solution for simply supported and multilayered MEE plates under surface load and internal load. Zhang et al. (2019) suggested a semi-analytical model to investigate vibration of laminated MEE composite plate. Further, Ngak et al. (2019) employed a semi-analytical three-dimensional method to analyze the dynamic response of multilayer MEE plate. Although analytical approaches can offer closed-form solutions, they unfortunately are limited to elementary geometries, certain kinds of gradation of material properties, particular types of boundary cases and specific loading conditions. To accomplish more common analysis, it is necessary to resort to numerical methods and particularly finite element method (FEM). Generally, the linear shell theories are inspected for shell structures showing small displacements under various types of loads. Nonlinear analysis of shells behavior, however, has been realized in the context when shells undertake large deformations. Several research works are obtainable in literature to analyze the nonlinear mechanical behavior of shell structures (Mellouli et al. 2019). Moreover, investigations of geometrically nonlinear behavior of smart shell structures have been carried out in the works of Rao et al. (2016); Marinkovic and Rama (2017), Rama et al. (2017, 2018); Balamurugan and Narayanan (2008); Kpeky et al. (2018) and Carrera et al. (2018).

Further, over the years, numerical methods are widely used to provide a way to approximate solutions to problems in many engineering disciplines such as aerospace, mechanical, and electrical, quickly and easily compared to analytic solutions and when analytical solutions aren't possible. In addition, finite element analysis (FEA) allows simulations to predict and understand the structure behavior under several physical conditions. It is a numerical approach to finding the real results and providing an approximate solution to different problems in mechanical engineering. A design optimization of dental implant geometrical characteristics enhancing primary stability is proposed by (Elleuch et al. 2021) using FEA of stress distribution around dental prosthesis. A numerical methodology to characterize aluminum behavior considering non-associated plasticity model coupled with isotropic ductile damage was carried out by (Bouhamed et al. 2021, 2022) in order to identify ductile damage constitutive equations for thin sheet metal applications.

The main contribution of this research is to examine for the first time in the literature, the geometrically nonlinear behavior of MEE shell enduring large deformations and finite rotations. The present model is based on first-order shear deformation theory (FSDT). Geometrically nonlinear analysis of MEE pinched cylinder with end diaphragms is conducted in order to outline the accuracy and performance of the present model.

2 Basic Formulation of MEE Cylindrical Shell

The basic formulation of the FSDT model is briefly reviewed in this section. For convenience of presentation, the Cartesian tensor notation is adopted. The position vector of a material point (q) of the shell structure placed at the distance z from the mid surface can be associated to the position vector of (p) in both reference C_0 and current C_t

configurations:

$$\begin{cases} \boldsymbol{X}_q\left(\xi^1, \xi^2, z\right) = \boldsymbol{X}_p\left(\xi^1, \xi^2\right) + z\boldsymbol{D}\left(\xi^1, \xi^2\right) \\ \boldsymbol{x}_q\left(\xi^1, \xi^2, z\right) = \boldsymbol{x}_p\left(\xi^1, \xi^2\right) + z\boldsymbol{d}\left(\xi^1, \xi^2\right) \end{cases} z \in [-h/2, h/2] \tag{1}$$

with $\boldsymbol{\xi} = \left(\xi^1, \xi^2, \xi^3 = z\right)$ representing the curvilinear coordinates. $\boldsymbol{D}$ and $\boldsymbol{d}$ are the shell director vectors in reference and current configurations, respectively. h denotes the thickness. The strain ε can be separated in in-plane and transverse shear strains as:

$$\begin{cases} \varepsilon_{\alpha\beta} = e_{\alpha\beta} + z\,\chi_{\alpha\beta} \\ 2\varepsilon_{\alpha 3} = \gamma_\alpha \end{cases} ;\ \alpha, \beta = 1, 2 \tag{2}$$

$\varepsilon_{\alpha\beta}$, $\chi_{\alpha\beta}$ and γ_α are the components of linearized membrane, bending and shear strains vectors given by.

$$\begin{aligned} &\delta e_{\alpha\beta} = 1/2\left(\boldsymbol{A}_\alpha.\delta\boldsymbol{x}_{,\beta} + \boldsymbol{A}_\beta.\delta\boldsymbol{x}_{,\alpha}\right); \\ &\delta\chi_{\alpha\beta} = 1/2\left(\boldsymbol{A}_\alpha.\delta\boldsymbol{d}_{,\beta} + \boldsymbol{A}_\beta.\delta\boldsymbol{d}_{,\alpha} + \delta\boldsymbol{x}_{,\alpha}.\boldsymbol{d}_{,\beta} + \delta\boldsymbol{x}_{,\beta}.\boldsymbol{d}_{,\alpha}\right);\ \alpha, \beta = 1, 2 \\ &\delta\gamma_\alpha = \boldsymbol{A}_\alpha.\delta\boldsymbol{d} + \delta\boldsymbol{x}_{,\alpha}.\boldsymbol{d} \end{aligned} \tag{3}$$

Using matrix notation, the membrane, bending, and shear strains vectors are

$$\mathbf{e} = \begin{Bmatrix} e_{11} \\ e_{22} \\ 2e_{12} \end{Bmatrix},\ \boldsymbol{\chi} = \begin{Bmatrix} \chi_{11} \\ \chi_{22} \\ 2\,\chi_{12} \end{Bmatrix}\ \boldsymbol{\gamma} = \begin{Bmatrix} \gamma_1 \\ \gamma_2 \end{Bmatrix} \tag{4}$$

The electric and magnetic fields are supposed to be constant over an element of the active MEE shell. The electric φ and magnetic potential ψ vary linearly through the thickness and can be expressed as

$$\begin{matrix} \boldsymbol{E} = -\varphi_{,\alpha} \\ \boldsymbol{H} = -\psi_{,\alpha} \end{matrix},\ \alpha = 1, 2, 3 \tag{5}$$

Strains, electric and magnetic field expressions shown respectively in Eqs. (4) and (5) can be substituted in the weak form of equilibrium equations as below:

$$G = \int_A \left(\boldsymbol{N}.\delta\boldsymbol{e} + \boldsymbol{M}.\delta\boldsymbol{\chi} + \boldsymbol{T}.\delta\boldsymbol{\gamma} + \widetilde{\boldsymbol{q}}.\delta\,\boldsymbol{E} + \widetilde{\boldsymbol{b}}.\delta\boldsymbol{H}\right)dA - G_{ext} = 0 \tag{6}$$

where the membrane $\boldsymbol{N}$, bending $\boldsymbol{M}$, and shear $\boldsymbol{T}$ resultants, the electric displacement $\widetilde{\boldsymbol{q}}$ resultant and the magnetic induction $\widetilde{b}$ can be written in the form:

$$\begin{aligned} &\boldsymbol{N} = \int_{-h/2}^{h/2} \begin{bmatrix} \sigma_{11} \\ \sigma_{22} \\ \sigma_{12} \end{bmatrix} dz,\ \boldsymbol{M}_k = \int_{-h/2}^{h/2} z \begin{bmatrix} \sigma_{11} \\ \sigma_{22} \\ \sigma_{12} \end{bmatrix} dz,\ \boldsymbol{T}_1 = \int_{-h/2}^{h/2} f(z) \begin{bmatrix} \sigma_{13} \\ \sigma_{23} \end{bmatrix} dz, \\ &\widetilde{\boldsymbol{q}} = \int_{-h/2}^{h/2} \boldsymbol{q}\,dz,\ \widetilde{\boldsymbol{b}} = \int_{-h/2}^{h/2} \boldsymbol{b}\,dz, \qquad k = 1, 2 \end{aligned} \tag{7}$$

$f(z) = \frac{5}{4}\left(1 - \frac{4z^2}{h^2}\right)$ is employed to insure parabolic shear strain through the thickness and $\boldsymbol{\sigma}$ is the stress tensor.

To obtain generalized resultant of stress $\boldsymbol{R}$ and $\boldsymbol{\Sigma}$ strain vectors can be written as

$$R = \left[N\ M\ T\ \widetilde{q}\ \widetilde{b}\right]_{14\times1},\ \boldsymbol{\Sigma} = \left[\ \boldsymbol{e}\ \boldsymbol{\chi}\ \boldsymbol{\gamma}\ \boldsymbol{E_m}\ \boldsymbol{H_m}\right]_{14\times1} \tag{8}$$

The linear constitutive equations of magnet-electro-elasticity expressing the coupling between the elastic, electric and magnetic fields relevant to present problem can be defined as:

$$\begin{cases} \boldsymbol{\sigma} = \boldsymbol{C}\,\boldsymbol{\varepsilon} - \boldsymbol{p}^T\,\boldsymbol{E} - \boldsymbol{\eta}^T\boldsymbol{H} \\ \boldsymbol{q} = \boldsymbol{p}\,\boldsymbol{\varepsilon} + \boldsymbol{k}\boldsymbol{E} + \boldsymbol{d}.\boldsymbol{H} \\ \boldsymbol{B} = \boldsymbol{\eta}\,\boldsymbol{\varepsilon} + \boldsymbol{d}.\boldsymbol{E} + \boldsymbol{\mu}\boldsymbol{H} \end{cases} \tag{9}$$

where $\boldsymbol{C}, \boldsymbol{p}, \boldsymbol{\eta}, \boldsymbol{k}, \boldsymbol{d}$, and $\boldsymbol{\mu}$ represent respectively the elastic matrix for a constant electric and magnetic field, the piezoelectric coupling matrix, the piezomagnetic coupling matrix, the dielectric permittivity matrix, the magneto electric coupling matrix and the magnetic permeability.

Using Eqs. (4), (5), (7) and (9), the resultant of stress R is related to the strain field $\boldsymbol{\Sigma}$ can be given as follows, with $\mathbf{H}_T$ is the linear coupling magnetoelectroelastic matrix expressed as:

$$\boldsymbol{R} = \boldsymbol{H}_T\boldsymbol{\Sigma},\ \boldsymbol{H}_T = \begin{bmatrix} \boldsymbol{H}_{11} & \boldsymbol{H}_{12} & 0 & \boldsymbol{H}_{14} & \boldsymbol{H}_{15} \\ & \boldsymbol{H}_{22} & 0 & \boldsymbol{H}_{24} & \boldsymbol{H}_{25} \\ & & \boldsymbol{H}_{33} & \boldsymbol{H}_{34} & \boldsymbol{H}_{35} \\ & & & \boldsymbol{H}_{44} & \boldsymbol{H}_{45} \\ Sym & & & & \boldsymbol{H}_{55} \end{bmatrix} \tag{10}$$

With

$$\begin{cases} (\boldsymbol{H}_{11}, \boldsymbol{H}_{12}, \boldsymbol{H}_{22}) = \int_{-h/2}^{h/2} \left(1, z, z^2\right) \boldsymbol{C} dz\,;\ \boldsymbol{H}_{33} = \int_{-h/2}^{h/2} (f(z))^2\, \boldsymbol{C}_\tau\, dz \\ (\boldsymbol{H}_{14}, \boldsymbol{H}_{24}) = \int_{-h/2}^{h/2} (1, z)\, \boldsymbol{p}_1^T\, dz;\ (\boldsymbol{H}_{15}, \boldsymbol{H}_{25}) = \int_{-h/2}^{h/2} (1, z)\, \boldsymbol{\eta}_1^T\, dz \\ \boldsymbol{H}_{34} = \int_{-h/2}^{h/2} f(z)\, \boldsymbol{p}_2^T dz;\ \boldsymbol{H}_{35} = \int_{-h/2}^{h/2} f(z)\, \boldsymbol{\eta}_2^T dz \\ \boldsymbol{H}_{44} = \int_{-h/2}^{h/2} \boldsymbol{k}\, dz;\ \boldsymbol{H}_{45} = \int_{-h/2}^{h/2} \boldsymbol{d}\, dz \\ \boldsymbol{H}_{55} = \int_{-h/2}^{h/2} \boldsymbol{\mu}\, dz \end{cases} \tag{11}$$

where $\boldsymbol{C}$ and $\boldsymbol{C}_\tau$ are in plane and out-of-plane linear elastic sub-matrices and $\boldsymbol{p}_1^T, \boldsymbol{p}_2^T$ and $\boldsymbol{k}$, represent the in plane and out-of-plane piezoelectric coupling sub-matrices and dielectric permittivity matrix, respectively. $\eta_1^T, \eta_2^T, \boldsymbol{d}$and $\boldsymbol{\mu}$, represent the in plane and out-of-plane piezomagnetic coupling sub-matrices, magneto electric coupling matrix and magnetic permeability matrix, respectively.

Therefore, the weak form of the equilibrium equation can be present as

$$G(\boldsymbol{\Phi}, \delta\boldsymbol{\Phi}) = \int_A \delta\Sigma^T.\boldsymbol{R}dA - G_{ext}(\boldsymbol{\Phi}, \delta\boldsymbol{\Phi}) = 0 \tag{12}$$

Next, by using the weak form of the equilibrium equation, the procedure of Newton iterative will be employed to solve the nonlinearity of the problem. The tangent operator will be divided into geometric $D_G G.\mathbf{\Delta\Phi}$ and material parts $D_M G.\mathbf{\Delta\Phi}$:

$$DG.\mathbf{\Delta\Phi} = D_G G.\mathbf{\Delta\Phi} + D_M G.\mathbf{\Delta\Phi}; \begin{cases} D_M G.\mathbf{\Delta\Phi} = \int_A \delta\mathbf{\Sigma}^T \mathbf{\Delta R} dA \\ D_G G.\mathbf{\Delta\Phi} = \int_A \left(\Delta\delta\Sigma^T.\mathbf{R}\right) dA \end{cases} \tag{13}$$

3 Numerical Results

In this section, a short cylinder subjected to mechanical loads is considered to assess the proposed method. Two opposite pinching vertical forces applied at the middle of the structure which is enhanced at the end by two rigid diaphragms in order to avoid the section's distortion. The severity of this test is shown in the complexity of the membrane and the bending stress. Because of the symmetry of the geometry of this model, only the octant of the cylinder will be modeled. The circumferential periphery is fully clamped.

This example, shown in Fig. 1, can be imagined as a pinched can of soda. The geometric properties of cylinder are gathered in Fig. 1 according to Simo et al. (1989) and Costa et al. (2013).

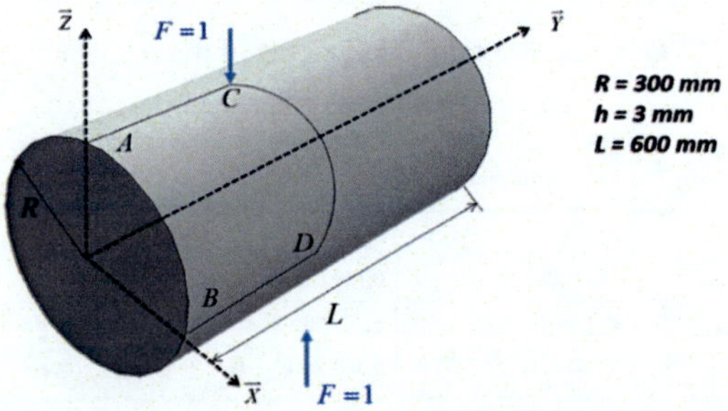

Fig. 1. Description of the pinched cylinder with end diaphragms.

On one hand, to determine the accuracy and convergence of the proposed method, this numerical example is considered with an isotropic material with the following material properties: Young's modulus $E = 3.10^6$ Pa and Poisson's ratio $\nu = 0.3$ according to Simo et al. (1989); Costa et al. (2013). The deflections are checked under pinching force $F = 1$N. The obtained results are compared with those made by Simo et al. (1989) which propose a shell finite element model with geometrically exact constraint-results. The solutions found by Costa el al. (2013), exploiting the meshless method "the Multiple Fixed Least-Squares (MFLS)" with a complete stress-resultant theory with Reissner-Mindlin kinematics based on an inextensible director, are also used for comparison. The radial displacements are also measured at the loading points, an analytical solution of the order of 1.82488×10^{-5} mm is reported in the work of Simo et al. (1989) and considered as a reference.

A convergence analysis is carried out in order to outline the validity and to determine the suitable mesh size to use for the simulation. Different numbers of nodes per side are considered, as listed in Table 1. The convergence study conducted indicates that the use 16x16 nodes is appropriate.

Table 1. Radial displacements of pinched cylindrical shell

Elements	Déflexion ($\times 10^{-5}$ mm)			Precision %
	Costa et al. (2013)	Simo et al. (1989)	Present model	
4 × 4	0.583	0.728	0.0678	3.71
8 × 8	0.820	1.390	1.3555	74.28
12 × 12	1.149	–	–	–
16 × 16	1.459	1.706	1.694	92.82

On the other hand, the effectiveness of the developed MEE shell model is assessed through analysis of static response of the cylindrical structure in Fig. 1 using magneto-electro-elastic material $BF50\%$ which material properties are listed in Table 2 according to Sladek et al. (2013).

Table 2. Material coefficients of the $BF50\%$

$c_{11}=c_{22}$ 213×10^9 Nm^{-2}	c_{12} 113×10^9Nm^{-2}	$c_{13}=c_{23}$ 112.8×10^9Nm^{-2}	c_{33} 207×10^9Nm^{-2}	$c_{44}=c_{55}$ 49.9×10^9 Nm^{-2}	μ_{33} 83.140Ns$^2/C^2$
c_{66} 50×10^9 Nm^{-2}	$e_{31}=e_{32}$ −2.710 Cm^{-2}	e_{33} 8.860 Cm^{-2}	$e_{24}=e_{15}$ 0.150 Cm^{-2}	$\eta_{31}=\eta_{32}$ 222.6 N/Am	$d_{11}=d_{22}$ 5.35×10^{-12} Ns/VC
η_{33} 292.010 N/Am	$\eta_{15}=\eta_{24}$ 185.130 N/Am	$\kappa_{11}=\kappa_{22}$ 0.710×10^{-9} C/Vm	κ_{33} 6.320×10^{-9} C/Vm	$\mu_{11}=\mu_{22}$ −192.200Ns^2/C^2	d_{33} $2751.4 \times 10^{-12} Ns/VC$

Figure 2 illustrates the variation of deflection at point C as function of the load. It is remarkable that the nonlinear curve starts with the same shape and values as the linear curve. Furthermore, the nonlinear curve take different values, at some points, higher than the linear one by keeping the same trend. It can be seen that there is a crucial load factor at which the linear solution begins to differ compared to its nonlinear one specially when subjected to large displacement.

Figure 3 shows the deflection of the pinched cylindrical shell as function of the R/h ratio. It is clear that the rise of this ratio provides a higher deflection. Indeed, the thickness of the MEE pinched cylinder reduced with the increase of the ratio R/h which allows higher flexibility of the structure.

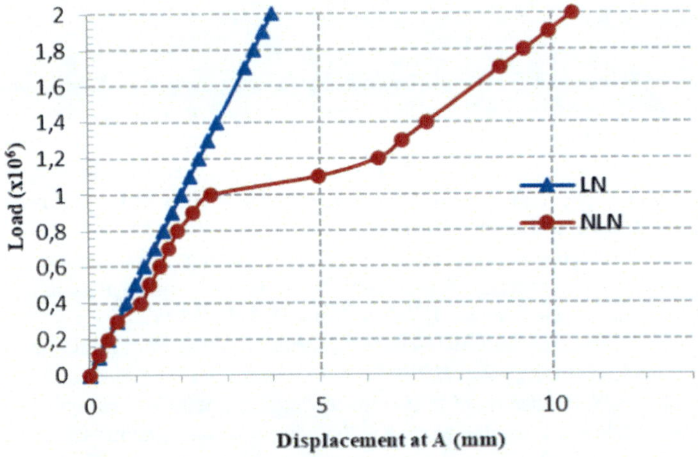

Fig.2. Variation of deflection as function of the load factor

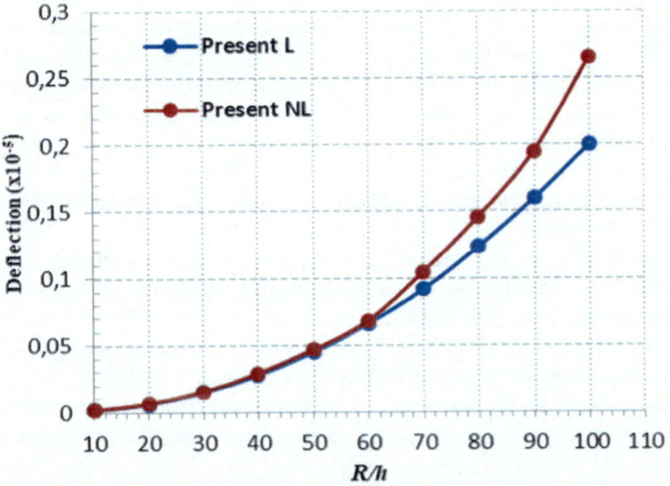

Fig.3. Effect of R/h ratio on the deflection of the pinched cylinder with end diaphragms.

4 Conclusion

An accurate finite element method based on the Mindlin's first-order shear deformation theory is proposed in this work, to study the geometrically nonlinear static response of 3D-MEE-shell structures. The validation of the present model is based on comparing obtained results to the literature for a pinched cylinder with end diaphragms. A good agreement is obtained between the actual results and the reference ones.

References

Balamurugan, V., Narayanan, S.: A piezolaminated composite degenerated shell finite element for active control of structures with distributed piezosensors and actuators. Smart Mater. Struct. **17**(3), 035031 (2008)

Bouhamed, A., et al.: Identification of fully coupled non-associated-Ductile damage constitutive equations for thin sheet metal applications: numerical feasibility and experimental validation. Thin-Walled Struct **176**, 109365 (2022)

Bouhamed, A., Mars, J., Jrad, H., Wali, M., Dammak, F.: Experimental and numerical methodology to characterize 5083-aluminum behavior considering non-associated plasticity model coupled with isotropic ductile damage. Int. J. Solids Struct. **229**, 111139 (2021)

Carrera, E., Nali, P.: Multilayered plate elements for the analysis of multifield problems. Finite Elem. Anal. Des. **46**(9), 732–742 (2010)

Carrera, E., Valvano, S., Kulikov, G.M.: Electro-mechanical analysis of composite and sandwich multilayered structures by shell elements with node-dependent kinematics. Int. J. Smart Nano Mater. **9**(1), 1–33 (2018)

Costa, J.C., Tiago, C.M., Pimenta, P.M.: Meshless analysis of shear deformable shells: the linear model. Comput. Mech. **52**(4), 763–778 (2013). https://doi.org/10.1007/s00466-013-0837-8

Elleuch, S., Jrad, H., Kessentini, A., Wali, M., Dammak, F.: Design optimization of implant geometrical characteristics enhancing primary stability using FEA of stress distribution around dental prosthesis. Comput Methods Biomech. Biomed. Eng. **24**(9), 1035–1051 (2021)

Ewolo Ngak, F.P., Ntamack, G.E., Azrar, L.: Dynamic analysis of multilayered magnetoelectroelastic plates based on a pseudo-Stroh formalism and Lagrange polynomials. J. Intell. Mater. Syst. Struct. **30**(6), 939–962 (2019)

Kondaiah, P., Shankar, K., Ganesan, N.: Pyroelectric and pyromagnetic effects on multiphase magneto–electro–elastic cylindrical shells for axisymmetric temperature. Smart Mater. Struct. **22**(2), 025007 (2012)

Kpeky, F., Abed-Meraim, F., Boudaoud, H., Daya, E.M.: Linear and quadratic solid–shell finite elements SHB8PSE and SHB20E for the modeling of piezoelectric sandwich structures. Mech. Adv. Mater. Struct. **25**(7), 559–578 (2018)

Marinkovic, D., Rama, G.: Co-rotational shell element for numerical analysis of laminated piezoelectric composite structures. Compos. B **125**, 144–156 (2017)

Mellouli, H., Jrad, H., Wali, M., Dammak, F.: Geometrically nonlinear meshfree analysis of 3D-shell structures based on the double directors shell theory with finite rotations. Steel Compos. Struct. **31**(4), 397–408 (2019)

Milazzo, A.: A one-dimensional model for dynamic analysis of generally layered magneto-electro-elastic beams. J. Sound Vib. **332**(2), 465–483 (2013)

Nan, C.W., Bichurin, M.I., Dong, S., Viehland, D., Srinivasan, G.: Multiferroic magnetoelectric composites: historical perspective, status, and future directions. J. Appl. Phys. **103**(3), 1 (2008)

Pan, E.: Exact solution for simply supported and multilayered magneto-electro-elastic plates. J. Appl. Mech. **68**(4), 608–618 (2001)

Rama, G.: A 3-node piezoelectric shell element for linear and geometrically nonlinear dynamic analysis of smart structures. Facta Universitatis Ser. Mech. Eng. **15**(1), 31–44 (2017)

Rama, G., Marinkovic, D., Zehn, M.: Efficient three-node finite shell element for linear and geometrically nonlinear analyses of piezoelectric laminated structures. J. Intell. Mater. Syst. Struct. **29**, 345–357 (2018)

Rao, M.N., Tarun, S., Schmidt, R., Schröder, K.U.: Finite element modeling and analysis of piezo-integrated composite structures under large applied electric fields. Smart Mater. Struct. **25**(5), 055044 (2016)

Simo, J.C., Fox, D.D., Rifai, M.: On a stress resultant geometrically exact shell model. Part II: the linear theory; computational aspects. Comput. Methods Appl. Mech. Eng. **73**(1), 53–92 (1989)

Sladek, J., Sladek, V., Krahulec, S., Pan, E.: The MLPG analyses of large deflections of magnetoelectroelastic plates. Eng. Anal. Bound. Elem. **37**(4), 673–682 (2013)

Zhang, P., Qi, C., Fang, H., Ma, C., Huang, Y.: Semi-analytical analysis of static and dynamic responses for laminated magneto-electro-elastic plates. Compos. Struct. **222**, 110933 (2019)

Analysis of Nonlinear Behavior of Smart MEE Composite Plate

Hajer Ellouz[1,2](✉), Hanen Jrad[1,2], Mondher Wali[1,2], and Fakhreddine Dammak[1,2]

[1] Laboratory of Electromechanical Systems (LASEM), National Engineering School of Sfax, University of Sfax, Route de Soukra Km 4, 3038 Sfax, Tunisia
hajer.ellouz@yahoo.fr, hanen.jrad@essths.u-sousse.tn, Fakhreddine.dammak@enis.tn

[2] Higher School of Sciences and Technology of Hammam Sousse, University of Sousse, Rue Lamine Abassi, 4011, Hammam Sousse, Tunisia

Abstract. With the advancement of technology, an intelligent mix of important material qualities is being employed advantageously in various engineering domains. Over the past few decades, considerable attention has been devoted to thin structures that combine piezoelectric material layers or patches, due to their many potential applications. Accurate modeling for multi-physics coupled situations is a significant challenge. This paper deals with geometrically nonlinear analysis of Magneto-electro-elastic composite materials (MEE) shell structure. In the presented model, the governing equations are developed via the high-order shear deformation theory (HSDT), which provides a high distribution of the displacement field by taking into account the effect of transverse shear deformations. Numerical studies are conducted to validate the numerical stability of the proposed model in order to predict the behavior of MEE composite shell structure. The large deflection behavior of MEE square plate with surface-bonded piezoelectric patches is investigated in order to outline the effectiveness of the developed model.

Keywords: MEE plate · HSDT · smart shell structure

1 Introduction

Magneto-electro-elastic materials (MEE) have been opening towards new interesting and actual applications in several technological fields, counting the next generation of transport vehicles, vibration control, structural health observation, medical instruments, sensor and actuator uses, robotics, energy harvesting and many others. In fact, the MEE materials have magneto-electric coupling effect that carry out the mutual conversion among the elastic, electric and magnetic form, which allows designing multi-functional components able of accomplishing specific tasks with great performance requirements, Nan et al. (2008); Carrera and Nali (2010); Kondaiah et al. (2012); Milazzo (2013).

To explore MEE materials, several researchers have used various theories and methods. Pan (2001) proposed an exact solution for simply supported and multilayered MEE

A. Akrout et al. (Eds.): ICAV 2022, ACM 22, pp. 52–60, 2023.
https://doi.org/10.1007/978-3-031-34190-8_7

plates under surface load and internal load. Zhang et al. (2019) suggested a semi-analytical model to investigate vibration of laminated MEE composite plate. Further, Ngak et al. (2019) employed a semi-analytical three-dimensional method to analyze the dynamic response of multilayer MEE plate. Despite analytical approaches can offer closed-form solutions, they unfortunately are limited to elementary geometries, certain kinds of gradation of material properties, particular types of boundary cases and specific loading conditions. To accomplish more common analysis, it is necessary to resort to numerical methods and particularly finite element method (FEM). Mainly, the linear shell theories are inspected for shell structures showing small displacements under various types of loads. Nonlinear analysis of shells behavior, however, has been realized in the context when shells undertake large deformations. Several research works are obtainable in literature to analyze the nonlinear mechanical behavior of shell structures. Moreover, investigations of geometrically nonlinear behavior of smart shell structures have been carried out in the works of Balamurugan and Narayanan (2008) Marinkovic and Rama (2017), Rama et al. (2017, 2018);

Further, over the years, numerical methods are widely used to provide a way to approximate solutions to problems in many engineering disciplines such as aerospace, mechanical, and electrical, quickly and easily compared to analytic solutions and when analytical solutions aren't possible. In addition, finite element analysis (FEA) allows simulations to predict and understand the structure behavior under several physical conditions. It is a numerical approach to finding the real results and providing an approximate solution to different problems in mechanical engineering. A design optimization of dental implant geometrical characteristics enhancing primary stability is proposed by (Elleuch et al. 2021) using FEA of stress distribution around dental prosthesis. A numerical methodology to characterize aluminum behavior considering non-associated plasticity model coupled with isotropic ductile damage was carried out by (Bouhamed et al. 2021, 2022) in order to identify ductile damage constitutive equations for thin sheet metal applications.

The purpose of this research is to examine for the first time in the literature, the geometrically nonlinear behavior of MEE shell enduring large deformations and finite rotations. It is based on double directors MEE shell model, which assumes the high order shear deformation theory (HSDT) and quasi-static behavior for the electric and magnetic fields. Geometrically nonlinear analysis of MEE square plate with surface-bonded piezoelectric layers is conducted in order to outline the accuracy and performance of the present model.

2 Basic Formulation of Double Director MEE Shell

The basic formulation of the double director shell theory, used in this research to simulate geometrically non-linear response of an MEE shell, is briefly reviewed in this section. For convenience of presentation, the Cartesian tensor notation is adopted, Auricchio and Sacco, (2003); Alaimo et al. (2013) and Bagheri et al. (2018). Parameterizations of the shell material points are realized in terms of curvilinear coordinates $\boldsymbol{\xi} = (\xi, \eta, z)$. $\boldsymbol{X}_p$ and $\boldsymbol{x}_p$ are denoted to be the position vectors of one point p of the mid surface in the reference configuration C_0 and the current configuration C_t, respectively and can be described as:

$$\begin{cases} \boldsymbol{X}_q(\xi,\eta,z) = \boldsymbol{X}_p(\xi,\eta) + z\boldsymbol{D}(\xi,\eta) \\ \boldsymbol{x}_q(\xi,\eta,z) = \boldsymbol{x}_p(\xi,\eta) + f_1(z)\boldsymbol{d}_1(\xi,\eta) + f_2(z)\boldsymbol{d}_2(\xi,\eta); \quad z \in [-h/2, h/2] \\ \quad f_1(z) = z - \frac{4z^3}{3h^2}; \; f_2(z) = \frac{4z^3}{3h^2} \end{cases} \tag{1}$$

where h is the thickness; $\boldsymbol{D}$ is the initial MEE shell director which is perpendicular to the undeformed mid-surface; $\boldsymbol{d}_1$ and $\boldsymbol{d}_2$ represent the double director MEE shell vectors in the deformed configuration.

The Lagrangian strain components can be written as follows:

$$\begin{cases} \varepsilon_{\alpha\beta} = e_{\alpha\beta} + f_1(z)\chi^1_{\alpha\beta} + f_2(z)\chi^2_{\alpha\beta} \\ 2\varepsilon_{\alpha 3} = f_1^{'}(z)\gamma^1_\alpha + f_2^{'}(z)\gamma^2_\alpha \end{cases} ; \; \alpha, \beta = 1, 2 \tag{2}$$

in which $e_{\alpha\beta}$, $\chi^k_{\alpha\beta}$ and γ^k_α denoted the membrane, the bending and the shear strains respectively. These components are given in the following form according to Mellouli et al. (2019):

$$\begin{cases} \delta\boldsymbol{e} = \boldsymbol{B}_m.\delta\boldsymbol{x}; \; \delta\boldsymbol{\chi}^k = \boldsymbol{B}_{bmk}\delta\boldsymbol{x} + \boldsymbol{B}_{bbk}\delta\boldsymbol{d}_k \\ \delta\boldsymbol{\gamma}^k = \boldsymbol{B}_{smk}\delta\boldsymbol{x} + \boldsymbol{B}_{sbk}\delta\boldsymbol{d}_k \end{cases} ; \; k = 1, 2 \tag{3}$$

Based on double director theory, and in order to insure the condition of zero transverse shear stress at top and bottom surfaces, $\delta\boldsymbol{\gamma}^2$ should be neglected in discrete form of the variational formulation.

Further, the proposed model consists of a core made of one magneto-electro-elastic and two materials with two integrated active layers. For that reason, one magnetic degree of freedom ψ is assumed for the MEE layer and one electrical degree of freedom φ is assumed for each of piezolayers. The linear distribution of the electric and magnetic field vector $\boldsymbol{E}$ and $\boldsymbol{H}$, which can be expressed as:

$$\begin{aligned} \boldsymbol{E} &= -\varphi_{,\alpha} \\ \boldsymbol{H} &= -\psi_{,\alpha} \end{aligned} \tag{4}$$

The variation of work done by external forces, can be described as:

$$G = \int_A \left(\boldsymbol{N}\delta\boldsymbol{e} + \sum_{k=1}^{2} \left(\boldsymbol{M}_k \delta\boldsymbol{\chi}^k \right) + \boldsymbol{T}_1 \delta\boldsymbol{\gamma}^1 + \tilde{\boldsymbol{q}}\delta\,\boldsymbol{E} + \tilde{b}.\delta\boldsymbol{H} \right) dA - G_{ext} = 0 \tag{5}$$

where $\boldsymbol{N}$, $\boldsymbol{M}_{\boldsymbol{k}}$ and $\boldsymbol{T}_1$ represent respectively the membrane, the bending and the shear stress resultants. $\delta\boldsymbol{e}$, $\delta\chi^k$ and $\delta\gamma^1$ denote the shell strains. $\overline{\boldsymbol{q}}$ and $\boldsymbol{E}$ are the resultant electric displacement and the electric field, respectively. $\boldsymbol{b}$ and $\boldsymbol{H}$ are the magnetic induction and the magnetic field.

Next, by using the weak form of the equilibrium equation, the procedure of Newton iterative will be employed to solve the nonlinearity of the problem. The tangent operator will be divided into geometric $D_G G.\boldsymbol{\Delta\Phi}$ and material parts $D_M G.\boldsymbol{\Delta\Phi}$:

$$DG.\boldsymbol{\Delta\Phi} = D_G G.\boldsymbol{\Delta\Phi} + D_M G.\boldsymbol{\Delta\Phi}; \begin{cases} D_M G.\boldsymbol{\Delta\Phi} = \int_A \delta\boldsymbol{\Sigma}^T \boldsymbol{\Delta R}\, dA \\ D_G G.\boldsymbol{\Delta\Phi} = \int_A \left(\boldsymbol{\Delta}\delta\boldsymbol{\Sigma}^T.\boldsymbol{R} \right) dA \end{cases} \tag{6}$$

where $\boldsymbol{R}$ and $\boldsymbol{\Sigma}$ represent the generalized resultant of stress and strain vectors, which are expressed as:

$$\begin{aligned} \boldsymbol{R} &= \left[\boldsymbol{N}\ \boldsymbol{M}_1\ \boldsymbol{M}_2\ \boldsymbol{T}_1\ \tilde{\boldsymbol{q}}\ \ \tilde{\boldsymbol{b}}\right]^T_{17\times 1}; \\ \boldsymbol{\Sigma} &= \left[\boldsymbol{e}\ \boldsymbol{\chi}^1\ \boldsymbol{\chi}^2\ \boldsymbol{\gamma}^1\ -\boldsymbol{E}\ \ -\boldsymbol{H}\right]^T_{17\times 1} \end{aligned} \tag{7}$$

After developing the weak form of equilibrium, the interpolation of the displacement, electric and stain fields, can be undertaken within a four nodes finite shell element. Therefore, the magneto-electro-elastic equilibrium equation can be seen as:

$$\boldsymbol{K}\Delta\boldsymbol{\Phi} = \boldsymbol{F} \tag{8}$$

where $\boldsymbol{F}$ is the contribution of both internal and external work. $\boldsymbol{K}$ denotes generalized tangent stiffness matrix englobing the material and geometric tangent matrix. $\boldsymbol{\Delta\Phi} = (\boldsymbol{\Delta u}, \boldsymbol{\Delta d}_1, \boldsymbol{\Delta d}_2, \boldsymbol{\Delta}\varphi, \boldsymbol{\Delta}\psi)$ symbolized the increments of the generalized displacement.

3 Numerical Results

In this section, the nonlinear static behavior of square plate with piezoelectric patches is performed to demonstrate the accuracy and the performance of the proposed MEE shell element.

3.1 Aluminum Square Plate with Piezoelectric Patches

Let's consider a fully clamped square aluminum plat of 200 × 200 mm^2 with a thickness of 8 mm. Four piezoelectric layers, made of PZT-5H and having dimensions of 40 × 40 mm^2 with a thickness of 1 mm, bonded to the lower and upper surfaces of the plate. The square plate with piezoelectric patches is shown in Fig. 1.

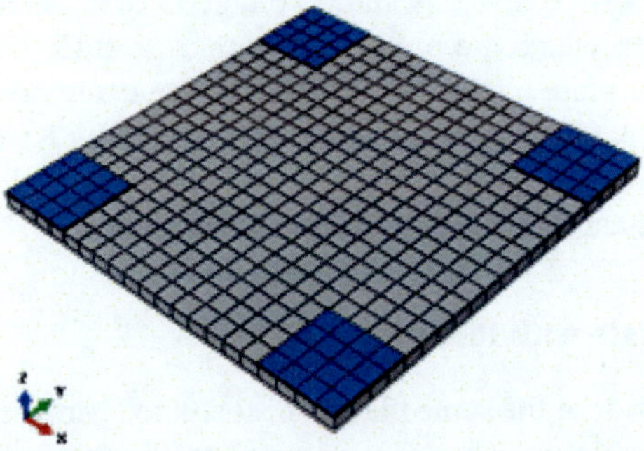

Fig. 1. Description of the square plate with piezoelectric patches

The material properties of the structure are listed in Table 1, according to Kpeky et al. (2018). The square aluminum plate is meshed using 20 × 20 elements and the

Table 1. Material properties

PZT-5H		Aluminum
$\rho = 7730$ kg.m^{-3} $C_{11} = C_{22} = C_{33} = 126$ GPa $C_{44} = C_{55} = C_{66} = 23$ GPa	$e_{15} = e_{24} = 17$ C/m^2 $e_{31} = e_{32} = -6.5$ C/m^2; $e_{33} =$ 23.3 C/m^2 $\kappa_{11} = \kappa_{22} = 1.503 \times 10^{-8}$ F/m; $\kappa_{33} = 1.3 \times 10^{-8}$ F/m	$\rho = 2690$ kg.m^{-3} $E = 70.3$ GPa $\nu = 0.345$

Table 2. Centerline deflection of square plate under uniform load and various values of voltage

Central line deflection ($\times 10^{-5}$/mm)	Voltage (V)	Position along x-direction (mm)					
		0	40	80	120	160	**200**
Kpeky et al. (2017)	0	0	−4.995	−8.856	−8.856	−4.995	0
	−10	0	−2.449	−4.427	−4.433	−2.438	0
	−20	0	−0.1613	−0.011	−0.016	−0.144	0
Present Model	0	0	−4.123	−9.145	−9.145	−4.123	0
	−10	0	−2.001	−4.340	−4.340	−2.001	0
	−20	0	−0.151	−0.027	−0.027	−0.151	0
Error %	0	–	21.15	3.16	3.16	21.15	–
	−10	–	22.38	2.00	2.14	21.84	–
	−20	–	6.82	59.26	40.74	4.63	–

PZT-5H patches are meshed using 4 × 4 elements, respectively, along axial and width directions.

A uniformly distributed mechanical load of 100 N.m^{-2} is applied over the whole surface of the active square plate. A static voltage is then given incrementally to the piezoceramic actuators, which are polarized in opposite sides. Static centerline deflections of the composite plate along the x-direction, are examined under different input voltages. The present predictions and solutions, provided by Kpeky et al. (2018), using solid shell elements SHB8PSE and SHB20E, are shown in Table 2. It can be seen that the results are in good agreement with the alternative solutions.

3.2 MEE Square Plate with Piezoelectric Patches

The mechanical behavior of the same plate is next studied using magneto-electro-elastic material instead of aluminum. The fully clamped MEE plate is made of $BF50\%$. Four piezoelectric layers are made of PZT-5, which are bonded to the top and bottom surfaces. Thicknesses of the square plate and the piezoelectric layers are, respectively, 8 and 1 mm. The material properties of the MEE plate is listed in Table 3, according to Sladek et al. (2013) and Daga et al. (2009).

Table 3. Material properties of MEE plate

BF50%	PZT-5
$C_{11} = C_{22} = 213$ GPa; $C_{33} = 207$ GPa $C_{12} = 113$ GPa; $C_{13} = C_{23} = 112.8$ GPa $C_{44} = C_{55} = 49.9$ GPa; $C_{66} = 50$ GPa $e_{15} = e_{24} = 0.150\ C/m^2$ $e_{31} = e_{32} = -2.710 C/m^2$; $e_{33} = 8.860 C/m^2$ $\kappa_{11} = \kappa_{22} = 0.24 \times 10^{-9}$ C/Vm; $\kappa_{33} = 6.37 \times 10^{-9}$ C/Vm $\eta_{31} = \eta_{32} = 222$ N/Am; $\eta_{33} = 292$N/Am $\eta_{15} = \eta_{24} = 185$N/Am $\mu_{11} = \mu_{22} = 2.01 \times 10^{-4} Ns^2/C^2$; $\mu_{33} = 0.839 \times 10^{-4} Ns^2/C^2$ $d_{11} = d_{22} = -5.23 x 10^{-12}$ Ns/VC; $d_{33} = 2750 \times 10^{-12}$ Ns/VC $\rho = 5550\ kg.m^{-3}$	$C_{11} = C_{22} = 99.2$GPa $C_{33} = 86.85$ GPa $C_{12} = 54.01$ GPa $C_{13} = C_{23} = 50.77$GPa $C_{44} = C_{55} = 21.1$ GPa $e_{15} = e_{24} = 12.32\ C/m^2$ $e_{31} = e_{32} = -7.2 C/m^2$ $e_{33} = 15.11 C/m^2$ $\kappa_{11} = \kappa_{22} = 1.53 \times 10^{-8}$ F/m $\kappa_{33} = 1.5 \times 10^{-8}$ F/m

Figure 2 illustrates the static response of the MEE plate under uniform load of 100 $N.m^{-2}$ and various values of voltage applied to the piezo-electric actuators. From Fig. 2, it can be seen that the curve of the centerline deflection has the same shape as its alternative obtained by the numerical model of Kpeky et al. (2018), for a square plate with piezoelectric patches subjected to electromechanical loads. Further, the centerline deflection of the MEE plate decreases with the increase of the intensity of the electric charges.

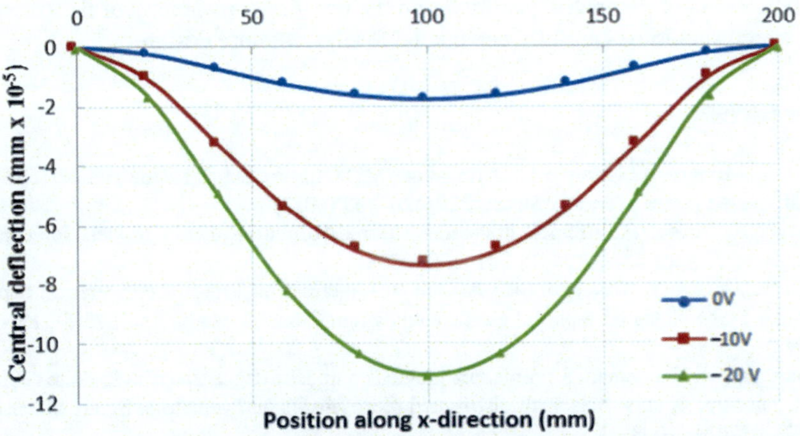

Fig. 2. Centerline deflection along x-direction

The effect of the ratio a/h on the deflection of the MEE plate is examined in Fig. 3. It can be observed that with the increase of the geometrical ratio *a/h*, the centerline deflection increases. Hence, the geometrical ratio *a/h* has a great impact on the smart magneto-electro-mechanical behavior of the studied structure. Indeed, with the rise of *a/h* ratio, the structure become thinner which allows more flexibility of the MEE plate.

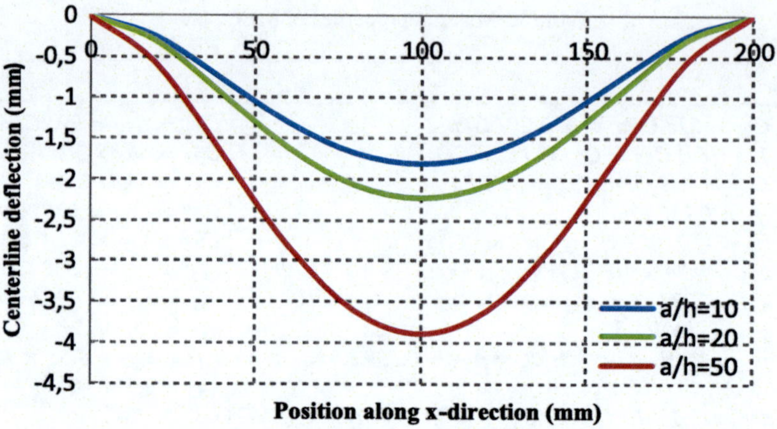

Fig. 3. Effect of geometrical ratio on centerline deflection along x-direction

4 Conclusion

In this paper, geometrically nonlinear behavior of square plate with surface-bonded piezoelectric patches is examined using a finite element procedure based on the high-order shear deformation theory. The effectiveness of the present method is proved by validating the obtained results against those of other studies available in the literature. The extension of this formulation for its application to smart MEE is proposed. A parametric study including the variation of geometrical parameter is performed. Numerical results reveal the effect of geometric parameter on the non-linear responses of the smart MEE plate, which should be carefully considered for the structure design.

References

Alaimo, A., Milazzo, A., Orlando, C.: A four-node MITC finite element for magneto-electro-elastic multilayered plates. Comput. Struct. **129**, 120–133 (2013)

Auricchio, F., Sacco, E.: Refined first-order shear deformation theory models for composite laminates. J. Appl. Mech. **70**(3), 381–390 (2003)

Bagheri, H., Kiani, Y., Eslami, M.R.: Asymmetric thermal buckling of temperature dependent annular FGM plates on a partial elastic foundation. Comput. Math. Appl. **75**(5), 1566–1581 (2018)

Balamurugan, V., Narayanan, S.: A piezolaminated composite degenerated shell finite element for active control of structures with distributed piezosensors and actuators. Smart Mater. Struct. **17**(3), 035031 (2008)

Bouhamed, A., et al.: Identification of fully coupled non-associated-Ductile damage constitutive equations for thin sheet metal applications: numerical feasibility and experimental validation. Thin-Walled Struct. **176**, 109365 (2022)

Bouhamed, A., Mars, J., Jrad, H., Wali, M., Dammak, F.: Experimental and numerical methodology to characterize 5083-aluminum behavior considering non-associated plasticity model coupled with isotropic ductile damage. Int. J. Solids Struct. **229**, 111139 (2021)

Carrera, E., Nali, P.: Multilayered plate elements for the analysis of multifield problems. Finite Elem. Anal. Des. **46**(9), 732–742 (2010)

Daga, A., Ganesan, N.A., Shankar, K.: Behaviour of magneto-electro-elastic sensors under transient mechanical loading. Sens. Actuators, A **150**(1), 46–55 (2009)

Elleuch, S., Jrad, H., Kessentini, A., Wali, M., Dammak, F.: Design optimization of implant geometrical characteristics enhancing primary stability using FEA of stress distribution around dental prosthesis. Comput. Methods Biomech. Biomed. Eng. **24**(9), 1035–1051 (2021)

Ewolo Ngak, F.P., Ntamack, G.E., Azrar, L.: Dynamic analysis of multilayered magnetoelectroelastic plates based on a pseudo-Stroh formalism and Lagrange polynomials. J. Intell. Mater. Syst. Struct. **30**(6), 939–962 (2019)

Kondaiah, P., Shankar, K., Ganesan, N.: Pyroelectric and pyromagnetic effects on multiphase magneto–electro–elastic cylindrical shells for axisymmetric temperature. Smart Mater. Struct. **22**(2), 025007 (2012)

Kpeky, F., Abed-Meraim, F., Boudaoud, H., Daya, E.M.: Linear and quadratic solid–shell finite elements SHB8PSE and SHB20E for the modeling of piezoelectric sandwich structures. Mech. Adv. Mater. Struct. **25**(7), 559–578 (2018)

Mellouli, H., Jrad, H., Wali, M., Dammak, F.: Geometrically nonlinear meshfree analysis of 3D-shell structures based on the double directors shell theory with finite rotations. Steel Compos. Struct. **31**(4), 397–408 (2019)

Milazzo, A.: A one-dimensional model for dynamic analysis of generally layered magneto-electro-elastic beams. J. Sound Vib. **332**(2), 465–483 (2013)

Nan, C.W., Bichurin, M.I., Dong, S., Viehland, D., Srinivasan, G.: Multiferroic magnetoelectric composites: historical perspective, status, and future directions. J. Appl. Phys. **103**(3), 1 (2008)

Pan, E.: Exact solution for simply supported and multilayered magneto-electro-elastic plates. J. Appl. Mech. **68**(4), 608–618 (2001)

Rama, G.: A 3-node piezoelectric shell element for linear and geometrically nonlinear dynamic analysis of smart structures. Facta Universitatis Series: Mech. Eng. **15**(1), 31–44 (2017)

Rama, G., Marinkovic, D., Zehn, M.: Efficient three-node finite shell element for linear and geometrically nonlinear analyses of piezoelectric laminated structures. J. Intell. Mater. Syst. Struct. **29**, 345–357 (2018)

Sladek, J., Sladek, V., Krahulec, S., Pan, E.: The MLPG analyses of large deflections of magnetoelectroelastic plates. Eng. Anal. Bound. Elem. **37**(4), 673–682 (2013)

Zhang, P., Qi, C., Fang, H., Ma, C., Huang, Y.: Semi-analytical analysis of static and dynamic responses for laminated magneto-electro-elastic plates. Compos. Struct. **222**, 110933 (2019)

Non-destructive Identification of Damage Mechanisms in Unidirectional Composites by Acoustic Emission and Machine Learning-Based Clustering

Mariem Ben Hassen[1,3](✉), Sahbi Tamboura[1], Joseph Fitoussi[2], and Hatem Mrad[3]

[1] Écolee Nationale d'Ingénieurs de Sousse, LMS, Pôle Technologique, Route de Ceinture, 4054 Sousse, Tunisie
mariem.benhassen@uqat.ca

[2] Arts et Métiers ParisTech (ENSAM), PIMM – UMR CNRS 8006, 151 Boulevard de l'Hôpital, 75013 Paris, France
joseph.fitoussi@ensam.eu

[3] Université du Québec en Abitibi-Témiscamingue (UQAT), 445 Bd de l'Université, Rouyn-Noranda J9X 5E4, Canada

Abstract. The objective of the present experimental work is to analyze and identify the damage mechanisms on various composite materials. The present paper deals with different types of composites that were subjected to static tensile tests in order to investigate their damage mechanisms. A progressive approach consisting of working on different, unidirectional composites with different fiber orientations. The specimens were tested under static tensile tests. All experimental tests carried out were monitored by acoustic emission (AE): a non-destructive technique to detect the appearance and identify the evolution of major damage mechanisms. An unsupervised clustering technique, based on pattern recognition and widely used in machine learning (ML), the k-means clustering technique, was used to classify the AE recorded events based on a temporal parameter such as amplitude range, the cumulative number of hits, duration, and the acoustic energy. Four classes of AE events were identified for most materials depending on the specimen configuration. A comparison was carried out between AE data sets processed using machine learning based clustering (MLC) method and commercial software.

Keyword: Acoustic emission · Damage · Machine learning · Clustering

1 Introduction

A major effort has been made in recent years to control the long-term behavior of composites materials and anticipate their breakage. For this purpose, it is essential to have the best possible knowledge, in terms of microstructure and mechanical properties to be able to monitor these properties under stress in order to identify the damage mechanisms involved, their initiation and development until failure.

A. Akrout et al. (Eds.): ICAV 2022, ACM 22, pp. 60–69, 2023.
https://doi.org/10.1007/978-3-031-34190-8_8

In this context, acoustic emission (AE) is a good candidate because of their advantages as it is a non-destructive technique, applicable on a structure in service, including in difficult environmental conditions. Acoustic emission (AE) technique has been used extensively for the characterization and identification of micro failure mechanisms in composites (Dahmene et al. 2015). Moreover, acoustic emission was carried out on glass or carbon fiber reinforced composites often involve mixed time-frequency analyses (Sause et al. 2012). Romhány (Romhány et al.2003) found that AE processing could be used to identify the failure mode sequence and to separate the damage mechanisms of flax fiber composites during static test. AE technique has also been performed for several natural fiber composites with different reinforcements and matrices (Kersani et al. 2015) and at different observation scales.

Haggui (Haggui et al. 2019) used AE technical to pick up the various damage mechanisms that develop under static and dynamic tests. As a result, four classes of acoustic events were identified. These are matrix microcracking, fiber-matrix debonding, fiber pull-out, and fiber breakage. Also, the combination between the microscopic analysis and AE processing in the study of Monti et al. (Monti et al. 2017) for different configurations of composites subjected to static testing show that the first registered defects are matrix cracking and fiber-matrix debonding. Then, fiber pulls out and fiber breakage accumulates to provoke the final failure.

The present paper deals with different types of composites that were subjected to quasi-static tensile loading, the tensile tests were monitored by acoustic emission to pick up the different damage mechanisms. This study aims to identify and classify the failure mechanisms occurring under the load for unidirectional laminates composites with different fibers orientation. After processing data, and recording acoustic emission events, an unsupervised clustering technique, based on pattern recognition and widely used in ML, was used to identify the recorded events.

2 Materials and Experimental Setup

2.1 Materials

The composite materials studied in this work are PPGF40: thermoplastic composites filled with short fibers, it is a polypropylene filled with glass fibers (PPGF 40) resulting from the mixture of polypropylene granules manufactured by Styron loaded with long glass fibers with a mass rate of 40% and a length of 6 to 12 mm in the material and pure polypropylene granules. It is obtained by injection. GMT50: It is a propylene reinforced with long glass fiber, cut into wicks of 4 to 5 cm length, performed by thermocompression. SMC, thermosetting matrix composites reinforced with calcium carbonate particles (CaCO3) obtained by the injection molding process (Sheet Molding Compounds (SMC)), two types of SMC used in our study are standard SMC and high-performance SMC. The standard SMC based on polyester reinforced with Glass fiber wick of approximately 25 mm (30% mass of the composite). The A-SMC, or Advanced SMC, is an SMC based on vinylester resin allowing the impregnation of a high percentage of glass fibers (>50% of the mass), which increases the mechanical performance.

Before conducting mechanical tests, the samples for different types of materials used in this work, were prepared as below.

The samples for GMT50 and A-SMC were sliced using a bar-shaped rotary blade cutting machine with a dimension of 130 mm in length, 19 mm in width and 3.3 mm in thickness. It has been ensured that all cut samples are perpendicular to the plane. The tensile tests were carried out on dumbbell-type samples for PPGF40 and on double dumbbell specimens for SMC-Standard. This geometry provides a uniform behavior, with a final fracture located in the useful area of the specimens.

For the composites studied in this work, three types of unidirectional specimens with fiber directions of 90°, 45° and 0° except the PPGF40 only two orientations of 45° and 0°.

2.2 Experimental Setup

For each type of sample, five rectangular specimens were examined under uniaxial loading. Two tensile machines of 10 and 50 kN load cell were used to perform the tests until failure. The experiments were conducted using a 5 mm/min displacement rate at room temperature. As an example, the load versus displacement curve for all GMT50 orientations were illustrated in Fig. 1(b). Moreover, the tests were followed by AE, in this study two sensors with 100 kHz to 1 MHz bandwidth and two preamplifiers with a 40 dB gain were used. The input of the sensors was clamped to the samples using a coupling agent (Fig. 1(a)), while their output was linked to the preamplifiers and then to an EPA (Euro Physical Acoustic) acquisition system. To check the system functionality and to set the level of amplitude to distinguish signal from noise, the pencil lead breaks procedure were executed and the threshold was set at 30 dB. Preliminary tests led to the selection of the following acquisition settings: Peak Definition Time PDT = 50 μs, Hit Definition Time HDT = 150 μs and Hit Lockout Time HLT = 300 μs.

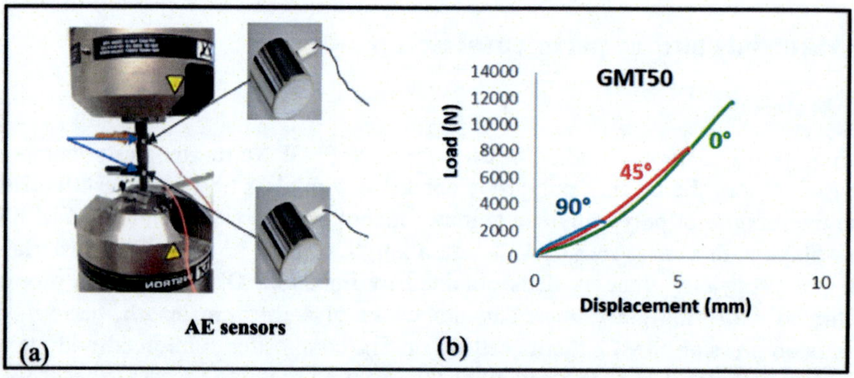

Fig. 1. (a): AE sensors installation for acoustic emission test and (b) GMT50 load curve

3 Results and Discussion

3.1 Acoustic Emission Data Processing Using Machine Learning Based Clustering

In order to examine and investigate the different damage mechanisms within the studied materials, it is very useful to apply a non-destructive technique during mechanical tests such as the AE. After recording the AE data, clustering analysis was performed to classify the different types of damage. The unsupervised clustering K-means method was used in this work. The choice of this algorithm was based on the high efficiency of the algorithm and the simplicity of the class separations by calculating the Euclidean distance between each x vector et the centers of k classes.

K-Means method (Likas et al. 2003) is part of the family of algorithms of "mobile centers" classification. It is an iterative method of partitioning data by minimizing within-group variance. The purpose of this method is to partition the recoded events into K clusters by considering five temporal parameters: amplitude, rise time, duration, number of counts to peak and energy.

A database of acoustic emission signals from tensile tests on the three composite materials stressed at different orientations with respect to the axis of the fibers was used. An acoustic emission signal is assimilated to a vector with five components. The five temporal parameters were considered as effective parameters for data clustering. Amplitude, rise time, duration, numbers of counts to peak and energy. They are described in detail below and presented in Fig. 2 (Ben Ameur et al. 2019).

- Amplitude: is the signal's peak value in an AE event, spanning the range of 30 dB to 100 dB.
- Threshold: is a defined level for distinguishing signal from noise.
- Rise time: is the amount of time it takes for the signal to reach amplitude after crossing the threshold.
- Duration: is the elapsed time between the event signal's first and last threshold crossing.
- Number of counts to peak: is the total number of times the signal passes through the threshold from peak amplitude to threshold.
- Energy: is the area under the amplitude-time curve for an event.

3.1.1 Acoustic Emission Data Clustering

The clustering methodology based on the five temporal parameters was applied to identify the damage mechanisms involved in the different samples and to monitor their temporal evolution.

After classification, the results were reported by plotting the number of vectors of hits as a function of time for different configurations.

Based on the bibliography, four types of damage can occur for unidirectional laminates and to assess and analyze the evolution of this mechanism four classes are identified, a damage mechanism is associated to each class. The signal distribution is relatively similar to the three types of materials and the five-time descriptors.

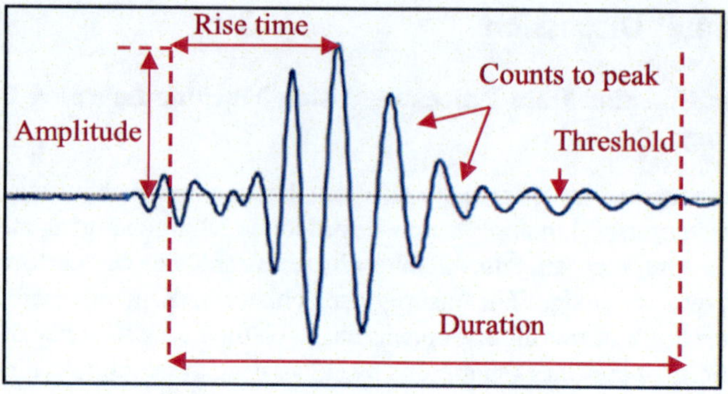

Fig. 2. Characteristics of AE signals

Class 0 includes the first events appearing in the specimen chronology. Therefore, Class 0 can be associated with matrix cracking. Class 1 can be attributed to the mechanisms of fiber/matrix debonding. Where Class 2 is estimated to be pseudo-delamination or delamination. The last Class 3 is attributed to the fiber breakage which generally contains few events with a high energy rate and a very high amplitude up to 90 dB. The clustering process based on the five temporal parameters was used to classify the damage mechanisms occurring in the specimens prepared and to follow their temporal evolution until failure.

Figure 3 shows the distribution of AE signal duration versus time during tensile tests for the PPGF40 in the two orientations of fibers 0° and 45°, it can be shown that the signal duration distribution reveals small intersections area for unidirectional specimens proving the adoption of a multi-parameter technique

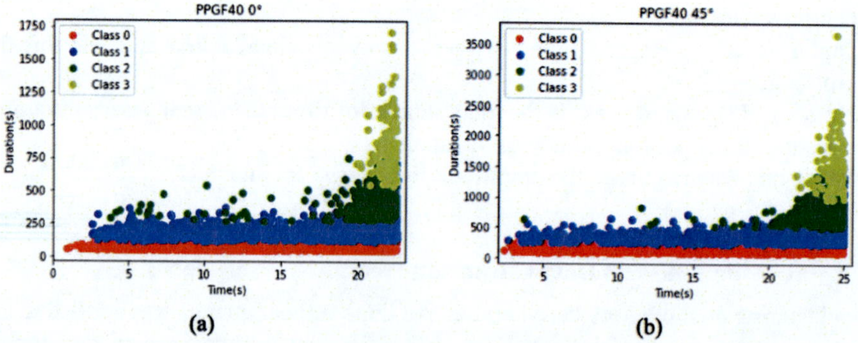

Fig. 3. Distribution of duration versus time during static tensile tests for: (a) PPGF40 0°and (b) PPGF40 45°

Figures 4 and 5 present the evolution of the cumulative number of vectors of hits versus time, for the different composites studied in this work, demonstrating the evolution of each damage mechanism with highlighting the chronology of their appearance throughout the tensile test. As illustrated, the matrix cracking events (class 0) have the

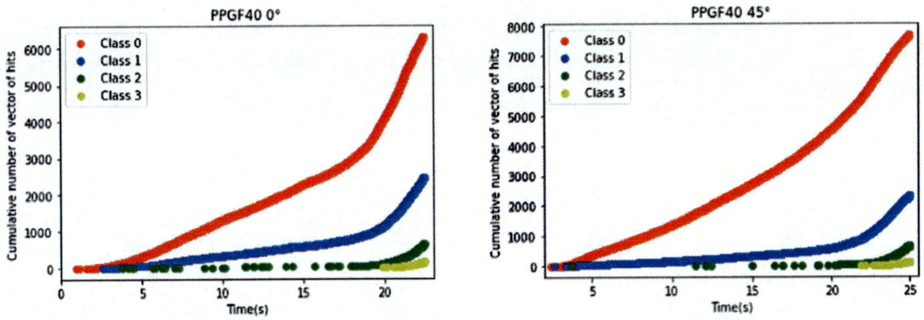

(a) (b)

Fig. 4. Distribution of cumulative number of vector of hits versus time during static tensile tests for (a) PPGF40 0° and (b) PPGF40 45°

important number of events for all specimens and it appears from the beginning of the test, while fiber-matrix debonding (class 1) has less activity and its importance varies for specimens and for fiber orientations. Delamination events (class 2) have fewer events compared to class 0 and class 1 and for the majority of specimens it appears from the middle of the test. Finally, fiber breakage, it occurs at the end of the test with fewer events than all the previous classes, and it's almost absent for some specimens.

For the A-SMC we note the differentiation in the appearance and the evolution of mechanisms from one orientation to another, for the SMC (Randomly Oriented) the last two events appear at the same time with a difference in the number of events while for the A-SMC (Highly Oriented) the fiber breakage events is present only for A-SMC 0° (parallel to the Mold Flow Direction (MFD)) and for the last two orientation's A-SMC 45° and A-SMC 90° (perpendicular to (MFD)), it's considered absent. Also, in comparison with the three A-SMC specimens used in this study, the closer the orientation of fibers is near to the perpendicular to the mold flow direction, the higher the number of events is and the shorter lifetime.

For the GMT50 the dominance of the matrix mechanism is very intense compared to the other mechanisms, while for the first specimens: GMT50 0° the last three classes are absent compared to GMT50 45° and GMT50 90°, we also notice that when the interface slope changes, the fiber-matrix debonding starts.

3.2 AE Data Processing Using Commercial Software

The PPGF40 damage mechanisms for the recorded AE signals were identified in this section using the AE classification approach. For AE data clustering, The NOESIS software based on the effective time parameters such as amplitude, rise time, duration, energy, and number of counts to peak at peak was used.The unsupervised algorithm, K-Mean (Likas, et al.2003) is employed in this study. Based on the relationship shown below, regarding the Davies and Bouldin R_{kl} coefficient's (D&B) minimum mean value, the ideal number of classes is determined (Davies and Bouldin 1979).

$$R_{kl}(D\&B) = \frac{1}{n}\sum_{k=1}^{n} max_k\left(\frac{d_k + d_l}{d_{kl}}\right) \quad (1)$$

where n, d_k, d_l and d_{kl} are respectively, the number of classes, the average distance in the kth, the average distance in the lth classes and the average distance between the two classes k and l.

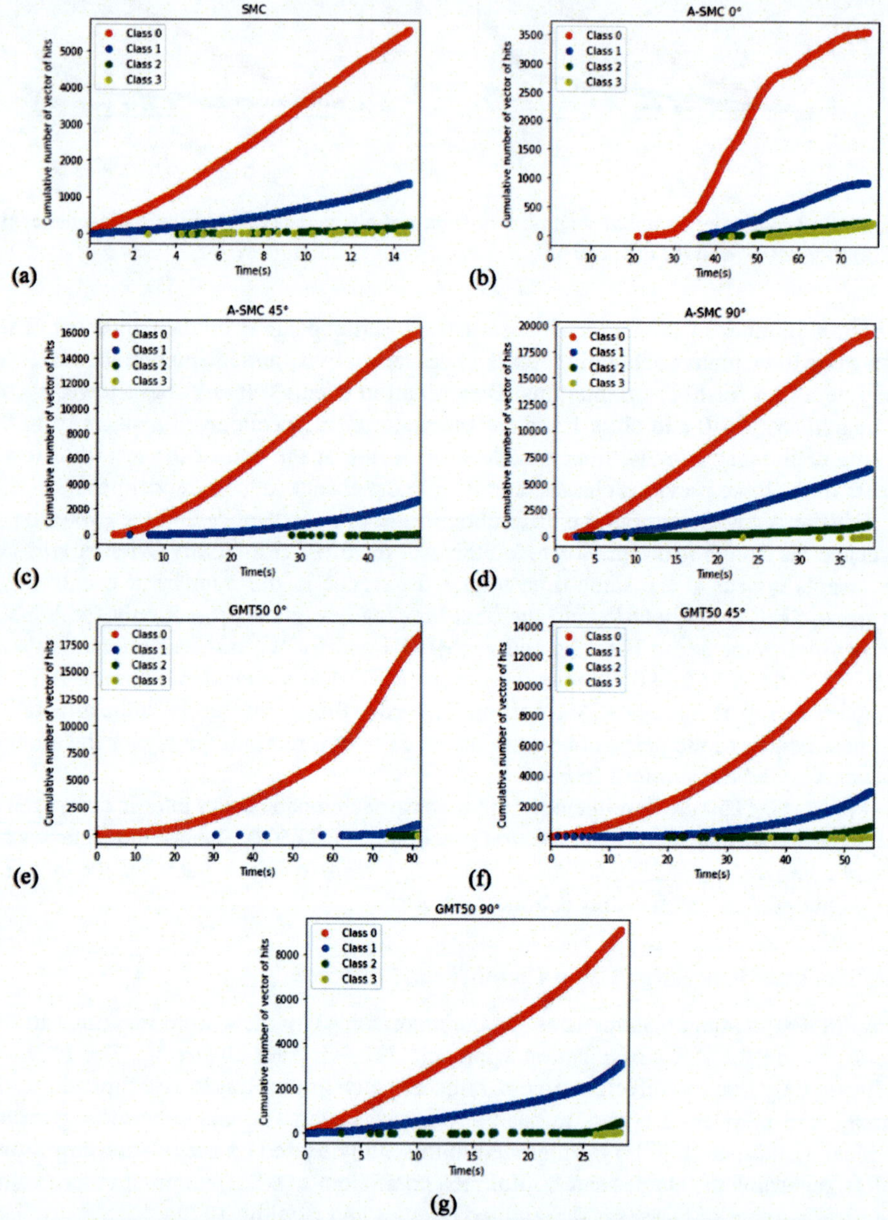

Fig. 5. Distribution of cumulative number of hits versus time for: a) SMC °, b) A-SMC 0°, c) A-SMC 45°, d) A-SMC 90°, e) GMT50 0°, f) GMT50°, and g) GMT50 90°

The acquisition threshold, was set at 30 dB, is the minimum amplitude for which a signal will be recorded. The results obtained for a PPGF 40 samples, after classification are presented by the amplitude curves and the cumulative number of vectors of as a function of time for different configurations (Fig. 6 and Fig. 7).

Four classes are identified for both PPGF40 orientations with a similar distribution. Class 0 contains the very first events in the specimen chronology. The amplitude is between [38–55 dB]. Class 0 can thus be referred to the matrix cracking mechanism. Both configurations also exhibit Class 1 at an amplitude within [54 and 65 dB]. That was associated to the mechanisms of fiber / matrix debonding. Where Class 2 is estimated to be pseudo-delamination or delamination. Rare event occurrences with a high energy rate and an amplitude up to 78 dB are typically seen in the last Class 3. Their occurrence indicates the specimen failure. Therefore, they are attributed to the rupture of the fiber breakage.

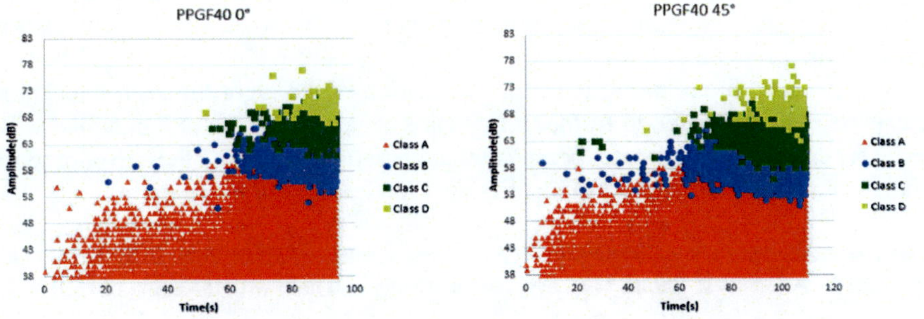

(a) (b)

Fig. 6. Distribution of amplitude versus time during static tensile tests for PPGF40 and PPGF40 45°

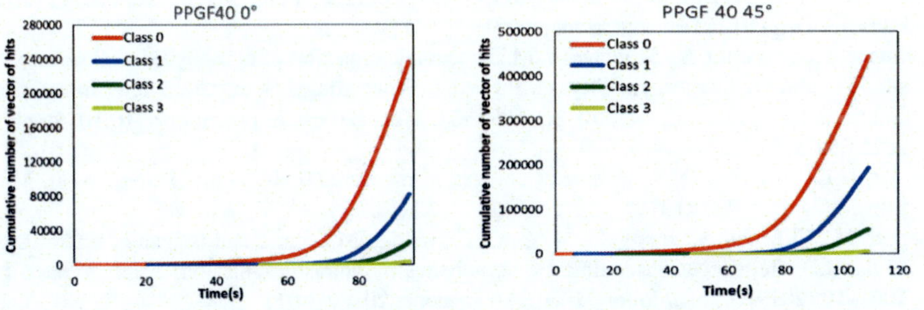

(a) (b)

Fig. 7. Distribution of cumulative number of vectors of hits versus time during static tensile tests for PPGF40 and PPGF40 45

4 Conclusions and Future Work

The objective of this work was to validate the use of a non-destructive control technique capable of distinguishing the main damage mechanisms occurring in different composite materials.

During the tests, several tools were used to identify, quantify, and qualify the damage mechanisms for different composites. First, a real-time multiparametric acoustic emission (EA) analysis was coupled to the mechanical static tensile tests. The classification of AE signals, using the K-means method, distinguished four or three classes depending on the nature of composites and on the fiber's orientations. The results obtained by using commercial software are more refined in terms of class separations, while the AE classification using MLC used in this work did not allow a perfect identification of every damage mechanism occurring during tensile tests as it shows overlap between classes that require refinements, due to the influence of the starting point chosen for the initialization of class centers and the algorithm results. Despite the imperfections, the results using MLC are reliable and satisfy the need of this work as it can cut costs to a great extent.

Therefore, in future work, after the application of the algorithm, other techniques or expertise must be used to further refine class separations. The tool obtained will be useful for crossed ply-laminates composites under fatigue tests with the aim of proposing a fatigue damage scenario and to identify the kinetics of each mechanism.

Acknowledgments. The authors would like to thank Zouhaier.JENDLI and Mondher HAGGUI from ESTACA-lab West Campus in Laval for the supply and help with the using of NOESIS.

References

Ameur, M.B., et al.: Investigation and identification of damage mechanisms of unidirectional carbon/flax hybrid composites using acoustic emission. Eng. Fract. Mech. **216**, 106511 (2019). https://doi.org/10.1016/j.engfracmech.2019.106511

Dahmene, F., Yaacoubi, S., Mountassir, M.E.: acoustic emission of composites structures: story, success, and challenges. In: Physics Procedia, Proceedings of the 2015 ICU International Congress on Ultrasonics, vol. 70, pp. 599–603. Metz (2015). https://doi.org/10.1016/j.phpro.2015.08.031

Davies, D.L., Bouldin, D.W.: A cluster separation measure. IEEE Trans. Pattern Anal. Mach. Intell. **1**(2), 224–227 (1979)

Haggui, M., El Mahi, A., Jendli, Z., Akrout, A., Haddar, M.: Static and fatigue characterization of flax fiber reinforced thermoplastic composites by acoustic emission. Appl. Acoust. **147**, 100–110 (2019). https://doi.org/10.1016/j.apacoust.2018.03.011

Kersani, M., Lomov, S.V., Vuure, A.W.V., Bouabdallah, A., Verpoest, I.: Damage in flax/epoxy quasi-unidirectional woven laminates under quasi-static tension. J. Compos. Mater. **49**(4), 403–413 (2015). https://doi.org/10.1177/0021998313519282

Likas, A., Vlassis, N., Verbeek, J.J.: The global k-means clustering algorithm. Pattern Recogn. **36**(2), 451–461 (2003). https://doi.org/10.1016/S0031-3203(02)00060-2

Monti, A., El Mahi, A., Jendli, Z., Guillaumat, L.: Experimental and finite elements analysis of the vibration behaviour of a bio-based composite sandwich beam. Compos. Part B: Eng. **110**, 466–475 (2017)

Romhány, G., Karger-Kocsis, J., Czigány, T.: Tensile fracture and failure behavior of technical flax fibers. J. Appl. Polym. Sci. **90**(13), 3638–3645 (2003). https://doi.org/10.1002/app.13110

Sause, M.G., Gribov, A., Unwin, A.R., Horn, S.: Pattern recognition approach to identify natural clusters of acoustic emission signals. Pattern Recogn. Lett. **33**, 17–23 (2012). https://doi.org/10.1016/j.patrec.2011.09.018

Mechanical Behavior and Damage of Advanced Iron-Based Metal Matrix Composite Under Shear

Manel Dammak(✉) and Monique Gaspérini

LSPM-CNRS, Université Paris 13, 99 Avenue Jean Baptiste Clément, 93430 Villetaneuse, France
manel.dammak@univ-paris13.fr

Abstract. This paper aims to analyse the microstructure and the mechanical behavior of a novel steel-based composite reinforced by TiB_2 ceramic particles, during simple monotonic and reverse shear tests. The microstructure was studied by Scanning Electron Microscopy and Backscattered Electron Diffraction (SEM-EBSD). TiB_2 particles are homogeneously distributed in the matrix through two distinct populations: small sized eutectic particles and coarse primary ones. A very pronounced morphological and crystallographic texture was evidenced. Significant damage occurs at around 10% of equivalent strain amount and appears first within the large-sized particles. It increases with the cumulated shear deformation. The analysis of the overall cracks orientation, in relationship with local stress state, shows that damage corresponds essentially to crack opening-mode ruptures of both mono and polycrystalline particles. Few particle /matrix debonding was observed, ssupporting the good interfacial cohesion conferred for this steel-based composite. After strain reversal, damage exhibits less open and more fragmented cracks and thus a better load transfer to the matrix, which may explain the later macroscopic cracking. For a given cumulative deformation amount, the composite microstructure is quite different after monotonic and reverse shear loading, showing more intragranular substructures within ferritic matrix grains in monotonic path loading.

Keywords: Metal matrix composite · simple shear · EBSD · plasticity · damage

1 Introduction

Metal matrix composites (MMCs) are excellent candidates for lightweight structures applications, in particular in transportation field, which has to meet more and more severe requirements in terms of Co_2 gas emissions. Composites with a metallic matrix are particularly advantageous for improving mechanical properties (Hadjem-Hamouche et al. 2018; Dammak et al. 2014) especially the specific modulus (i.e. the ratio between the elastic modulus E and the density ρ), which is classically aimed in this industry. The attractive physical and mechanical properties characterizing particulate reinforced MMC have made them interesting candidate materials for aerospace, automotive, reactor vessels, and several other engineering applications (Jrad et al. 2018; 2019; Bouhamed et al.

A. Akrout et al. (Eds.): ICAV 2022, ACM 22, pp. 70–77, 2023.
https://doi.org/10.1007/978-3-031-34190-8_9

2019). Further, the addition of stiff lightweight ceramic particles increases the specific stiffness; however, the reinforcement of ductile materials by fragile particles requires the analysis of the mechanisms of plasticity and damage during different mechanical loading paths in order to control the risks of ruin and to optimize the use properties. A methodology to characterize aluminium behavior considering non-associated plasticity model coupled with isotropic ductile damage was carried out by Bouhamed et al, (2021, 2022) in order to identify Ductile damage. Composites with Aluminum alloy matrix have been the subject of several studies but more recently in the last decade, steel-based composites have found a considerable interest (Lartigue et al. 2012; Hadjem-Hamouche et al. 2012; Dammak et al. 2014; Gaspérini et al. 2017; Dorhmi et al. 2020). In particular Fe–TiB_2 composites have attracted consequent interest, since TiB_2 has a high Young's modulus E = 583 GPa (Okamoto et al. 2010) and lower density than steel and thus TiB_2 are often considered as the best reinforcement for steel matrix composites.

The present work is based on experimental microstructural and mechanical characterizations of a novel titanium diboride (TiB_2)-reinforced steel composite, developed by ArcelorMittal for automotive applications and prepared by in situ precipitation of the TiB_2 particles during eutectic solidification. This continuous casting production process leads to an increase in specific stiffness (higher than 20%), and to a better strength to ductility compromise. The objective here is to analyze the statistically representative aspects of mechanical behavior up to large plastic strain and under both monotonic and reverse load path, useful in the context of sheet metal forming. Damage mechanisms are studied by SEM -EBSD observations and quantitative image analysis. Monotonous and reverse simple shear tests up to large cumulated plastic strains are achieved, and differences induced on micro/macro scales are discussed.

2 Experimental Procedure

2.1 Material

The material considered here is developed by ArcelorMittal and prepared by in-situ precipitation of TiB_2 particles, during eutectic solidification (Arcelor Research Group 2008). The choice of titanium diboride (TiB_2) have been motivated due to its very high Young's modulus (585 GPa) and its low density ($\rho = 4.52$ g/cm^3). The as cast composite is then hot rolled.

2.2 Simple Shear Tests

The simple shear device, developed at LSPM Laboratory and adapted to a tensile machine, makes it possible to characterize the mechanical behavior of metal sheets for large plastic deformations (Bouvier et al. 2006). Rectangular samples with the dimensions of length L = 38 mm, width w = 23 mm and Thickness T = 3 mm were used, thereby providing a useful sheared zone being: L = 38 mm, width b = 2,5 mm and Thickness T = 3 mm. The reference frame for the tests was (SD, NS, ND), with SD = Shear Direction, NS = Normal to the Shear plane, ND = Normal Direction to the sheet. Tests were carried up to the onset of macroscopic fracture, as well as up to an equivalent deformation ε = 3%, 10%, 30% in monotonous loading, and up to a cumulated strain of 30% in reverse loading after a pre-strain of 10%. The shear direction initially coincided with the transverse direction of the material.

2.3 Microstructural Characterizations

The investigations were carried out by Scanning Electron Microscopy (SEM), equipped for the Electron Backscattered diffraction (EBSD) measurements. EBSD is useful to study local crystallographic orientation in the microstructure. This microscopic observations required specific surface preparation involving careful mechanical polishing using specific diamond grinding discs (diamond size down to 1 μm) and commercial colloidal silica suspension (abrasive particles size 0.25 μm). The initial microstructures were characterized from (RD, TD), (RD, ND) and (TD, ND) plane sections, whereas (SD,NS) and (SD,ND) plane sections were used for analysis of the deformed samples.

3 Results

3.1 Characterization of the Initial Microstructure

The overall aspect of the initial microstructure obtained by SEM-EBSD, characterized from both longitudinal (parallel to the rolling direction RD) (TD,RD), and transversal (TD, ND) plane sections, is illustrated in Figs. 1 and 2.

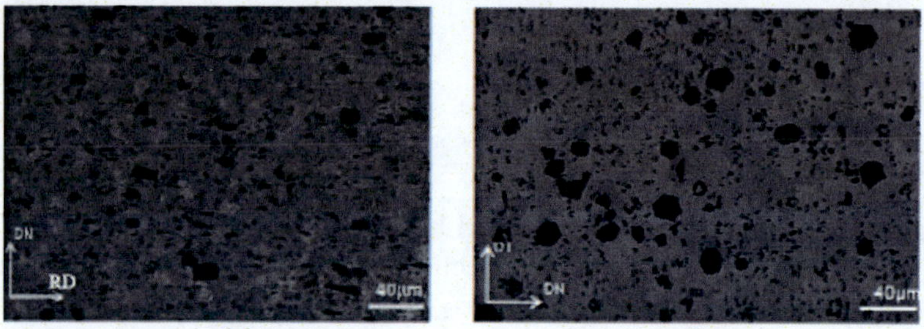

(a) (b)

Fig. 1. SEM micrographs of Fe–TiB_2 composite of (a) TD normal section and (b) RD normal section (TiB_2 particles appear with dark contrast)

The reinforcement particles are almost homogeneously dispersed in the matrix. Two TiB_2 populations of distinct size and shape are evidenced: large-sized primary particles, mainly hexagonal or prismatic sections (Fig. 1a and b), and small particles with round corners and various shapes. TiB_2 particles are generally oriented along the rolling direction (Fig. 1a). The quantitative morphological characteristics determined after statistical image processing are specified in Table 1. From this analyse on (RD,TD) plane, the total area fraction of TiB_2 was estimated to 11%. The mean size of large particles and small eutectic ones is of 13,5 μm and 2,2 μm respectively.

The average matrix grain size is around 20 μm. The crystallographic texture of the matrix is very weak, as illustrated by Fig. 2a and confirmed by global measurements X-ray diffraction. On the other hand, the particles exhibits a very marked crystallographic texture, with a preferential orientation of the c axis along the rolling direction (Fig. 2b).

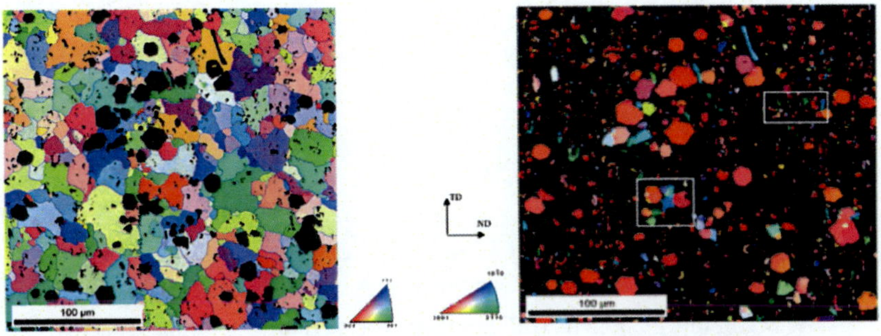

(a) matrix **(b) particles**

Fig. 2. Inverse pole figure mapping (RD direction) of (TD,ND) sections

Table 1. Particle sizes and corresponding area fractions (f) from quantitative image processing analysis. d is the equivalent diameter of the supposed spherical particles.

	d (μm)	*f (%)*
	(RD,ND)	*(RD,ND)*
Small particles (d < 8 μm)	2,2	6,5
Coarse particles (d > 8 μm)	13,5	4,5

3.2 Mechanical Behavior in Simple Shear

Figure 3 shows the stress-strain curves corresponding to monotonic and reverse simple shear tests up to 40% cumulated equivalent strain (corresponding to shear strain $\gamma = 40.\sqrt{3} \approx 70\%$ according to our isotropic hypothesis). A remarkable ductility is noted for this composite (up to $\gamma = 50\%$ of shear strain in monotonic path), higher than those of MMC's systems in the literature (Bacon et al. 2013), which is particularly appreciable for forming processes. A Bauschinger effect and a temporary stagnation of work hardening appear after inversing the loading direction, these effects of strain path change have been already evidenced in different mono or multi-phase steels (Hu et al. 1992; Dillien et al. 2010). The curves show that the reverse loading leads to later macroscopic cracking comparing to the monotonous loading.

3.3 Microstructural Observations of Damage

SEM-EBSD investigations show that after $\varepsilon = 3\%$ of equivalent strain, the microstructure is rather similar to the initial state. Significant damage level is observed for $\varepsilon = 10\%$ and

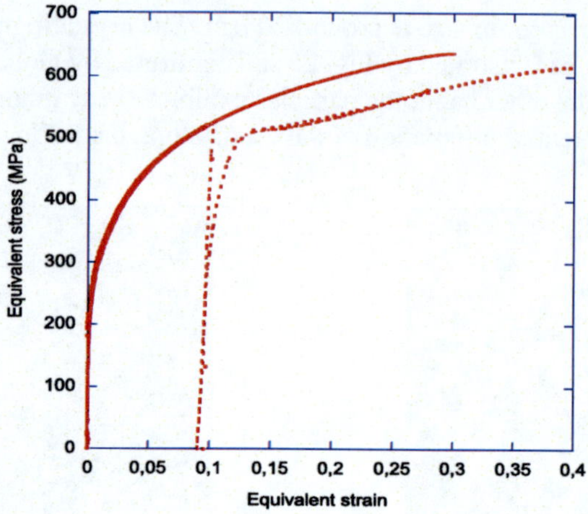

Fig.3. Simple monotonic (full line) and reverse (dotted line) shear curves

increases with strain. Figure 4 shows the appearance of broken particles after 30% of cumulative equivalent strain in monotonic and reverse loading and exhibits more open and rectilinear cracks in monotonous loading. The particle/matrix interface appears very resistant even after large plastic deformations of the matrix, which is consistent with the HR-TEM observations and analysis of the interface structure of these materials, carried by (Cha et al. 2012).

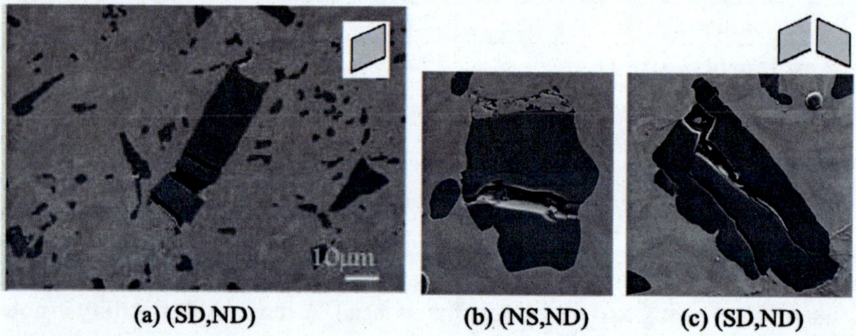

Fig. 4. Damaged particles after $\varepsilon = 30\%$ in monotonic (a,b) and reverse (c) loading, on different sections with SD = Shear Direction, NS = Normal to the Shear plane, ND = Normal direction to the sheet.

The damage appears firstly within the large particles family, and is mainly produced by ruptures of mono or polycrystalline particles, as could be noticed in Fig. 5a et b. The matrix plastic deformation induces a heterogeneous intergranular structuration accompanied by walls weakly disoriented, and then accentuated with the deformation. After reverse loading, the crystallographic textures differ little from the initial state, while that of monotonic loading changed significantly, for the same cumulative strain of 30%, as illustrated by Fig. 5c, d, e and f. Further analysis of intra-granular strain mechanisms in

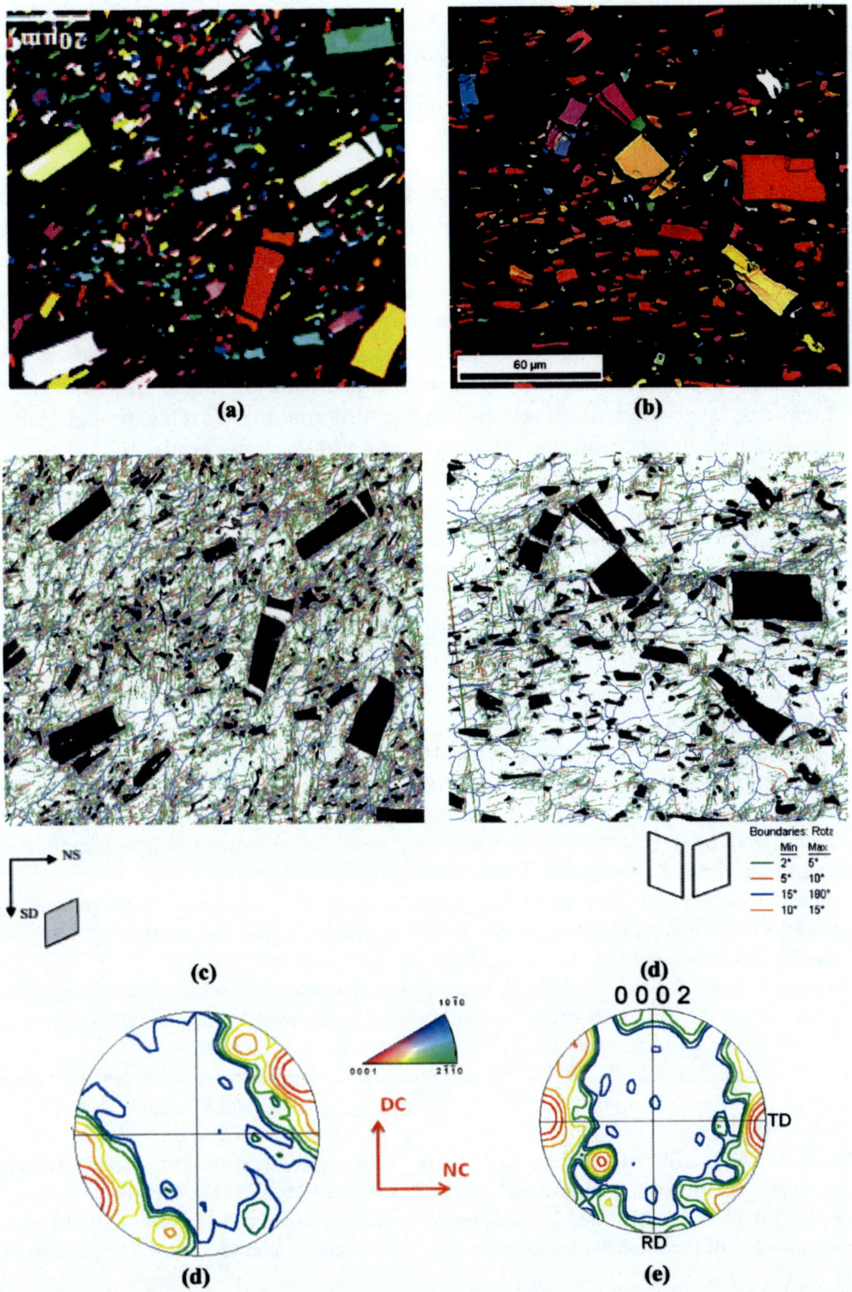

Fig. 5. EBSD analyses after 30% cumulative strain in simple shear by monotonic (a,c,d) and reverse (b,d,e) loading: (a,d) IPF (RD) map and (c,d) disorientation maps in the matrix. (d,e) 0002 pole figures in the particles.

the matrix and effects induced by the presence of particles leads to better understanding of their relationship with macroscopic behavior.

4 Conclusion

The study highlighted the statistically representative microstructural characteristics of an advanced industrial hot rolled Fe-TiB_2 composite and exposed their evolution during simple shear tests up to large plastic strain. Some major conclusions can be drawn: (i) TiB_2 particles present very pronounced morphological and crystallographic textures and are distributed in two types of populations: coarse primary particles and small eutectic particles. Some Particles are polycrystalline. (ii) The composite offers a high ductility, very appreciated for its cold forming applications. Significant Bauschinger effect was displayed after strain reversal, suggesting important kinematic hardening, to be considered when modeling the strain hardening of these materials.

Acknowledgements. This work is supported by ANR project n°ANR-09-MAPR-0001–05. The authors also gratefully acknowledge the helpful comments and suggestions of the reviewers, which have improved the presentation.

References

Animesh, A., Bandyopadhyay, T.K., Das, K.: Synthesis and characterization of TiB2-reinforced iron-based composites. J. Mat. proc. Tech. **172**, 70–76 (2006)

ARCELOR Research Group. Patent EP 1 897 963 A1, Bulletin 2008/ 11, p. 20 (2008)

Bouhamed, A., Jrad, H., Mars, J., Wali, M., Gamaoun, F., Dammak, F.: Homogenization of elasto-plastic functionally graded material based on representative volume element: application to incremental forming process. Int. J. Mech. Sci. **160**, 412–420 (2019)

Bouhamed, A., et al.: Identification of fully coupled non-associated-Ductile damage constitutive equations for thin sheet metal applications: Numerical feasibility and experimental validation. Thin-Walled Structures **176**, 109365 (2022)

Bouhamed, A., Mars, J., Jrad, H., Wali, M., Dammak, F.: Experimental and numerical methodology to characterize 5083-aluminium behavior considering non-associated plasticity model coupled with isotropic ductile damage. Int. J. Solids Struct. **229**, 111139 (2021)

Cha, L., Lartigue-Korinek, S., Walls, M., Mazerolles, L.: Interface structure and chemistry in a novel steel-based composite Fe–TiB2 obtained by eutectic solidification. Acta Mater **60**, 6382–6389 (2012)

Dammak, M., Gaspérini, M., Barbier, D.: Microstructural evolution of iron based metal–matrix composites submitted to simple shear. Mater Sci Eng A **616**, 123–131 (2014)

Bacon, D., Edwards, L., Moffatt, J.E., Fitzpatrick, M.E.: Fatigue and fracture of a 316 stainless steel metal matrix composite reinforced with 25% titanium diboride. Int. J. Fatigue **48**(2013), 39–47 (2013)

Dorhmi, K., Morin, L., Derrien, K., Hadjem-Hamouche, Z., Chevalier, J.P.: A homogenization-based damage model for stiffness loss in ductile metal-matrix composites. J. Mech. Phys. Solids **137**, 103812 (2020)

Gaspérini, M., Dammak, M., Franciosi, P.: Stress estimates for particle damage in Fe-TiB2 metal matrix composites from experimental data and simulation. Eur. J. Mech. A. Solids **64**, 85–98 (2017)

Hadjem-Hamouche, Z., Chevalier, J.P., Cui, Y., Bonnet, F.: Deformation behavior and damage evaluation in a new titanium diboride (TiB2) steel-based composite. Steel Res. Int. **83**, 538–545 (2012)

Hadjem-Hamouche, Z., Derrien, K., Héripré, E., Chevalier, J.P.: In-situ experimental and numerical studies of the damage evolution and fracture in a Fe-TiB2 composite. Mater. Sci. Eng. A **724**, 594–605 (2018)

Ibrahim, I., Mohamed, F., Lavernia, E.: Particulate reinforced metal matrix composites: a review. J. Mater. Sci. **26**, 1137–1156 (1991)

Jrad, H., Mars, J., Wali, M., Dammak, F.: An extended finite element meth-od for modeling elastoplastic FGM plate-shell type structures. Struct. Eng. Mech. Int. J. **68**(3), 299–312 (2018)

Jrad, H., Mars, J., Wali, M., Dammak, F.: Geometrically nonlinear analysis of elastoplastic behavior of functionally graded shells. Eng. Comput. **35**(3), 833–847 (2018). https://doi.org/10.1007/s00366-018-0633-3

Okamoto, N.L., Kusakari, M., Tanaka, K., Inui, H., Otani, S.: Anisotropic elastic constants and thermal expansivities in monocrystal CrB2, TiB2, and ZrB2. Acta Mater. **58**, 76–84 (2010)

Bouvier, S., Haddadi, H., Levée, P., Teodosiu, C.: Simple shear tests: experimental techniques and characterization of the plastic anisotropy of rolled sheets at large strains. J. Mater. Proc. Tech. **172**, 96–103 (2006)

Dillien, S., Seefeldt, M., Allain, S., Bouaziz, O., Van Houtte, P.: EBSD study of the substructure development with cold deformation of dual phase steel. Mater. Sci. Eng. A **527**, 947–953 (2010)

Hu, Z., Rauch, E.F., Teodosiu, C.: Work-hardening behavior of a mild steel under stress reversal at large strains. Int. J. Plasticity **8**, 839–856 (1992)

Vibration Characteristics of Porous Functionally Graded Cylindrical Shells

Sameh Elleuch[1(✉)], Hanen Jrad[1,2], Mondher Wali[1,2], and Fakhreddine Dammak[3]

[1] Laboratory of Electromechanical Systems (LASEM), National Engineering School of Sfax, University of Sfax, Route de Soukra Km 4, 338 Sfax, Tunisia
sameh.elleuch@enis.tn, hanen.jrad@essths.u-sousse.tn

[2] École superieure des Sciences et de la Technologie de Hammam Sousse, University of Sousse, Rue Lamine Abassi, 411 Hammam Sousse, Tunisia

[3] Laboratory of Electrochemistry and Environment (LEE), National Engineering School of Sfax, ENIS, University of Sfax, Sfax, Tunisia
fakhreddine.dammak@enis.tn

Abstract. In this research paper, the free vibration of a functionally graded porous (FGP) cylindrical shell is numerically examined. It is assumed that the modulus of elasticity of the porous composite is graded in the thickness direction. Material properties are defined according to the coordinates points in a user defined material UMAT subroutine implemented with the commercial code ABAQUS / Standard. First order-shear deformation theory (FSDT) is employed for analyzing the free vibration of cylindrical shell. Two types of porosity distribution (Even and Uneven types) are considered in the present work. This study examined the effect of material gradient index, porosity arrangements and porosity coefficient of free vibration characteristics of functionally graded porous (FGP) cylindrical shell. The numerical simulation illustrates very close results to solutions in the literature, assessing the accuracy and the effectiveness of the actual implementation. The proposed solution procedure is significantly efficient from the computational point of view.

Keywords: Free vibration · cylindrical shell · porosity distribution · FGM

1 Introduction

Cylindrical shells are widely used as common structural components in a variety of engineering fields, like nuclear reactors, aircraft engineering and aerospace engineering. Previous studies have examined the mechanical properties of laminated and isotropic composite cylindrical shells.

Over the years, numerical methods are widely used to provide a way to approximate solutions to problems in many engineering disciplines such as aerospace, mechanical, and electrical, quickly and easily compared to analytic solutions and when analytical solutions aren't possible. In addition, finite element analysis (FEA) allows simulations to predict and understand the structure behavior under several physical conditions. It is a

A. Akrout et al. (Eds.): ICAV 2022, ACM 22, pp. 78–84, 2023.
https://doi.org/10.1007/978-3-031-34190-8_10

numerical approach to finding the real results and providing an approximate solution to different problems in mechanical engineering. A design optimization of dental implant geometrical characteristics enhancing primary stability is proposed by (Elleuch et al. 2021) using FEA of stress distribution around dental prosthesis. A numerical methodology to characterize aluminum behavior considering non-associated plasticity model coupled with isotropic ductile damage was carried out by (Bouhamed et al. 2021, 2022) in order to identify ductile damage constitutive equations for thin sheet metal applications.

Metal matrix composites (MMCs) are excellent candidates for lightweight structures applications, indeed, the attractive physical and mechanical properties characterizing particulate reinforced MMC has made them interesting candidate materials for aerospace, automotive, reactor vessels, and several other engineering applications Rawal (2001); Kainer (2006); Macke (2012); Kim et al. (2017); Singh et al. (2020) et Mandal et al. (2022). Furthermore, functionally graded materials are inhomogeneous composite materials, as their mechanical properties vary continually and smoothly from one surface to the other in the preferred direction and this is obtained by gradually the volume fraction of the constituent materials. Numerous research studies are available in the literature to analyze the mechanical behavior of FGM shell structures. (Matsunaga (2008); Ghanned and Nejad (2013); Shariyat and Alipour (2014); Kim and Lee (2016); Liu et al. (2017) and Eliseeva et al. (2019)). Functionally graded porous material (FGP) is a new FGM in which porous materials are characterized by the graduated distribution of internal pores in the microstructure. Thus, local density can act as a design variable to improve structural performance (Zhao et al. (2009); Sofiyev (2010); Thai and Vo (2012) and Ramu and Mohanty (2014)). Recently, functionally graded porous materials (FGP) have received increasing attention due to their advantages, such as efficient energy capacity and low specific weight in several areas like manufacture and engineering design, Wang and Wu (2017). It should be observed from the aforementioned literature study that there is still a dearth of publications discussing the mechanical behavior of porous FGM constructions. There is a lack of research on FGM shell geometry and the majority of assessments focus on rectangular plates.

In this study, the ABAQUS software's UMAT and USDFLD subroutines are used to determine the material characteristics of a functionally graded porous cylindrical shell in accordance with the coordinates of the integration points. It is determined that gradient index, a porosity parameter, plays a major role in the frequency responses of cylindrical porous FGM shells by examining the impacts of both uniform and uneven distributions of porosity and the reliance on porosity. The effects of geometric factors, porosity distribution type, and porosity coefficient on the free vibration properties of a porous cylindrical shell are investigated using numerical technique.

2 Functionally Graded Porous (FGP) Cylindrical Shell

A functionally graded porous (FGP) cylindrical shell is presented in this work in Fig. 1. The length, thickness and radius of the cylindrical shell correspond to L, h, and R. The FGP cylindrical shell has a thickness to radius ratio h/R = 0.002 and length to radius ratio L/R = 20.

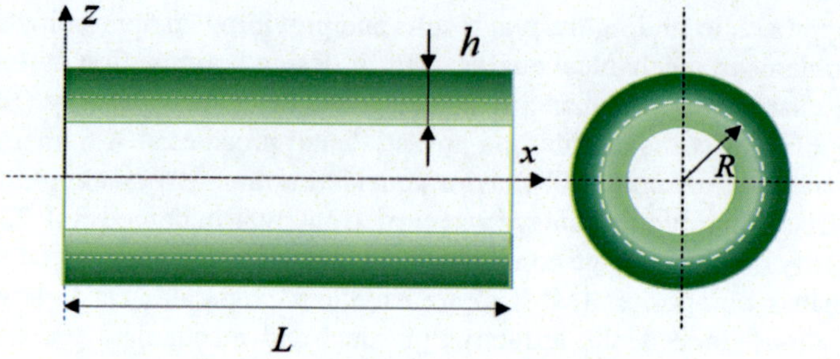

Fig. 1. Representation of the cylindrical shell

A functionally gradient porous cylindrical shell composed of metallic and ceramic components is considered where the structure considers two types of porosity distributions, namely type-I (even porosity) and type-II (uneven porosity) as shown in Fig. 2.

$$E(z,p) = \begin{cases} (E_c - E_m)V_c + E_c - \frac{p}{2}(E_c + E_m) & \text{Even porosity} \\ (E_c - E_m)V_c + E_c - \frac{p}{2}(E_c + E_m)(1 - \frac{2|z|}{h}) & \text{Uneven porosity} \end{cases} \quad (1)$$

The continuous variation as the Young's modulus E can be given in function of the porosity parameter (p) and the variable z as mentioned in Eq. (1) according to Wattanasakulpong et al. (2012); Wattanasakulpong and Ungbhakorn (2014); Wattanasakulpong and Chaikittiratana (2015), where $0 \prec p \prec 1$ is the porosity coefficient which corresponds to the ratio between the volume of voids and the total volume. For ($p = 0$), the FGP cylindrical shell is considered perfect. E_m and E_c correspond to the Young's modulus of metal and ceramic portions. V_c denotes the volume fraction of the ceramic, which can be defined as follows, Mellouli et al. (2019):

$$V_c = \left(\frac{1}{2} + \frac{z}{h}\right)^n \; z \in \left[-\frac{h}{2}, \frac{h}{2}\right] \quad (2)$$

$$V_m + V_c = 1 \quad (3)$$

In Eq. (2), n is the material gradient index which varies from zero to infinity. h is the structure thickness and z denotes the coordinate measured along the thickness direction. When ($n = 0$), the FGP cylindrical shell is ceramic-rich and when n tends to infinity, the FGP cylindrical shell is metallic-rich.

The numerical simulation is performed with the commercial software ABAQUS. Material properties are implemented according to coordinates of the integration points to avoid discontinuity of stress. UMAT subroutine was performed to implement the elastic mechanical behavior along the thickness using the number of integration points.

3 Numerical Results

Numerical results of free vibration with shear-diaphragm boundary condition of FGP cylindrical shell is performed in this section. The FGP cylindrical shell has a thickness to radius ratio set at 0.002 and a length to radius ratio set at 20. Materials properties for

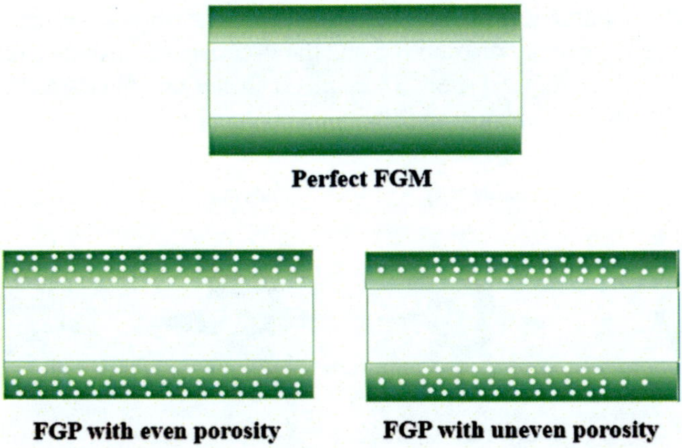

Fig. 2. Perfect FGM and FGP with even and uneven distribution

cylindrical shell structure are nickel and stainless steel with ($E_m = 205.098$ GPa, $E_{c,} =$ 207.788 GPa, $\nu_m = 0.31$, $\nu_c = 0.317756$, $\rho_m = 8900$ kg.m^{-3}, $\rho_c = 8166$ kg.m^{-3}).

For free vibration, the formulations take the form of an eigenvalue problem:

$$([\boldsymbol{K}] - \mathrm{w}^2[\boldsymbol{M}])\boldsymbol{\Delta} = 0 \tag{4}$$

where $[\boldsymbol{K}]$ the structural stiffness matrix, w is denoted the eigen-frequencies, $[\boldsymbol{M}]$ is the mass matrix and $\boldsymbol{\Delta}$ denote the vector from the displacement functions which represents the modal shapes of cylindrical shell. The discretization of the cylindrical shell is performed using the S4 shell element.

The computational code and the free vibration formulations are verified by comparing the present results of FGM cylindrical shell with the results of Wali et al, (2015). The comparison result is presented in Table1, which illustrate the natural frequencies (Hz) with the circumferential wave number.

Table 1. Frequencies (Hz), in relation to the number of circumferential waves

Wave number		Stai.st	n = 0.5	n = 1	n = 5	Nickel
2	Present	4.5922	4.5251	4.5207	4.5102	4.378
	Wali et al. 2015	4.6719	4.5955	4.4838	4.4531	4.4455
4	Present	7.2415	7.1259	7.1182	7.133	6.8881
	Wali et al. 2015	7.2416	7.1136	7.0546	6.9412	6.8736
6	Present	16.989	16.713	16.694	16.660	16.159
	Wali et al. 2015	16.909	16.605	16.467	16.203	16.048
8	Present	30.870	30.367	30.332	30.272	29.362
	Wali et al. 2015	30.604	30.057	29.801	29.325	29.046
10	Present	48.904	48.107	48.051	47.957	46.514
	Wali et al. 2015	48.243	47.373	46.977	46.226	45.716

In this section, numerical results are shown to demonstrate the effect of power gradient index, type of porosity distribution and the variation of porosity coefficient on the frequencies of FGP cylindrical shell. Some mode shapes are illustrated in Fig. 3 for the shear-diaphragm FGP cylindrical shell with even porosity type (n = 0.1 and p = 0.1).

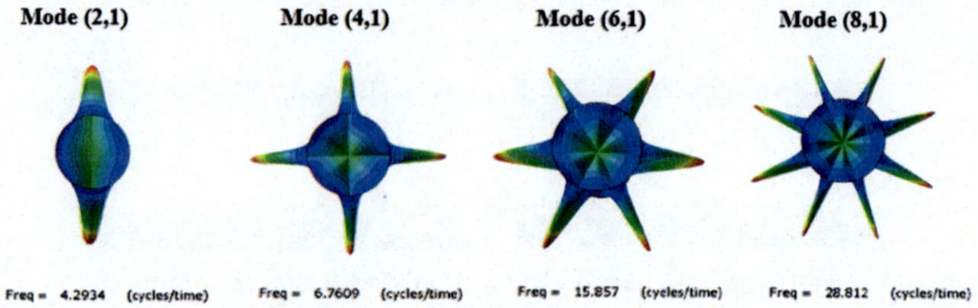

Fig. 3. Mode shapes of the FGP cylindrical shell with even porosity type (n = 0.5, p = 0.1)

Table 2 presents frequencies of FGP cylindrical shell. Two types of porosity distributions (Even and Uneven) are included. Results illustrate that the difference between the lowest and the highest values of even cylindrical shell are more evident when compared to Uneven cylindrical shell. As shown in Table 2, the decrease in the highest and the lowest values at n = 0.5 correspond to 18.31% for the even cylindrical shell and 4.02% for uneven cylindrical shell. This means that the frequency is more susceptible to the porosity coefficient for even FGP cylindrical shell than to the uneven type.

It is important to note that the frequencies decrease as the power gradient index and the porosity coefficient increase for Even and Uneven cylindrical shell.

Table 2. Frequencies (Hz) of cylindrical shell with various porosity coefficients.

Even porosity type				
p	n = 0.5	n = 1	n = 5	n = 10
0.1	6.7609	6.7529	6.7378	6.7312
0.2	6.3749	6.3667	6.3512	6.3444
0.3	5.9641	5.9554	5.9395	5.9325
0.4	5.5227	5.5136	5.4970	5.4898
Uneven porosity type				
p	n = 0.5	n = 1	n = 5	n = 10
0.1	7.0339	7.0261	7.0113	7.0048
0.2	6.9407	6.9329	6.9180	6.9115
0.3	6.8463	6.8384	6.8235	6.8169
0.4	6.7505	6.7425	6.7276	6.7210

4 Conclusion

The first order-shear deformation theory was used in the current study to create a robust computational model for the free vibration of porous cylindrical FGM shells. The suggested model is credible, as shown by a comparison with earlier findings. The research looked at how certain FGP cylindrical shell parameters affected certain vibrational properties. Results indicate that the vibration characteristics of a porous, functionally graded cylindrical shell are significantly influenced by the power gradient index, porosity distribution types, and porosity coefficient.

References

Bouhamed, A., et al.: Identification of fully coupled non-associated-Ductile damage constitutive equations for thin sheet metal applications: numerical feasibility and experimental validation. Thin-Walled Struct **176**, 109365 (2022)

Bouhamed, A., Mars, J., Jrad, H., Wali, M., Dammak, F.: Experimental and numerical methodology to characterize 5083-aluminum behavior considering non-associated plasticity model coupled with isotropic ductile damage. Int J Solids Struct **229**, 111139 (2021)

Eliseeva, O.V., et al.: Functionally graded materials through robotics-inspired path planning. Mater. Des. **182**, 107975 (2019)

Elleuch, S., Jrad, H., Kessentini, A., Wali, M., Dammak, F.: Design optimization of implant geometrical characteristics enhancing primary stability using FEA of stress distribution around dental prosthesis. Comput Methods Biomech. Biomed. Eng. **24**(9), 1035–1051 (2021)

Ghanned, M., Nejad, M.Z.: Elastic solution of pressurized clamped-clamped thick cylindrical shelles made of functionally graded materials. J. Theor. Appl. Mech. **51**, 1067–1079 (2013)

Kainer, K.U.: Basics of metal matrix composites. Metal Matrix Composites: Custom-made Materials for Automotive and Aerospace Engineering, 1–54 (2006)

Kim, C.-S., Cho, K., Manjili, M.H., Nezafati, M.: Mechanical performance of particulate-reinforced Al metal-matrix composites (MMCs) and Al metal-matrix nano-composites (MMNCs). J. Mater. Sci. **52**(23), 13319–13349 (2017). https://doi.org/10.1007/s10853-017-1378-x

Kim, N.I., Lee, J.: Geometrically nonlinear isogeometric analysis of functionally graded plates based on first-order shear deformation theory considering physical neutral surface. Compos. Struct. **153**, 804–814 (2016)

Liu, Z., Meyers, M.A., Zhang, Z., Ritchie, R.O.: Functional gradients and heterogeneities in biological materials: design principles, functions, and bioinspired applications. Prog. Mater Sci. **88**, 467–498 (2017)

Macke, A., Schultz, B.F., Rohatgi, P.: Metal (2012)

Mandal, V., Tripathi, P., Kumar, A., Singh, S.S., Ramkumar, J.: A study on selective laser melting (SLM) of TiC and B4C reinforced IN718 metal matrix composites (MMCs). J. Alloy. Compd. **901**, 163527 (2022)

Matsunaga, H.: Free vibration and stability of functionally graded plates according to a 2-D higher-order deformation theory. Compos. Struct. **82**(4), 499–512 (2008)

Mellouli, H., Jrad, H., Wali, M., Dammak, F.: Geometrically nonlinear meshfree analysis of 3D-shell structures based on the double directors shell theory with finite rotations. Steel Compos. Struct. **31**(4), 397–408 (2019)

Ramu, I., Mohanty, S.C.: Buckling analysis of rectangular functionally graded material plates under uniaxial and biaxial compression load. Procedia Eng. **86**, 748–757 (2014)

Rawal, S.P.: Metal-matrix composites for space applications. Jom **53**(4), 14–17 (2001)
Shariyat, M., Alipour, M.M.: A novel shear correction factor for stress and modal analyses of annular fgm plates with non-uniform inclined tractions and non uniform elastic foundations. Int. J. Mech. Sci. **87**, 60–71 (2014)
Singh, L., Singh, B., Saxena, K.K.: Manufacturing techniques for metal matrix composites (MMC): an overview. Adv. Mater. Process. Technol. **6**(2), 441–457 (2020)
Sofiyev, A.: Buckling analysis of FGM circular shells under combined loads and resting on the Pasternak type elastic foundation. Mech. Res. Commun. **37**(6), 539–544 (2010)
Thai, H.T., Vo, T.P.: Bending and free vibration of functionally graded beams using various higher-order shear deformation beam theories. Int. J. Mech. Sci. **62**(1), 57–66 (2012)
Wali, M., Hentati, T., Jarraya, A., Dammak, F.: Free vibration analysis of fgm shell structures with a discrete double directors shell element. Compos. Struct. **125**, 295–303 (2015). https://doi.org/10.1016/j.compstruct.2015.02.032
Wang, Y., Wu, D.: Free vibration of functionally graded porous cylindrical shell using a sinusoidal shear deformation theory. Aerosp. Sci. Technol. **66**, 83–91 (2017)
Wattanasakulpong, N., Chaikittiratana, A.: Flexural vibration of imperfect functionally graded beams based on timoshenko beam theory: Chebyshev collocation method. Meccanica **50**(5), 1331–1342 (2015)
Wattanasakulpong, N., Ungbhakorn, V.: Linear and nonlinear vibration analysis of elastically restrained ends fgm beams with porosities. Aeros. Sci. Technol. **36**, 111–120 (2014)
Wattanasakulpong, N., Prusty, B.G., Kelly, D.W., Hoffman, M.: Free vibration analysis of layered functionally graded beams with experimental validation. Mater. Des. **1980–2015**(36), 182–190 (2012)
Zhao, X., Lee, Y.Y., Liew, K.M.: Free vibration analysis of functionally graded plates using the element-free kp-Ritz method. J. Sound Vib. **319**(3–5), 918–939 (2009)

Damping Behavior of Water-Aged Bio-Based Sandwich with Auxetic Core

Zeineb Kesentini[1,2](✉), Abderrahim El Mahi[1], Jean Luc Rebiere[1], Rachid El Guerjouma[1], Moez Beyaoui[2], and Mohamed Haddar[2]

[1] Laboratoire d'Acoustique de l'Université du Mans (LAUM) UMR CNRS 6613, Av. O. Messiaen, 2085 Le Mans Cedex 9, France
Zeineb.Kesentini@univ-lemans.fr

[2] Laboratoire de Modélisation et Production Mécanique (LA2MP), Ecole Nationale d'Ingénieurs de Sfax, Université de Sfax, BP N ° 1173-3038, Sfax, Tunisie

Abstract. The advantages of bio-based composites have led to a significant increase in their application in recent years. Sandwich with auxetic structures made up of periodic cells is likewise being pursued by industrial enterprises. The excellent characteristics of these materials have been highlighted in several experiments. This study aims to examine how water aging impacts the vibration behavior of sandwich composites made of bio-based materials. A polylactic acid (PLA) tape reinforced with flax fiber filament is employed in this study. It is made of a biobased, recyclable, and biodegradable fabric. The sandwich's core is an auxetic core with three cells. The investigated sandwich construction is created using a 3D printer and submerged in tap water at room temperature. The water absorption is evaluated at first by measuring the mass gain at each time interval of immersion. Second, using a free vibration test, the vibration behavior of aged materials is examined for various percentages of absorption. When measured over time, water absorption results in a drop in stiffness and an increase in loss factor. The findings show that the sandwich is sensitive to humidity, which results in a decline in stiffness with an increase in the loss factor.

Keywords: auxetic structure · bio-based sandwich · water absorption · vibration · 3D printing

1 Introduction

Sandwich panels have been utilized in many industries such as aerospace, automobile industry and also sports and leisure industries due to its excellent energy absorption, strength-to-weight ratio, and bending stiffness [1]. A composite sandwich structure is made by joining two thin and rigid skins to a light and solid core. Although the sandwich composite has a higher thickness than the core material, which is often low in strength, it nevertheless has a low overall density.

The sandwich structure's mechanical characteristics are determined by the fundamental material and fiber reinforcement employed in its fabrication, as well as the core structural design. In this research, auxetic structures with a negative Poisson's ratio are

A. Akrout et al. (Eds.): ICAV 2022, ACM 22, pp. 85–91, 2023.
https://doi.org/10.1007/978-3-031-34190-8_11

used as the architectural cellular structure in sandwich composites. It has been demonstrated that sandwich composites with auxetic cores perform better in terms of load bearing, energy absorption, and bending deflection. [2, 3]. The mechanical qualities and damping properties of a 3D printed bio-based sandwich were also investigated [4, 5]. The sophisticated architectural and geometrical properties of these materials can now be managed thanks to recent advancements in 3D printing [6, 7]. Today's environmental issues need the use of bio-based products rather than synthetic materials [8, 9]. These materials' advantages include biodegradability, low cost, high specific strength, and recyclability. Flax fibers are the most studied natural fibers due to their ability to improve the mechanical strength and stiffness of composite materials [10, 11].

Using bio-based composites is anticipated to be greatly impeded by the challenges of evaluating their capabilities in a humid environment, which is a significant barrier that also slows down manufacturing [12, 13]. Water sensitivity can be explained by their biological structure and content [14]. The sandwich materials implemented using as a core agglomerated cork and laminated flax fiber skins and the SR greenpoxy resin is studied in a hostile environment [15]. They demonstrated how the amount of water absorbed affects the static behavior of sandwich materials. Sandwich materials' flexural modulus decreases, while their loss factor rises with both frequency and immersion time.

Studying the impact of humidity on the vibrational behavior of a sandwich composite material with an auxetic core made entirely of bio-based materials is the goal of this research.

2 Materials and Methods

The suggested sandwich composites are made of polylactic acid (PLA) reinforced with flax fibers having a density of 1000 kg.m^{-3} (core and skins).

The composite filaments with a 1.75 mm diameter that NANOVIA provided are utilized in this experiment. The 3D printer is a MakerBot Replicator2 Desktop.

The construction platform temperature is set to 55 °C, and the extrusion temperature is set to 210 °C, both in accordance with the manufacturer's instructions.

Solidworks is used to create the samples, and the MakerBot Replicator2 Desktop program is used to convert the designs into printable instructions. The samples are submerged in room temperature tap water. Five samples of a $25 \times 25 \times 7$ mm^3 measurement were dipped in water to evaluate the percentage of water absorption of flax/PLA. The samples are weighed with an accuracy of 10^{-4} g using a SARTORIUS balance. Regularly, samples are taken out of the water to be weighed.

According to ASTM E-756 [16], flexural vibrations are utilized to evaluate the materials' damping properties. In a clamp-free arrangement, specimens with dimensions of 230 mm, 25 mm, and 7 mm are held horizontally (Fig. 1). 40 mm has been chosen as the clamping length. The specimens are stimulated using a PCB084A14 impact hammer over three free lengths in order to achieve different frequencies (230 mm, 200 mm, and 170 mm).

A signal dynamic analyzer digitizes and analyzes the excitation and response signals. This analyzer which is linked to a PC collects signals, keeps track of the acquisition settings, and examines the data it has gathered (Fourier transform, frequency response, mode shapes, etc…). To take into consideration the allegedly dispersed qualities of these natural materials, a minimum of 5 beams are tested each time. Using MATLAB software, the beam frequency response function (FRF) is then produced from the data [10, 11, 17]. To detect resonance peaks, each FRF is given an automated method. Figure 2 shows how a half-power bandwidth (HPB) system and a Matlab software optimization application are used to calculate the resonance frequency fi and modal loss factor η_i of each mode.

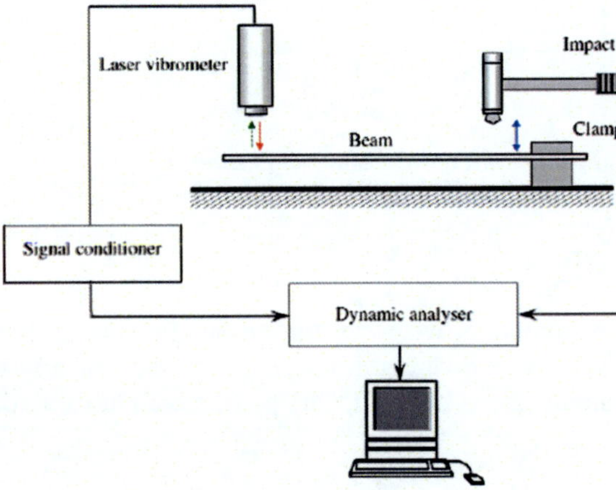

Fig. 1. Experimental set up of vibration

The modal loss factor is calculated using equation Eq. (1). The loss factor η_i is the ratio between the bandwidth frequencies where Δf_i the amplitude resonance decreases by 3db, split up by the resonance frequency f_i.

$$\eta_i = \frac{\Delta f_i}{f_i} = \frac{f_2 - f_1}{f_i} \tag{1}$$

The equivalent stiffness of the sandwich for each bending mode can be calculated by equation Eq. (2)

$$(EI)_{eq} = (\frac{m}{L})(2\pi f_n)^2(\frac{l^2}{(\beta_n l)^2})^2 \tag{2}$$

where m is the mass of the sandwich, L is the total length of the specimen, f_n is the resonance frequency of the nth flexural mode, l is the free length of the specimen in the vibration direction and $(\beta_n l)$ is a coefficient for the nth mode of a clamped-free specimen, with:

$\beta_1 l = 1.8751$,$\beta_2 l = 4.66940$, $\beta_3 l = 7.8547$,$\beta_4 l = 10.9955$, $\beta_5 l = 14.1371$ and $\beta_n l = (\pi/2).(2n-1)^2$ for $n\rangle 5$[14].

These experiments are performed on a dry specimen as well as various aged specimens to determine the influence of humidity on vibration behavior. For each absorption percentage, five specimens are evaluated to account for the variability of findings owing to experimental circumstances.

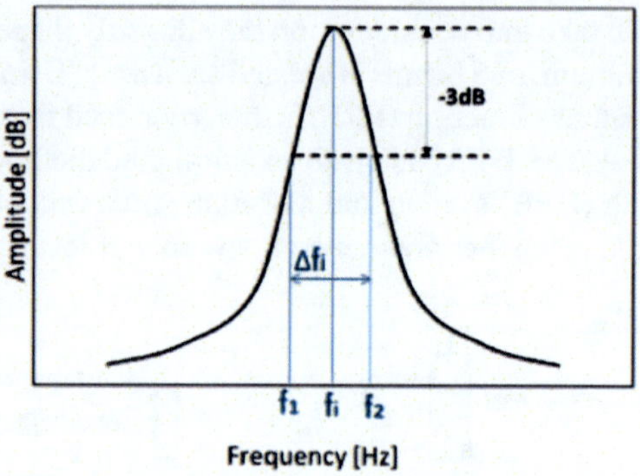

Fig. 2. Half power bandwidth method [16]

3 Water Absorption

Figure 3 depicts the development of water absorption as a function of immersion time. The amount of water that the sandwich absorbs first increases linearly with time and then slows down until saturation is attained. The panel's weight saturation is roughly 6%.

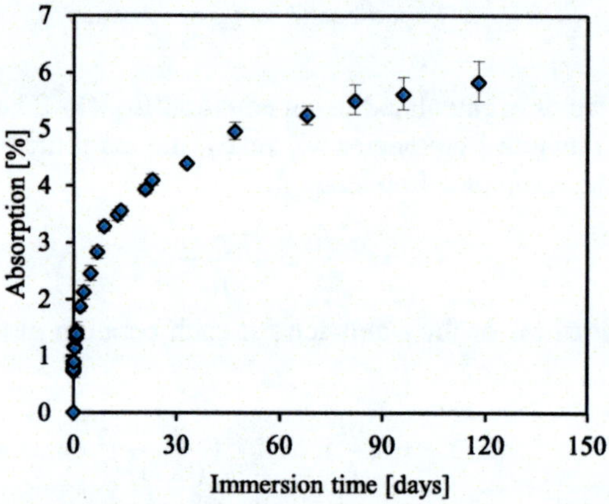

Fig. 3. Evolution of water absorption of the sandwich

The chemical composition of the bio-based composite as well as the void created by the auxetic core's structure may be responsible for this large weight rise. Between the auxetic cells, where it is trapped, water joins the other molecules in the bio-composite. Amorphous lignin and hemicelluloses [18–22] in an amorphous matrix are the components of bio-fibers, which are cellulose fiber-reinforced hybrid textiles. The main responsible are the hydroxyl groups -OH, which trap water.

4 Aging Effect on the Vibration Behavior of the Sandwich

Vibration studies are carried out for a range of immersion times to investigate how the aging of the water affects the behavior of a sandwich with an auxetic core.

Figures 4 and 5 depict how the vibration characteristics change with frequency for three different immersion times.

The results show that the equivalent stiffness of the sandwich (Fig. 4) decreases with increasing frequency and immersion period.

The loss of mechanical characteristics is caused by the weakening of the hydrogen bonds that connect the polar groups of adjacent macromolecular chains to a water molecule in bio-composition, composites [23]. The stiffness of specimens decreases with high frequency and significant water absorption.

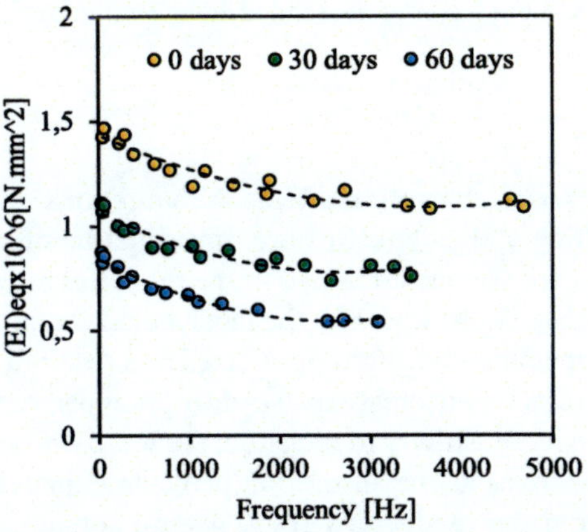

Fig. 4. Equivalent stiffness of the sandwich for 3 immersion times

As seen in Fig. 5, the loss factor increases as a function of frequency. The strong damping properties of composite flax fibers may be attributed to their architecture, which promotes energy dissipation through friction between cell walls and cellulose and hemicelluloses [23]. Non-aged materials have a low failure factor because they have few flaws and the interactions between the fiber and matrix are still not damaged.

As the absorption % increases, the equivalent stiffness and the loss factor show a conflicting trend.

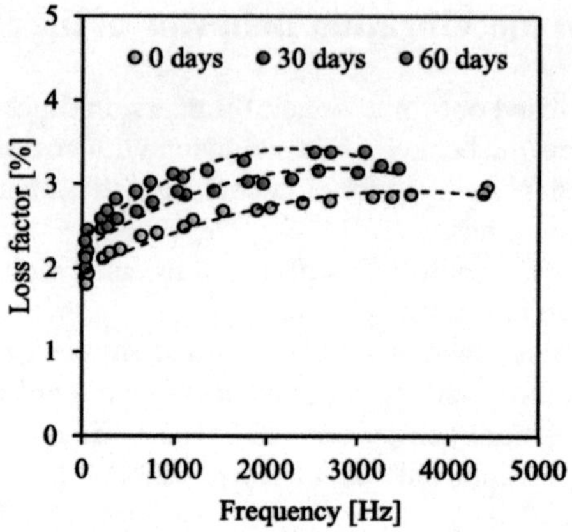

Fig. 5. Loss factor of the sandwich for 3 immersion times

5 Conclusion

It is investigated how water absorption affects the vibrational behavior of a sandwich created using 3D printing and an auxetic core. Investigation of absorption comes first. Second, vibration tests are conducted to determine the different water absorption rates. The resulting data are then contrasted with the mechanical characteristics of a reference, unaged composite. A specimen becomes plasticized as a result of water absorption. The mechanical characteristics will decrease as the absorption percentage rises as a result of this occurrence. Due to its sensitivity to moisture, the bio-based sandwich loses rigidity, although the loss factor rises as the absorption percentage increases. Additionally, the hydrophilic properties of the sandwich's 100% natural components and the voids that reentrant auxetic cells formed contribute to the high percentage of water absorption at saturation.

References

1. Schaedler, T.A., Carter, W.B.: Architected cellular materials. Annu. Rev. Mater. Res. **46**, 187–210 (2016)
2. Wang, P., Zhang, Y., Chen, H., Zhou, Y., Jin, F., Fan, H.: Broadband radar absorption and mechanical behaviors of bendable over-expanded honeycomb panels. Compos. Sci. Technol. **162**, 33–48 (2018)
3. Evans, K.E.: Auxetic polymers: a new range of materials. Endeavour **15**(4), 170–174 (1991)
4. Essassi, K., Rebiere, J.L., El Mahi, A., Ben Souf, M.A., Bouguecha, A., Haddar, M.: Dynamic characterization of a bio-based sandwich with auxetic core: experimental and numerical study. Int. J. Appl. Mech. **11**(2), 1950016 (2019)
5. Essassi, K., Rebiere, J.L., El Mahi, A., Ben Souf, M.A., Bouguecha, A., Haddar, M.: Experimental and numerical analysis of the dynamic behavior of a bio-based sandwich with an auxetic core. J. Sandwich Struct. Mater., 1099636219851547 (2019)
6. Wang, Q., Yang, Z., Lu, Z., Li, X.: Mechanical responses of 3D cross-chiral auxetic materials under uniaxial compression. Mater. Des. **186**, 108226 (2019)

7. Yu, S., Hwang, Y.H., Hwang, J.Y., Hong, S.H.: Analytical study on the 3D-printed structure and mechanical properties of basalt fiber-reinforced PLA composites using X-ray microscopy. Compos. Sci. Technol. **175**, 18–27 (2019)
8. Allagui, S., El Mahi, A., Rebiere, J.L., Beyaoui, M., Bouguecha, A., Haddar, M.: Effect of recycling cycles on the mechanical and damping properties of flax fibre reinforced elium composite: experimental and numerical studies. J. Renew. Mater. **9**(4), 695–721 (2021)
9. Monti, A., El Mahi, A., Jendli, Z., Guillaumat, L.: Experimental and finite elements analysis of the vibration behaviour of a bio-based composite sandwich beam. Compos. B Eng. **110**, 466–475 (2017)
10. Daoud, H., Rebiere, J.L., Makni, A., Taktak, M., El Mahi, A., Haddar, M.: Numerical and experimental characterization of the dynamic properties of flax fiber reinforced composites. Int. J. Appl. Mech. **08**(05), 1650068 (2016)
11. Daoud, H., El Mahi, A., Rebiere, J.L., Taktak, M., Haddar, M.: Characterization of the vibrational behaviour of flax fibre reinforced composites with an interleaved natural viscoelastic layer. Appl. Acoust. **128**, 23–31 (2017)
12. Hill, C.A., Norton, A., Newman, G.: The water vapor sorption behavior of natural fibers. J. Appl. Polym. Sci. **112**(3), 1524–1537 (2009)
13. Azwa, Z.N., Yousif, B.F., Manalo, A.C., Karunasena, W.A.: Review on the degradability of polymeric composites based on natural fibres. Mater. Des. **47**, 424–442 (2013)
14. Yan, L., Chouw, N., Jayaraman, K.: Flax fibre and its composites–a review. Compos. B Eng. **56**, 296–317 (2014)
15. Malloum, A., El Mahi, A., Idriss, M.: The effects of water ageing on the tensile static and fatigue behaviors of greenpoxy–flax fiber composites. J. Compos. Mater. **53**(21), 2927–2939 (2019)
16. Standard test method for measuring vibration-damping properties of materials. ASTM International. American Society For Testing and Materials (2010)
17. El Mahi, A., Assarar, M., Sefrani, Y., Berthelot, J.M.: Damping analysis of orthotropic composite materials and laminates. Compos. B Eng. **39**(7–8), 1069–1076 (2008)
18. Khalil, A., Rozman, H.D., Ahamd, N.N., Ismail, H.: Acetylated plant-fibre-reinforced polyester composites: a study of mechanical, hydrothermal and ageing characteristics. Polym. Plast. Technol. Eng. **39**(4), 757–781 (2000)
19. Tröger, F., Wegener, G., Seemann, C.: Miscanthus and flax as raw material for reinforced particleboards. Ind. Crops Prod. **8**(2), 113–121 (1998)
20. Dittenber, D.B., GangaRao, H.V.: Critical review of recent publications on use of natural composites in infrastructure. Compos. A Appl. Sci. Manuf. **43**(8), 1419–1429 (2012)
21. Charlet, K., Baley, C., Morvan, C., Jernot, J.P., Gomina, M., Breard, J.: Characteristics of Hermès flax fibres as a function of their location in the stem and properties of the derived unidirectional composites. Compos. A **38**(8), 1912–1921 (2007)
22. Baley, C.: Analysis of the flax fibres tensile behaviour and analysis of the tensile stiffness increase. Compos. A **33**(7), 939–948 (2002)
23. Duc, F., Bourban, P.E., Manson, J.A.: The role of twist and crimp on the vibration behaviour of flax fibre composites. Compos. Sci. Technol. **102**, 94–99 (2014)

Experimental and Numerical Characterization of the Acoustic Behavior of a Roller Shutter Box

Soraya Bakhouche[1], Walid Larbi[1](✉), Jean-François Deü[1], and Philippe Macquart[2]

[1] Conservatoire National des Arts et Métiers (Cnam), Laboratoire de Mécanique des Structures et des Systèmes Couplés (LMSSC), 2 Rue Conté, 75003 Paris, France
{soraya.bakhouche2,walid.larbi,jean-francois.deu}@lecnam.net

[2] Union des Fabricants de Menuiseries (UFME), 39 Rue Louis Blanc, 92038 Courbevoie, France
p.macquart@ufme.fr

Abstract. Roller shutter boxes are facade elements of paramount importance in the thermal and acoustic insulation of buildings. Their sound transmission is generally characterized by carrying out laboratory tests according to standards. These tests have the disadvantage of their high cost and their lack of repeatability and reproducibility between the experimental measurements, especially in the low frequency range. The purpose of this work is to propose a numerical model able to predict the vibro-acoustic response of roller shutter boxes. The proposed numerical approach is based on the application of the finite element method for modeling the studied structure which will be excited by a diffuse field. Its sound radiation is calculated by applying the infinite elements method. The experimental protocol used to measure the sound transmission of an element in a laboratory is exposed. Two cases are studied with and without insulation for the roller shutter box. The results of the proposed numerical model are validated by comparison to the results of laboratory tests.

Keywords: roller shutter box · experimental test · numerical simulation · porous material

1 Introduction

Roller shutter boxes play a significant role in the acoustic insulation of building facades. In this work, we are interested in the sound transmission loss of a roller shutter box according to the standards of tests in laboratories which considers a wall dividing an emission room from a receiving room. This wall has an opening where the test element is installed. One of the major problems of these tests is the reproducibility of the acoustic measurements. Indeed, differences between different laboratories have been observed (Cops et al. 1987), (Kihlman and Nilsson 1972). Also, the position of the curtain has an impact on the transmission loss. The curtain can be rolled or unrolled. The material used for the sound insulation plays an important role. The combination of melamine foam and heavy masses gives the best insulation. In this work, we propose a reliable method based on numerical analyzes and experimental measurements able to predict the acoustic behavior of roller shutter box.

A. Akrout et al. (Eds.): ICAV 2022, ACM 22, pp. 92–101, 2023.
https://doi.org/10.1007/978-3-031-34190-8_12

The literature is very poor in bibliographic references concerning the acoustic study of the roller shutter boxes. F. Asdrubali, C. Buratti present in (Asdrubali and Buratti 2005) an experimental investigation on different prototypes of high sound insulation ventilating windows with and without rolling shutter boxes. C. Díaz and A. Pedrero (Díaz and Pedrero 2009, Díaz and Pedrero 2013) present a summary of the studies carried out on the sound insulation from airborne noise in several types of windows (double side-hung casement and double horizontal sliding sash) with built-on shutter and prefabricated box. For each type of window, an analysis was made of the effects of the interior finish of the shutter box and the shutter position (whether fully retracted or extended). The authors knowledge, there is no bibliographical reference on the application of numerical methods for the acoustic characterization of roller shutter boxes, which constitutes the originality of this work. However, we can cite in the same context the recent work of the authors in the numerical prediction and the experimental calibration of the sound transmission of windows and double glazing (Soussi et al. 2021).

This paper is organized as follows: In a first part, we are interested in (i) the characterization of the sound transmission loss in laboratory according to the standard (ISO 14010 2010) and (ii) the acoustic measures to determine the acoustic performances of a tested sample. The second part is dedicated firstly to the choice justification of the numerical tools used to obtain the power radiated by the roller shutter box in the reception room. For this purpose, three methods will be compared: Rayleigh integral, infinite elements and finite elements methods. The modal effect of the receiving room and the influence of its absorption coefficients are particularly analyzed. Finally, a comparison between numerical and experimental results of sound transmission loss factor of a roller shutter box is proposed. The effect of the introduction of porous materials and heavy mass in the acoustic insulation of the box is analyzed.

2 Characterization of the Sound Transmission Loss

2.1 Laboratory Measurement

The measurements of the airborne sound insulation of building elements are performed according to the standard ISO 14010. The test laboratory consists of two adjacent reverberation rooms. It is recommended that the volume of the two rooms differ by at least 10%. The ratios of the dimensions of each test room are chosen so that the natural frequencies in the low frequency bands are spaced as evenly as possible. The rooms should have a volume of at least 50 m^3.

Between these two rooms, there is an opening to introduce the sample to be tested. The sample can be a window, an air inlet, or a roller shutter box. The reverberation time of the rooms should not exceed 2 s. It is important to specify the temperatures and relative humidity of the two rooms. The relative humidity should be at least 30%. Room temperatures should be between 17 °C and 23 °C.

In laboratories used for sound loss index measurements, the sound transmitted through any indirect path must be negligible compared to the sound transmitted through the sample.

For this work, the configuration of the emitting and receiving rooms used for experimental results is presented in Fig. 1. The transmitting chamber has a volume of 78 m^3

and the receiving chamber has a volume of 62.3 m^3. In the receiving room, the walls noted 1 are made of concrete of thickness 200 mm and are separated by a layer of mineral wool noted 2 for insulation while the wall 3 is made of concrete block of thickness 100 mm. The receiving room is placed on a spring box noted 7. In the emitting room, two steel sheets, of thickness 2 mm noted 4 and thickness 6 mm noted 5, respectively are introduced. These sheets are separated by mineral wool. Finally, the separating wall is noted 6 and the tested box is noted C.

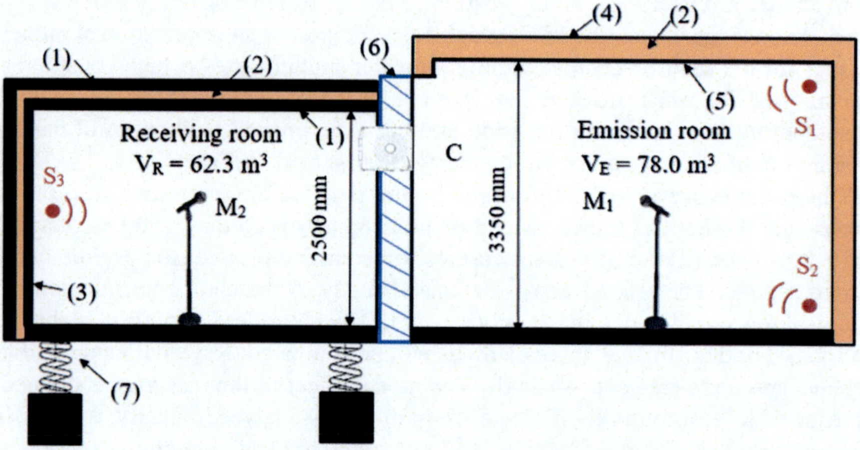

Fig. 1. Experimental set-up of acoustic experimental test

In the emitting chamber, two monopole sources (S_1 and S_2) are placed in the corners of the room to generate several acoustic wave reflections on the walls to create a sufficiently diffuse field. The power incident on the sample is deduced from the sound pressure. A rotating microphone and a sound amplifier in the emitting chamber are used in this room. In the receiving room, we have the presence of a rotating microphone (M_2) and a source (S_3) to measure the reverberation time of the room. The microphone in the receiving room also measures the sound level of the room transmitted by the tested element. Finally, there is the control room where we find a real time spectrum analyzer allowing us to give the results in third octave band, a calibrator and a computer for the post-processing of results.

2.2 Acoustic Measures

Transmission loss (TL) describes the cumulative decrease in energy intensity of a waveform as it propagates outward from a source, or as it propagates through a certain area or type of structure. Mathematically, the transmission loss is measured on the decibels (dB), and it can be defined using the following formula:

$$TL = 10\log\left|\frac{W_i}{W_t}\right| \tag{1}$$

where W_i is the power of the incident wave and W_t is the power of the transmitted wave. The sound insulation between rooms is the difference in the sound pressure level between the emission room and the reception room. We distinguish two types of indicators between rooms: (i) the gross sound insulation index between rooms in dB defined by:

$$D = L_1 - L_2 \tag{2}$$

and (ii) the normalized sound insulation index between the rooms in dB defined as follows:

$$D_n = L_1 - L_2 - 10\log\left(\frac{A}{A_0}\right) \tag{3}$$

where A_0 is a reference equivalent absorption area, taken as 10 m^2 and A is the equivalent absorption area of the receiving room in m^2 for a wall made of different surface materials S_i defined by:

$$A = \sum_i \alpha_i . S_i \tag{4}$$

where α_i is the absorption coefficient of each wall. For a room of volume V, the equivalent absorption area A is related to the reverberation time T_R by Sabine's formula (Vigran 2008).

$$T_R = \frac{0.16V}{A} \tag{5}$$

The corrective term used corresponds to an evaluation of the level difference in the receiving room when the equivalent absorption area varies from A to A_0. This normalized index is used for laboratory measurements.

3 Numerical and Experimental Results

3.1 Sound Radiation of a Rectangular Plate in a Cavity

Let's consider the vibration of an elastic plate, of dimensions: 0.3 m × 1.5 m and a thickness of 5 mm, made in aluminum ($E_s = 71$ GPa, $\nu_s = 0.33$, $\rho_s = 2770$ kg/m^3, $\eta_s = 0.01$) coupled to an acoustic cavity of dimensions 1.3 m × 2 m × 1.5 m and filled with air (velocity 340 m/s, density 1.225 kg/m^3 and fluid damping 0.02). The plate is simply supported at its edges and excited by a unit nodal force applied at its center.

The resolution of the associated acoustic radiation problem involving bounded domains by the finite element method has the disadvantage of the numerical cost which can be exorbitant especially to reach high frequencies. Other methods can be used as an alternative to reduce the numerical cost. In this section, the sound radiated by the vibrating plate coupled to the acoustic cavity in the frequency range between 100 Hz and 500 Hz is calculated using five configurations for modelling the acoustic cavity: (i) Rayleigh integral method, (ii) infinite element method using a half-sphere domain of radius 1 m (Fig. 2a), (iii) infinite element method using a rectangular parallelepiped domain with the same dimensions of the acoustic cavity (Fig. 2b), (iv) finite element method with rigid boundaries (Fig. 2c), and (v) finite element method with absorbing coefficient 0.4 applied to all walls of the cavity except the one in contact with the plate and damping coefficient of 0.02 in the cavity (Fig. 2d). A quadratic element (Quad4) of size 0.03 m is used for the mesh of the plate and hexahedral element of size 0.1 m for the acoustic cavity.

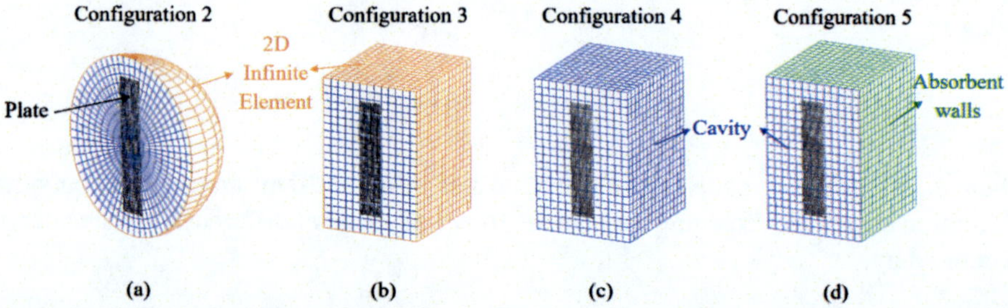

Fig. 2. Configurations of interest: (a) infinite elements with a half-sphere domain, (b) infinite elements with rectangular parallelepiped domain, (c) finite elements with rigid boundaries, and (d) finite elements with absorbent walls

Regarding the infinite element method, a standard finite element mesh is applied around the vibrating structure in conjunction with infinite elements used for modelling the far field region. To eliminate sound wave reflections on the boundary of the truncated acoustic domain, the infinite element method uses special shape functions ϕ with the following form:

$$\Phi(r, \theta, \varphi) = e^{-jkr} \sum_{n=1}^{m} \frac{F_n(\theta, \varphi)}{r^n} \quad (6)$$

where (r, θ, φ) are the ellipsoidal reference coordinate, m is the radial interpolation order and $F_n(\theta, \varphi)$ is a continuous regular function in the infinite fluid domain (Coyette and Van den Nieuwenhof 2000). For this work, the interpolation order m is 10 for configurations 2 and 3.

The radiated sound power from the plate was calculated in decibels (dB) in narrow band for the five presented configurations (Fig. 3). The simulations were performed with the software Actran.

We note from Fig. 3, that the radiated acoustic power calculated from the finite element modeling of the acoustic cavity without absorptions (configuration 4) diverges compared to the other configurations. This is because in the low frequency range the acoustic modes of the cavity exert an influence on the vibratory behavior of the plate. The introduction of absorption in its walls (configuration 5) makes the results of the finite element model converge towards the other models. The Rayleigh integral method is the least expensive method. However, its field of application is limited to plane baffled structures radiating in an unbounded or bounded acoustic domain with absorbent walls.

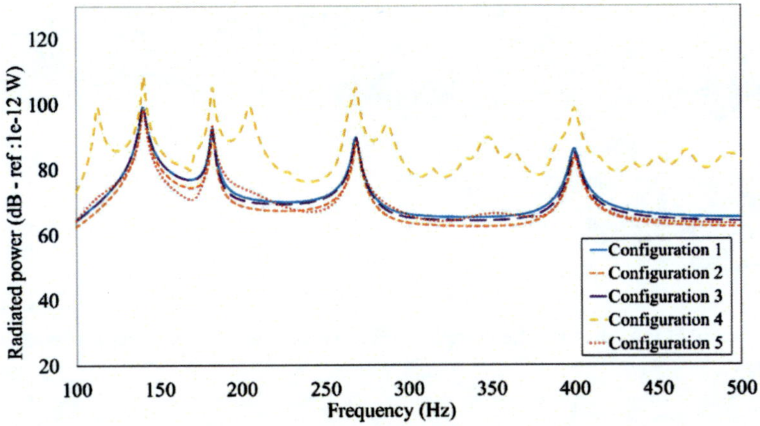

Fig. 3. Sound radiation of the plate in a cavity with different methods in narrow band

The infinite elements method, despite its high cost compared to the Rayleigh integral, has the advantage of being applied to any geometric shape of structure, even non-plane, radiating in a bounded acoustic domain or unbounded. It is less expensive than the finite element method. For the next simulations, we cannot use the Rayleigh integral because the roller shutter box is not mounted in a baffled plane structure. Also, we will not modelized the entire receiving room for calculation costs. Compared to configuration 2 using also infinite elements, configuration 3 allows a better optimization of the mesh and presents the advantages of the use of regular finite elements shape and a compatible mesh at the interfaces between the different parts of the system. So, we will keep configuration 3 and the application of the infinite element method for the rest of the numerical simulations.

3.2 Sound Transmission Through Roller Shutter Box

This paragraph concerns the numerical prediction of the sound transmission through a roller shutter box. The numerical results found from the application of configuration 3 will be compared to the results of laboratory measurements. The effect of introducing acoustic treatment with a melamine foam and heavy mass will be studied.

The studied box (Fig. 4a) is made of PVC (PolyVinyl Chloride) with the following mechanical properties: $E_b = 2.8$ GPa, $\nu_b = 0.35$, $\rho_b = 1460$ kg/m^3 and $\eta_b = 0.04$. Its dimensions are: 210 mm × 250 mm × 1450 mm. The walls of the box are composed of stiffened panels of thickness 1.42 mm as shown in (Fig. 4b). The end caps of the box, with 5 mm thick, are made of ABS (Acrylonitrile Butadiene Styrene) which is an industrial thermoplastic polymer with the following mechanical properties: $E_c = 2.9$ GPa, $\nu_c = 0.42$ $\rho_c = 1100$ kg/m^3 and $\eta_c = 0.01$. The rolled curtain of diameter 163 mm is represented as rigid cylinder. The boundary conditions are those used in the laboratory tests and relate to the embedding of surfaces 1, 2 and 3 of the box and caps as shown in Fig. 5a. Finally, we consider that the roller shutter box is filled with air (c = 340 m/s, ρ = 1.225 kg/m^3 and $\eta = 0.01$).

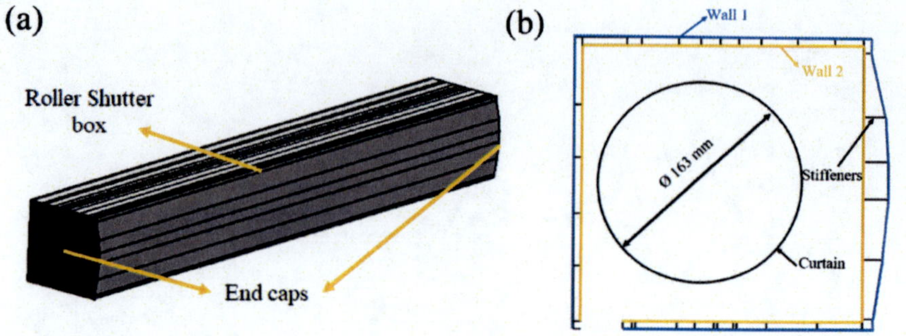

Fig. 4. The CAD model of (a) the complete roller shutter box and (b) the profile of the roller shutter box

3.2.1 Resonance Frequencies of the Roller Shutter Box

Table 1 presents the first six resonance frequencies of the roller shutter box calculated by modeling its walls with the stiffeners and compared to a simple box model where its walls are modeled by a single plate (without stiffener) of thickness 6.5 mm. The objective of this comparison is to validate a box model with simple geometry to minimize the cost of acoustic calculation. The results of this comparison show a good agreement between the two models allowing to validate the geometry without stiffener which will be used for the acoustic simulations.

Table 1. Modal analysis of the roller shutter box.

Real box (Hz)	Simple box (Hz)	Relative gap (%)
94.9	101.6	7%
119.7	126.2	5%
150.4	155.2	3%
182.2	188.5	3%
220.2	224.6	2%
259.5	271.7	5%

3.2.2 Numerical and Experimental Analysis of Sound Transmission

In this section, results of laboratory measurements of the sound transmission through the roller shutter box described above are compared to results of numerical simulations using configuration 3 of Sect. 3.2. For the numerical model, to avoid modeling the emitting chamber, only a diffuse sound field is applied to the structure as an acoustic excitation (Soussi et al. 2021) (Fig. 5). The diffuse waves are defined statistically using the Power Spectral Density matrix (PSD) by applying a Cholesky decomposition (Van den Nieuwenhof et al. 2010). The receiving chamber is replaced by a parallelepiped-shaped domain of dimensions 820 mm × 1030 mm × 1850 mm meshed with hexahedral elements of size 25 mm and filled with air ($c_{IE} = 340$ m/s and $\rho_{IE} = 1.225$ kg/m^3 and $\eta_{IE} = 0.02$). At the border of this domain, 2D quadrilateral infinite elements of size 25 mm are applied to avoid acoustic wave reflections.

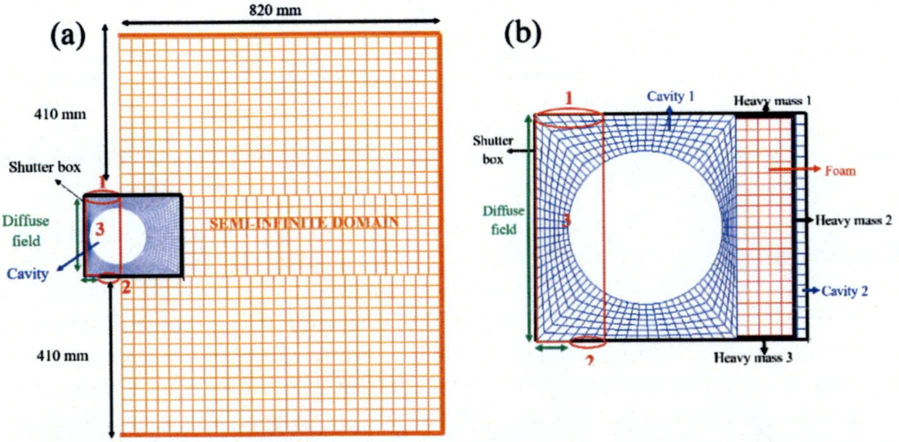

Fig. 5. Numerical model used for acoustic simulation without (a) and with (b) insulation

Figure 6a shows a comparison of the sound transmission loss (in dB) measured in laboratory and calculated numerically in third octave band. A good agreement between these two methods can be observed with an average relative deviation of 9%.

3.2.3 Sound Insulation Using Porous Materials and Heavy Mass

To reduce the sound transmission through the roller shutter box, porous materials (melamine foam) in combination with heavy viscoelastic masses are introduced in the system (Fig. 6b). The used melamine of dimensions 202 mm × 60 mm × 1450 mm is characterized by the following parameters: $E_f = 160$ kPa, $\nu_f = 0.44$, $\rho_f = 8.35$ kg/m^3, $\eta_f = 0.06$, $\sigma = 12600$ N.m^{-4}.s, $\phi = 0.99$, $\alpha = 1$, $\Lambda = 78$ µm and $\Lambda' = 192$ µm. It's modeled using Biot theory (Sgard et al. 2000). The heavy mass, of 4 mm thickness, has the following properties: $E_h = 120$ MPa, $\nu_h = 0.43$, $\rho_h = 1600$ kg/m^3 and $\eta_h = 0.65$. Figure 6b shows a comparison between numerical and experimental results. We can observe a good agreement between the two methods which validate the proposed numerical approach. The introduction of sound proofing materials significantly improves the acoustic insulation of the roller shutter box.

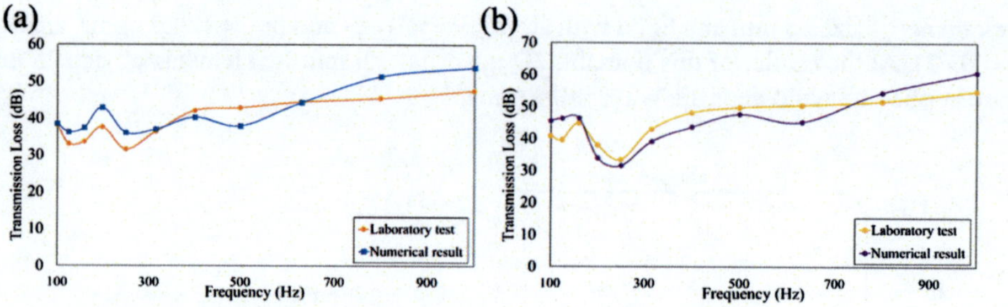

Fig. 6. Correlation between test and numerical calculation without (a) and with (b) sound insulation

4 Conclusion

We were able to develop a numerical model to characterize the acoustic behavior of roller shutter boxes with and without insulation. We have exposed different methods to calculate the sound radiation power for a structure coupled with an air domain. In our case, we have chosen to work with infinite element method. We have presented the test results from the laboratory tests and compared them to the numerical results with a simplified box model. Both results show a good agreement. Future work will focus on performing sensibility analysis to find the most influential parameters on the sound transmission loss.

Acknowledgements. The authors would like to express gratitude to CODIFAB (Comité professionnel de Développement des Industries Française de l'Ameublement et du Bois) and UFME (Union des Fabricants de Menuiseries) for their cooperation and financial support.

References

Asdrubali, F., Buratti, C.: Sound intensity investigation of the acoustics performances of high insulation ventilating windows integrated with rolling shutter boxes. Appl. Acoust. **66**, 1088–1101 (2005)

Cops, A., Minten M., Myncke, H.: Influence of the design of transmission rooms on the sound transmission loss of glass-intensity versus conventional method. Noise Control Eng. J. 121–129 (1987)

Coyette, J.P., Van den Nieuwenhof, B.: A conjugated infinite element method for half-space acoustic problems. J. Acoust Soc. Am. **108**(4), 1464–1473 (2000)

Díaz, C., Pedrero, A.: An experimental study on the effect of rolling shutters and shutter boxes on the airborne sound insulation of windows. Appl. Acoust. **70**, 369–377 (2009)

Díaz, C., Díaz, A., Navacerrada, M.A.: An experimental study on the effect of rolling shutters on the field measurements of airborne sound insulation of façades. Appl. Acoust. **74**, 134–140 (2013)

Kihlman, T., Nilsson, A.: The effects of some laboratory designs and mounting conditions on reduction index measurements. J. Sound Vib. 349–364 (1972)

Lloret Gavila, M.: Prediction of the airborne sound transmission through a car front end model including poroelastic acoustic treatments. Dissertation, University of Magdeburg (2018)

Sgard, F.C., Atalla, N., Nicolas, J.: A numerical model for the low frequency diffuse field sound transmission loss of double-wall sound barriers with elastic porous linings. J. Acoust Soc. Am. **108**(6), 2865–2872 (2000)

Soussi, C., Aucejo, M., Larbi, W., Deü, J.F.: Numerical analyses of the sound transmission at low frequencies of a calibrated insulating glazing unit. Appl. Acoust. **179**, 108065 (2021)

Soussi, C., Aucejo, M., Larbi, W., Deü, J.F.: Numerical analyses of the sound transmission at low frequencies of a calibrated domestic wooden window. Proc. Inst. Mech. Eng. C J. Mech. Eng. Sci. **235**(14), 2637–2650 (2021)

Van den Nieuwenhof, B., Lielens, G., Coyette, J.P.: Modeling acoustic diffuse fields: updated sampling procedure and spatial correlation function eliminating grazing incidences. In: Proceedings of ISMA2010 Including USD2010, vol. 3, pp. 4723–4736 (2010)

Vigran, T.E.: Building Acoustics. Taylor & Francis, London (2008)

Structural Performance Evaluation of the Settling Station of the Gafsa Phosphate Company

Majdi Yangui(✉), Ahmed Samet, Moez Beyaoui, and Mohamed Haddar

Laboratoire de Mécanique, Modélisation et Productique (LA2MP), École Nationale d'Ingénieurs de Sfax, Sfax, Tunisia
yanguimajdi@gmail.com, mohamed.haddar@enis.rnu.tn

Abstract. The primary aim of this study is to create a prototype that incorporates a coagulation, flocculation, and settling station. The prototype is intended to optimize the water recovery rate within storage basins and reduce the surface area required for sludge storage. To achieve this, the study will involve the design of treatment and de-watering systems. The production flows of clarified water will be determined based on the initial flow of phosphate sludge. Finally, the settling station will be sized according to the specific requirements of the Eurocode 3 standard. The sizing is performed using numerical simulations based on the finite elements method using SolidWorks simulation. To model the sedimentation basin structure made up of sheet metal parts and structural members, a mixed meshing is developed based on triangular shell elements and beam elements. The development of this prototype is expected to enhance the efficiency and sustainability of wastewater treatment processes, while also reducing the environmental impact of industrial activities.

Keywords: Settling station · Dewatering system · Structural performance evaluation · Finite Element Modeling

1 Introduction

Faced with the lack of water resources, the depletion of underground water tables and the successive years of drought, the reuse of water contaminated by clay sludge is now of major importance, especially in the arid regions of the mining basin of Gafsa. As a matter of fact, the discharge into nature of more than 20 million m^3/year of sludge, by the Company of Phosphate of Gafsa CPG, which essentially contains only water and a clay suspension constitutes a waste of a vital resource that can be treated and used for agricultural development. In addition to the disfigurement of the natural landscape by the sludge which spreads over hundreds of km. In this context, a Federated Research Project has been funded to provide a simple method of water treatment that could lead to its use in the irrigation of palm groves and other fruit trees. Currently, the discharge of phosphate process water has a negative impact on the quality of the water table (Batarseh and El-hassen 2009); Othman and Al-Masri 2007). Thus, the efforts presented by the

A. Akrout et al. (Eds.): ICAV 2022, ACM 22, pp. 102–108, 2023.
https://doi.org/10.1007/978-3-031-34190-8_13

CPG in collaboration with national and international research laboratories are constantly evolving with the aim to reduce this environmental problem.

Due to the large size of the sedimentation basin in the water treatment process, several investigations tried to optimize the basin geometry (McLaughlin et al. 2009). Kang et al. (2014) showed that the sedimentation basin can greatly improve the obtained water quality. However, the sedimentation station's large size remains a notable limitation in the cited works.

In this work, we aimed to improve the efficiency of sludge settling in water treatment by designing a sludge settling station with a screw press system. This system was developed based on the water treatment methods and processes commonly used in the industry. Our design selectively removes only the solid part of the settling sludge, resulting in a significant reduction in sedimentation basin volume. This innovation has the potential to decrease the operational costs of water treatment facilities while maintaining high-quality standards.

2 Sludge Settling Station Presentation

As presented in Fig. 1, the developed sludge treatment station consists of four main stations.

The sludge rich in water is pumped to the coagulation station. A quick stirring system was designed to promote the chemical reaction between the small particles in the sludge and the coagulant.

To ensure the flocculation process of the sludge, a small stirring velocity must be adjusted in the second station.

The sedimentation basin is the most important structure in terms of size and role in the water treatment process. This step generates a lot of wasted time, where to have pure water, the precipitation of the mud must be performed. In the conventional settling systems, emptying the basin of sedimentation causes the loss of water and coagulant remaining in the sludge. For this reason, a screw press system is designed to clear out only the solid part of the settling sludge.

The obtained clear water from the sedimentation basin can be used in the irrigation, but the filtered water will be injected once again in the coagulation station. The use of the filtered water will reduce the amount of the chemical used to ensure the coagulation-flocculation processes.

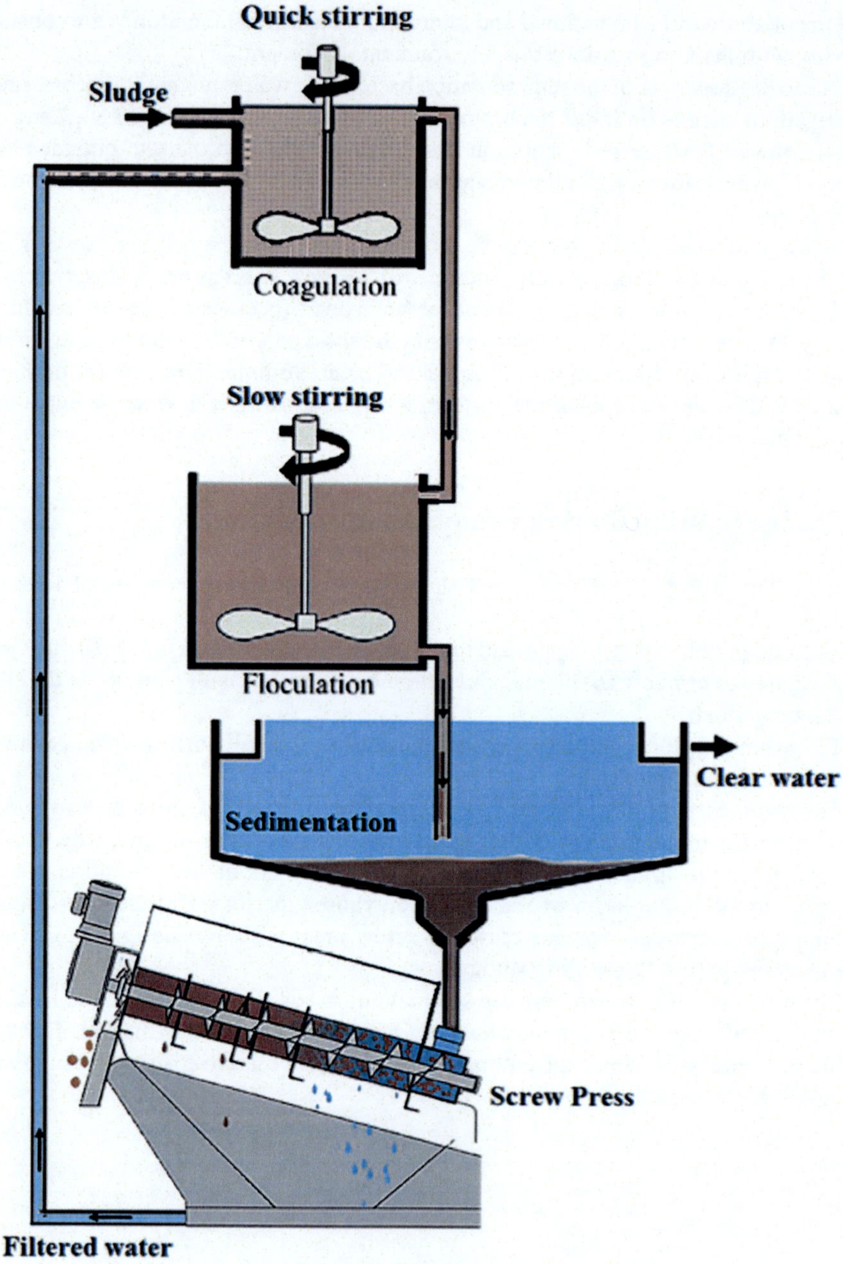

Fig. 1. Sludge settling station

3 Design and Sizing of the Sedimentation Basin

Based on already existing systems as well as on mechanical considerations a first design of the sedimentation basin, the most important structure in the presented water treatment process, was created using SolidWorks software (Fig. 2). The vertical components of the supporting structure have a length of 3.6 m. The height of the basin varies linearly from 1 m to 3.6 m in order to facilitate the degradation of the sludge. The selected materials are the most widely used weldable low carbon manganese structural steels, S235 for the sheet metal parts and S275 for the supporting structure components.

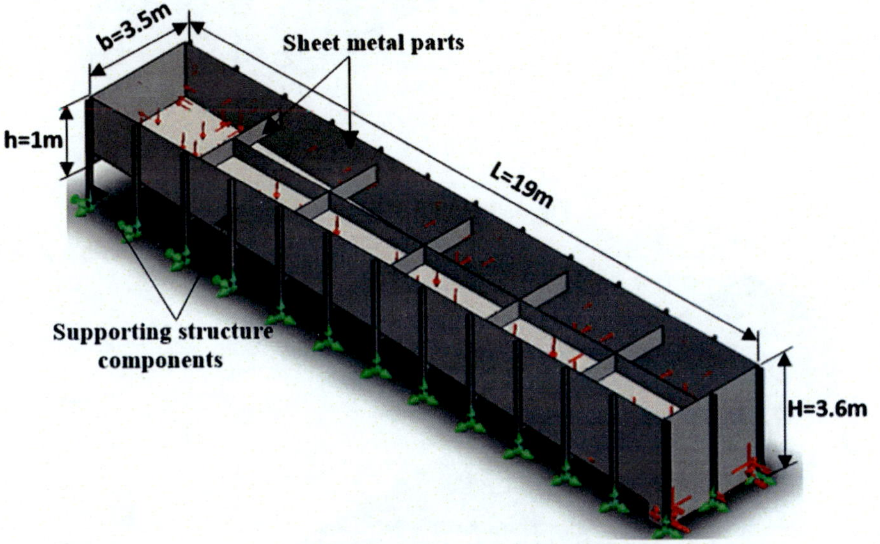

Fig. 2. Design of the sedimentation basin

The sizing is carried out using numerical simulations based on the finite elements method (FEM). For this, a simple elastic stress analysis was performed. In Fig. 2, the loading situation and the used boundary conditions are presented. The loading is applied as hydrostatic pressure, which varies according to the depth of the sludge in the basin. The density of the sludge has been estimated to be 1800 kg/m^3. Figure 3 presents the generated FE-model. A mixed meshing is carried out based on triangular shell elements for modelling the sheet metal parts and beam elements to model the structural members.

The model consists of 8602 elements, and the finite element size of 200 mm was determined through meticulous application of numerical convergence, guaranteeing the accuracy and precision of the simulations.

In Fig. 4 the distribution of the Von Mises equivalent stress in the designed sedimentation basin is illustrated. As shown, the maximum value of this stress does not exceed 211 MPa located at the supporting beams structure. Thus, a safety coefficient of 1.3 is guaranteed which is greater than 1.25 the minimum value of the partial coefficient mentioned by the Eurocode 3 standard (Standard 1993).

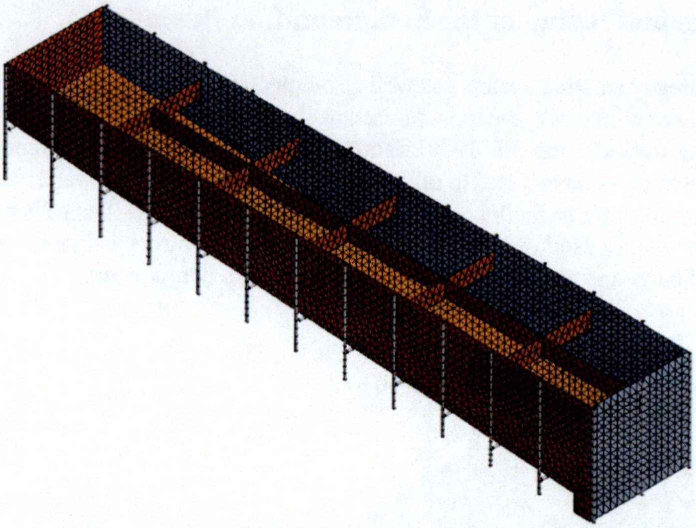

Fig. 3. Generated FE-model

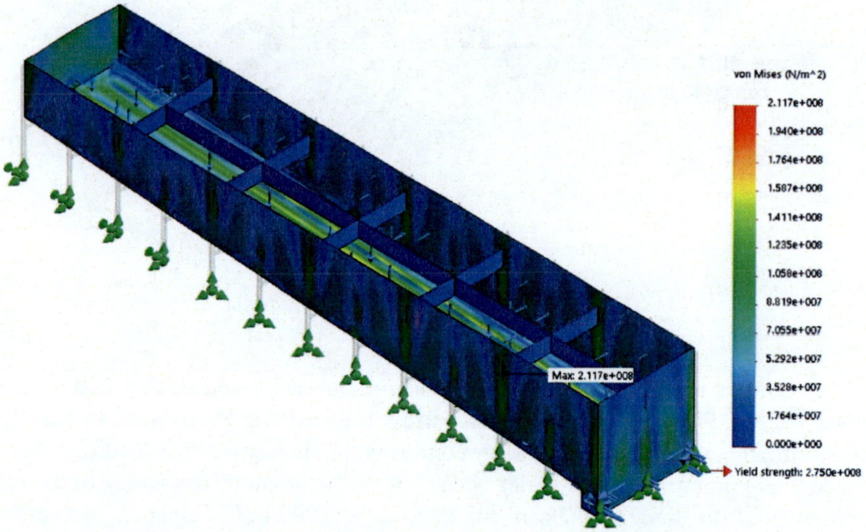

Fig. 4. Calculated distribution of the equivalent stress in the basin

In addition, the analysis of the displacements in the considered structure (Fig. 5) does not exceed the value of 55 mm.

In fact, the Eurocode 3 standard emphasizes the displacements of the supporting structure. The permissible displacement δ mentioned by standard of a column of height H can be calculated as follows:

$$\delta = \frac{H}{150} = 24\,\text{mm} \tag{1}$$

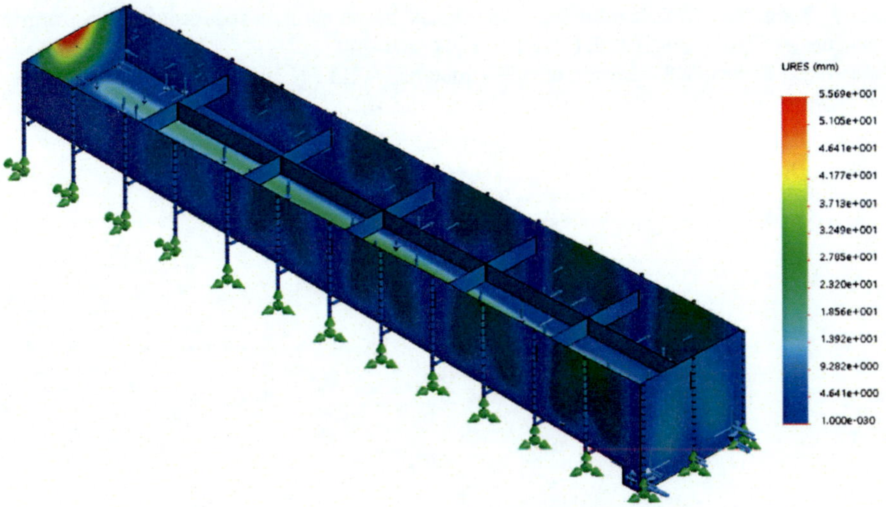

Fig. 5. Sedimentation structure displacements

Figure 5 shows that the maximum displacement of the columns is about 23 mm, what agrees with the applicable standard.

4 Conclusion

The current study introduced a novel sludge treatment station design that offers the benefit of decreasing the sedimentation basin volume. To develop the numerical model of the sedimentation basin, both shell and beam elements were employed while considering the nodal connectivity between these two types of elements. Through numerical simulations, the structure components were optimized in accordance with the specific requirements of the Eurocode 3 standard. However, it is important to note that only the sedimentation station was addressed in this research, and therefore, there is potential to extend this study by optimizing the size of other stations as well.

Acknowledgements. The authors gratefully acknowledge the Project "PRF 2019-D6P1" funded by the Tunisian Ministry of Higher Education and Scientific Research.

References

Batarseh, M., El-Hassen, T.: Toxic element levels in the phosphate deposits of central Jordan. Soil Sediment. Contam. **18**, 205–215 (2009)

Othman, I., Al-Masri, M.S.: Impact of phosphate industry on the environment: a case study. Appl. Radiat. Isotopes **65**, 131–141 (2007)

McLaughlin, R.A., Hayes, S.A., Clinton, D.L., McCaleb, M.M., Jennings, G.D.: Water quality improvements using modified sediment control systems on construction sites. Trans. ASABE **52**(6), 1859–1867 (2009)

Kang, J., King, S.E., McLaughlin, R.A.: Impacts of flocculation on sediment basin performance and design. Trans. ASABE **57**(4), 1099–1107 (2014)
Standard, B.: Eurocode 3—Design of steel structures—. BS EN 1993-1, 1, 2005 (2006)

Fluid-Structure Interaction: Application to Segmented Wind Turbine Blades

Ahmed Samet, Majdi Yangui(✉), Mohamed Amine Ben Souf, and Mohamed Haddar

Laboratoire de Mécanique, Modélisation et Productique (LA2MP), École Nationale d'Ingénieurs de Sfax, Sfax, Tunisie
yanguimajdi@gmail.com, mohamed.haddar@enis.rnu.tn

Abstract. The interaction of a moving fluid with a structure present nowadays a particular interest in academic and industrial projects. This interaction gives rise to several physical phenomena such as the reaction of tall structures to winds and the vibration of turbine blades. Modeling both the structure and the fluid helps to understand these physical phenomena. In this work, the Computational Fluid Dynamics CFD code ANSYS Fluent was used to model the aerodynamic behavior of the flow around a Segmented Wind Turbine Blades (SWTB) and to determine the pressure load on the blades. Then, a structural analysis study was also performed used the FEA (Finite Element Analysis) model implemented in ANSYS Structural module to find the performance parameter such as the total displacement. The interface of CFD and FEA is based on a one-way coupling, in which aerodynamic loads calculated from CFD modelling are mapped to FEA modelling as load boundary conditions.

Keywords: Fluid Structure Interaction · Wind turbine · Segmented blade · CFD · FEM · Aerodynamic behavior · Structural analysis

1 Introduction

The depletion of fossil fuels and the emission of greenhouse gases are the major problems of the earth's descendants. To solve these problems, the use of renewable energy such as wind power has become the target of global attention [1]. In fact, improving the capabilities of wind turbines continues to attract particular interest from academia and industry.

Recently, research is focused on aerodynamic [2, 3] and structural [4] modeling of wind turbines and their interaction has increased significantly in recent years. In fact, Fluid-Structure Interaction (FSI) is a common phenomenon during wind turbine operation. Many numerical studies have been developed in this context. Bazilevs et al. [5] developed a fully coupled fluid-structure interaction (FSI) simulation methodology for wind turbine rotors. Roul et al. [6] numerically studied the fluid-structure interaction in wind turbines using CFD and FEM modeling. Wang et al. [7] studied the influence of wind speed on the structural behavior of the blade using CFD and FEM modeling.

A. Akrout et al. (Eds.): ICAV 2022, ACM 22, pp. 109–115, 2023.
https://doi.org/10.1007/978-3-031-34190-8_14

Note that most of the researches in the literature have not dealt with a segmented blade which has a good advantage in ease of assembly. In this context, the main novelty of this paper is to study the FSI in segmented wind turbine blades. The CFD code ANSYS Fluent was used to determine the pressure load on the segmented blades. Then, a structural analysis was also performed using the ANSYS Structural FEA code to determine the total displacement of the blade.

This paper is organized as follows: Sect. 2 discusses the FSI analysis, the design of the segmented blade, the CFD and FEM modelling and Sect. 3 discusses the numerical results.

2 FSI Analysis

The FSI analysis consist of modelling the interaction between the structural and fluid flow.

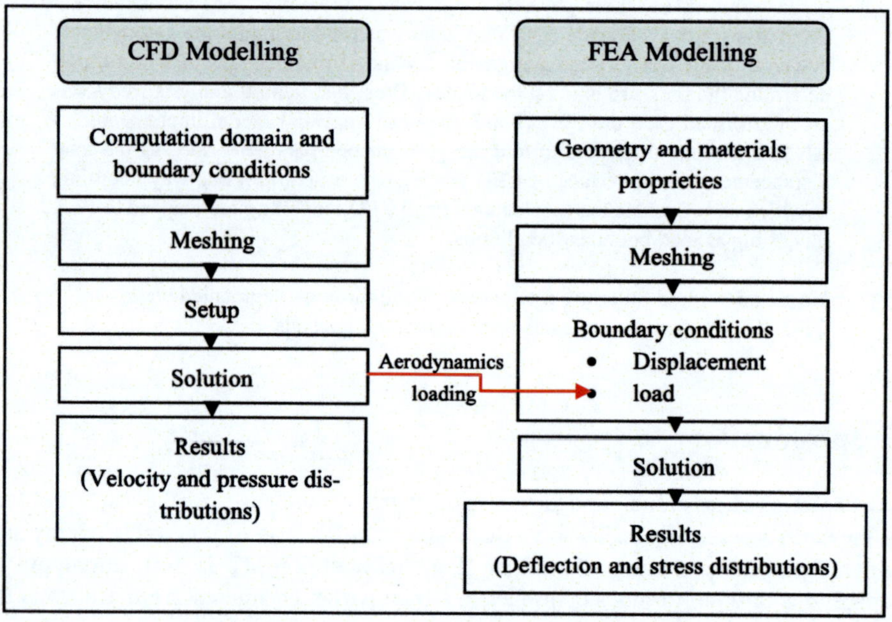

Fig. 1. The numerical methodology of FSI analysis.

In this paper, the performance of the SWTB is the result of the contact between the aerodynamics of the wind flow and the structural model. This physical phenomenon is modelling using ANSYS software. In fact, the numerical methodology of FSI analysis is presented in Fig. 1. As presented, two computational areas have been considered, which are studied separately according to their mesh and significant equations.

2.1 Segmented Wind Turbine Blade Design

The system under study is a SWTB for aerodynamic power utilization. The 3D geometry of the turbine is presented in Fig. 2. The turbine diameter is 1 m and it has three blades [10].

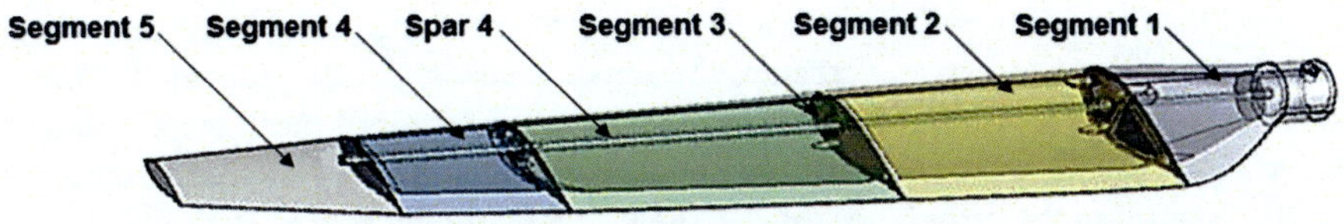

Fig. 2. Three dimension of the SWTB.

The material properties of the SWTB components are presented in the following Table 1.

Table 1. Material properties of the blade components [4].

Parameters	Material	Density (kg/m^3)	Poisson's ratio	Elastic modulus (GPa)
Blade segments	PC-ABS	1070	0.3879	2.25
Spar	Steel	7850	0.3	210

2.2 CFD Modeling

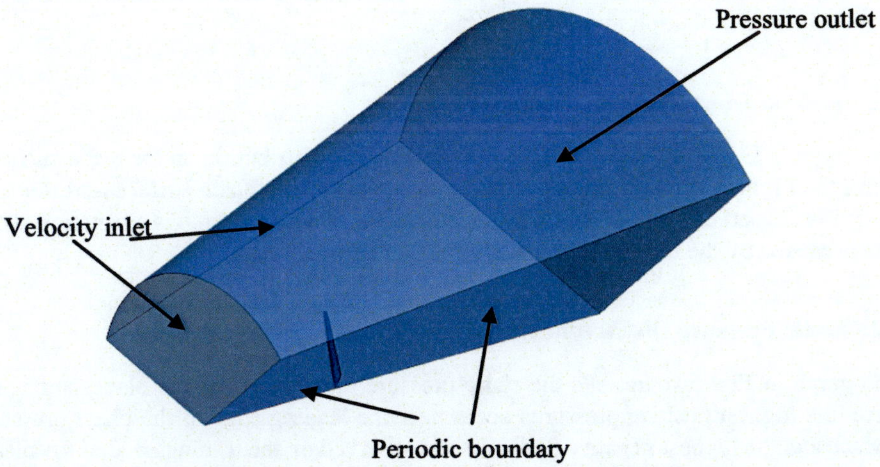

Fig. 3. Computational domain and boundary conditions.

This section deal with the CFD parameters applied to model the aerodynamics behavior of the flow around the segmented wind turbine blade.

The 3D-flow model was built using the commercial CFD code ANSYS Fluent. The operating conditions used for the 3D-CFD simulations are as follows: the rated wind speed is 5 m/s, the pitch angle is 0° and the rotational velocity of the turbine is 25 rad/s. The procedure to follow for the CFD simulation consists of two steps starting with the creation of the computational domain and define the boundaries conditions, as presented in Fig. 3. Once the geometry is created, the mesh is subsequently generated. This step requires care for the CFD simulation. To this end, several sizes of mesh were tested from a coarse mesh to a refined one until the numerical results no longer depend on the size of mesh. Regarding the turbulence model, the k-ω SST model is used in this study, since it has been used in this area of study and gives favorable results [8, 9].

2.3 FEA Modeling

This section discusses the FEA parameters applied to model the structural behavior of the segmented wind turbine blade. The 3D model was built using the commercial FEA code ANSYS Static structural. The operating conditions used for the 3D FEA simulations are as follows: the aerodynamic loads determined from the CFD modeling are applied as the load boundary condition to the FEA modeling. Also, there are two other sources of loads generated by the centrifugal loads, caused by the blade rotation, and the optimal assembly load, estimated to 65 N, investigated by Yangui et al. [10]. In fact, in this work, rotational speed and aerodynamic loads were considered as static loads.

3 Results and Discussions

In this section, the numerical simulation results of the FSI model for the segmented wind turbine blades are presented. The interaction between the structural and fluid flow allow the visualization in the first steps of the velocity and pressure distributions and the displacement distribution in the second steep.

3.1 Velocity Magnitude Distributions

The graph in Fig. 4 represents the magnitude velocity distribution on the blade surface. Here, the highest velocity occurs near the upstream of the blade surface and the low velocity is observed on the downstream of the blade near the rotor axis of rotation. This fact is caused by the turbine rotation.

3.2 Static Pressure Distributions

The graph in Fig. 5 represents the static pressure distribution on the blades surfaces. Here, the highest positive pressures occur near the leading edge of the blade pressure surface and the highest negative pressures are observed on the leading edge of the blade suction surface near the rotor axis of rotation.

Fig. 4. Velocity magnitude distributions.

Fig. 5. Static pressure distributions.

3.3 Total Displacement Distributions

The graph in Fig. 6 represents the total displacement distribution on the blade surface. Here, the maximum deformation occurs at the tip of the blade. This fact is due to the highest velocity occurs near the upstream of the blade surface.

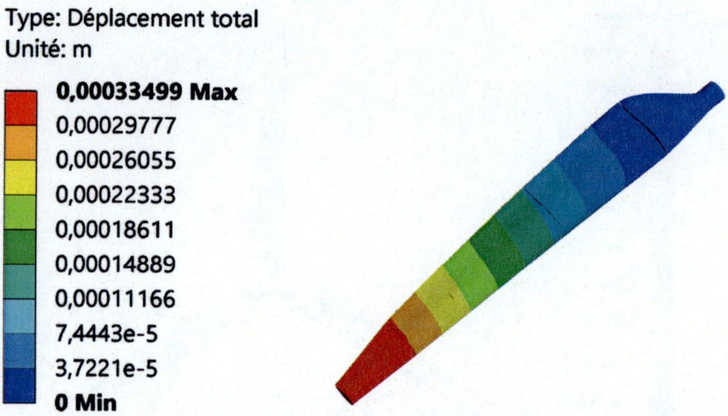

Fig. 6. Total displacement distributions.

4 Conclusion

In this paper, a numerical study of aerodynamic and structural analysis of segmented wind turbines by combining Computational Fluid Dynamics (CFD) and Finite Element Analysis (FEA) is presented. The Fluid structure interaction strategy is based on one-way coupling, in which the aerodynamic loads calculated by CFD modelling are mapped to FEA modelling as load boundary conditions. In this work, the CFD code ANSYS Fluent is used to model the aerodynamic behavior of the flow around a SWTB and to determine the pressure load on the blades. Then, a structural analysis study is also performed used the FEA code ANSYS Structural to find the performance parameters such as the total displacement.

Acknowledgements. This work has been funded by the research project PRF 2019 D1-P2 entitled "Design, modeling and diagnosis of wind turbines for sustainable energy efficiency".

References

1. Jonkman, J.M., Buhl Jr., M.L.: FAST user's guide. Technical report NREL/EL-500-38230, National Renewable Energy Laboratory, Golden, CO (2005)
2. Gómez-Iradi, S., Steijl, R., Barakos, G.N.: Development and validation of a CFD technique for the aerodynamic analysis of HAWT. J. Solar Energy Eng. **131**, 031009–1–13 (2009)
3. Zahle, F., Sørensen, N.N., Johansen, J.: Wind turbine rotor- tower interaction using an incompressible overset grid method. Wind Energy **12**, 594–619 (2009)
4. Yangui, M., Bouaziz, S., Taktak, M., Haddar, M.: Experimental updating of a segmented wind turbine blade numerical model using the substructure method. J. Strain Anal. Eng. Design **56**(2), 67–75 (2020)
5. Bazilevs, Y., Hsu, M.-C., Kiendl, J., Wüchner, R., Bletzinger, K.-U.: 3D simulation of wind turbine rotors at full scale. Part II: fluid–structure interaction modeling with composite blades. Int. J. Numer. Methods Fluids **65**, 236–253 (2011)
6. Roul, R., Kumar, A.: Fluid-structure interaction of wind turbine blade using four different materials: numerical investigation. Symmetry **12**(9), 1467 (2020)

7. Wang, L., Quant, R., Kolios, A.: Fluid structure interaction modelling of horizontal-axis wind turbine blades based on CFD and FEA. J. Wind Eng. Ind. Aerodyn. **158**, 11–25 (2016)
8. Sørensen, N.N., Michelsen, J., Schreck, S.: Navier–Stokes predictions of the NREL phase VI rotor in the NASA Ames 80 ft× 120 ft wind tunnel. Wind Energy **5**, 151–169 (2002)
9. Mo, J.-O., Lee, Y.-H.: CFD Investigation on the aerodynamic characteristics of a small-sized wind turbine of NREL PHASE VI operating with a stall-regulated method. J. Mech. Sci. Technol. **26**, 81–92 (2012)
10. Yangui, M., et al.: Numerical assessment of the structural performance of a segmented wind turbine blade. In: Hammami, A., Heyns, P.S., Schmidt, S., Chaari, F., Abbes, M.S., Haddar, M. (eds.) Modelling and Simulation of Complex Systems for Sustainable Energy Efficiency. MOSCOSSEE 2021. Applied Condition Monitoring, vol. 20, pp. 1–7. Springer, Cham (2021). https://doi.org/10.1007/978-3-030-85584-0_1

Effects of LRB Isolators and Viscous Dampers on Seismic Isolated Irregular Reinforced Concrete Buildings

Abed El Rahman Yaktine[1], Magdalini Titirla[2], and Walid Larbi[2](✉)

[1] Department of Civil Engineering, ISSAE CNAM LIBAN, Beirut, Lebanon
https://isae.edu.lb/accesGmail/

[2] Laboratory of Structural Mechanics and Coupled Systems (LMSSC), Conservatoire National des Arts et Métiers (CNAM), Paris, France
{magdalini.titirla,walid.larbi}@lecnam.net

Abstract. The aim of this paper is to present the effects of lead rubber bearings (LRB) or viscous dampers (VD) in irregular in plan reinforced concrete buildings. The buildings (a three-story building, an eight-story building, and a twenty-story building) had been studied under earthquake recorded accelerograms, three of them are real and the other four are artificial, by nonlinear dynamic time-history analyses. In this paper the optimal design of LRB bearings and viscous dampers is described, focusing the design on minimizing certain parameters: (i) the maximum displacement (top of the structures), (ii) the torsion of the buildings, (iii) the base shear loads and, (iv) the maximum horizontal interstory drift. The comparative results show the effectiveness of the VD in terms of minimizing the torsion effects for all the buildings, and mostly in the low rise one. However, the use of LRB provides a significant reduction for all the buildings in terms of minimizing the rotation, the maximum acceleration, the max-story drift, the maximum displacement and the base shear loads.

Keywords: Base isolation · LRB · Viscous dampers · Irregularity in plan · Optimization · multi-story buildings · seismic response

1 Introduction

Torsional irregularity is one of the most important factors, which causes damages in the majority of the structures. In the literature, a large number of studies investigated various aspects of torsional irregularity including geometric asymmetry (Duan and Chandler 1997; Demir et al. 2010, Alla et al. 2022) in pushover and non-linear dynamic analyses (Martinez-Romero, 1993; Penelis and Kappos, 2002, Ayoub et al. 2022, Forcellini and Kalfas, 2023) as well as experimental and analytical studies (Jeong and Elnashai, 2006, Amendola et al. 2016, Russillo et al 2022). In addition, some of them are focused their studies on the seismic vulnerability of irregular systems in plan or elevation (Chandler et al. 1996; De Stefano and Pintucchi, 2002; Bosco et al. 2015; Stathi et al. 2015; Das and Nau, 2003).

A. Akrout et al. (Eds.): ICAV 2022, ACM 22, pp. 116–124, 2023.
https://doi.org/10.1007/978-3-031-34190-8_15

Dampers that belong to passive energy dissipation devices, such as viscous dampers (VD), friction dampers (FD) and base isolation with lead rubber bearings (LRB), have been used in order to control the seismic response of civil engineering structures. The main idea of passive energy dissipation systems is to reduce the structural damages, by absorbing the structural vibratory energy and dissipating energy through their hysteretic behaviour. As a result, many bridges and buildings were strengthening with them (Benzoni and Casarotti, 2009; Fabbrocino et al. 2017; Titirla et al. 2017, Titirla et al. 2018, Mrad et al 2021; Jaissee et al. 2021; Gao et al 2022).

Which is the appropriate selected solution to prevent the failure of an under-designed reinforced concrete frame has been studied by previous researchers, taking into account various parameters (Valente and Milani, 2018; Requena-García-Cruz et al. 2019). In this study, the effectiveness of lead rubber bearings (LRB) and viscous dampers (VD) on seismic isolated irregular reinforced concrete buildings is investigated in order to minimize the torsional irregularity. A comparable presentation of the effect of lead rubber bearings (LRB) or viscous dampers (VD) is illustrated. The 3 buildings are similar in plan but the number of the floor is varied from 3 to 20. The comparative results show the effectiveness of the VD in terms of minimizing the torsion effects for all the buildings, and mostly in the low rise one. However, the use of LRB provides a significant reduction for all the buildings in terms of minimizing the rotation, the maximum acceleration, the max-story drift, the maximum displacement and the base shear loads.

2 Buildings Description

The three investigated buildings are irregular in plan and have the shape L, with 42.00 m long in the longitudinal direction, while in the traversal directions the long varies from 11.00 m to 21.00 m, as illustrated in Fig. 1.

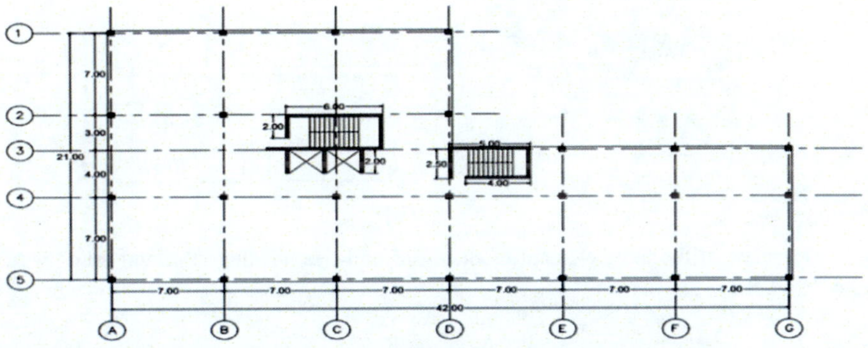

Fig. 1. Description of the three investigated buildings (in m).

The height of a single story is equal to 3.50 m, while the number of them is varied. The first building, named as « Building_Low», consists of 3 stories with the ground floor, the second one, mentioned from now on as « Building_Mid», consists of 8 stories, while the third building, mentioned as « Building_High», consists of 20 stories. The concrete material belongs to category C35/45, while the steel rebar's to S500B.

3 FE Modeling of the Investigated Buildings

In this study, walls were modelled as shell elements, columns and beams were modelled as frame elements with the appropriate selected rectangular cross section. Gravity and lateral loads were taken into account based on the provisions of EC1. The building has a fixed base at the foundation level. In addition, a rigid floor diaphragm has been modelled in each story. The viscous dampers (VD) were modelled by non-linear link elements, presented the performance low of Maxwell model. VDs were positioned in brace elements, simple diagonal configuration, which were modelled by frame element (see Fig. 2a). The bearings (LRB) were modelled by non-linear link elements (see Fig. 2b), which describe the corresponding translational and rotational stiffness of each bearing according to Section 7.5 of Eurocode 8 Part 2 (2005). The details of the dampers parameters were carried out thought the optimal design that is presented in Sect. 4.

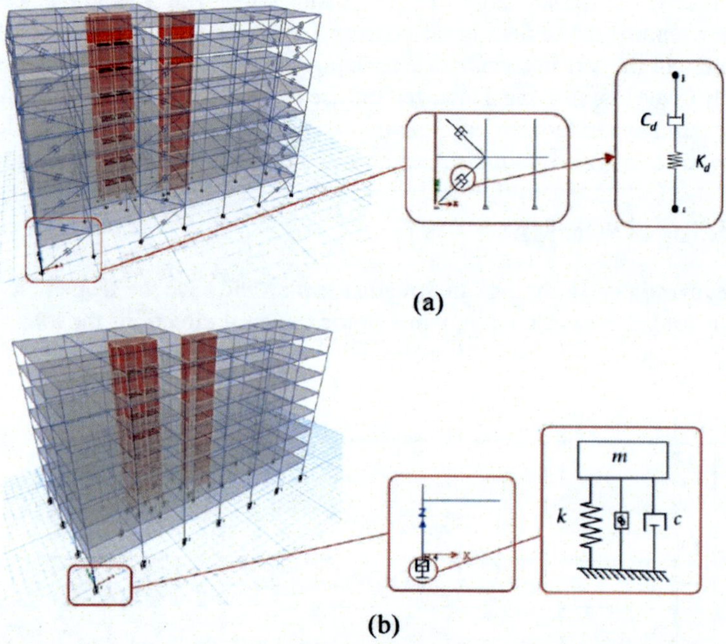

Fig. 2. Overview of the finite element modeling of mid-rise building "Building-Mid" with (a) VD and, (b) LRB

The three building have been studied for seven (real and artificial) accelerograms that were compatible to ground type B (seismic zone V according to the French national annex 1998). The accelerograms that have been selected from the word database, respect the provisions of Eurocode 8 Part 1. The direct integration, known β-Newmark method, was used. Bi-directional recorded accelerograms (3 of them real and 4 artificial) were used in the dynamic time history analysis of this study.

4 Optimization

4.1 Optimization of Viscous Dampers (VDs)

For the viscous dampers, the force $P(t)$ of each damper presented by the well-known mathematical formula of Seleemah and Constatinou (1997), presented in Eq. 1:

$$P(t) = Cd \cdot |\dot{u}(t)|^{a} \cdot sgn[\dot{u}(t)] \tag{1}$$

So $P(t)$ depends on the damping coefficient $C(d)$, the velocity across the damper $\dot{u}(t)$ and the coefficient a based on the piston head design and viscosity properties of fluid.

Coefficient a, based on the piston head design, tooks values between 0.3 to 1.0 in earthquake resistance structures, because the design aims to increase the forces as well as to minimize shocks for high velocities (Taylor Devices 2020). Silicone fluid is used that is completely non-toxic and is cosmetically and chemically inert. In this study a value equal to 0.3 has been selected for the coefficient a based to the results of previous research (Hwang et al. 2013, Mrad et al. 2022; Li et al. 2022). Damping coefficient $C(d)$, which is also one of the most important parameters of the optimal design of viscous dampers, selected based to desired effective damping attributed of each structure but also in a way that the damping ratio-repair cost remains low. In this study, a selected

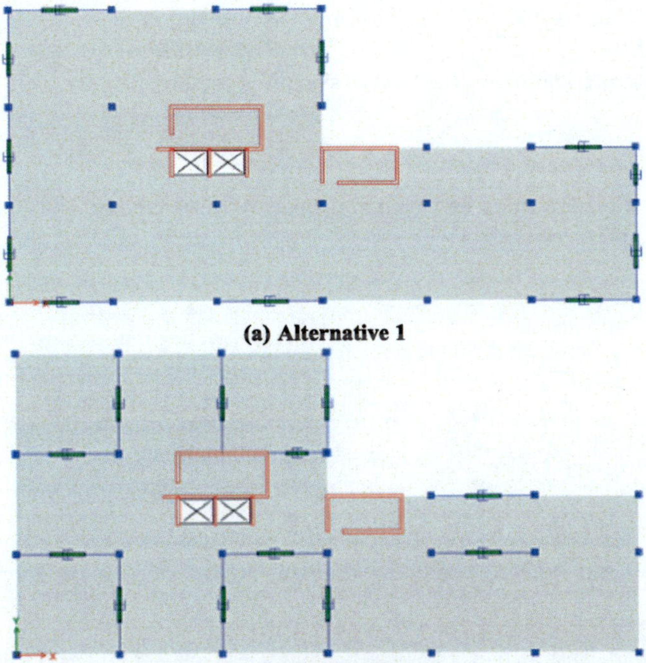

(a) Alternative 1

(b) Alternative 2

Fig. 3. Alternative 1 and 2 of viscous dampers placement.

damping coefficient achieved an average effective damping ξ_{eff} equal to 35% (30% for "Building-Low", 35% for "Building-Mid" and 40% for "Building-High").

It was important to figure out that the dampers are located in plan and height with a way that does not increase the eccentricity of the buildings. In this study, many configuration of VD's placement were studied, but only two of them are presented in the Fig. 3. For each solution, three dampers were located in each direction and on each side of the center mass of each building. The suitable placement has been chosen based on the following parameters: the fundamental period as well as the rotation ration (both directions). As a result, alternative solution 1 is chosen for all the buildings.

4.2 Optimization of Lead Rubber Bearings (LRB)

An advanced bearing N-link element is selected and positioned in each column as the representative lead rubber bearing system (LRBs). Bilinear elastomeric bearing model, which assumes a stable hysteresis for LRBs, showed that LRBs had various degenerations (Ye et al. 2019; Kumar et al 2014, Gupta et al. 2022). These various degenerations include coupled bi-directional horizontal motion effects, vertical and horizontal motion effects, gravitation and post-cavitation behavior in tension, strength deterioration in cyclic tensile loading due to cavitation, and variation in critical buckling load capacity due to sideways displacement.

To avoid additional eccentricity of the isolated buildings, it was necessary to limit the eccentricity between the center of mass and the stiffness. After parametric study, the appropriate selected system of each building is presented in the following figures (see Figs. 4, 5 and 6).

The nonlinear equation of that type of system is presented by the well-known Eq. 2:

$$M \cdot \ddot{v}_b + \sum m_n \cdot \ddot{v}_n + c_b \cdot v_b + F_b = -M \cdot \ddot{u}_g \tag{2}$$

where [m] and [c] are mass and damping matrices respectively, and v and F are the displacement and force vectors.

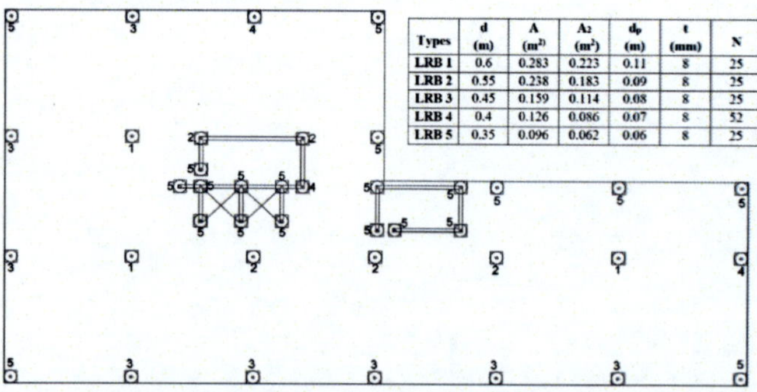

Types	d (m)	A (m²)	A₂ (m²)	d_p (m)	t (mm)	N	t_r (m)	t_s (mm)	H (m)
LRB 1	0.6	0.283	0.223	0.11	8	25	0.2	2	0.298
LRB 2	0.55	0.238	0.183	0.09	8	25	0.2	2	0.298
LRB 3	0.45	0.159	0.114	0.08	8	25	0.2	2	0.298
LRB 4	0.4	0.126	0.086	0.07	8	52	0.2	2	0.298
LRB 5	0.35	0.096	0.062	0.06	8	25	0.2	2	0.298

Fig. 4. Distribution of low-rise building's lead rubber bearings (LRB).

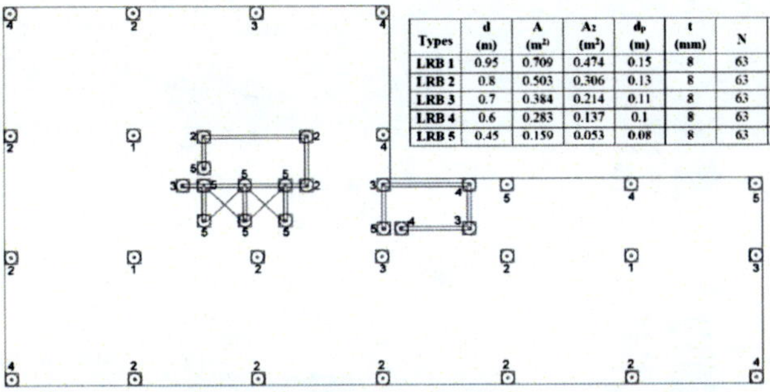

Types	d (m)	A (m^2)	A_2 (m^2)	d_p (m)	t (mm)	N	t_r (m)	t_s (mm)	H (m)
LRB 1	0.95	0.709	0.474	0.15	8	63	0.504	2	0.678
LRB 2	0.8	0.503	0.306	0.13	8	63	0.504	2	0.678
LRB 3	0.7	0.384	0.214	0.11	8	63	0.504	2	0.678
LRB 4	0.6	0.283	0.137	0.1	8	63	0.504	2	0.678
LRB 5	0.45	0.159	0.053	0.08	8	63	0.504	2	0.678

Fig. 5. Distribution of middle-rise building's lead rubber bearings (LRB).

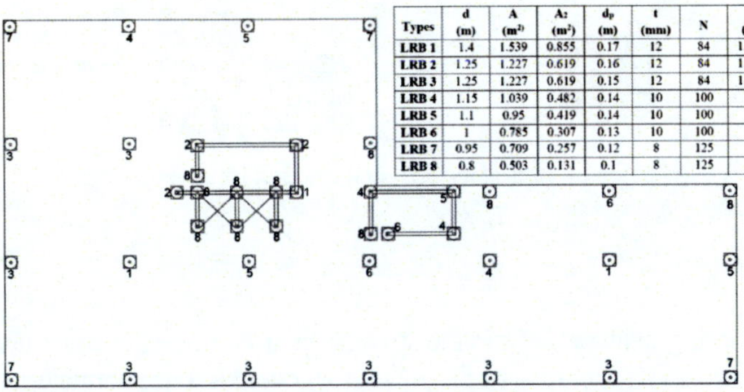

Types	d (m)	A (m^2)	A_2 (m^2)	d_p (m)	t (mm)	N	t_r (m)	t_s (mm)	H (m)
LRB 1	1.4	1.539	0.855	0.17	12	84	1.008	3	1.307
LRB 2	1.25	1.227	0.619	0.16	12	84	1.008	3	1.307
LRB 3	1.25	1.227	0.619	0.15	12	84	1.008	3	1.307
LRB 4	1.15	1.039	0.482	0.14	10	100	1	3	1.347
LRB 5	1.1	0.95	0.419	0.14	10	100	1	3	1.347
LRB 6	1	0.785	0.307	0.13	10	100	1	3	1.347
LRB 7	0.95	0.709	0.257	0.12	8	125	1	3	1.422
LRB 8	0.8	0.503	0.131	0.1	8	125	1	3	1.422

Fig. 6. Distribution of high-rise building's lead rubber bearings (LRB).

5 Results and Conclusions

Figure 6 illustrates the percentage of reduction of the inter-story horizontal displacement and the rotation in the longitudinal and tranversal direction of each building equipped with LRB or VD compared with the structure without dampers. Blue columns describe the buildings ("Building-Low", "Building-Mid" and "Building-High") with LRB, while red columns the buildings with VDs respectively.

By evaluating the mean value of percentage reduction in terms of interstory-drift in both directions which is equal to 63.53% and 59.91% with viscous dampers, 60.64% and 58.76% with LRB, while reduction of rotation 49.6% and 80.3% respective, it can be seen that LRB dampers perform better in the low-rise buildings ("Building-low"). According to the results of the mid-rise buildings ("Building-Mid") (Fig. 6 c, d), it can be seen that the 2 investigated solutions, with LRB or VD, perform well under the seismic records. On the other hand, the percentage of reduction for the high-rise building ("Building-High") equipped with VD reaches a maximum of 63.53% in the longitudinal direction and a maximum of 60% in the transversal direction, which is higher that the reduction of LRBs in terms of inter-story displacement (Fig. 7).

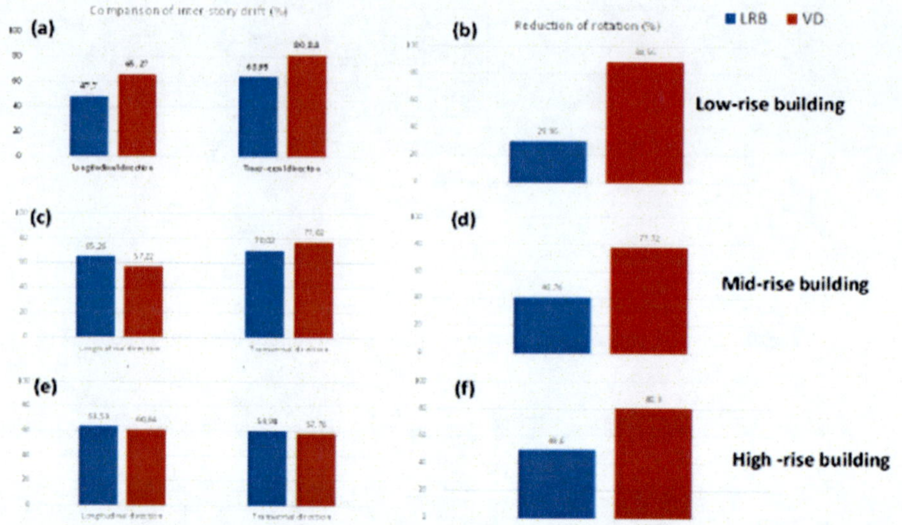

Fig. 7. Reduction of inter-story drift and rotation for all buildings.

6 Conclusion

Which is the appropriate selected solution to strengthen an engineering construction has been studied by previous researchers, by taking into account various parameters (Valente and Milani, 2018; Requena-García-Cruz et al. 2019). In this study, we do not try to found the best solution for irregular in plan reinforced concrete buildings, but to present the effect of lead rubber bearings (LRB) or viscous dampers (VD). The buildings (a three-story building, an eight-story building, and a twenty-story building) had been studied under earthquake recorded accelerograms, three of them are real and the other four are artificial, by nonlinear dynamic time-history analyses. The optimal design of LRB bearings and viscous dampers is described, focusing the design on minimizing certain parameters: (i) the maximum displacement (top of the structures), (ii) the torsion of the buildings, (iii) the base shear loads and, (iv) the maximum horizontal interstory drift. The comparative results show the effectiveness of the VD in terms of minimizing the torsion effects for all the buildings, mostly in the low rise building ("Building-Low") with effectiveness equal to 88%. However, the use of LRB provides a significant reduction for all the buildings in terms of rotation, minimizing the maximum acceleration, the max-story drift, the maximum horizontal displacement and the base shear loads. In order to have more accurate conclusions, a further parametric study is necessary, by increasing the number of the buildings, studying new dampers and more types of irregularities.

References

Alaa, K.M., El-Kashif, K.F., Salem, H.M.: New definition for torsional irregularity based on floors rotations of reinforced concrete buildings. J. Eng. Appl. Sci. **69**(1), 1–35 (2022). https://doi.org/10.1186/s44147-021-00061-5

Amendola, A., et al.: Experimental response of additively manufactured metallic pentamode materials confined between stiffening plates. Compos. Struct. **142**, 254–262 (2016)

Ayoub, N., Deü, J.-F., Larbi, W., Pais, J., Rouleau, L.: Application of the POD method to nonlinear dynamic analysis of reinforced concrete frame structures subjected to earthquakes. Eng. Struct. **270**, 114854 (2022)

Benzoni, G., Casarotti, C.: Effects of Vertical Load, strain rate and cycling on the response of lead-rubber seismic isolators. J. Earthq. Eng. **13**, 293–312 (2009)

Bosco, M., Ferrara, G.A.F., Ghersi, A., Marino, E., Rossi, P.: Seismic assessment of existing R.C framed structures with inplan irregularity by nonlinear static methods. Earthq. Struct. **8**(2), 401–422 (2015)

CEN. Eurocode 1 (EC1–2001): Actions on structures. EN 1991–1–1. European Committee for Standardization, Brussels (2001)

CEN. Eurocode 8 (EC8–2004): Design of earthquake resistance structures.. EN 1998–1. European Committee for Standardization, Brussels 2004

Chandler, A., Duan, X., Rutenberg, A.: Seismic torsional response: assumptions, controversies and research progress. Eur. Earthq. Eng. **10**(1), 37–51 (1996)

Das, S., Nau, J.M.: Seismic design aspects of vertically irregular reinforced concrete buildings. Earthq. Spectra **19**(3), 455–477 (2003)

De Stefano, M., Pintucchi, B.: A review of research on seismic behaviour of irregular building structures since 2002. Bull. Earthq. Eng. **6**(2), 285–308 (2008)

Demir, A., Demir, D., Erdem, R., Bagci, M.: Torsional irregularity effects of local site classes in multiple storey structures. Int. J. Res. Rev. Appl. Sci. 258–262 (2010)

Duan, X.N., Chandler, A.M.: An optimized procedure for seismic design of torsional unbalanced structures. Earthq. Eng. Struct. Dynam. **26**(7), 737–757 (1997)

Fabbrocino, F., Titirla, M., Amendola, A., Benzoni, G., Fraternali, F.: Innovative devices for the base isolation of existing buildings. In: COMPDYN 2017, 6th ECCOMAS Thematic Conference on Computational Methods in Structural Dynamics and Earthquake Engineering, Rhodes Island, Greece (2017). https://doi.org/10.7712/120117.5732.17771

Forcellini, D., Kalfas, K.N.: Inter-story seismic isolation for high-rise buildings. Eng. Struct. **275**, 115–175 (2023)

Gao, J., Yuan, Y., Qiu, T., Wang, C., Qu, Z.: Performance optimization and loading rate-dependency of friction dampers with non-metallic friction materials .J. Build. Eng. **54**, 104609 (2022)

Gupta, P., Ghosh, G., Kumar, V., Paramasivam, P., Dhanasekaran, S.: Effectiveness of LRB in curved bridge isolation: a numerical study. Appl. Sci. **12**, 11289 (2022). https://doi.org/10.3390/app122111289

Jeong, S.H., Elnashai, A.S.: New three-dimensional damage index for RC buildings with planar irregularities. J. Struct. Eng. ASCE **132**(9), 1482–1490 (2006)

Hwang, J.S., Lin, W.C., Wu, N.: Comparison of distribution methods for viscous damping coefficients of buildings. Struct. Infrastruct. Eng. **9**(1), 28–41 (2013)

Kumar, M., Whittaker, A.S., Constantinou, M.C.: An advanced numerical model of elastomeric seismic isolation bearings. Earthq. Eng. Struct. Dyn. **43**, 1955–1974 (2014). https://doi.org/10.1002/eqe.2431

Li, L., LiangY, G.C., Yang, D.: Simultaneous layout and size optimization of nonlinear viscous dampers for frame buildings under stochastic seismic excitation. Eng. Struct. **273**(5), 115067 (2022). https://doi.org/10.1016/j.engstruct.2022.115067

Martinez-Romero, E.: Experiences on the use of supplemental energy dissipators on building structures. Earthq. Spectra **9**, 581–624 (1993)

Mrad, C., Titirla, M.D., Larbi, W.: Comparison of strengthening solutions with optimized passive energy dissipation systems in symmetric buildings. Appl. Sci. **11**, 10103 (2021)

Mrad, C., Titirla, M.D., Larbi, W.: Optimal design of viscous and friction dampers in symmetric reinforced concrete buildings. In: Proceedings of the 7th World Congress on Civil, Structural, and Environmental Engineering, CSEE 2022, online, Lisbon, Portugal (2022). https://doi.org/10.11159/icsect22.149

Penelis, G., Kappos, A.: 3D pushover analysis: the issue of torsion. In: 12th European Conference on Earthquake Engineering (2002)

Requena-Garcia-Cruz, M.V., Morales-Esteban, A., Durand-Neyra, P.: Assessment of specific structural and ground-improvement seismic retrofitting techniques for a case study RC building by means of a multi-criteria evaluation. Eng. Struct. **38**, 265–278 (2019). https://doi.org/10.2495/ERES19008

Russillo, A.F., Failla, G., Amendola, A.: Luciano R (2022) on the free vibrations of non-classically damped locally resonant metamaterial plates. Nanomaterials **12**(3), 541 (2022). https://doi.org/10.3390/nano12030541

Seleemah, A., Constantinou, M.: Investigation of Seismic Response of Buildings with Linear and Nonlinear Fluid Viscous Dampers; Report No. NCEER 97–0004; National Center for Earthquake Engineering Research, State University of New York at Buffalo: New York, NY, USA (1997)

Stathi, C., Bakas, N., Lagaros, N., Papadrakakis, M.: Ratio of torsion (ROT): an index for assessing global induced torsion in plan irregular buildings. Earthq. Struct. **9**(1), 145–171 (2015)

Jaisee, S., FengYue, and Yi Hao Ooi,: A state-of-the-art review on passive friction dampers and their applications. Eng. Struct. **235**, 112022 (2021). https://doi.org/10.1016/j.engstruct.2021.112022

Taylor Devices (2020) Fluid viscous dampers. Third Edition. General Guidelines for Engineers

itirla, M., Katakalos, K., Zuccaro, G., Fabbrocino, F.: On the mechanical response of an innovative energy dissipation device. Int. J. Earthq. Eng. Ing. Sismica (2), 126–138 (2017)

Titirla, M., Papadopoulos, P., Doudoumis, I.: Finite element modelling of an innovative passive energy dissipation device for seismic hazard mitigation. Eng. Struct. **168**, 218–228 (2018). https://doi.org/10.1016/j.engstruct.2018.04.081

Titirla, M., Papadopoulos, P.: Finite element investigation of a new seismic energy absorption device through simultaneously yield and friction. In: 5th International Conference on Computational Methods in Structural Dynamics and Earthquake Engineering, COMPDYN 2015, Greece (2015). https://doi.org/10.7712/120115.3678.1109

Valente, M., Milani, G.: Alternative retrofitting strategies to prevent the failure of an under-designed reinforced concrete frame. Eng. Fail. Anal. **89**, 271–285 (2018)

Ye, K., Xiao, Y., Hu, L.: A direct displacement-based design procedure for base-isolated building structures with lead rubber bearings (LRBs). Eng. Struct. **197**, 109402 (2019). https://doi.org/10.1016/J.ENGSTRUCT.2019.109402

The Effect of Flights Delayed on Passenger Load and Utilization of Airbus A320 Aircraft

Khaled Aljaly[1](✉), Omar Ayadi[2], Salem Sultan[3], and Faouzi Masmoudi[2]

[1] Afriqiyah Airways Company and University of Tripoli, Tripoli, Libya
kaljaly0@gmail.com

[2] University of Sfax, Mechanics, Modelling, and Production Research Laboratory, National School of Engineering of Sfax, Sfax, Tunisia

[3] College of Engineering Technology, Janzour, Libya

Abstract. This study attempts to evaluate the impacts of delayed flights, passenger-carrying factors, and aircraft utilization of the Airbus A320 of Afriqiyah Airways. Flight delays cause schedule disruptions and discomfort for passengers, but they also decrease productivity, raise capital costs, reallocate flight crews and aircraft, and add to crew costs. Since flight delays require the consumption of additional labor, capital, and other inputs needed in the process. We infer from an economic perspective.

Domestic and international departure routes were examined to address this issue. The results showed that the departure flight delay had negative effects on revenue, Based on econometric estimates.

On-time performance at the airport is related to and is easily affected by the prevalence of delays from previous operations of flights using the airport. In this paper, we use the methodology of scientific research in collecting and analyzing data through the impact of three variables, namely flight delays, passenger loads, and the utilization of the aircraft.

The results showed that flight delays represent the largest percentage, which negatively affects both the passenger load and the utility of the aircraft.

Keywords: flight delay · passenger load · utilization of flights · Airbus · A320 · Afriqiyah airways

1 Introduction

In recent years, airlines have grown from simple contract mail carriers to intellectually appealing businesses. The airlines compete with the conceit of speed and urgency around the world to increase travel for business and leisure. As a result, it becomes an intense competition between companies. Airlines need to be more demanding of customers to be high-return. Therefore, airlines must become more competitive to achieve service quality. The first logical solution is to fly their aircraft as much as possible: an aircraft on the ground means real profit losses (Alves et al. 2010).

The airplanes were working to reduce margins and carrying millions of people facing pressure from competitors. Therefore, there is an urgent need to measure the impact of

A. Akrout et al. (Eds.): ICAV 2022, ACM 22, pp. 125–139, 2023.
https://doi.org/10.1007/978-3-031-34190-8_16

the flight delay, passenger load factor, and utilization process in order to obtain the best trade-off between safety and costs (Brueckner et al. 2022). The cost of a flight delay, passenger load, and utility leads to remarkable spending, and that effect the challenge with other airlines (Gilbert 2021). Aircraft flight delay, passenger load, and utilization have the biggest contributors to the success of any airline (Brueckner et al. 2022) (Laik et al. 2021) (Gilbert 2021).

The punctuality of airlines has become a key factor for performance. The flight departure and arrival times are very important for the cargo and passengers. And one way to increase the operating profit is to pay more attention to On-Time Performance (OTP). The structure of this paper consists of an introduction to the reliability of flight delay, Passenger load factor, and the concept of flight utilization.

2 Reliability Flight Delay

Delays that occur prior to an aircraft taking off are numerous and some of them may cause an unexpected delay (Brueckner et al. 2022). Sometimes there is only one reason for a delay or more, airlines use delay codes that are established by IATA to track and report flight delays. Appendix A has more details on these codes. in line with IATA.

2.1 Delay Codes

IATA (International Air Transport Association) standardized the flight delay reporting format by employing "IATA delay codes," which identify the cause and source of the delay. Delay code attributes cover nine category sets for the delay. Each category group can be described by either a two-digit numeric number or a two-letter alphanumeric code.

- Passengers and Baggage Handling (code 11–18)
- Cargo and Mail (code 21–29)
- Aircraft and Ramp Handling (code 31–39)
- Technical and Aircraft Equipment (code 41–47)
- Damage to Aircraft and Automated Equipment Failure/EDP (computer system) (code 51–57)
- Flight Operations and Crewing (code 61–69)
- Weather (code 71–77)
- Air Traffic Control Restrictions and Airport or Governmental Authorities (code 81–89)

Reactionaries Reasons and Miscellaneous (code 91–99) (Sridhar et al. 2008) (Alves et al. 2010) (Appendix A).

3 Passenger Load Factor

The number of passengers carried is a key indicator of how well the airline system is performing. Selling tickets allow for the recovery of the aircraft's relatively high capital expenses (investment), and a high passenger load factor (full occupancy, 100%) is required (Laik et al. 2021).

As the passenger's load rises, the performance of the airline gets better. As flight on-time rates rise, the passenger load rises as well (Monmousseau et al. 2020).

A high load factor means that an airline's aircraft is fully loaded with passengers sitting in the majority of the seats. Each flight is connected with significant fixed expenses for airlines. Every flight needs to have a complete flight crew and support staff, a fuel-efficient, well-maintained aircraft, and amenities that will amuse and soothe passengers. The airline is not making as much money as it could be flying a full plane if only half of the seats on a flight are occupied (Safiuddin et al. 2019). Investors may find the load factor useful in understanding how the airline makes money and pays its bills. Low load factors can be problematic and may be a sign that an airline is losing money (Li et al. 2021).

4 The Concept of Flight Utilization

By maximizing aircraft usage, which includes short turnaround times at the gates, an airline may enhance the return on the sizeable financial investment it has made in its planes (Senturk et al. 2010). An airline's fleet planning, schedule planning, passenger reservations, flight operations, ground operations, and airplane maintenance systems must work closely together as well as with air traffic controllers and airport authorities to ensure efficient airplane usage. Particularly for short-haul airlines, even a slight reduction in turnaround time at the gate can have significant positive effects (Zhang et al. 2013) (Gilbert 2021).

5 Evaluation of Delay African Airways of the Fleet A320

The Libyan African Aviation Holding Company is the owner of Afriqiyah Airways, a registered Libyan airline with its main office in Tripoli. The company was founded in the year 2007. The fleet of the company is dependent on Airbus planes. Members of the International Air Transport Association (IATA), African Airlines Association (AFRAA), Arab Air Transport Organization (AACO), and International Civil Aviation Organization include the business (ICAO).

Three company-owned A330s and nine A319/A320 aircraft are used by the company to fly domestic and international routes. In this research paper, we will examine the A320 fleet from the time of purchase to 2019 and analyze the data from Table 1 regarding the A320 fleet (African airways 2022).

Table 1. AAW Fleet data

A/C Type	No. of A/C	A/C Reg	A/C seat Capacity	A/C Received date	Remarks
A320	7	5A-ONA	140	14/08/2007	-
		5A-ONB		29/08/2007	-
		5A-ONJ		28/01/2010	In 2019 the airplane carried 180 passengers
		5A-ONL		21/10/2010	
		5A-ONM		26/11/2010	-
		5A-ONN		23/11/2012	-
		5A-ONO		11/01/2013	-
Total	7				

6 Statistic Evaluation

- **Data collection**

The data on flight delays, passenger loads, and aircraft utilization were collected from Afriqiyah Airways for the years 2007 to 2019 (excluding 2011) using the Airbus software kinds AMSIS - KEOPS for aircraft type A320.

- **The data analysis for flight delays**

The objective of this analysis is to ascertain the typical flight delay of the Airbus A320. The following mathematical Eq. (1) is used to determine the flight delay, which is organized in Table 2 and Fig. 1.

$$\text{Delay Rate} = \frac{\text{Delay Flight}}{\text{Actual Total Flight}} \times 100 \tag{1}$$

Flight delays have negative impacts, mainly economic, for passengers, airlines, and airports. Given the unpredictability of their occurrence, travelers typically plan to leave for their appointments many hours before than necessary, raising the expense of their journey. However, airlines must pay penalties, fines, and other operating expenses like keeping staff and aircraft at airports.

The data which had been collected for Airbus type A320 of delayed flights is illustrated in Table 2 and also plotted graphically in Fig. 1. Which indicated the relationship between the actual total flights and flight delay.

- **data analysis of passenger load factor.**

The dimensionless percentage ratio of the actual number of passengers to the number of available seats on an airplane for a particular flight is known as the passenger load factor, the following mathematical equation is used (2) to determine passenger load, and it is structured in Table 3 & Fig. 2.

$$\text{Passenger load} = \frac{\text{Actual Number of Passenger}}{\text{Aircraft Capacity}} \times 100 \tag{2}$$

Table 2. Delay Rate of type A320

Years	Actual total flights	Flight delay	Delay Rate
2007	1052	363	35%
2008	3564	789	22%
2009	3081	541	18%
2010	4993	3622	73%
2012	4977	3356	67%
2013	8411	3890	46%
2014	5848	3514	60%
2015	4106	898	22%
2016	3979	554	14%
2017	3396	1098	32%
2018	2792	1013	36%
2019	3855	3420	89%
Total	50054	23058	46%
Average rate			43%

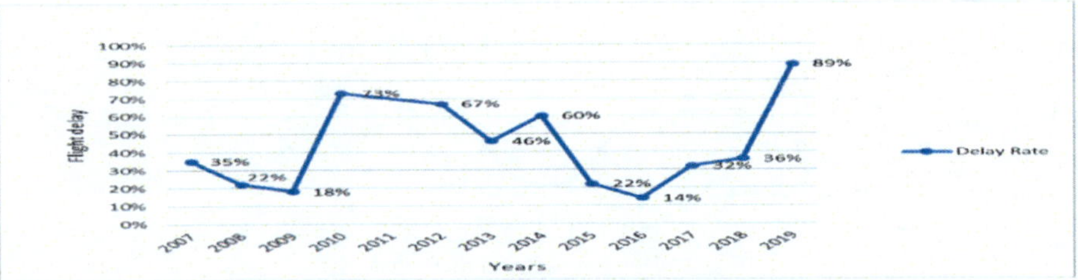

Fig. 1. Delay Rate of type A320

The passenger load factor is equal to 100 divided by the number of passengers actually on board.

The passenger load factor gauges how well an airline is using its available aircraft capacity. It is used to evaluate how well an airline produces fare money and fills seats.

The passenger load factor is a crucial component in determining how well the airline system is doing. Because airplanes have substantial capital expenses (investments) that can be recouped through ticket sales, a high passenger load factor (full seats) is required.

As the load factor rises, the airline performs better. As flight on-time rates rise, the load factor rises as well.

The collected data of Airbus type A320 of Passenger Load is illustrated in Table 3. And also plotted graphically in Fig. 2. Which indicated the relationship between the aircraft capacity and the actual number of passengers.

Table 3. Passenger Load of type 320

Years	Aircraft Capacity	The actual number of passengers	Passenger Load
2007	147280	31938	22%
2008	498960	92102	18%
2009	431340	181610	42%
2010	699020	284410	41%
2012	696780	261714	38%
2013	1177540	396595	34%
2014	818720	3888	0.5%
2015	574840	25186	4%
2016	557060	253014	45%
2017	475440	341673	71%
2018	390880	193858	50%
2019	627420	188516	30%
Total	7095280	2254504	32%
Average rate			33%

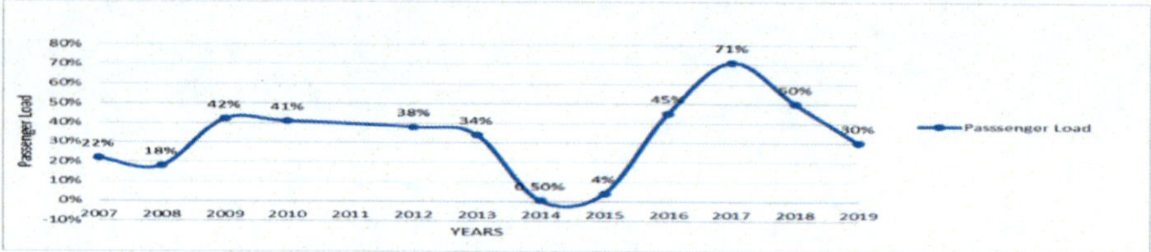

Fig. 2. Passenger Load of type A320

- **data analysis flight utilizations.**

The term "aviation usage" refers to an aircraft's typical daily flying cycles or hours. The analysis's objective is to determine the airline's typical usage of Airbus A320-type aircraft, the following mathematical equation is used (3) to determine flight utilization, and it is structured in Table 4 & Fig. 3.

$$\text{Aircraft Utilization} = \frac{\text{Block Hours}}{\text{Aircraft Time}} \times 100 \tag{3}$$

Table 3 illustrates data collected from Afriqiyah airways for the aircraft utilization of airbus type A320 from the year 2007 to 2019 and also indict the relationship between the aircraft time and block hours.

Equation 3 is used to calculate the total percentage of aircraft utilization.

Figure 3 shows the maximum benefits of the aircraft were achieved at the year 2013 whereas the minimum was achieved at the year 2017.

Table 4. Aircraft Utilization of type 320

Years	Aircraft time	Block hours	Aircraft Utilization
2007	20496	2458.85	12%
2008	61320	8849.58	14%
2009	61320	8087.14	13%
2010	61320	12472.8	20%
2012	61320	8884.6	14%
2013	61320	13752.9	22%
2014	61320	3024.9	5%
2015	61320	1637.68	3%
2016	61320	1847.67	3%
2017	61320	1118.5	2%
2018	61320	2447.85	4%
2019	61320	2689.18	4%
Total	695016	67271.6	10%
Average ratio			10%

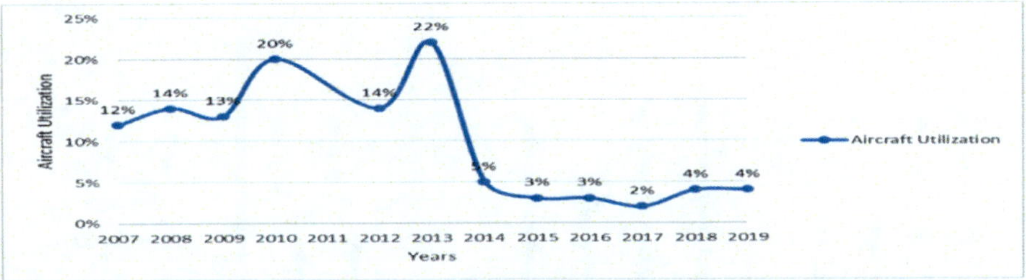

Fig. 3. Aircraft Utilization of type 320

- **Comprising flight delay, passenger load, and utilization.**

Comprising have been carried out between flight delay, passenger load, and utilization for Afriqiyah Airways of aircraft type Airbus A 320, the obtained results are illustrated in the Table 5 & Fig. 4.

Table 5 shows the comparison of the percentages between the flight delay, passenger load, and utilization of the aircraft.

Figure 4 illustrates the chart for the comparison between the flight delay, passenger load, and utilization for aircraft type A320.

Table 5. Comprising flight delay, passenger load, and utilization of type 320

Years	Flight Delay	Passenger Load	Aircraft Utilization
2007	35%	22%	12%
2008	22%	18%	14%
2009	18%	42%	13%
2010	73%	41%	20%
2012	67%	38%	14%
2013	46%	34%	22%
2014	60%	0.5%	5%
2015	22%	4%	3%
2016	14%	45%	3%
2017	32%	71%	2%
2018	36%	50%	4%
2019	89%	30%	4%
Average ratio	43%	33%	10%

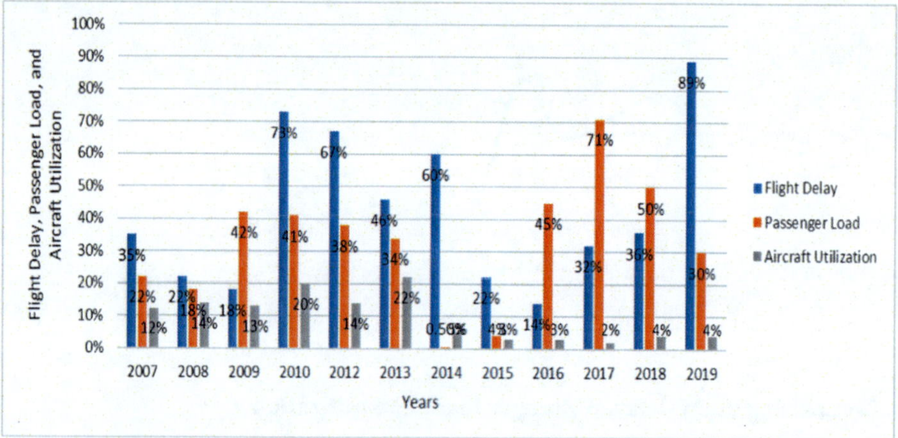

Fig. 4. Comprising flight delay, passenger load, and utilization of type 320

7 Conclusions

The aim of this study is to discuss the delay of air transport flights in Libya, specifically African Airlines, for the Airbus 320 as a case study. The extent of its impact on the factors of flight delays in passenger load and benefits from them.

It concludes from the results obtained for the period from the year 2007 to the year 2019, excluding year 2011, show that the average aircraft flight delay for Af-riqiyah

Airways is very low (A320 = 43%), the passenger load is very low (A320 = 33%). As well the utilization of flights is very low (A320 = 10%).

Comparisons were made between flight delay, passenger load factor, and utilization. The results showed that with an increase in flight delay resulting, in fact, leads to a decrease in passenger load, the utilization will decrease, which indeed leads to a decrease in benefits.

Due to the absence of an analysis of this data, the researcher conducted an analysis of the data on flight delays and passenger loads to reduce the delay in departure times, which causes passengers to be upset, which leads to their loss and thus leads to the company's economic loss.

The Statistic evaluation is to perspectives a parametric study of more influencing delay types.

Appendix A

Illustrates the Standard of IATA Delay Codes (Tabe 6).

Table 6. Standard IATA Delay Codes (AIRBUS)

Company internal codes (Always use SI to explain the reasons as internal codes are not known by everyone)			
01	IT	CREW TRANSPORTATION	Flight crew transportation delay
02	IC	IMMIGRATION OFFICER	Absence of an immigration officer to complete boarding procedures
03	IS	SECURITY MEMBER	Security member of the flight delayed
04	IA	A/C TOWING/ TAXING	Aircraft position change
Others			
06	OA	NO GATE/STAND AVAILABILITY	Extra time to board passengers as the aircraft was out of gate
09	SG	SCHEDULED GROUND TIME	Scheduled ground time less than declared minimum ground time
Passenger and Baggage			
11	PD	LATE CHECK - IN	Acceptance after deadline
12	PL	LATE CHECK - IN	Congestion in check-in area
13	PE	CHECK – IN ERROR	Error with passenger and/or baggage details
14	PO	OVERSALES	Booking errors
15	PH	BOARDING	Discrepancies and paging, missing checked-in passenger

(continued)

(*continued*)

Passenger and Baggage			
16	PS	COMMERCIALPUBLICITY AND PASSENGER CONVENIENCE	VIP, press, ground meals and missing personal items
17	PC	CATERING ORDER	Late or incorrect order given to supplier
18	PB	BAGGAGE PROCESSING	Late or incorrect baggage sorting

Cargo and mail (if delays caused by mail handing can be identified, use codes 27,29 otherwise use codes 21–26)			
21	CD	DOCUMENTATION	Late or incorrect documentation for booked cargo
22	CP	LATE POSITIONING	Late delivery of booked cargo to Airport/Aircraft
23	CC	LATE ACCEPTANCE	Acceptance of cargo after deadline
24	CI	INADEQUATE PACKING	Repackaging and/or re-labelling of booked cargo
25	CO	OVERSALES	Booked load in excess of load capacity(weight or volume), resulting in reloading or off-load
26	CU	LATE PREPARATION IN WAREHOUSE	Late preparation of cargo load from warehouse
Mail only			
27	CE	DOCUMENTATION, PACKING	Incomplete and/or inaccurate documentation/packaging of mail
28	CL	LATE POSITIONING	Late delivery of mail to Airport/Aircraft
29	CA	LATE ACCEPTANCE	Acceptance of mail after deadline

Aircraft and Ramp Handling			
31	GD	AIRCRAFT DOCUMENTATION LATE/ INACCURATE	Late or inaccurate Mass and Balance documentation, general declaration, passenger manifest, etc
32	GL	LOADING / UNLOADING	Bulky items, special load, cabin load, lack of loading staff
33	GE	LOADING EQUIPMENT	Lack of and/ or breakdown, e.g. container pallet loader, lack of staff
34	GS	SERVICING EQUIPMENT	Lack of and/ or breakdown, lack of staff, e.g. steps

(*continued*)

(*continued*)

Aircraft and Ramp Handling			
35	GC	AIRCRAFT CLEANING	Lack completion of aircraft cleaning
36	GF	FUELLING / DEFUELLING	Lack delivery of fuel from fuel supplier; excludes late request
37	GB	CATERING	Lack and/ or breakdown, lack of staff, e.g. steps
38	GU	ULD	Lack of and/ or incomplete delivery; late loading
39	GT	TECHNICAL EQUIPMENT	Lack and/ or breakdown, lack of staff, includes GPU, pushback tug, de-icing

Technical and Aircraft Equipment			
41	TD	AIRCRAFT DEFECTS	Aircraft defects including items covered by MEL
42	TM	SCHEDULED MAINTENANCE	Lack release from maintenance
43	TN	NON-SCHEDULED MAINTENANCE	Special checks and/or additional works beyond normal maintenance schedule
44	TS	SPARES AND MAINTENANCE EQUIPMENT	Lack of spares, lack of and/ or breakdown of specialist equipment required for defect rectification
46	TA	AOG SPARES	Awaiting AOG spare(s) to be carried to another staion
46	TC	AIRCRAFT CHANGE	For technical reasons, e.g. a prolonged technical delay
47	TL	STANDBY AIRCRAFT	Lack of planned standby aircraft for technical reasons
48	TV	SCHEDULED CABIN CONFIGURATION/VERSION ADJUSTMENT	Due to change required for cabin configuration, e.g. change from three-class to two-class configuration, moving curtain etc

Damage to Aircraft			
51	DF	DAMAGE DURING FLIGHT OPERATIONS	Bird or lightning strike, turbulence, heavy or overweight landing, collision during

(*continued*)

(*continued*)

Damage to Aircraft			
52	DG	DAMAGE DURING GROUND OPERATIONS	Collisions (other than during taxiing), loading/ off- loading damage, contamination, towing, extreme weather conditions

EDP Automated Equipment Failure			
55	ED	DEPARTURE CONTROL	Failure of automated systems, including check-in; load control systems producing mass and balance
56	EC	CARGO PREPARATION/ DOCUMENTATION	Failure of automated cargo system for cargo preparation/ documentation
57	EF	FLIGHT PLANS	Failure of automated flight planning systems
58	EO	OTHER AUTOMATED SYSTEM	Can be related to operations control, crew rostering systems

Flight Operations and Crewing			
61	FP	FLIGHT PLAN	Late completion or change of flight documentation
62	FF	OPERATIONAL REQUIREMENTS	Late alteration to fuel or payload
63	FT	LATE CREW BOARDING OR DEPARTURE PROCEDURES	Late flight deck, or entire crew, other than connection and standby; late completion of flight deck crew checks
64	FS	FLIGHT DECJ CREW SHORTAGE	Sickness, awaiting standby, flight time limitations, valid visa, health documents, etc
65	FR	FLIGHT DECJ CREW SPECIAL REQUEST	Requests not within operational requirements
66	FL	LATE CABIN CREW BOARDING OR DEPARTURE PROCEDURES	Late cabin crew other than connection and standby; late completion of cabin crew checks
67	FC	CABIN CREW SHORTAGE	Sickness, awaiting standby, flight time limitations, valid visa health documents, etc.
68	FA	CABIN CREW ERROR OR SPECIAL REQUEST	Requests not within operational requirements
69	FB	CAPTAIN REQUEST FOR SECURITY CHECK	Extraordinary request outside mandatory requirements

Weather			
71	WO	DEPARTURE STATION	Below operating limits
72	WT	DESTINATION STATION	Below operating limits
73	WR	EN-ROUTE OR ALTERNATE	Below operating limits
75	WI	DE-ICING OF AIRCRAFT	Removal of ice and/or snow; frost prevention, excluding unserviceability of equipment
76	WS	REMOVAL OF SNOW, ICE, WATER, AND SAND FROM AIRPORT	Runway, taxiway conditions
77	WG	GROUND HANDLING IMPAIRED BY ADVERSE WEATHER CONDITIONS	High winds, heavy rain, blizzards, monsoons etc

Air Traffic Flow Management Restrictions			
81	AT	ATFM DUE TO ATC EN-ROUTE DEMAND/ CAPACITY	Standard demand/ capacity problems
82	AX	ATFM DUE TO ATC STAFF / EQUIPMENT EN-ROUTE	Reduced capacity caused by industrial action or staff shortage or rquipment failure, extraordinary demand due to capacity reduction in neighbouring area
83	AE	ATFM DUE TO RESTRICTION AT DESTINATION AIRPORT	Airport and/ or runway closed due to obstruction, industrial action, staff shortage, political unrest, noise abatement, night curfew, special flights
84	AW	ATFM DUE TO WEATHER AT DESTINATION	Airport and/ or runway closed due to extreme weather, e.g. sever wind, fog, heavy snow causing stand still
Airport and Governmental Authorities			
85	AS	MANDATORY SECURITY	Passengers, baggage, crew, etc.
86	AG	IMMIGRATION, CUSTOMS, HEALTH	Passengers, crew
87	AF	AIRPORT FACILITIES	Parking stands, ramp congestion, lighting, buildings, gate limitations, etc.
88	AD	RESTRICTION AT AIRPORT OF DESTINATION	Airport and/ or runway closed due to obstruction, industrial action, staff shortage, political unrest, noise abatement, night curfew, special flights

(*continued*)

(continued)

Airport and Governmental Authorities			
89	AM	RESTRICTIONS AT AIRPORT OF DEPARTURE WITH OR WITHOUT ATFM RESTRICTIONS	Including Air Traffic Services, start-up and pushback, airport and/ or runway closed due to obstruction or weather (restriction due to weather in case of ATFM regulation only, else refer to code 71 (WO)) industrial action, staff shortage, political unrest, noise abatement, night curfew, special flights

Reactionary			
91	RL	LOAD CONNECTION	Awaiting load from another flight
92	RT	THROUGH CHECK-IN ERROR	Passenger and baggage check-in error at originating station
93	RA	AIRCRAFT ROTATION	Late arrival of aircraft from another flight or previous sector
94	RS	CABIN CREW ROTATION	Awaiting cabin crew from another flight
95	RC	CREW ROTATION	Awaiting flight deck or entire crew from another flight
96	RO	OPERATIONS CONTROL	Re-routing, diversion, consolidation, aircraft change for reasons other than technical

Miscellaneous			
97	MI	INDUSTRIAL ACTION WITH OWN AIRLINE	Industrial action related to own airline disputes
98	MO	INDUSTRIAL ACTION OUTSIDE OWN AIRLINE	Industrial action (excluding Air Traffic Services)
99	MX	MISCELLANEOUS	No suitable code; explain reason(s) in plain text as SI

References

Alves, V.A., Ghobbar, A.A.: A predictive model to evaluate and improve punctuality of fleet of wide-body aircraft maintenance and reliability. In: Air Transport and Operations: Proceedings of the First International Air Transport and Operations Symposium 2010, p. 323. IOS Press (2010)

African airways (2022)

Brueckner, J.K., Czerny, A.I., Gaggero, A.A.: Airline delay propagation: a simple method for measuring its extent and determinants. Transp. Res. Part B: Methodol. **162**, 55–71 (2022). https://doi.org/10.1016/j.trb.2022.05.003

Alves, V.A.: A predictive model to evaluate and improve punctuality of fleet of wide-body aircraft maintenance and reliability. Mestre thesis, technical university of Lisbon (2010)

Gilbert, E.J.: An analysis of the utilization rates for eastern kentucky university's fleet of training aircraft (2021)

Laik, M.N., Choy, M.: Airport check-in counters assignment using predicted passenger load: a case study. In: 11th Annual International Conference on Industrial Engineering and Operations Management (2021)

Li, X., de Groot, M., Bäck, T.: Using forecasting to evaluate the impact of COVID-19 on passenger air transport demand. Decis. Sci. (2021). https://doi.org/10.1111/deci.12549

Senturk, C., Kavsaoglu, M.S., Nikbay, M.: Optimization of aircraft utilization by reducing scheduled maintenance downtime. In: 10th AIAA Aviation Technology, Integration, and Operations (ATIO) Conference, p. 9143 (2010). https://doi.org/10.2514/6.2010-9143

Safiuddin, S.K.: Passenger load factor and financial health-a study of select airline companies. Think India J. **22**(10), 356–365 (2019)

Sridhar, K.: Delays–how to mitigate it. In: Maintenance Reliability & Cost analysis Seminar, pp. 2097–2114 (2008)

Monmousseau, P., Marzuoli, A., Feron, E., Delahaye, D.: Putting the air transportation system to sleep: a passenger perspective measured by passenger-generated data (2020). arXiv preprint arXiv:2004. P.14372. https://doi.org/10.48550/arXiv.2004.14372

Zhang, Z., Wang, P., Wang, P.: Calculation model of airspace utilization rate based on flight time and the space and capacity of airspace. In: 2013 IEEE 8th Conference on Industrial Electronics and Applications (ICIEA), pp. 760–765 (2013). https://doi.org/10.1109/ICIEA.2013.6566469

A Wavelet-Based Statistical Control Chart Approach for Monitoring and Detection of Spur Gear System Faults

Rasheed Majeed[1,3](✉), Maroua Haddar[1,2], Fakher Chaari[1], and Mohamed Haddar[1]

[1] Laboratory of Mechanics, Modeling, and Production (LA2MP), National School of Engineers of Sfax, BP 1173–3038, Sfax, Tunisia
rasheed-majeed-ali.al-jorani@enis.tn

[2] National School of Engineers of Sousse, University of Sousse, 4003, Sousse, Tunisia

[3] Ministry of Construction, Housing, Municipalities and Public Works, Diyala Governorate, IQ32001 Diyala, Iraq

Abstract. Gears are one of the essential transmission systems in various industry sectors. The gear faults are widespread due to the severe conditions of their work. Therefore, gear failure diagnosis has become an important topic. This paper proposes a method for fault diagnosis in the dynamic model of a spur gear system using one of the statistical control charts (SCC) known as an exponentially weighted moving average (EWMA). The original simulation vibration signals are first obtained by dynamic modeling. Second, additive white Gaussian noise (AWGN) is added to the modeling signals to simulate the working environment. Thirdly, the discrete wavelet transform (DWT) technique is used to analyze the vibration signals into eight independent levels of detail. Then the relative wavelet energy (RWE) is extracted as a statistical feature for all levels of the DWT. Fourth, the univariate EWMA scheme was created based on the RWE features under healthy gear conditions. Finally, an EWMA control scheme tests the defective gear signal using the same limits as the scheme for the typical case to detect gear breakage. The performance of the proposed EWMA scheme proved effective in diagnosing and distinguishing between abnormal shifts of healthy and defective gears.

Keywords: Spur Gear · Gear Fault · Additive White Gaussian Noise · Discrete Wavelet Transform · Relative Wavelet Energy · Statistical Control Charts (SCC)

1 Introduction

Gearbox systems that find considerable use in many sectors, such as power plants, transmission systems, production, and industrial machines, comprise one of the most critical mechanical equipment. However, the challenge lies in the problem of failure of the gear teeth due to different causes, including manufacturing and assembly errors, insufficient lubrication, overload, and material (Chaari et al. 2009). To overcome this problem, developing new methods to monitor gear systems to detect gear teeth faults early

A. Akrout et al. (Eds.): ICAV 2022, ACM 22, pp. 140–152, 2023.
https://doi.org/10.1007/978-3-031-34190-8_17

before a sudden gear breakage occurs has become the most significant aim nowadays. If the gear flaws cannot be discovered in a timely manner, the health of the gear will continue to deteriorate, which may result in a significant loss of production or system shutdown. Early failure diagnosis and detection enable it to schedule shutdowns in such a way as to prevent a major loss, which, in turn, leads to a safer operation and a better cost reduction (Liang et al. 2018).The modeling of malfunctions in a gearbox is a promising and helpful approach to understanding the dynamic response properties of a gearbox while it is faulty. Moreover, simulations can simply be performed with a good dynamic model to monitor different types and levels of gear failures (Tian et al. 2012).

In the past twenty years, several studies (Endo et al. 2009; Wei et al. 2016; Palermo et al. 2013; Kubur et al. 2004; Yuan et al. 2019) have been published on the dynamic models of spur gears; however, these models do not consider the faults diagnosis. The primary focus of these studies is on modeling the tooth distortion that occurs due to contact, static, and dynamic evaluation, in addition to the effects of assembly and manufacturing errors on dynamic behavior. The dynamic models of the spur gear systems with faults have been the subject of several studies as of late, all with the goal of better comprehending the mechanics by which gear faults are generated and then developing more efficient fault detection and diagnosis techniques. The authors (Wu et al. 2008; Tian et al. 2012) studied several statistical indicators of the time domain method to investigate the fault detection of spur gear systems. The findings concluded that the strongest statistical features were root mean square (RMS)and kurtosis. In contrast, presented (Chen and Shao 2011) spectrum analysis as a powerful approach for detecting the influences of tooth cracking; outcomes showed that sidebands produced from crack faults are more sensitive than the gear mesh frequency and its harmonics. Similarly, (Yang et al. 2019) introduced frequency domain analysis methods. The results indicated that frequency domain characteristics, such as fast Fourier transform (FFT) spectra and sidebands, can serve as indicators of gear defects to some extent. Time-domain and frequency-domain vibration analysis methods can effectively identify errors but are less effective in condition monitoring. Further, the frequency components are canceled out during the time domain analysis. Time-frequency domain methods have been applied to overcome these limitations. The discrete wavelet transform (DWT) is one of the most important time-frequency techniques known for its ability to provide signal information simultaneously in both time and frequency domains. The DWT provides fast computation since it is based on sub-band coding. Therefore, it is easy to implement and its calculation time is short. Furthermore, the DWT helps extract important features of the stationary or non-stationary signals with a time-frequency resolution (Peng and Chu 2004). Relative wavelet energy (RWE) based on DWT analysis was revealed as a good indicator that reflects the prevalence of early crack defects when combined with well-trained neural network (NN) models (Li et al. 2015). Statistical Control Charts (SCC) have a number of advantages compared to NN models, such as the ability to use less computational capacity with fewer data, detect faults in less time, have a high capacity, and easily apply. Control charts are one of the basic techniques of the statistical process control (SPC) approach. They consist of three control limits are center limit (CL), upper control limit (UCL), and lower control limit (LCL) (Montgomery and Runger 2010). However, the control charts have not been widely used to detect faults in gear systems except in very few recent studies. Among these studies, (Lal and Kane 2019) applied an EWMA chart to detect tooth fracture in the spur gear system with fracture levels of 37.5, 75, and 100% by the signals analyzed in the time domain. (Maraş et al. 2021) employed the X-S chart with time-domain features to design a system for monitoring and detecting gear wear faults in the spur gear system. Nevertheless, these two studies focused on using only features and time domain methods to analyze the experimental vibration signal. Moreover, obtaining vibration data from field measurements is costly due to supplying the test rigs and gears (Mohammed et al. 2015). The difference between this study and the published literature is that a fault detection approach is based on simulation vibration data of the model of the spur gear system using an exponentially weighted moving average (EWMA) chart. Furthermore, the extracted relative wavelet energy (RWE) from DWT analysis was adopted as a statistical indicator to design the univariate EWMA chart. Finally, an EWMA chart was generated using the feature of RWE that is most sensitive to detecting the breakage fault. The performance of EWMA control schemes was compared by detecting the deterioration of gear breakage Fig. 1 describes a flowchart of the proposed methodology steps to the EWMA control chart in this study, which is explained in the following sections.

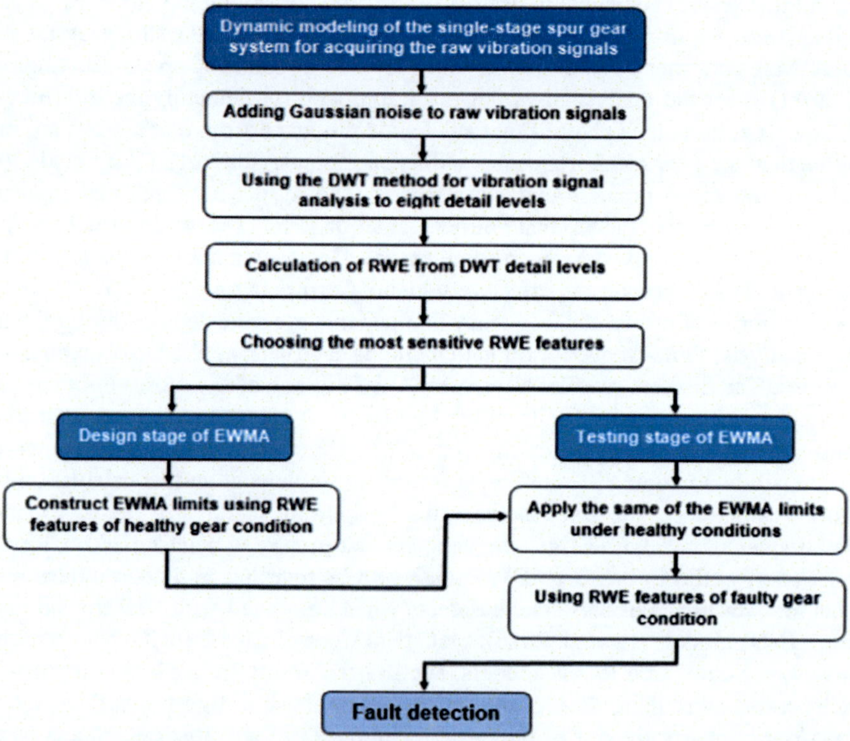

Fig. 1. EWMA framework for detecting the gear tooth breakage using simulated vibration signals

2 Dynamic Modeling of Spur Gear System

The four Degrees Of Freedom (4 DOF) dynamic model is used to simulate the linear behavior of the modeling of a one-stage spur gear system developed by (Chaari et al. 2008) as illustrated in Fig. 2. It is consisting of two blocks pinion has 20 teeth and a wheel has 30 teeth. The 4 DOF assigned to each block are two translations, and two rotations both of the pinion and wheel. The two gear bodies are assumed as rigid spur discs and the shafts are presented by torsional stiffness. The bearings are used to support the shafts, each of which is represented by two linear springs. A linear Springs moving along the motion line of the meshing teeth represents the gear mesh stiffness. The description of the displacement on the path of action is expressed by

$$\delta(t) = (x_1 - x_2)\sin(\alpha) + (y_1 - y_2)\cos(\alpha) + \theta_{12} rb_{12} + \theta_{21} rb_{21} \tag{1}$$

x_i and y_i are the translations of block i (i = 1, 2). θ_{ij} is thc angular displacement of the wheel j in block i $(i, j = 1, 2)$. rb_{12} and rb_{21} are respectively the base radius of the pinion and the wheel and α is the pressure angle.

The equation of motion of the system is acquired by applying Lagrangian formulation as follows by:

$$Mq + K(t)q = F \tag{2}$$

M is a diagonal mass matrix, F is the external applied torques vector given by

$$q = \{x_1.y_1.x_2.y_2.\theta_{11}.\theta_{12}.\theta_{21}.\theta_{22}\}^T$$

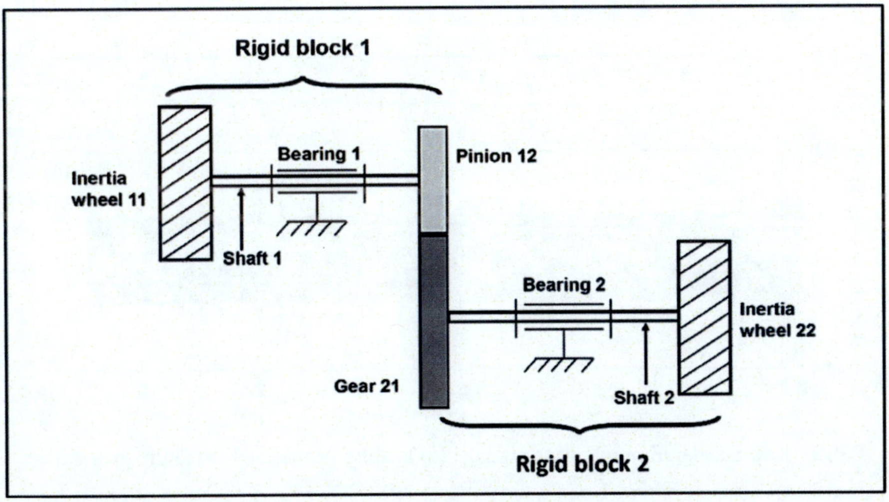

Fig. 2. Model of a one-stage spur gear system

3 Simulation of Gaussian Noise for Raw Signals

This section demonstrates the simulation process of adding Gaussian noise to initial acceleration vibration signals for the spur gear system under healthy and faulty conditions, as illustrated in Fig. 3. Additive white Gaussian noise (AWGN) was added with different signal-to-noise ratios to simulate the operating conditions of gear systems in real practice. The study considered examples of commonly used noise levels, with values of 5,10 and 15 used by (Zhao et al. 2016; Zhang et al. 2015). The SNR values were computed using an equation provided by (Li et al. 2017).

$$SNR = 10 \times \log10(\frac{P_x}{P_\varepsilon}) \tag{3}$$

where $P_x = \sum_{n=1}^{N}[x_{(n)}]/N$ is the average power of the signal and $P_\varepsilon = \sum_{n=1}^{N}[\varepsilon_{(n)}]/N$ is the average power of the noise.

A higher signal-to-noise ratio (SNR) value implies that the fault signal power is greater than the power of the background noise; consequently, the fault will be simple to identify. As a result, the appropriate noise level was selected to be 15 dB since it gave the highest signal-to-noise ratio(SNR) value, as shown in Table 1. Therefore, it is added for healthy and faulty signals, as shown in Fig. 4.

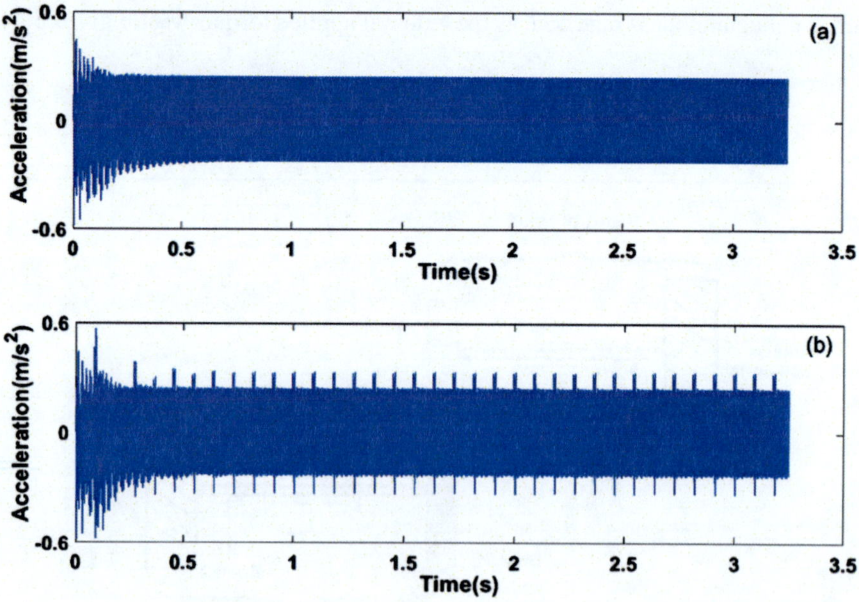

Fig. 3. Raw acceleration vibration signals: (a) healthy gear signal (b) faulty gear signal

Table 1. Noise Simulation of signals based on the signal-to-noise ratio (SNR)

SNR ratios for signals(in dB)	5	10	15
Healthy gear signal	11.4682	22.9823	34.4638
Faulty gear signal	11.5148	23.0795	34.4555

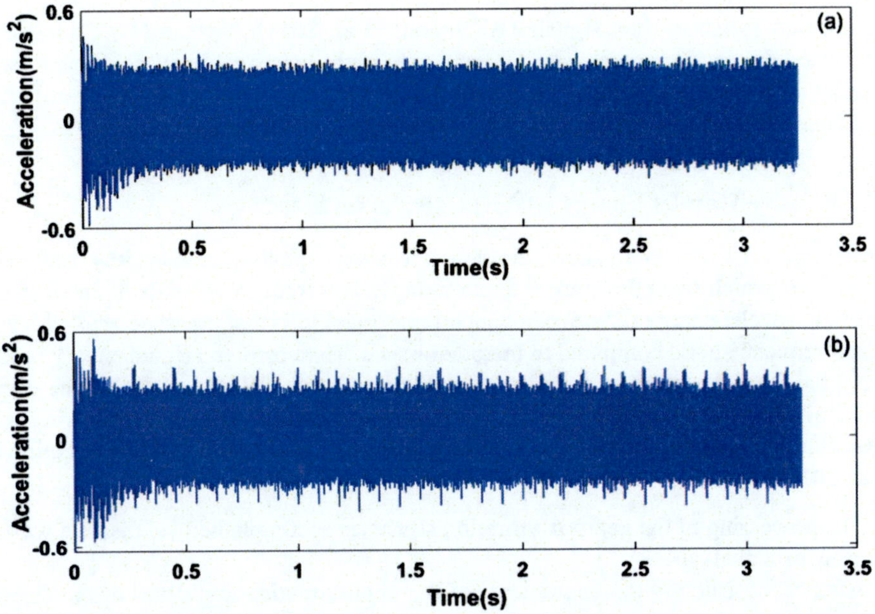

Fig. 4. Noisy acceleration vibration signals: (a) healthy gear signal (b) faulty gear signal

4 Discrete Wavelet Transform (DWT) Analysis

The discrete wavelet transform (DWT) is one of the Wavelet Transform (WT) and it is used for vibration signal processing in the time-frequency domain. It is known as a powerful tool in signal feature extraction. The object of DWT analysis is to decompose the signal into independent and different frequency bands by the use of filters with different frequencies. In this way, the signal is passed through a high-pass filter for the purpose of the high-frequency analysis called coefficients detail (cd), and for low-frequency analysis, the signal is passed through a low-pass filter called coefficients approximate (ca) and can be applied as in Eqs. (4) and (5). As a result, a fluctuating representation of the vibration signal is obtained.

$$cd_j(k) = \sum\nolimits_{i=0}^{I} g[i]a_j - 1[2k - i] \quad (4)$$

$$ca_j(k) = \sum\nolimits_{i=0}^{I} h[i]a_j - 1[2k - i] \quad (5)$$

where $cd_j(k)$ coefficient detail, and $ca_j(k)$ is the coefficient approximations are the outcomes of the g high-pass filter and h low-pass filter respectively.

At the start, the signals are analyzed using DWT, each vibration signal under healthy and faulty conditions with a total time of 3.247 s is divided into 35 segments as short time signals of 0.09 s. Considering it is the period for every two cycles of pinion shaft rotation that are associated with the occurrence of two periodic pulses of the fault. The main purpose of the signal partitioning prior to proceeding with a DWT analysis is due to the fact that the frequency bands of DWT are better and more accurate in fault localization for short-time signals (Al-Badour et al. 2011). Next, a DWT analysis is performed for each segment (short signal) into eight levels of detail coefficient using wavelet DB8. Since the detail coefficients contain the majority of the signal information, so adopted and used in the signal analysis (Fan et al., 2018) as shown in Fig. 5.

4.1 Relative Wavelet Energy (RWE) Calculation Steps

Depending on DWT analysis of the gear vibration signals under healthy and faulty conditions, which were decomposed into independent frequency bands of detail levels. Relative wavelet energy (RWE) can contain important information associated with each detail frequency band compared to the other bands. Therefore, the signal energy in each detail band can be used as a statistical indicator of the EWMA control chart design for monitoring the gearbox operating condition and fault detection. The following steps demonstrate the calculation of the RWE as a feature of gear breakage fault detection (Li et al. 2015):

1. The processing of the gearbox vibration signals is accomplished via discrete wavelet transform analysis.
2. After reconstructing the coefficients of the approximation and detail of the discrete wavelet transformation, the signals are extracted into a number of different frequency bands. The RWE is selected based on the detail coefficients because it contains the major part of the signal information.
3. For each coefficient of the detail coefficients, computed the wavelet energy Ej. After that, calculated the total energy ET, as shown in Eqs. (6), and (7).

$$E_j = \sum\nolimits_{k=1}^{n} |x_{jk}|^2 \tag{6}$$

$$E_T = \sum\nolimits_{j=1}^{J} E_j \tag{7}$$

   where k= 1,2,3.... n is defined as the number of discrete variables for each frequency level j. J is the number of levels of decomposed, and x_{jk} represents the amplitude of the discrete variables.
4. Find the relative wave energy, which is the result of splitting the signal energy (E_j) in each frequency band on the total energy (E_T), as described in the Eq. (8).

$$RWE = \frac{E_j}{E_T} \tag{8}$$

The RWE feature revealed a relatively significant increase at detail levels D4, D5, and D6 under healthy and faulty gear conditions as shown in Fig. 6. This difference was highlighted by the fact that DWT analysis levels can contain signal detail in a specific time-frequency range. Thus, the magnitude of the energy level can vary greatly. Therefore, the RWE values prevalent at these three levels were adopted as a variable to design the EWMA control chart. In contrast, their small values in the remaining detail levels are ignored because they can be easily contaminated with measurement noise.

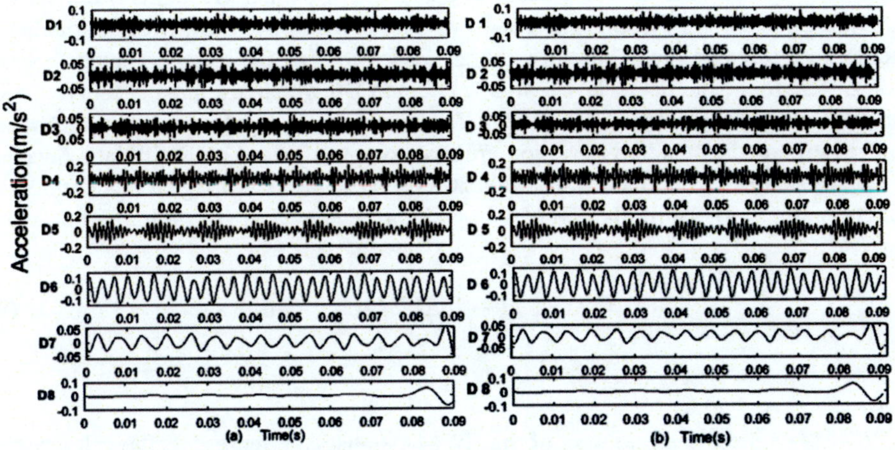

Fig. 5. DWT analysis of the spur gear system signals: (a) Healthy signal (b) Faulty signal

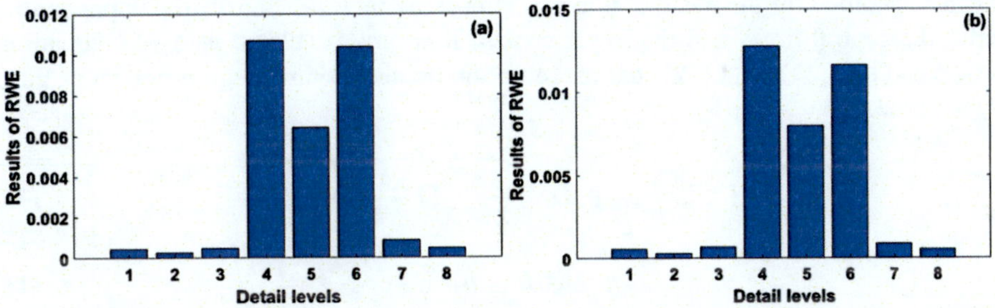

Fig. 6. Relative Wavelet Energy features of the gear system a) healthy conditions and b) faulty conditions

5 Statistical Control Charts (SCC) for Fault Detection

Statistical control charts (SCC) is a widely used technology in manufacturing. The SCC major goal is to detect abnormal changes in processes based on process variables including vibrations, forces, and dimensions (Douglas Montgomery 2009) (chapter 9). The exponentially weighted moving average (EWMA) control chart is one of the SCC. This section includes an explanation and evaluation of the results of the design and testing stages of the EWMA control chart.

5.1 Design Stage of the EWMA Control Chart

The EWMA chart is effective in detecting small shifts of the process mean, because it takes into account all the past and current data in the process log to be fully accounted for it. Moreover, it provides a little weight to the previous data, while the weight of the new data is greater. For this reason, the EWMA chart is seen as a time-weighted average of all the observations (Mohsin et al. 2016). It was first used by (Roberts 2000), and its use has been widely used in time-series analyses. The chart was applied in many different disciplines by scientists and engineers to detect abnormalities (Douglas Montgomery 2009) (chapter 9). The EWMA chart is univariate, so it depends on only one variable for the purpose of its design. Therefore, it was designed using the relative wavelet energy (RWE) feature of the detail levels at D4, D5, and D6 which are most sensitive under the health conditions of the gear system. To implement this procedure, the target value (the mean) and the standard deviation are first computed based on RWE data. Second, the set of samples from the three dominant RWEs for each short signal is identified as subgroups (n = 3). Third, the EWMA statistic is calculated as shown in the following Eq. (9).

$$EWMA_i = \lambda x_i + (1 - \lambda)EWMA_{i-1}, \qquad \text{if } 1 < \text{t} \tag{9}$$

$$EWMA_0 = \mu_0 \qquad \text{if t} = 0$$

$EWMA_i$ represents the output of the EWMA statistic, x_i represents the value corresponding to the observation from the real-time monitored process, and $\lambda(0 < \lambda < 1)$ is the smoothing parameter (Douglas Montgomery 2009). The value $EWMA_0$ in starting is often specified for the mean of normal (fault-free) data, μ_0. Fourth, the upper control limit, the central limit, and the lower control limit are calculated using the equations shown in (10), (11), and (12) that make up the process control region known as the 3 sigma zone.

$$UCL = \mu_0 + L\sigma\sqrt{\frac{\lambda}{2-\lambda}[1-(1-\lambda)^{2i}]} \tag{10}$$

$$CL = \mu_0 \tag{11}$$

$$LCL = \mu_0 - L\sigma\sqrt{\frac{\lambda}{2-\lambda}[1-(1-\lambda)^{2i}]} \tag{12}$$

where L is a multiplier of the EWMA standard deviation σ.

The parameters L and λ are necessary for the design of control plots in EWMA, so they were chosen as used in (Lal and Kane 2019) and are $L = 3$ and $\lambda = 0 \cdot 2$. Finally, the Minitab19 software draws the control chart after providing it with the required information, which is above explained.

Investigating the likelihood of erroneous errors is an essential component of the performance evaluation of control charts for fault detection. It is common knowledge that utilizing control charts can result in two distinct sorts of errors, namely Type-I and

Type-II errors. A Type-I error, often known as a false positive, indicates that a fault has occurred when none does. A type-II error, often known as a false negative error, does not indicate the presence of a problem even while the fault is present (Yang et al. 2018). Figure 7 shows an EWMA chart of the healthy gearbox state designed based on RWE features. It is seen that the behavior of the observed statistics by abnormal patterns appearance, represented by more than one point outside the upper control. As a result, the performance of the EWMA chart points to the occurrence of a type I error, often known as a false positive. This error suggests that a fault happened even though it did not occur. This performance can be explained by the increase in the vibration amplitude that coincided with the commencement of the rotational speed acceleration during the first 0.27 s, as shown in Fig. 3. This acceleration occurred during the first 0.27 s of the experiment. Then, the shift starts moving in the direction of the controlled variables to become stable within the limits of the control region. After that, the shift changes into constant amplitude since the motor's rotational speed has reached the rated speed. Thus, the working condition of the gear system has become fully stable.

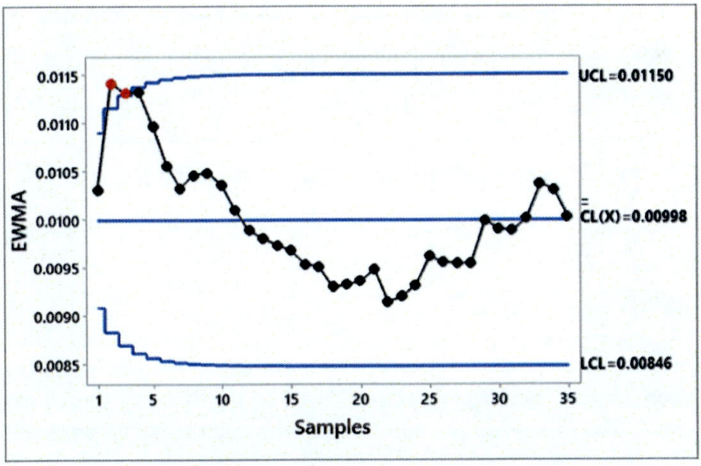

Fig. 7. Design EWMA control chart of the healthy gear state

5.2 Test Stage of the EWMA Control Chart

Following the completion of the design phase of the EWMA chart for the gear system in its healthy state, it is being tested for condition monitoring of the gear breakage. The RWE features values of faulty signal are moved to Minitab 19 software, then mean and standard deviation values are computed. Next, the same upper control limits and lower control limits are used for the EWMA chart in the healthy state. The performance of the EWMA chart in detecting fault is tested, depending on the SCC rules, which are special rules associated with SCC used to predict the presence of an abnormal operation (Douglas Montgomery 2009) (chapter 9). These rules indicate that the appearance of any non-random patterns in the monitoring process, such as the upward or downward trend of eight or more consecutive points on one side of the control limits, or the exit of more than one point is an indication of the gear failure occurrence. In Fig. 8, the EWMA statistics show a non-random behavior represented in the upward direction of consecutive observation RWE samples above the center line (CL) and as well as to the exit of six samples of RWE outside the control area limits.

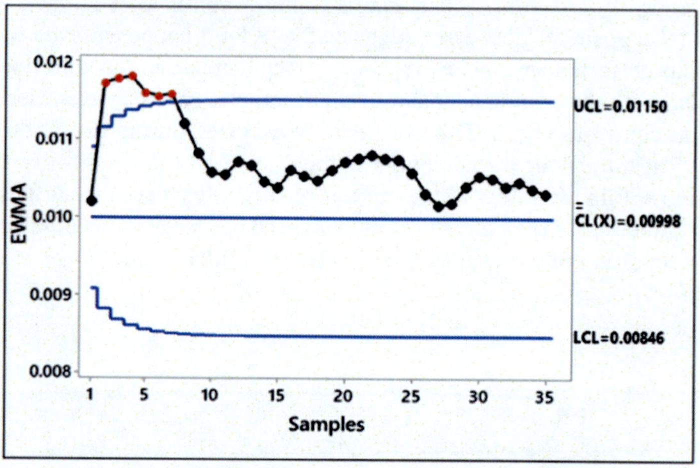

Fig. 8. Testing EWMA control chart of the faulty gear state

6 Conclusion

The main objective of this paper is to investigate the use of the statistical control charts (SCC) approach for gear breakage detection based on the dynamic model of a one-stage spur gear system. The following is a summary of the outcomes that were obtained from the study:

i. A simulation was performed to add white Gaussian noise (AWGN)to the initial signals acquired from the dynamic model using different noise levels at 5,10, and 15 to simulate the working environment of the gears in the real world. Then, the appropriate noise level is chosen to equal 15 based on the highest signal-to-noise ratio (SNR).
ii. The relative wavelet energy, that is obtained from the detailed levels has been suggested for the DWT decomposition as a univariate control indicator to construct control limits for the EWMA chart. In addition to this, the RWE delivered significant details regarding the relative levels of energy that were associated with the different frequency bands.
iii. The statistical control charts (SCC) technique is applied using a univariate chart known as an exponentially weighted moving average (EWMA), depending on the RWE statistics. EWMA chart was created by design parameters ($\lambda = 0.2$ & $L = 3$). The results showed the adequate performance of the EWMA chart in detecting the

gear defect that occurs by increasing the number of points out of control. Moreover, control chart efficiency indicates the alarm issuance of the prediction of the gear breakage, which happens by observing the sudden upward pattern of the observed variables. As a result, it can be concluded that the EMWA statistical control chart effectively distinguished between healthy and defective gear cases and any abnormal changes that could occur during gear work.

iv. Referring to the previous studies presented, this study is the first to investigate the SCC approach for gear failure detection using simulation data derived from a dynamic model of a single-stage spur gear system. As a result, further research into the use of SCC technology as a monitoring and detection strategy for continuous defects under different conditions and types of dynamic models of gear systems or similar industrial applications is required in the future.

References

Al-Badour, F., Sunar, M., Cheded, L.: Vibration analysis of rotating machinery using time-frequency analysis and wavelet techniques. Mech. Syst. Signal Process. **25**(6), 2083–2101 (2011). https://doi.org/10.1016/j.ymssp.01.017

Chaari, F., Baccar, W., Abbes, M.S., Haddar, M.: Effect of spalling or tooth breakage on gearmesh stiffness and dynamic response of a one-stage spur gear transmission. Eur. J. Mech. A/Solids **27**(4), 691–705 (2008). https://doi.org/10.1016/j.euromechsol.2007.11.005

Chaari, F., Fakhfakh, T., Haddar, M.: Analytical modelling of spur gear tooth crack and influence on gearmesh stiffness. Eur. J. Mech. A/Solids **28**(3), 461–468 (2009). https://doi.org/10.1016/j.euromechsol.2008.07.007

Chen, Z., Shao, Y.: Dynamic simulation of spur gear with tooth root crack propagating along tooth width and crack depth. Eng. Fail. Anal. **18**(8), 2149–2164 (2011). https://doi.org/10.1016/J.ENGFAILANAL.2011.07.006

Montgomery, D.C.: Introduction to Statistical Quality Control, 6th edn. Wiley, USA (2009)

Endo, H., Randall, R.B., Gosselin, C.: Differential diagnosis of spall vs. cracks in the gear tooth fillet region: experimental validation. Mech. Syst. Signal Process. **23**(3), 636–651 (2009). https://doi.org/10.1016/J.YMSSP.2008.08.015

Fan, W., Zhou, Q., Li, J., Zhu, Z.: A wavelet-based statistical approach for monitoring and diagnosis of compound faults with application to rolling bearings. IEEE Trans. Autom. Sci. Eng. **15**(4), 1563–1572 (2018). https://doi.org/10.1109/TASE.2017.2720177

Kubur, M., Kahraman, A., Zini, D.M., Kienzle, K.: Dynamic analysis of a multi-shaft helical gear transmission by finite elements: Model and experiment. J. Vibr. Acoust .Trans. ASME **126**(3), 398–406 (2004). https://doi.org/10.1115/1.1760561

Li, C., Zhang, W., Peng, G., Liu, S.: Bearing fault diagnosis using fully-connected winner-take-all autoencoder. IEEE Access **6**(92), 6103–6115 (2017). https://doi.org/10.1109/ACCESS.2017.2717492

Li, Z., Ma, Z., Liu, Y., Teng, W., Jiang, R.: Crack fault detection for a gearbox using discrete wavelet transform and an adaptive resonance theory neural network. Strojniski Vestnik/J. Mech. Eng. **61**(1), 63–73 (2015). https://doi.org/10.5545/sv-jme.2014.1769

Liang, X., Zuo, M.J., Feng, Z.: Dynamic modeling of gearbox faults: a review. Mech. Syst. Signal Process. **98**, 852–876 (2018). https://doi.org/10.1016/J.YMSSP.2017.05.024

Maraş, S., Arslan, H., Birgören, B.: Detection of gear wear and faults in spur gear systems using statistical parameters and univariate statistical process control charts. Arab. J. Sci. Eng. **46**(12), 12221–12234 (2021). https://doi.org/10.1007/s13369-021-05930-y

Mohammed, O.D., Rantatalo, M., Aidanpää, J.O.: Dynamic modelling of a one-stage spur gear system and vibration-based tooth crack detection analysis. Mech. Syst. Signal Process. **54**, 293–305 (2015). https://doi.org/10.1016/j.ymssp.2014.09.001

Mohsin, M., Aslam, M., Jun, C.H.: A new generally weighted moving average control chart based on Taguchi's loss function to monitor process mean and dispersion. Proc. Inst. Mech. Eng. Part B: J. Eng. Manuf. **230**(8), 1537–1547 (2016). https://doi.org/10.1177/0954405415625477

Montgomery, D.C., Runger, G.C.: Applied Statistics and Probability for Engineers. Wiley, Hoboken (2010)

Palermo, A., Mundo, D., Hadjit, R., Desmet, W.: Multibody element for spur and helical gear meshing based on detailed three-dimensional contact calculations. Mech. Mach. 6213–6230 (2013). https://doi.org/10.1016/j.mechmachtheory.2012.11.006

Peng, Z.K., Chu, F.L.: Application of the wavelet transform in machine condition monitoring and fault diagnostics: a review with bibliography. Mech. Syst. Signal Process. **18**(2), 199–221 (2004). https://doi.org/10.1016/S0888-3270(03)00075-X

Lal, H., Kane, P.V.: Gearbox fault detection using exponentially weighted moving average control charts. Mach. Mech. Robot. 39–47 (2019). ISBN 978-981-10-8596-3

Roberts, S.W.: Control chart tests based on geometric moving averages. Technometrics **42**(1), 97–101 (2000). https://doi.org/10.1080/00401706.2000.10485986

Tian, Z., Zuo, M.J., Wu, S.: Crack propagation assessment for spur gears using model-based analysis and simulation. J. Intell. Manuf. **23**(2), 239–253 (2012). https://doi.org/10.1007/s10845-009-0357-8

Wei, J., Zhang, A., Gao, P.: A study of spur gear pitting under EHL conditions: theoretical analysis and experiments. Tribol. Int. **94**, 146–154 (2016). https://doi.org/10.1016/j.triboint.2015.08.037

Wu, S., Zuo, M.J., Parey, A.: Simulation of spur gear dynamics and estimation of fault growth. J. Sound Vib. **317**(3–5), 608–624 (2008). https://doi.org/10.1016/j.jsv.2008.03.038

Yang, H.H., Huang, M.L., Lai, C.M., Jin, J.R.: An approach combining data mining and control charts-based model for fault detection in wind turbines. Renew. Energy **115**, 808–816 (2018). https://doi.org/10.1016/j.renene.2017.09.003

Yang, Y., Xia, W., Han, J., SongY, W.J., Dai, Y.: Vibration analysis for tooth crack detection in a spur gear system with clearance nonlinearity. Int. J. Mech. Sci. **157–158**, 648–661 (2019). https://doi.org/10.1016/j.ijmecsci.2019.05.012

Yuan, B., Chang, S., Liu, G., Chang, L., Liu, L.: Quasi-static analysis based on generalized loaded static transmission error and dynamic investigation of wide-faced cylindrical geared rotor systems. Mech. Mach. Theory **134**, 74–94 (2019). https://doi.org/10.1016/j.mechmachtheory.2018.12.019

Zhang, Y., Tang, B., Liu, Z., Chen, R.: An adaptive demodulation approach for bearing fault detection based on adaptive wavelet filtering and spectral subtraction. Meas. Sci. Technol. **27**(2), 25001 (2015). https://doi.org/10.1088/0957-0233/27/2/025001

Zhao, M., Lin, J., Miao, Y., Xu, X.: Detection and recovery of fault impulses via improved harmonic product spectrum and its application in defect size estimation of train bearings. Meas. J. Int. Meas. Confed. **91**, 421–439 (2016). https://doi.org/10.1016/j.measurement.2016.05.068

A Study on Temperature Evolution During Milling of FRP Composites

Sami Ghazali(✉)

Mechanical and Materials Engineering Department, Faculty of Engineering, University of Jeddah, P.O. Box 80327, Jeddah 21589, Saudi Arabia
sghazali@uj.edu.sa

Abstract. This paper investigates temperature variation in fiber-reinforced polymers (FRP) during milling operations by using thermocouples (TCs) embedded within the core of the cut specimen. TCs are connected to a temperature difference (T.D.) heat transfer device to log temperature variation during milling. The effects of process parameters (depth of cut, feed rate, and rotational speed) on temperature generation within the material are discussed by cutting six specimens from a bulk FRP panel using a laser machine. Holes were drilled on the trim plane side of the specimens. These three holes allow the interstation of type K thermocouples connected to the acquisition system. It was observed that increasing the initial depth of the cut has a directly proportional relationship with the amount of heat generated within the specimen. Furthermore, increasing the feed rate also had a similar trend with heat generation within the material, apart from fifty-five mm/min. However, a contrary effect was observed and attributed to the preexisting middle-milled channel effects were heat escapes more rapidly. Finally, the impact of increasing the rotational rates of the tool was discussed, and it was revealed that the temperature developed within the workpiece rises proportionally to the rotational speed.

Keywords: FRPs · Milling · Thermocouples · Heat-Transfer · Temperature

1 Introduction

Fiber Reinforced Polymers (FRPs) are transforming many industries due to their economic competence and superior mechanical properties. FRPs have several unique mechanical properties over conventional materials. For example, FRPs are lightweight, have good impact resistance, a high strength-to-weight ratio compared to traditional materials, have high damping ability, and have low density (Gilpin 2009). These advantages make them competitive in the aerospace, military, and aerospace sectors. Milling is frequently utilized in fabricating FRP parts. FRPs are non-homogeneous, anisotropic materials, making them difficult to machine (Qi et al. 2015). Based on experimental studies, researchers have recognized the importance of fiber orientation, tool geometry, and other cutting parameters in milling operations. Furthermore, it was concluded that the principal cutting depends strongly on fiber arrangement (Puw et al. 1995). Composites often require the removal of excess material to account for tolerance, and thus

A. Akrout et al. (Eds.): ICAV 2022, ACM 22, pp. 153–161, 2023.
https://doi.org/10.1007/978-3-031-34190-8_18

milling is used to manufacture a well-defined component with high-quality surfaces (Jahanmir et al. 1999). Nevertheless, many challenges arise when milling FRPs: surface damages, rough surfaces, accidental fiber extractions, different fiber types, and matrix failures. The fibers embedded within the matrix will determine the suitability of the cutting tools required for the milling operation (Davim and Reis 2005). Therefore, the presence of milling defects could affect the specimen's life. Furthermore, it is vital to control temperature distribution and heat generation within the workpiece to optimize the cutting process during dry cutting operations applications since refrigerants will alter the chemical characteristics of the material (Liu et al. 2014).

(KoPlev et al. 1983) investigated the influence of cutting orientation on the developed cutting force and concluded that a noticeably more cutting force is required for perpendicular fiber orientation. Furthermore, it was explained that chip formations, during the machining of FRP, were not subjected to large plastic deformation as metal chips. Also, it was observed that the depth of the cut has a significant impact on the surface roughness, delamination dagame, and cutting speed (Sreenivasulu 2013).

(Ramulu et al. 1993; Sreejith et al. 2000) observed that surface roughness is directly related to the feeding rate and inversely proportional to cutting speeds. Furthermore, investigations of varying cutting parameters on different fiber composites revealed that jute fiber reinforcement composites had a high surface roughness and more delamination damage than natural fiber composites and glass-reinforced polymers (Babu et al. 2013). Milling tools are a vital component that significantly affects the working piece's dimensional precision and mechanical properties. It was also observed that coated types of milling tools provided better results than uncoated tools (Raj et al. 2012). Hence, it was observed that less delamination damage occurred when machining was carried out with a high number of milling flutes (Kiliçkap et al. 2015). Another study by Erkan et al. (Erkan et al. 2013) confirms that less delamination damage occurs when milling with an increased number of flutes and at a lower feed rate. Furthermore, it was observed that tool life depends on feed rate and cutting speed for glass-reinforced composites (Azmi et al. 2013). Jia et al. (Jia et al. 2017) investigated the failure of the polycrystalline diamond tool during milling operations. Temperature development within the workpiece is another crucial aspect of milling operations. (Wang et al. 2016) observed that the cutting forces decreased with increasing fiber orientation angle. Furthermore, researchers concluded that cutting temperature is linearly proportional to cutting speed in carbon fiber-reinforced polymers. (Chibane et al. 2017) measured workpiece temperature and cutting forces. They concluded that temperature increases as cutting speed and chip thickness increase. Also, it was observed that vibration increases as the feed rate and the cutting depth increase. To put it concisely, the works done on FRPs machinability are a function of many different process parameters and the effects on the surface roughness conditions and dimensional accuracy of the workpiece desired. In comparison, this paper investigates temperature within FRP during the dry milling process using a K-type thermocouple positioned within the substrate of the workpiece. The study focuses on understanding the effects of initial depth, feeding rate, and rotational speed on the workpiece's fracture, delamination, and temperature development.

2 Experimental Procedure

The experimental procedure is composed of specimen design and milling tests as follows:

2.1 Specimen Design

Six pieces were cut using a laser machine from a bulk FRP workpiece. Each piece had a length of 100 mm, a width of 50 mm, and a thickness of 10 mm, as shown in Fig. 1. Three holes of 5 mm diameter and 25 mm depth were drilled within the core of the workpiece and placed 25 mm apart, as shown in Fig. 2. These three holes allow the interstation of type K thermocouples connected to a TD heat transfer device to measure the temperature within the specimen while milling.

2.2 Milling Tests

Three holes were drilled on the sides of the workpieces to allow the interstation of type K thermocouples. The holes were drilled 25 mm apart, as shown in Fig. 1 and Fig. 2. The hole's diameter and depth are 5 mm and 25 mm, respectively. The workpiece is placed on a milling machine with a Tungsten carbide tool of 12 mm diameter, and through the thermocouples embedded within and a TD heat transfer interface, temperature propagation during milling throughout the workpiece is determined. A sample of a finished workpiece is shown in Fig. 3.

2.2.1 Temperature Measurement vs. Depth of Cut

The workpiece is placed on the milling machine, and the thermocouples connections are established between the workpiece and the TD heat transfer interface. It then determines the coordinates needed to mill at the depths of (0.2, 0.4, 0.6, 0.8, 1, 1.4, 1.8, 2, 2.5, and 2.7) mm, respectively, while holding the feed rate at 92 mm/min and rotational speed at 495 rpm. Once the milling process starts, a high-resolution camera monitors the temperature elevation within the workpiece. This process is repeated at the abovementioned depths until a final depth of 2.7 mm is reached to ensure it is not cutting through the embedded thermocouples.

2.2.2 Temperature Measurement vs. Feed Rate

The workpiece is placed on the milling machine, and the thermocouples connections are established between the workpiece and the TD heat transfer interface. Now by fixing the penetration depth at 2 mm, and rotational speed at 495 rpm. The rates are set to be $f_1 = 30, f_2 = 55$, and $f_3 = 92$ mm/min respectively. The repeats were performed on the same workpiece but 4 mm apart. f_1 being in the middle, while f_2 and f_3 are 4 mm distance apart to the right and left of f_1, respectively. In doing so and while waiting for a few minutes between different runs, effects of unequal heat transfer effects could be neglected. Also, this ensures proper capture of heat generation during the milling process. Once the milling process starts, a high-resolution camera monitors the temperature elevation within the workpiece.

2.2.3 Temperature Measurement vs. Rotational Speed

The workpiece is placed on the milling machine, and the thermocouples connections are established between the workpiece and the TD heat transfer interface. Now by fixing the penetration depth at 2mm, and feed speed at 24 mm/min. The rotational rates are set to be $\omega_1 = 195$, $\omega_2 = 340$, and $\omega_3 = 495$ mm/min respectively. The repeats were performed on the same workpiece but 4mm apart. ω_1 being in the middle, while ω_2 and ω_3 are 4 mm distance apart to the right and left of ω_1 respectively. In doing so and while waiting for a few minutes between different runs, effects of unequal heat transfer effects could be neglected. Also, this ensures proper capture of heat generation during the milling process. Once the milling process starts, a high-resolution camera monitors the temperature elevation within the workpiece. Results, Discussion, and conclusion.

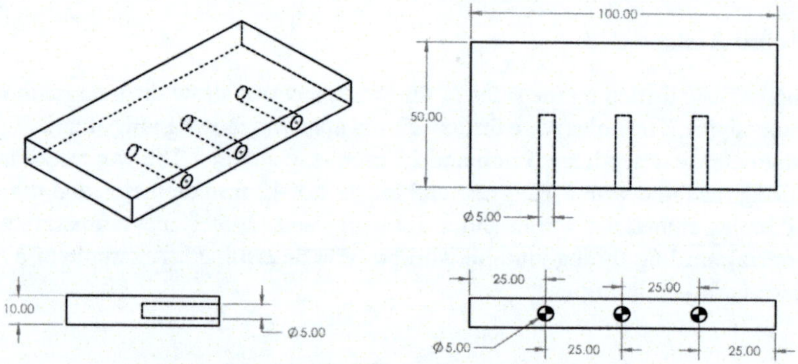

Fig. 1. FRP design and dimensions

Fig. 2. FRP with holes and thermocouples inserted

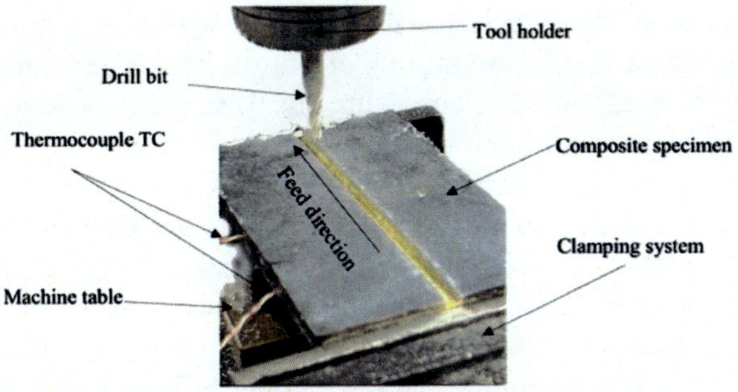

Fig. 3. Sample of a workpiece after milling

3 Results and Discussion

Collected data were extracted manually by utilizing and capturing a video of the experiments in the form of temperature versus time. Thus, various plots were constructed.

3.1 Discussion on Temperature Variation Versus Depth of the Cut

The experiment's outcome confirms that temperature increase is directly proportional to the depth of the cut in an almost linear fashion while fixing feed rate and rotational speed

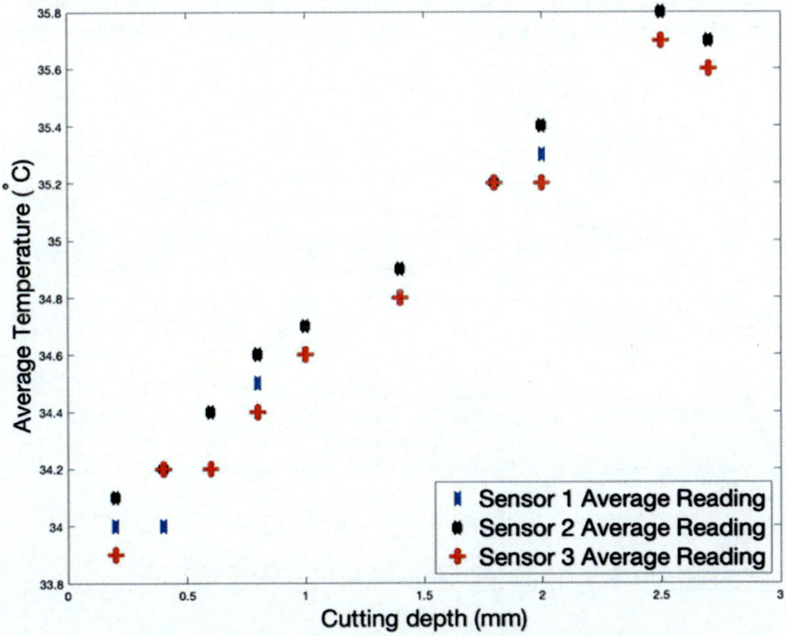

Fig. 4. Temperature change versus cutting depth

parameters. Data discrepancies between the three sensors are insignificant. An explanation for such disparity is that during milling operation, heat is generated and propagates toward sensors one, two, and three consecutively. Therefore, average temperature values are considered at each cutting depth, as illustrated in Fig. 4. It is worth noting that the time needed to reach the maximum temperature is a function of the cutting depth. For example, it took sensor-one thirty-three seconds to get a maximum temperature at 1.4 mm depth. In contrast, it took about 25 s to maximize temperature at a depth of 2.7 mm, which is owed to the proximity of the sensor at a deeper cutting depth. Similar observations were made for sensors two and three. Thus, the study of cutting depth versus temperature evolution concludes that the amount of heat generated is proportional to the cutting depth and the heat rate dissipation.

3.2 Discussion on Temperature Measurement Versus Velocity

Outcomes of varying feed rates on temperature propagation have been analyzed. An average maximum temperature reading of 36 °C is observed for 30 mm/min feed rate at the middle of the workpiece. A repeat is performed on the same workpiece to the right of the middle-milled channel using a feed rate of 55 mm/min. An increase in the kinetic energy of the indenter is expected to increase the temperature within the workpiece. However, increasing the feed rate to 55 mm/min led to an average maximum temperature reading of 34.5 °C. This decrease in average temperature reading is not expected. It is attributed to the existence of the middle-milled channel, were heat escapes more rapidly during the experiment. Conversely, a continuous increase of feed rate to 92 mm/min on the same workpiece to the left of the middle-milled channel led to a rise in the average maximum temperature reading of 38.6 °C which is expected. Therefore, higher feed

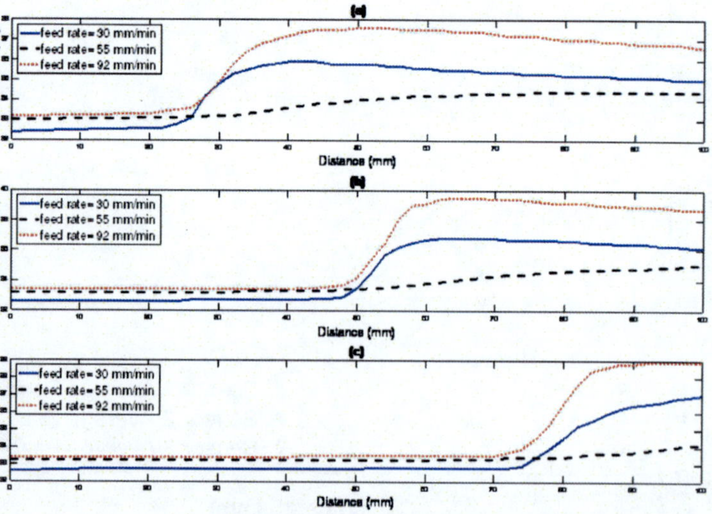

Fig. 5. Sensors' readings at various feed rates versus distance (a) sensor one readings, (b) sensor two readings, and (c) sensor three readings

rates lead to a higher sensor temperature reading, which explains the sudden jumps in temperature readings seen in distances of 25 mm, 50 mm, and 75 mm in Fig. 5. Please note that sensors one, two, and three are placed 25 mm apart.

3.3 Discussion on Temperature Measurement Versus Rotational Speed

Results of varying rotational speeds of the indenter on temperature propagation have been examined at 195, 340, and 495 rpm. A rotational speed of 195 rpm on the middle of the workpiece led to an average maximum temperature reading of 36 °C. Now two repeats are performed on the same workpiece to the right and left of the middle-milled channel using rotational speeds of 340 rpm and 495 rpm, respectively. Rotational velocities of 340 rpm and 495 rpm led to average maximum temperature readings of 35 °C and 45 °C, respectively. Therefore, higher rotational speeds lead to a higher sensor temperature reading, which explains the sudden jumps in temperature readings seen in distances of 25 mm, 50 mm, and 75 mm in Fig. 6. Please note that sensors one, two, and three are placed 25 mm apart.

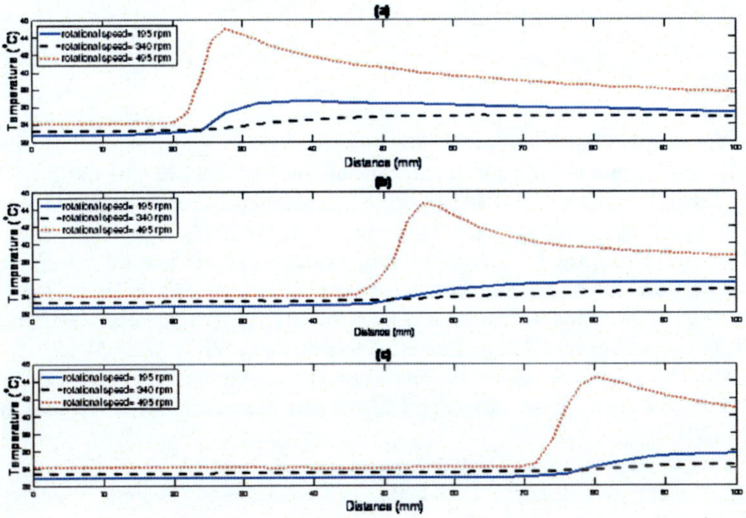

Fig. 6. Sensors' readings at various rotational speeds versus distance (a) sensor one readings, (b) sensor two readings, and (c) sensor three readings

4 Conclusion

This paper investigates temperature variation in FRP composites during milling operations by using several type K thermocouples embedded within the core of the FRP that is connected to a T.D. heat transfer device to log temperature variation within the specimen during the milling procedure. The effects of process parameters (cutting depth, feeding velocity, and rotational speed) on temperature generation within the workpiece are summarized as follows:

- Increasing the initial cutting depth of the indenter while fixing feed rate and rotational speed is proportional to the amount of heat generated during the milling operation on the workpiece in an almost linear relationship.
- The effects of increasing feed rates from 30 to 92 (mm/min) increased the temperature of the workpiece, except at a feed rate of 55 mm/min where a decrease of the workpiece temperature is observed. This decrease in average temperature reading is not expected. It is probably owed to the existence of the middle-milled channel were heat escapes more rapidly or a reading error during the experiment.
- The effects of increasing rotational rates of the indenter from 195 to 495 rpm increased the temperature of the workpiece.

Acknowledgments. The University of Jeddah partially supports this work. The author also gratefully acknowledges the support of Dr. Ali Mkaddem, Dr. Abdessalem Jarraya, Eng. Mohamed Hamdi and UJ students. The author also acknowledges the helpful comments and suggestions of the reviewers, which have improved the presentation.

References

Azmi, A.I., Lin, R.J.T., Bhattacharyya, D.: Machinability study of glass fibre-reinforced polymer composites during end milling. Int. J. Adv. Manuf. Technol. **64**, 247–261 (2013)

Babu, G.D., Babu, K.S., Gowd, B.U.M.: Effect of machining parameters on milled natural fiber-reinforced plastic composites. J. Adv. Mech. Eng. **1**, 1–12 (2013)

Chibane, H., Serra, R., Leroy, R.: Optimal milling conditions of aeronautical composite material under temperature, forces and vibration parameters. J. Compos. Mater. **51**, 3453–3463 (2017)

Davim, J.P., Reis, P.: Damage and dimensional precision on milling carbon fiber-reinforced plastics using design experiments. J. Mater. Process. Technol. **160**, 160–167 (2005)

Erkan, Ö., Işık, B., Çiçek, A., Kara, F.: Prediction of damage factor in end milling of glass fibre reinforced plastic composites using artificial neural network. Appl. Compos. Mater. **20**, 517–536 (2013)

Gilpin, A.: Tool solutions for machining composites. Reinf. Plast. **53**, 30–33 (2009)

Jahanmir, S., Ramulu, M., Koshy, P.: Machining of Ceramics and Composites. Marcel Dekker (1999)

Jia, Z., Su, Y., Niu, B., et al.: Deterioration of polycrystalline diamond tools in milling of carbon-fiber-reinforced plastic. J. Compos. Mater. **51**, 2277–2290 (2017)

Kiliçkap, E., Yardimeden, A., Çelik, Y.H.: Investigation of experimental study of end milling of CFRP composite. Sci. Eng. Compos. Mater. **22**, 89–95 (2015)

KoPlev, A.A., Lystrup, A., Vorm, T.: The cutting process, chips, and cutting forces in machining CFRP. Composites **14**, 371–376 (1983)

Liu, J., Chen, G., Ji, C., et al.: An investigation of workpiece temperature variation of helical milling for carbon fiber reinforced plastics (CFRP). Int. J. Mach. Tools Manuf. **86**, 89–103 (2014)

Puw, H.Y., Hocheng, H., Kuo, H.C.: Anisotropic chip formation models of cutting of FRP. In: Proceedings of ASME Symposium on Material Removal and Surface Modification Issues in Machining Processes, New York, pp 259–282 (1995)

Qi, Z., Zhang, K., Cheng, H., Wang, D., Meng, Q.: Microscopic mechanism based force prediction in orthogonal cutting of unidirectional CFRP. Int. J. Adv. Manuf. Technol. **79**(5–8), 1209–1219 (2015). https://doi.org/10.1007/s00170-015-6895-7

Raj, P.P., Perumal, A.E., Ramu, P.: Prediction of surface roughness and delamination in end milling of GFRP using mathematical model and ANN (2012)

Ramulu, M., Wern, C.W., Garbini, J.L.: Effect of fibre direction on surface roughness measurements of machined graphite/epoxy composite. Compos. Manuf. **4**, 39–51 (1993)

Sreejith, P.S., Krishnamurthy, R., Malhotra, S.K., Narayanasamy, K.: Evaluation of PCD tool performance during machining of carbon/phenolic ablative composites. J. Mater. Process. Technol. **104**, 53–58 (2000)

Sreenivasulu, R.: Optimization of surface roughness and delamination damage of GFRP composite material in end milling using Taguchi design method and artificial neural network. Procedia Eng. **64**, 785–794 (2013)

Wang, H., Sun, J., Zhang, D., et al.: The effect of cutting temperature in milling of carbon fiber reinforced polymer composites. Compos. Part A Appl. Sci. Manuf. **91**, 380–387 (2016)

Ball Bearing Diagnosis Using Data Hybridisation in Supervised Machine Learning

Souleymane Sow[1,2](✉), Xavier Chiementin[1], Lanto Rasolofondraibe[2], and Olivier Cousinard[1]

[1] Institute of Thermic, Mechanics, Material (ITheMM), University of Reims Champagne-Ardennes, Reims, France
souleymane.sow@univ-reims.fr

[2] Center of Research for Information and Communication Science and Technology (CReSTIC), University of Reims Champagne-Ardennes, Reims, France

Abstract. In the context of maintenance, the use of digital twins is a topic with huge prospects. A digital twin is a virtual and dynamic representation of an asset, but it is built on a numerical model based on hypothesis. For bearing numerical models, taking into account the flexibility of the bearings is still a challenge. In this study, a numerical model of a test bench composed of two bearings (one is healthy and the other one has a defect on its outer ring with different severities), is proposed while considering the flexibility of the bearings. From the numerical model, data are generated and added to the data acquired from the test bench to build a homogenous hybrid data model. The data are labelled according to their belonging to a severity and are then used to train two machine learning algorithms, KNN and Decision Tree. The tests was performed in the two machine learning algorithms by using the experimental data that were not used in the training. This study judged the effectiveness of a data hybridisation model through artificial intelligence diagnosis.

Keywords: digital twin · bearing · bearings flexibility · fault diagnosis · hybrid data model

1 Introduction

In today's industrial world, maintenance holds an important place. It allows to intervene as a numerator of productivity, by helping to increase the quantity produced and a denominator, by helping to optimize the expenses used for production (manpower, tools, energy spent, etc.). As a maintenance tool, vibration analysis is an important tool for system diagnosis and monitoring. It consists of detecting possible failures and dysfunctions of the rotating machine by following the evolution of indicators with the only goal of planning an intervention (Li et al. 1999; Karacay and Akturk 2009; Patidar and Soni 2013). In addition, the migration to industry 4.0 has introduced the notion of the digital twin which is one of the key pillars of this new form of industrialization. The notion of the digital twin has offered a fusion of physical and virtual spaces by creating

A. Akrout et al. (Eds.): ICAV 2022, ACM 22, pp. 162–170, 2023.
https://doi.org/10.1007/978-3-031-34190-8_19

cyber-physical systems. Its use in the maintenance context has been the subject of several researches in recent years because it can overcome a few blockages such as difficult or impossible instrumentation for complex systems and reduce the risk of damage to some expensive parts (Fadda and Moussaoui 2018). Using machine learning algorithms adds a touch of artificial intelligence to traditional diagnostic tools. In the diagnostic context, machine learning algorithms such as SVM (Samanta 2004); KNN (Guo et al. 2003); Neural Network (Paya et al. 1997) are used to perform a diagnosis by classifying different states of a system.

However, the process of developing a digital twin is based on hypotheses. In their study (Patil *et al.*, 2010), developed a numerical model of a cracking default on the outer race of a bearing and introduce an increase radial clearance when each rolling element passes through the defective area. In his work, (Farhat *et al.*, 2021) proposed a numerical model a rotor-ball bearing with an update of the generated data **to** the experimental data. In addition, the combination of vibration analysis, digital twin and artificial intelligence faces a challenge regarding the accuracy of the diagnosis made. This accuracy on the reliable diagnosis depends on several parameters considered during the analysis such as the model used for the training. It is already proved that machine learning algorithms are more efficient when they are trained with more data. (Ngandu and Kalala 2022), showed the efficiency of using a hybrid of homogeneous and heterogeneous data for training KNN and SVM machine learning algorithms in a mining context with a ball mill.

In this paper, a numerical model of a test bench composed bearings is developed by integrating the flexibilities of bearings. The final numerical model is a combined finite element (FE) and discrete element model (DE). Then a strategy of homogenous hybridisation model is proposed and used by adding numerical data generated and experimental data. These data are a function of the different severity on the outer race.

The paper is divided into two sections in addition to the introduction and conclusion. Section 2 presents the methods used in this study. In Sect. 3, the results of the application of the method are presented.

2 Methods

2.1 Global Methodology

The approach proposed in this study is presented in Fig. 1. It starts with an acquisition of experimental data of different operating modes, for which temporal and frequency indicators are calculated and then reduced to the most relevant using a selection algorithm. These data are labelled and divided in two according to a percentage, part of which will be used for hybridization and the other for testing. A numerical model with concentrated parameters is created to be able to generate signals in the same operating modes. Indicators are also calculated, and this data is labelled and fully used for hybridisation. The hybrid data set is used to train the MLAs. Finally, the diagnosis is made by classifying the different operating modes.

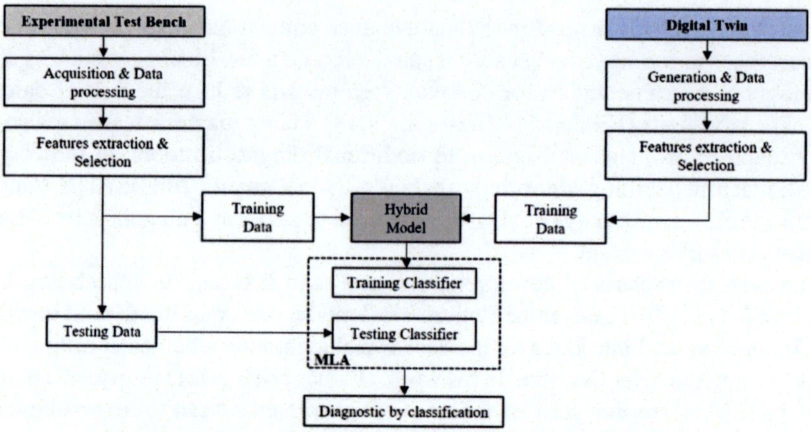

Fig. 1. Methodology based on diagnosis by classification

2.2 Test bench and Acquisition

The test bench of the bearing module presented in Fig. 2 is composed of a box, a steel shaft on which is mounted two ball bearings type 6206 BC 32 × 64 × 18 of the SKF brand. The shaft is coupled by a synchronous electric motor with a maximum power of 10 KW and a hydraulic cylinder connected to a freewheel pulley mounted on the shaft by a steel cable. The rotation frequency of the shaft is controlled by a variable speed drive via a PLC. A manual adjustable load is applied radially. Experimental data are collected by using two uniaxial piezoelectric accelerometers DJB A/120/VT, one in the radial direction and the other in the axial direction, on the bearing of the test bearing; a data collector type "OROS 36" set at a sampling frequency of 51.2 kHz and a radial load of 350 daN will be applied on the rotation shaft at 1000 rpm. The signals acquired on the test bench are a function of defects on the outer ring (1 mm, 2 mm, 3 mm, 4 mm, 5 mm) as presented in Fig. 3. The detailed characteristics of the test bench are given in Table 1.

Fig. 2. Bearing defect test bench

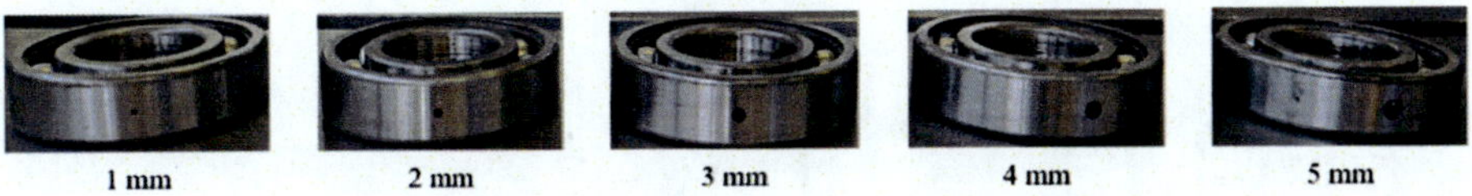

Fig. 3. Severity of defects on the outer ring

Table 1. Test bench characteristics

Parameters	Motor	Frame	Shaft	Bearing
Mass (kg)	8		2.5	
Moment of inertia (kg m^2)	0.01125		0.0108	
Young's modulus (N m^{-2})		2×10^{11}	2×10^{11}	
Poisson ratio		0.3	0.3	
Density (Kg m^{-3})		7900	7900	
Number of balls				9
Inner race radius (mm)				30
Outer race radius (mm)				62
Backlash γ (µm)				18
Bearing stiffness k (N m^{-1})				8.5×10^9

The signals collected (Fig. 4) on the test bench are acquired in four rounds of 4 s of time acquisition for each severity with a bearing change after each acquisition. These collected data are processed by reducing the noise on the signals by using SWT (Stationary Wavelet Transform) wavelets (Fowler 2005). Then cut into 40 signals for each severity, and calculate the indicators in the time domain such as: RMS, Peak, Crest Factor, Kurtosis, K Factor, Skewness, Talaf and the THIKAT and in the frequency domain: centre of frequencies (FC), standard deviation of frequencies, entropy, SPRO (Spectrum Peak Ratio Outer race), SPRI (Spectrum Peak Ratio Inner race), SPRB (Spectrum Peak Ratio Ball) (Darong and Peng 2012). The indicators are reduced to the most relevant ones using the SBS selection algorithm (Reeves and Zhe 1999), the selected ones are **RMS, FC, Entropy.**

The signals shown have the signature of a defect on the outer race as the theoretical defect frequency described in Eq. 1 at 59.37 Hz.

$$BPFO = \frac{N_b}{2}\left[1 - \left(\frac{d}{D} * \cos\theta\right)\right]f_r \tag{1}$$

where N_b is the number of balls, d is the ball diameter, D is the ball centre diameter, θ radial clearance and f_r the cage speed.

As the study is based on supervised learning, the data are labelled and then collected in the same database. Figure 5 shows a spatial representation according to the most relevant indicators of the experimental database.

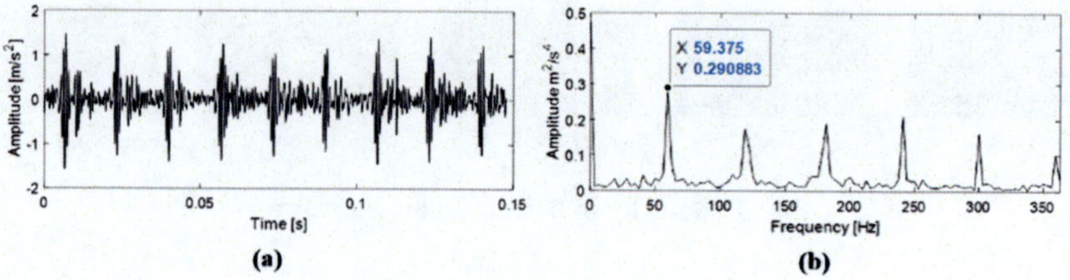

Fig. 4. Signal from the experimental test bench: (a) Time domain; (b) Envelope spectra (case of severity 1 = 1 mm)

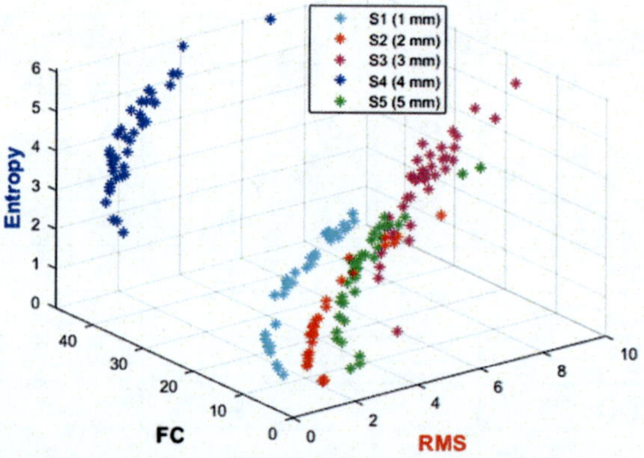

Fig. 5. Spatial representation of experimental data depending on the severities

2.3 Numerical Model

In his work, (Farhat *et al.* 2021), developed a discrete elements model of the test bench (DE). This model called "meso model", shown in Fig. 6 (a) has nine (9) nodes. Each having three degrees of freedom (DOF), which are: flexion in the $\vec{v}$ direction; flexion along the $\vec{w}$ direction and a torsion in $\vec{u}$ direction. The shaft and motor mass matrices, the shaft and bearing stiffness matrices are extracted, then, the bearing excitation forces are determined. In addition to his work, a finite element (FE) model is developed in a software calculation package (Abaqus) as shown in Fig. 6 (b). From this FE model, called the macro model, the mass and stiffness matrices of the frame are extracted, as well as the stresses on the rings caused by the load applied to the shaft radially along the $\vec{v}$ direction. Giving that, the digital twin is a virtual and dynamic representation of a physical asset, the extracted parameters from the FE model will be used in the DE model. Then the different parameters are assembled to reconstruct the equation of motion as given in Eq. 2, by using Lagrange's formalism. Where *q* is the generalized vector of coordinates that are defined by the degrees of freedom of each node $q = \{v_n, w_n, \theta_n\}$, $n \in (1...9)$, , [M] and [K] represent respectively the mass matrix and the global stiffness matrix, they are obtained by combining the elementary matrices of the system components. [C] is the global damping matrix. It is calculated proportionally to the mass matrix [M] and the average value of the stiffness matrix [K] as given thus Eq. 3 (Parker et al. 2000). F(t) includes the motor torque $C_m(t)$, gravity forces, total nonlinear forces generated by the rolling bearings.

$$[M]\ddot{q} + [C]\dot{q} + [K]q = F(t) \tag{2}$$

$$[C] = \alpha[M] + \beta[K] \tag{3}$$

α and β two real Rayleigh coefficients calculated such that the damping is viscous with $\alpha = 0.6$ et $\beta = 6e^{-4}$ (Parker et al. 2000).

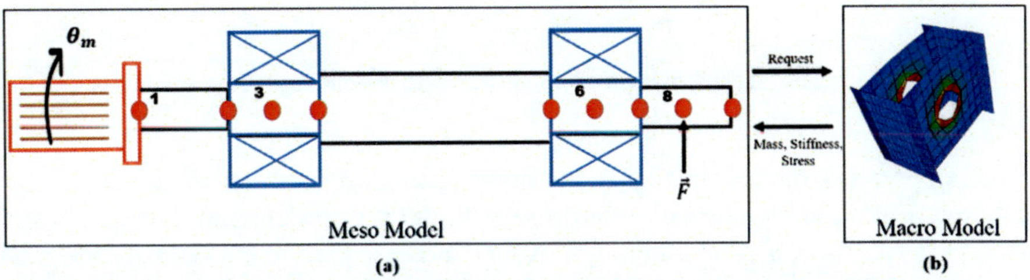

Fig. 6. Numerical model of the test bench: (a) meso model; (b) macro model

From the numerical model, signals are generated by solving the equation of motion (Fig. 7). These signals are function of each severity of defects initially acquired. Then the indicators selected during the experimental phase are calculated from the signals that have been generated. In Fig. 8, a spatial representation according to the most relevant indicators of the numerical database is given, then depending on the mode of operation describing the severity, all the data are labelled and assembled in a global digital database.

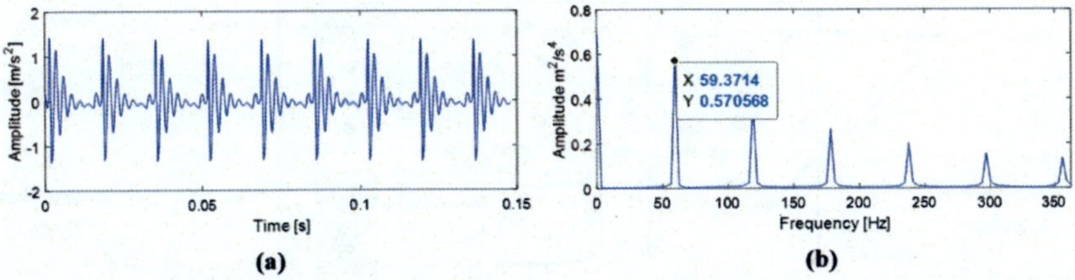

Fig. 7. Signal generated by the numerical model: (a) Time domain; (b) Envelope spectra (case of severity 1 = 1 mm)

2.4 Homogenous Hybridization Model

A hybrid model is a fusion of the experimental data and the data generated by the numerical model in the one database. This hybrid model can be homogeneous in the case of data used for the fusion of the experimental and numerical databases that reflect

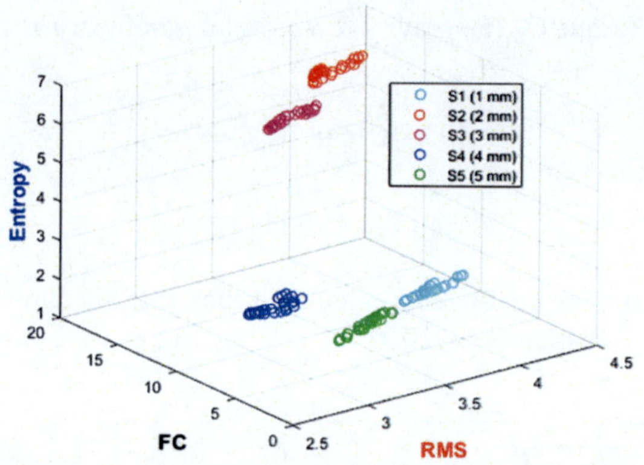

Fig. 8. Spatial representation of numerical data depending on the severities

the same operating modes. In Fig. 9, the homogeneous hybridisation model chosen is illustrated. First, the experimental database with the acquired signals for the different severities is split in two at each iteration. A first database used to build the hybrid model from 10% to 90% of the data of each severity. It is added to the total numerical database and used to train the two MLAs (KNN and Decision Trees). A second database is used for the test with MLAs ranging from 90% to 10% of the data for each severity contained in the experimental database.

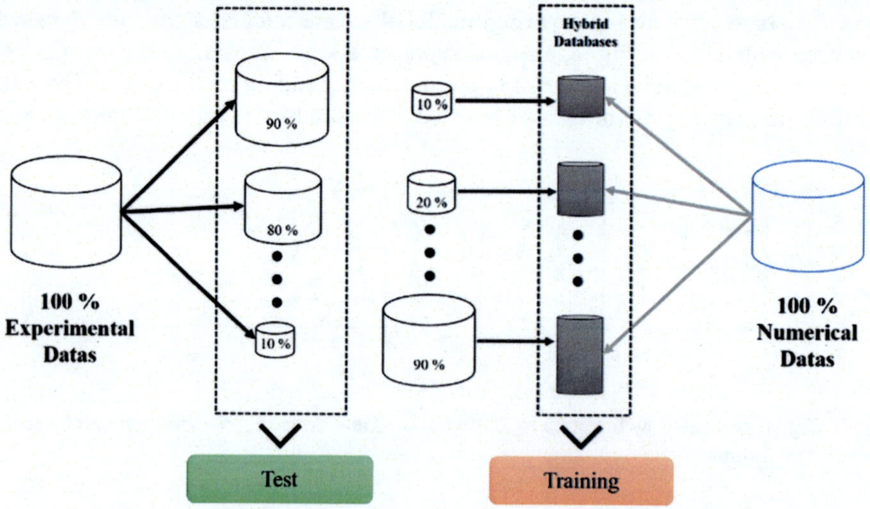

Fig. 9. Construction of the homogenous hybrid data-base methodology

3 Results and Discussions

Based on the hybridization model proposed in Sect. 2, In Fig. 10 are presented the accuracy of the test depending on the rate of experimental data added to the numerical database. As the contribution of experimental data increases, the accuracy of the diagnosis increases with the hybridisation method. With KNN algorithm, the accuracy average is over 92.3%, while, with the Decision Tree algorithm, the average accuracy is 88.1%. In Table 2, results at 10% and 50% experimental data used for hybridisation are compared, the results show that, at 10% accuracies are over 81.5% and 75% respectively for KNN and Decision Tree. With 50% of experimental data available the accuracies reached 95% and 100% respectively for KNN and Decision Tree.

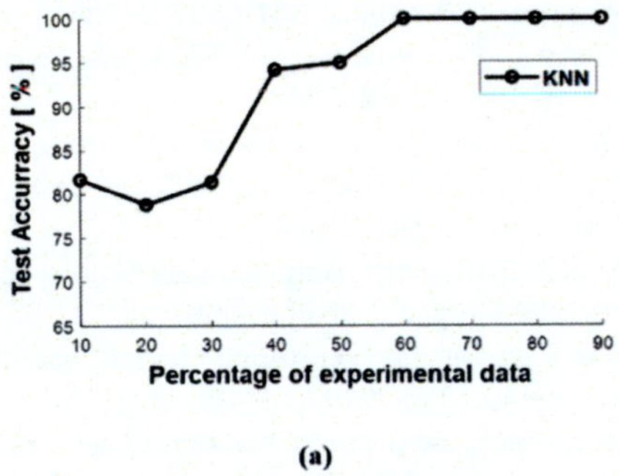

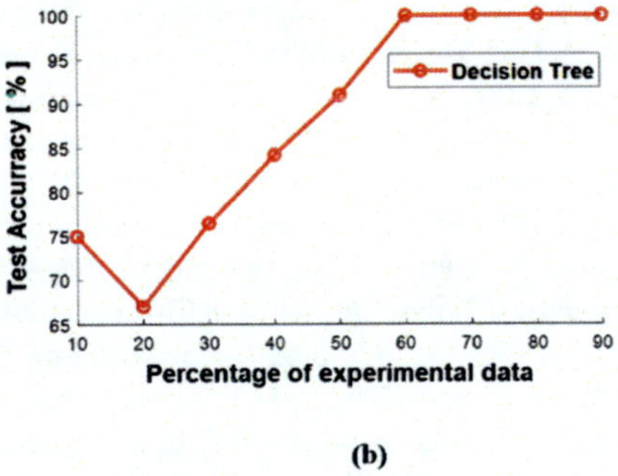

Fig. 10. Classification accuracy: (a) KNN; (b) Decision Tree

Table 2. Accuracy of classification for 10% and 50% of experimental data used

Accuracy		
Algorithm	10%	50%
KNN	81.5%	95%
Decision Tree	75%	91%

4 Conclusion

In this study, a numerical model of a test bench composed of bearings using a combination of a finite elements model and a discrete elements model is proposed. A diagnostic approach of bearing faults with different severities on the outer ring, using training on the hybrid database and the test performed on the rest of the experimental database, has been used. Two machine learning algorithms are trained and tested, with an accuracy averaging 92.3% for the KNN and 88.1% for the Decision Tree. To improve this result, in future work, the numerical model will be update on the test bench, using classical update techniques such as: eigenmode update, data update via extracted indicators and geometric parameter update.

Acknowledgements. This work is financed by the University of Reims Champagne-Ardenne and to the Region Grand Est.

References

Darong, H., Peng, W.: Grid-based DBSCAN algorithm with referential parameters. Phys. Procedia **24**, 1166–1170 (2012). https://doi.org/10.1016/j.phpro.2012.02.174

Fadda, M.L., Moussaoui, A.: Hybrid SOM–PCA method for modeling bearing faults detection and diagnosis. J. Braz. Soc. Mech. Sci. Eng. **40**(5), 1–8 (2018). https://doi.org/10.1007/s40430-018-1184-7

Farhat, M.H., et al.: Digital twin-driven machine learning: ball bearings fault severity classification. Meas. Sci. Technol. **32**(4), 044006 (2021). https://doi.org/10.1088/1361-6501/abd280

Fowler, J.E.: The redundant discrete wavelet transform and additive noise. IEEE Signal Process. Lett. **12**(9), 629–632 (2005). https://doi.org/10.1109/LSP.2005.853048

Guo, G., Wang, H., Bell, D., Bi, Y., Greer, K.: KNN model-based approach in classification. In: Meersman, R., Tari, Z., Schmidt, D.C. (eds.) OTM 2003. LNCS, vol. 2888, pp. 986–996. Springer, Heidelberg (2003). https://doi.org/10.1007/978-3-540-39964-3_62

Karacay, T., Akturk, N.: Experimental diagnostics of ball bearings using statistical and spectral methods. Tribol. Int. **42**(6), 836–843 (2009). https://doi.org/10.1016/j.triboint.2008.11.003

Li, Y., et al.: Dynamic prognostic prediction of defect propagation on rolling element bearings. Tribol. Trans. **42**(2), 385–392 (1999). https://doi.org/10.1080/10402009908982232

Ngandu Kalala, G.: Strategy for diagnosing gear defects by digital twin-assisted supervised classification on a ball mill drivetrain. In: proceedings of International Conference on Noise and Vibration Engineering. International conference on uncertainty in structural dynamics (ISMA-USD), Katholieke Univ Leuven (2022)

Parker, R.G., Vijayakar, S.M., Imajo, T.: Non-linear dynamic response of a spur gear pair: modelling and experimental comparisons. J. Sound Vibr. **237**(3), 435–455 (2000). https://doi.org/10.1006/jsvi.2000.3067

Patidar, S., Soni, P.K.: An overview on vibration analysis techniques for the diagnosis of rolling element bearing faults. Int. J. Eng. Trends Technol. (IJETT) **4**(5), 1804–1809 (2013)

Patil, M.S., et al.: A theoretical model to predict the effect of localized defect on vibrations associated with ball bearing. Int. J. Mech. Sci. **52**(9), 1193–1201 (2010). https://doi.org/10.1016/j.ijmecsci.2010.05.005

Paya, B.A., Esat, I.I., Badi, M.N.M.: Artificial neural network based fault diagnostics of rotating machinery using wavelet transforms as a pre-processor. Mech. Syst. Signal Process. **11**(5), 751–765 (1997). https://doi.org/10.1006/mssp.1997.0090

Reeves, S.J., Zhe, Z.: Sequential algorithms for observation selection. IEEE Trans. Signal Process. **47**(1), 123–132 (1999). https://doi.org/10.1109/78.738245

Samanta, B.: Gear fault detection using artificial neural networks and support vector machines with genetic algorithms. Mech. Syst. Signal Process. **18**(3), 625–644 (2004). https://doi.org/10.1016/S0888-3270(03)00020-7

Optimal Design Parameters of a Tuned Mass Damper for Offshore Wind Turbines

Helmi Mahmoudi[1,2], Maroua Hammami[1,3](✉), Nabih Feki[1,2], Olfa Ksentini[1,3], Mohamed Slim Abbes[1], and Mohamed Haddar[1]

[1] Laboratory of Mechanics, Modelling and Manufacturing (LA2MP), National Engineering School of Sfax, University of Sfax, Sfax, Tunisia
{MarouaHammami123,olfa.ksountini}@issig.u-gabes.tn, nabih.feki@issatso.u-sousse.tn, slim.abbes@enis.tn, Mohamed.haddar@enis.rnu.tn

[2] Higher Institute of Applied Science and Technology of Sousse, University of Sousse, Sousse, Tunisia

[3] Higher Institute of Industrial Systems of Gabes (ISSIG), University of Gabes, Gabes, Tunisia

Abstract. This article is dedicated to the development of a passive damping system for the reduction of vibrations of the Nacelle/Rotor of offshore wind turbines in the case of aerodynamic and hydrodynamic charges. The presence of vibration causes deflection at the top of the wind turbine structure, degradation of the blades and reduction of producing energy. A damping system using a passive damper, the Tuned Mass Damper (TMD), is designed and inserted in the Nacelle/Rotor of a monopile wind turbine (ref. NREL 5 MW). The dynamic behavior of the system is described by a differential equation of motion which is solved through the Newmark algorithm. The design parameters of the Tuned Mass Damper absorber are optimized by the interval method and the genetic algorithm method to achieve vibration attenuation. The performance of TMD is evaluated using the percentage reduction of the root mean square (RMS) of the displacement and acceleration of the Nacelle/Rotor.

Keywords: Tuned mass damper · offshore wind turbines · interval method · Genetic algorithm · vibration attenuation

1 Introduction

With the rapid growth of the offshore wind industry, the need for a reliable wind turbine tower is essential for the safety of the structure and power generation. As the wind resource is more stable (that means less turbulence) and more sustainable (that means more convertible wind resource) at higher altitudes (typically over 100 m), wind turbines grow to extract more energy. However, like all other technologies in the world, there is always a compromise to be made. The quest for ever larger turbines brings with it research challenges. To give an illustration, the tallest wind turbine tower suffers from very intensive loads directly from the airflow and indirectly from the nacelle.

A. Akrout et al. (Eds.): ICAV 2022, ACM 22, pp. 171–179, 2023.
https://doi.org/10.1007/978-3-031-34190-8_20

The rotor/nacelle system is subjected to wind blade loads which can often create vibrations or impose an ultimate change in load amplitude. In addition the wind turbines implemented in the sea are subject to hydrodynamic loads which can cause the bending of the tower and influence the stability of the nacelle.

The vibrations and ultimate loads transferred to the offshore wind turbine reduce the lifetime of the tower and lead to many other problems such as cracks due to fatigue damage and influence on the proper functioning of the different components constituting the nacelle (gearbox, generator shaft…). To overcome this research challenge, a damping system integrated inside the wind turbine offers a suitable solution, in particular a passive damping system that can reduce vibrations.

The installation of a tuned mass damper (TMD) system for optimal performance consists of optimizing the mass and stiffness while minimizing the vibration of the whole system caused by wind and wave charges.

The first theoretical study of tuned mass dampers (TMDs) was conducted by (Hartog and Ormonyard 1928). In his work "Mechanical Vibrations," initially published in 1940, Hartog (Hartog 1956) discovered optimal absorber settings. Passive, semi-active, and active control systems have been devised (Hartog and Ormonyard 1928, Hartog 1956). In the last decade, passive control of offshore wind turbines has been a major research topic. Murtagh et al. (Murtagh et al. 2008) investigated the use of a passive tuned mass damper (TMD) to control wind turbines along wind vibration.

As the performance of a TMD depends on its parameters, including mass, stiffness and damping, different methods have been proposed to determine the optimal value of TMD parameters according to different design criteria. Some methods are proposed for linear structures, considering the minimization of the root mean square (RMS) value of the displacement and acceleration of the structure. Gerges et al. (Gerges and Vickery 2005) identified the best design formula for planar pendulum tuned mass dampers (PTMDs) under white noise wind and seismic excitations. Roffel et al. (Roffel et al. 2013) developed the optimum design formula for the PTMD to account for the space motion of the PTMDs, to minimize the acceleration root mean square (RMS) of slender tall structures under wind loading.

Metaheuristic methods such as genetic algorithm (GA) have been applied to solve different optimisation problems. A wide application of a genetic algorithms for TMD parameter setting has been done in the studies of Hadi and Arfiadi (Hadi and Arfiadi 1998) and Singh et al. (Singh et al. 2002).

In this paper, optimal design parameters for a passive TMD should be found. Low vibration rates are achieved throughout the decreasing displacement and acceleration of the system. A dynamic model is proposed for the offshore wind turbine is proposed. Two optimization methods which are the interval method and the genetic algorithm were used to get an efficient TMD.

2 Structural model

The National Renewable Energy Laboratory (NREL) of the U.S. Department of Energy (DOE) establishes the detailed specifications of multimegawatt turbines that are suitable in deep waters ((Jonkman 2009), (Bir and Jonkman 2008). In this work, the "NREL offshore 5-MW baseline wind turbine" is chosen and its characteristics are summarized in Table 1. Figure 1 is a schematic of the NREL offshore 5-MW wind turbine configuration analyzed in this paper.

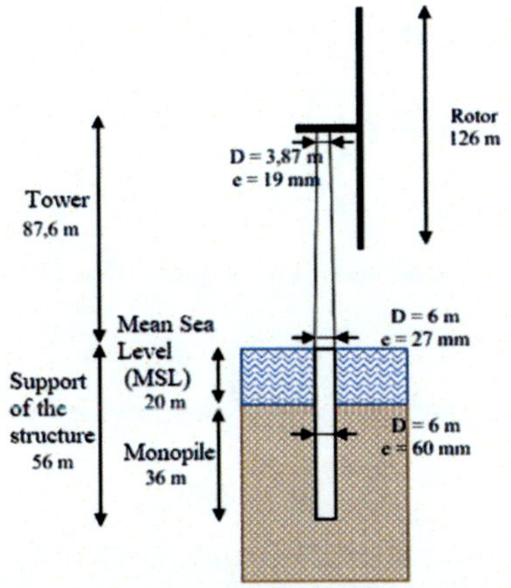

Fig. 1. Schematic configuration of the NREL offshore 5-MW wind turbine

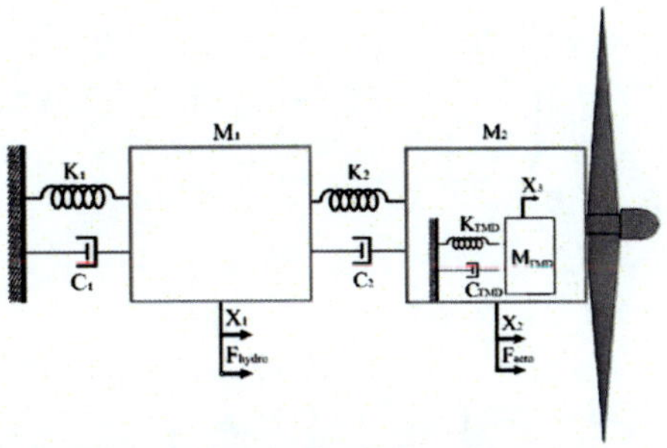

Fig. 2. Three degrees of freedom dynamic model including the linear absorber (TMD)

A three-degree-of-freedom model is proposed for the monopile Offshore Wind Turbine (OWT) design to control the excessive vibrations in the nacelle/rotor components. Figure 2 demonstrates the structural model of the OWT with TMD installed in the nacelle. The model parameters are illustrated as follows:

- M_1, M_2, M_{TMD} represent the mass of the tower system, the Rotor/Nacelle system and the tuned mass damper system.
- K_1, K_2 and K_{TMD} represent the spring stiffnesses of the tower system, the Rotor/Nacelle system and the tuned mass damper.
- C_1, C_2 and C_{TMD} represent the damping coefficients.

Table 1. Characteristics of the NREL 5MW reference wind turbine ((Jonkman 2009), (Bir and Jonkman 2008)).

	Diameter	Thickness	Length	Mass	Stiffness
Unit	*m*	*mm*	*m*	*Kg*	*N/m*
Tower	Top: 3,87 Bottom: 6	Top: 19 Bottom: 27	87,6	347460	15×10^6
Support of the structure	6	60	56	489070	–
Nacelle/Rotor	126	–	–	350000	4 × 106

The equation of motion of this system is given by:

$$[M]\{\ddot{x}\} + [C]\{\dot{x}\} + [K]\{x\} = \{F_{ext}\} \tag{1}$$

$\{x\} = \{X_1, X_2, X_3\}^T$, $\{x\}$, $\{\dot{x}\}$ and $\{\ddot{x}\}$ represent the generalized displacement, velocity and acceleration vectors respectively; $[M]$ is the mass matrix of the system with $[M] = \begin{bmatrix} M_1 & 0 & 0 \\ 0 & M_2 & 0 \\ 0 & 0 & M_{TMD} \end{bmatrix}$; $[C]$ is the damping matrix of the system with $[C] = \begin{bmatrix} C_1 + C_2 & -C_2 & 0 \\ -C_2 & C_2 + C_{TMD} & -C_{TMD} \\ 0 & -C_{TMD} & C_{TMD} \end{bmatrix}$; $[K]$ is the stiffness matrix of system with $[K] = \begin{bmatrix} K_1 + K_2 & -K_2 & 0 \\ -K_2 & K_2 + K_{TMD} & -K_{TMD} \\ 0 & -K_{TMD} & K_{TMD} \end{bmatrix}$ and $\{F_{ext}\}$ is the external force vector. applied to the system where $\{F_{ext}\} = \begin{Bmatrix} F_{hydro} \\ F_{aero} \\ 0 \end{Bmatrix}$: These forces are considered as harmonic excitations.

The proposed system may be integrated the time domain using a Newmark method, with parameters $\beta = 1/6$ and $\gamma = 1/2$, thus ensuring an unconditionally stable system.

2.1 Environmental Loads Applied on the Offshore Wind Turbine

The Floating Offshore Wind Turbine (OWT) Installation is subjected to several environmental conditions like wind action, hydrodynamic forces, earthquakes and ice and snow accumulation.

2.1.1 Aerodynamic Forces

The towers as well as the rotor blades are induced by airflow. The following equation is to be used for calculating the wind drag forces on the exposed structural components of the OWT.

$$F_{aero} = F_{rotor} + F_{tower} \tag{2}$$

The rotor aerodynamic load generated by wind perpendicular to the swept area of the blades can therefore be written as follow:

$$F_{rotor} = \frac{1}{2}\rho_a S C_t(\lambda)\, V_{ref}^2 \tag{3}$$

While ρ_a is the air mass density [kg/m^3], S is the swept area of the blades [m^2], C_t is the thrust coefficient, V_{ref} is the mean wind speed at hub height [m/s] and λ is the coefficient of the specific speed (see Eq. 4).

$$\lambda = \frac{\omega_{rotor} R_{rotor}}{V_{ref}} \tag{4}$$

Based on the guide for building and classing floating offshore wind turbine installations (ABS 2014), the aerodynamic load applied on the tower defines the wind loading normal to the considered surface can be described as follow:

$$F_{tower} = \frac{1}{2}\rho_a S\, C_s\, V_{ref}^2 \tag{5}$$

This equation of the wind load is substantially similar to that described in Eq. (3). Only two parameters were modified which are S is the tower projected area of windage on a plane normal to the direction of the considered force and the C_s is the shape coefficient which is equal to 0,5 for cylindrical shape.

2.1.2 Hydrodynamic Forces

The tower is the submerged part of the offshore wind turbines that will be subject to hydrodynamics forces caused by wave and sea currents. Morison's equation (Van Der Tempel 2006) is used to calculate hydrodynamic loading applied on the tower which is induced by the waves. The total hydrodynamic load on the cylinder is the sum of drag and inertia load as shown in the equations below.

$$F_{Morrison} = F_{inertia} + F_{drag} \tag{6}$$

with; $\begin{cases} F_{inertia} = \frac{\pi}{4}\,\rho_{water}\, C_i\, D^2\, \dot{v} \\ F_{drag} = \frac{1}{2}\,\rho_{water}\, C_d\, D|v|v \end{cases}$

Considering that C_i is the hydrodynamic inertia coefficient [-], C_d is the hydrodynamic drag coefficient, v is the water particle velocity [m/s], $\dot{v}$ is the water particle acceleration [m/s^2], ρ_{water} is the density of water [kg/m^3] and D is the diameter of the cylinder section [m].

For a monopile based wind turbine, the typical values for the inertia and drag coefficients are $C_i = 2.0$ and $C_d = 0.7$ ([(Van Der Tempel 2006), (Kühn 2012)). Furthermore, in Morison's Eq. (6), only waves are taken into account, to consider the effects of the current wave, the velocity, as well as the acceleration of the current wave, must be added which gives the following equations.

$$\begin{cases} F_{inertia} = \frac{\pi}{8\,K_{wave}}\,\rho_{water}\, C_i\, D^2\, H_{wave}\, \omega_{wave}^2 \\ F_{drag} = \frac{\pi}{16\,K_{wave}}\,\rho_{water}\, C_d\, D\, H_{wave}^2\, \omega_{wave}^2 \times \frac{sh(2\,K_{wave}\, d_{water}) + 2\,K_{wave}\, d_{water}}{ch(2\,K_{wave}\, d_{water}) - 1} \end{cases} \tag{7}$$

where K_{wave} is the wave number equal to $2\pi/\lambda_{wave}$ [m^{-1}], d_{water} is the water depth [m], H_{wave} is the wave amplitude [m] and ω_{wave} is the pulsation of the wave [rad/s].

Since considering the phase angle of each force, the overall hydrodynamic force is defined as (Kühn 2012):

$$F_{hydro} = F_{drag}|\cos(\omega_{wave}\, t)|\cos(\omega_{wave}\, t) - F_{inertia}\sin(\omega_{wave}\, t) \tag{8}$$

3 Optimization Design Methods

The design optimization of the TMD parameters has an objective to minimize the RMS of the displacement and the RMS of the acceleration of the Nacelle/Rotor system.

3.1 Interval Method Optimization

A numerical research approach is used to obtain the optimal design which consists of mass and stiffness that ensures minimum vibration. For this purpose, the root mean square (RMS) values of the displacement and acceleration are displayed to provide us with the values of the dynamic parameters ensuring reduced acceleration and displacement, because the attenuation of the displacement and acceleration response is the most important concern for wind turbines. A mono-objective approach is used based on an iterative interval method, used for each objective separately which is the RMS of the acceleration reduction or the RMS of the displacement reduction. In Figs. 3 and 4, the dark blue area is the area that contains the optimal values to be used to achieve the best efficiency of the TMD and to minimise vibration of the Nacelle/Rotor system.

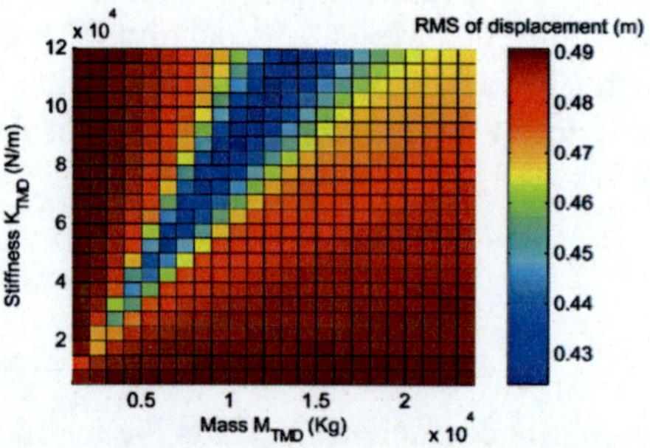

Fig. 3. RMS of the displacement of the Nacelle/Rotor system using different mass and stiffness values of TMD

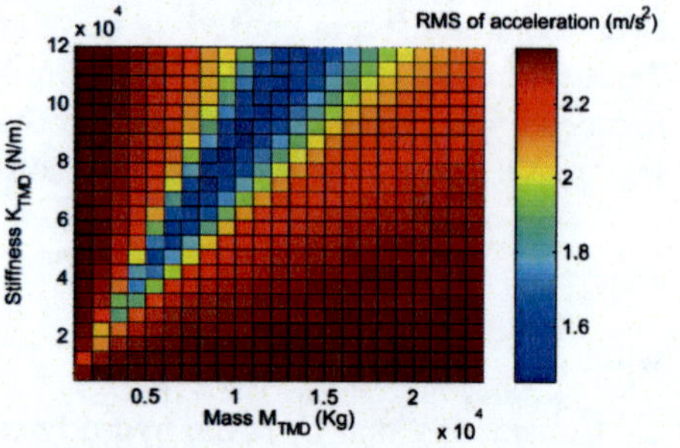

Fig. 4. RMS of the acceleration of the Nacelle/Rotor system using different mass and stiffness values of TMD

3.2 Genetic Algorithm Optimization

The GA optimization is carried out by hierarchically subdividing the parameters of the TMD in primary and secondary design TMD parameters. An in-house built GA toolbox computes the TMD optimal parameters of the 3DOF model. Then optimal couples of values for M_{TMD} and K_{TMD} are found.

The Pareto front for the introduced issue is provided by Multi-Objective Genetic Algorithm (MOGA). The minimum distance selection method (TMDSM) (Sun et al. 2011) is used to determine the most satisfactory choice from the Pareto front. The 'knee point' shown in Fig. 5, indicates the closest solution to the utopia point among the others Pareto results. The coordinates of the knee point represent the optimal values of both RMS of acceleration and displacement. This latter point is associated to the optimal TMD parameters which are $M_{TMD} = 10620.4233$ Kg and $K_{TMD} = 91467.3267$ N/m.

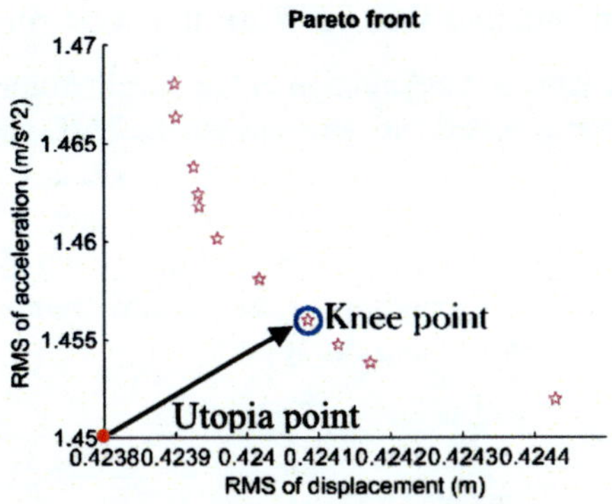

Fig. 5. Pareto front for a multi-objective optimization

4 Results and Discussions

4.1 Frequency Response with and Without Vibration Absorber

The difference in response between the structural systems with and without TMD represented in the previous two models can be visualized in the graphs in Figs. 6, and 7, which present the root mean square (RMS) values of displacement and acceleration for the stabilized harmonic nacelle/rotor system respectively. Based on these results, the reduction in the resonant amplitude of the displacement and acceleration is clear.

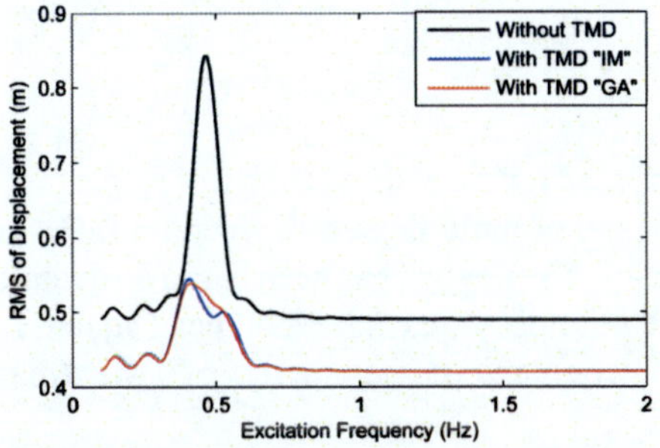

Fig. 6. Root mean value of displacement of the Nacelle/Rotor system against frequency

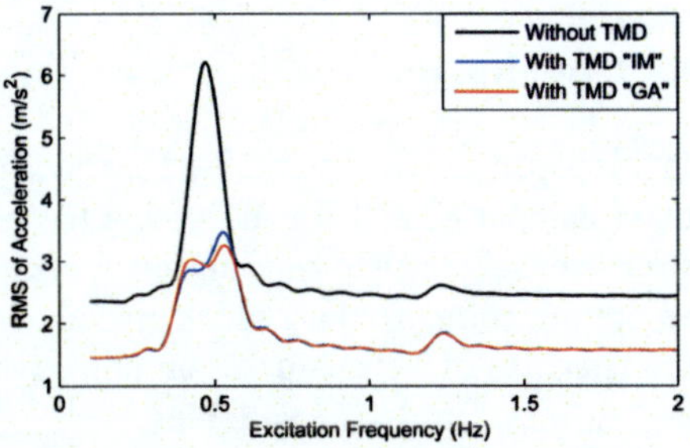

Fig. 7. Root mean value of acceleration of the Nacelle/Rotor system against frequency

For the choice of TMD parameters based on "IM" the root mean square values of displacement are reduced by almost 39% and the root mean square values of acceleration are reduced by almost 51% for the exciting frequency of 0.46 Hz. The RMS of displacement is decreased from 0.84 m to 0.5053 m and the RMS of the acceleration is reduced from 6.131 m/s^2 to 3.02 m/s^2. For the choice of TMD parameters based on "GA", the root mean square values of displacement are reduced by almost 40% and the root mean square values of acceleration are reduced by almost 52% for the exciting frequency of 0.46 Hz. The RMS of the displacement is decreased from 0.84 m to 0.5035 m and the RMS of the acceleration is reduced from 6.131 m/s^2 to 2.92 m/s^2.

4.2 Dynamic Response with and Without Vibration Absorber

To track our objective of vibration attenuation of the nacelle/rotor system, the dynamic responses of the system were plotted with and without the TMD absorber in acceleration (see Fig. 8) and the Fast Fourier transform (FFT) of the output acceleration signal (see Fig. 9). It can be seen that there are periodic oscillation packets with smaller maximum amplitudes, and relatively faster attenuation in the 3ddl model (With TMD). Figure 9 shows that the amplitudes corresponding to the natural frequencies are potentially reduced by the absorber implemented in the system.

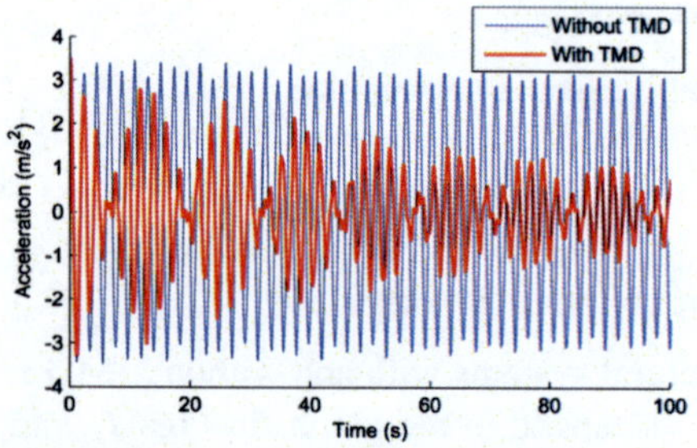

Fig. 8. Dynamic response of the output acceleration signal without TMD and with TMD

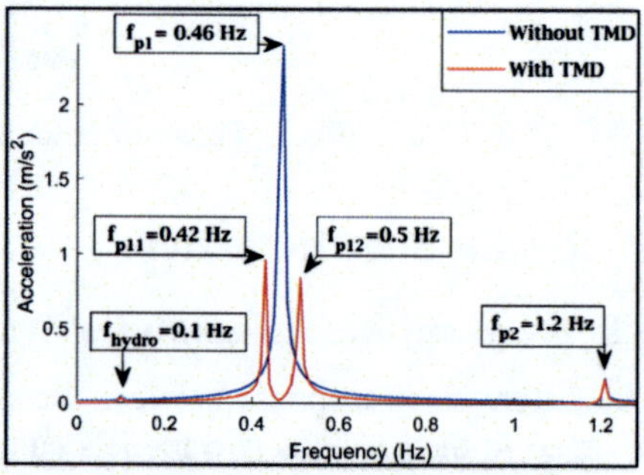

Fig. 9. Fast Fourier transform (FFT) of the output acceleration signal without TMD and with TMD.

5 Conclusion

In this paper, an effective method has been proposed for the optimal design of single tuned mass damper STMD for mitigating the response of multi degree of freedom (MDOF) structures subjected to wind and wave excitations. The method has been based on defining an optimization problem which considers the minimization of the root mean square for the response of a structure as an objective function and the parameters of TMD (mass and stiffness) as design variables.

1. The Interval method has been used to approach the optimal design parameters for a single objective (RMS of displacement, RMS of acceleration). This method has minimized the RMS of displacement value by 39% and the RMS of acceleration by 51%.
2. Multi objective genetic algorithm has been used to minimize the RMS of displacement and acceleration simultaneously. This method has minimized the RMS of displacement value by 40% and the RMS of acceleration by 52%.
3. The MOGA outperformed the iterative IM method, by optimizing the design based on a multi-objective approach likewise it enhanced the percentage of the vibration attenuation.

An acceptable reduction of vibration is achieved by the passive control strategy (TMD) with a plausible choice of design parameter based MOGA algorithm.

References

Ormondroyd, J., Den Hartog, J.P.: The theory of dynamic vibration absorber. Trans. ASME, APM **50–7**, 9–22 (1928)

Den Hartog, J.P.: Mechanical Vibrations, 4th edn. McGraw-Hill, New York (1956)

Murtagh, P.J., Ghosh, A., Basu, B., Broderick, B.M.: Passive control of wind turbine vibrations including blade/tower interaction and rotationally sampled turbulence. Wind Energy: Int. J. Prog. Appl. Wind Power Convers. Technol. **11**(4), 305–317 (2008). https://doi.org/10.1002/we.249

Roffel, A.J., Narasimhan, S., Haskett, T.: Performance of pendulum tuned mass dampers in reducing the responses of flexible structures. J. Struct. Eng. **139**(12), 04013019 (2013). https://doi.org/10.1061/(ASCE)ST.1943-541X.0000797

Gerges, R.R., Vickery, B.J.: Optimum design of pendulum-type tuned mass dampers. Struct. Design Tall Spec. Build. **14**(4), 353–368 (2005)

Hadi, M.N.S., Arfiadi, Y.: Optimum design of absorber for MDOF structures. J. Struct. Eng.-ASCE **124**(11), 1272–1280 (1998)

Singh, M.P., Singh, S., Moreschi, L.M.: Tuned mass dampers for response control of torsional buildings. Earthquake Eng. Struct. Dynam. **31**(4), 749–769 (2002). https://doi.org/10.1002/eqe.119

Jonkman ,J., Butterfield, S., Musial, W., Scott, G.: Definition of a 5-MW reference wind turbine for offshore system development. National Renewable Energy Laboratory NREL/TP-500- 38060 (2009)

Bir, G., Jonkman, J.: Modal dynamics of large wind turbines with different support structures. National Renewable Energy Laboratory NREL/CP-500–43045 (2008)

ABS, American Bureau of Shipping, Guide for building and classing offshore wind turbine installations (2014)

Van Der Tempel, J.: Design of support structures for offshore wind turbines. Dissertation, Delft technical university (2006)

Kühn, M.: Wind Power Plants: Fundamentals, Design, Construction and Operation, pp.520–539. Springer, Cham (2012)

Sun, G., Li, G., Zhou, S., Li, H., Hou, S., Li, Q.: Crashworthiness design of vehicle by using multiobjective robust optimization. Struct. Multidiscip. Optim. **44**(1), 99–110 (2011). https://doi.org/10.1007/s00158-010-0601-z

The Dependence of the Characteristics of the Dispersion Curves on the Orientation Angle of the CARALL Structures

Driss Hana[1,2](✉), El Mahi Abderrahim[1], Bentahar Mourad[1], Beyaoui Moez[2], and Haddar Mohamed[2]

[1] Laboratoire d'Acoustique de l'Université du Maine (LAUM), UMR CNRS 6613, Université du Maine, Av. O. Messiaen, 72085 Le Mans Cedex 9, France
Hana.Driss.Etu@univ-lemans.fr
[2] Laboratoire de Mécanique, Modélisation et Production (LA2MP), Ecole Nationale d'Ingénieurs de Sfax (ENIS), Université de Sfax, Route de Soukra, 3038 Sfax, Tunisia

Abstract. The purpose of this study is to present a parametric analysis of how the fiber orientation angle affects elastic wave dispersion relations in the case of the CARALL composite. The CARALL composite is made up of multiple plies of carbon fiber and aluminum plies bonded together. The material consists of 14 layers, where the layers are constructed from aluminum alloy and carbon/epoxy resin fibers. The overall configuration of the CARALL composite is [Al, θ5, Al]s, where the angle parameter is varied between 0° and 90°. The stiffness matrix approach is used to determine the dispersion curves, and the "Dispersion Calculator" computer program is employed in the analysis. The results indicate that the difference in fiber orientation angle significantly affects the shift in phase velocity between the primary symmetric mode and the horizontal shear mode. However, the sensitivity of the other modes to the angle is negligible. This study provides valuable insights into the impact of fiber orientation angle on elastic wave dispersion relations in CARALL composites and could inform future research on composite material design and optimization.

Keywords: Dispersion curve · CARALL · Lamb wave

1 Introduction

Existing analytical and semi-analytical modeling approaches exist for analyzing guided wave propagation in anisotropic composites such as multi-layer wind turbine blade assemblies and tidal plant hydrofoils. Dispersion curves, which characterize the frequency dependence of guided wave velocities in a certain structure, must be computed to measure wave propagation velocity.

[Muc et al. 2021] used neutron scattering and ultrasonic techniques to measure the propagation velocities of acoustic waves through the alloy. They then used these measurements to determine the elastic constants of the material. The results of the study showed that the acoustic wave propagation velocities in the alloy varied with the direction

A. Akrout et al. (Eds.): ICAV 2022, ACM 22, pp. 180–188, 2023.
https://doi.org/10.1007/978-3-031-34190-8_21

of propagation and the polarisation of the waves. The authors also determined that the elastic constants of the material varied as a function of the carbon concentration in the alloy. [Sasso et al. 2019] This study provides a detailed characterization of the mechanical behavior of a CARALL FML composite material at high strain rates. The results could be useful for the design of impact-resistant structures, particularly in aerospace and defense applications. [Bellini et al. 2019]This study provides a detailed assessment of the performance of carbon and aluminum fibre-reinforced metal laminates as a function of their composition and structure. The results could be useful for the design of lightweight and strong structures in a variety of applications, such as aerospace, automotive, and construction.

The majority of dispersion curves are computed using analytical or semi-analytical finite element techniques. [Nayfeh 1991] investigated the propagation of harmonic elastic waves in n-layer anisotropic plates. The solutions for every layer were identified using the transfer matrix method and expressed in terms of wave amplitudes. The wave amplitudes are replaced by stresses and displacements at the interfaces between the layers, and the complete transfer matrix is produced [Nayfeh 1991]. To take into consideration the attenuation and change in propagation direction brought on by the composite's anisotropic structure, Hosten and Castaings [1993] modified the transfer matrix formula. According to the transfer matrix technique, four waves with the same frequency and spatial characteristics should occur at each interface at the border of each layer of the N-layer composite laminate. An iterative root-seeking approach can yield an endless number of wave numbers at each frequency. To use matrix approaches for anisotropic layer simulation, the matrix must be expanded to six dimensions [Lowe 1995]. The disadvantage of this strategy is that it gets unstable when the FD is huge (frequency multiplied by the plate thickness). [Knopoff 1964] advocated the use of the global matrix approach for high frequencies or thick plates. However, when the number of components increases, the matrix gets more complicated, and the approach becomes slower [Lowe 1995; Knopoff 1964].

A0, SH0, and S0 modes refer to different types of wave modes that can propagate in a material. These modes are important for the study of dispersion curves, which describe how the speed of these waves depends on their frequency and direction of propagation. The A0 mode is a longitudinal wave mode. This mode is characterized by the movement of the material particles moving in phase with each other in the direction of propagation of the wave. This means that the particles expand and contract in unison as the wave passes through the material.

The SH0 mode is a transverse wave mode also known as the "horizontal shear" mode. The SH0 mode is a transverse wave mode also known as the "horizontal shear" mode. In this mode, the motion of the particles in the material is perpendicular to the direction of wave propagation. This means that the particles move from side to side as the wave travels through the material, with no up-and-down movement. The SH0 mode is useful for studying the shear properties of a material. The S0 mode is also a transverse wave mode, but it is known as the "shear" or "Rayleigh" mode. In this mode, the motion of the particles in the material is both perpendicular to the direction of wave propagation and in the plane of the material surface. This means that the particles move circularly as the wave passes through the material.

The S0 mode is useful for studying the surface properties of a material. Overall, understanding the different modes that can propagate in a material is important for studying its mechanical and acoustic properties, and for designing and optimizing materials for specific applications. The dispersion curves of these modes can provide valuable information about how the material responds to stress and strain, and can also be used to identify changes in the structure of the material.

For the computation of dispersion curves, the global matrix model has been incorporated in the commercial program "DISPERSE" [Pavlakovic et al. 1997]. Nayfeh and Chimenti [1988] developed the equations of motion for a transverse isotropic plate connected to a fluid and offered a continuous mixing theory for uniaxial fibrous composites with transverse isotropy. For a fluid-coupled composite plate, total transmission curves were calculated. In the instance of circumferential wave propagation, Towfighi et al. used coupled differential equations to resolve the issue of generating dispersion curves for curved anisotropic plates. Based on the Fourier series expansion of unknown values, they presented a method for an organized and comprehensive solution [Towfighi et al. 2002]. Karpfinger et al. (2008) proposed a spectrum-based approach that employs spectral differentiation matrices to discretize the underlying wave equations and solves the relevant equations as a generalized eigenvalue problem. The eigenvalues correspond to the wave numbers of the different modes at a given frequency. The advantage of this strategy is that it solves the generalized eigenvalue problem without the need for special functions.

As a result, it is straightforward to employ in circumstances where traditional root-finding methodologies are highly limited or difficult to apply due to attenuating, anisotropic, or poroelastic media.

2 Techniques for Generating Dispersion Curves

There are three methods to plot the dispersion curves: the first is based on numerical simulations, the second on data collected by experiment, and the third on a specialized coding program like Matlab.

[Sorohan et al. 2011] describe a method for getting all dispersion curves using simple numerical simulation and typical commercial finite element software. The approach is simply a series of modal assessments for a representative piece of the analyzed structure. Hora and Ervená [2012] provide Fourier transform methods for obtaining Lamb wave dispersion curves (FT). Propagating Lamb waves are sinusoidal in both the frequency and spatial domains. As a consequence, the temporal FT may be used to transition from the time domain to the frequency domain, and then the spatial FT can be used to transition to the frequency-wavenumber domain, where individual mode amplitudes and wavenumbers can be measured. [Schöpfer et al. 2013] provide a method for calculating dispersion curves using laser vibrometer measurements. After Fourier transforms the measurement data into the wavenumber domain, the matrix pencil technique is applied.

Harb and Yuan [2015] present a non-contact hybrid device for profiling the A0 Lamb wave dispersion of an isotropic aluminum plate that consists of an air-coupled Transducer (ACT) and a Laser Doppler Vibrometer (LDV). The ACT produces ultrasonic pressure on the plate's surface. Part of the pressure waves is re-fracted into the plate. The LDV is used to compute the out-of-plane velocity of the excited Lamb wave mode along the plate at various distances.

[Packo et al. 2014] demonstrate how to calculate dispersion curves and analyze numerical models for directed waves. The proposed technique utilizes the wave equation and through-thickness-only discretization of anisotropic, multilayer plates to determine the Lamb wave properties. An alternative strategy is proposed by [Honarvar et al. 2009]. [Honarvar et al. 2009] propose an alternative method for deriving the frequency equation's solution in the form of dispersion curves from its three-dimensional representation. To begin, a three-dimensional illustration of the real roots of the frequency equation is displayed. The dispersion curves, which are the numerical solutions to the frequency equation, are obtained by making a correct cut in the velocity frequency plane. Many researchers employ iterative strategies to solve frequency equations, such as linear Schwab and Knopoff [1970] and [Mal 1988] or quadratic Schwab and Knopoff [1970] interpolation [Haskell 1953] and [Press et al. 1961] or extrapolation [Lowe 1995] algorithms that are extremely quick on a single root. When two roots are near each other, such as at the intersections of longitudinal mode dispersion curves, the function changes sign twice, rendering such schemes unstable. To solve the frequency equations, slower but safer iteration techniques such as Newton-Raphson, Bisection, and Mueller [Press et al. 1987] might be utilised. These techniques, however, are difficult, slow, and time-consuming due to the huge number and diversity of operations, particularly in multilayered media [Honarvar et al. 2009].

3 Investigated CARALL Structure

The dispersion curves for the CARALL composite of the following general configuration of corner layers, namely [Al, +θ, +θ, +θ, +θ, +θ Al]s, are calculated in this study. The composite under consideration has 14 layers, four of which are AluminumAlloy1100 [Callister 2002] and the remainder are Carbon/Epoxy prepreg [Rokhlin et al. 2011]. Table 1 shows the mechanical and physical parameters of the materials employed. CARALL's entire thickness is constant and equal to t = 10 [mm]. The aluminum layers have a thickness of tAL = 4 [mm], while the carbon/epoxy layers have a thickness of tCE = 6 [mm].

Table 1. Mechanical properties of the used materials

Material	E1 [GPa]	E2 [GPa]	G12 [GPa]	υ12	υ23	ρ [g/cm^3]
AluminumAlloy1010	69	69	25.9	0.33		2710
Carbon/Epoxy	150.95	12.80	8	0.46	0.45	1610

Figure 1 depicts a multicouche composite material with a local (stratified) coordinate system (x′1, x′2, x′3) and a global coordinate system (x1, x2, x3). Assume that the elastic wave propagates along the axis x1 of the global coordinate system. The angle specifies the orientation of the fiber as well as the local coordinate system (layer) for each layer

(Fig. 1). The dispersion relationships are computed for the following coordinates: θ: = 0°, 30°, 45°, 60° et 90°.

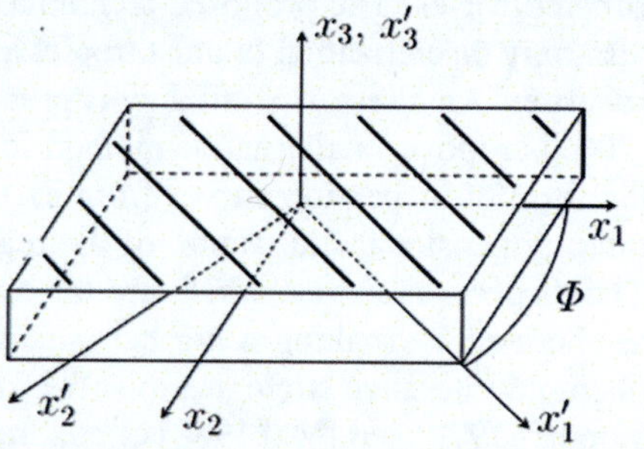

Fig. 1. Layered plate for investigated CARALL with local (x′1, x′2, x′3) and global coordinate systems (x1, x2, x3), where x1 is the direction of elastic wave propagation.

4 Results

4.1 Numerical Examples

In this part, we give our calculations on the Al-fiber matrix (carbon/epoxy) system using the Dispersion Calculator program. Eq. (1) shows the stiffness matrix and the density of carbon-epoxy at 1610 kg/m^3 (in GPa) is given by [Rokhlin et al. 2011].

$$C_{\frac{cabon}{epoxy}} = \begin{pmatrix} 162 & 11.8 & 11.8 & 0 & 0 & 0 \\ & 17 & 8.2 & 0 & 0 & 0 \\ & & 17 & 0 & 0 & 0 \\ & sym & & 4.4 & 0 & 0 \\ & & & & 8 & 0 \\ & & & & & 8 \end{pmatrix} \quad (1)$$

First, it is shown how to calculate the various mode families on laminates with a thickness of 10 mm of [Al,0$_5$,Al]s and [Al,90$_5$,Al]s. Then, to demonstrate the method's capability on the orientation angle variation, the dispersion curves corresponding to [Al,30$_5$, Al]s, [Al,45$_5$,Al]s, [Al,60$_5$,Al]s are computed. Wave propagation is always along 0 for coherence considerations.

4.2 Dispersion Diagrams

Figures 2 and 3 illustrate the dispersion curves produced for fiber orientation angles of 0° and 90°, respectively. The assumed frequency f and phase velocity Vp ranges are 0 [kHz] ≤ f ≤ 200 [kHz] and 0 [m/ms] ≤ Cp ≤ 20 [m/ms], respectively. The phase velocity of the antisymmetric fundamental mode A0 (bending wave) is substantially dependent on frequency in both circumstances, as illustrated for the comparatively low-frequency f < 30 [KHz]. As a result, this mode is quite dispersive. The basic horizontal shear mode SH0, on the other hand, is essentially constant. In both cases, the phase velocity of the SH0 mode Vp = 2.733 [m/ms] up to frequency f = 200 [KHz]. For = 0° and = 90°, the phase velocity Vp of the fundamental symmetric wave mode S0 (pressure or longitudinal wave) is equal to (f = 10 [KHz]) Vp = 7.51 [m/ms] and Vp = 4.35 [km/s]. The initial greater shear horizontal mode SH1 appears at the same frequency f 80 [KHz] in both laminates.

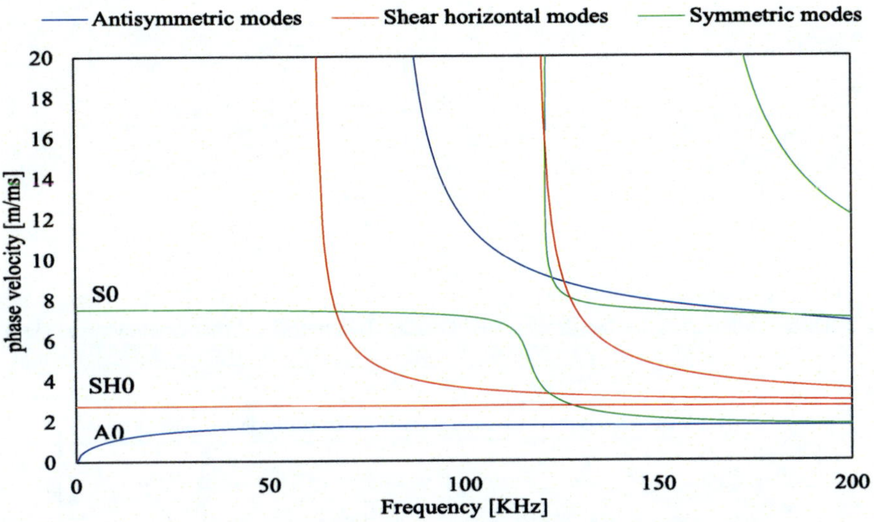

Fig. 2. Dispersion curves for the investigated CARALL composite "fiber orientation angle $\theta = 0°$"

Figure 4 displays the fundamental elastic wave modes A0, SH0, and S0 for 30°, 45°, and 60° fiber orientation angles. To improve clarity, the frequency range has been reduced to 0 [kHz] ≤ f ≤ 80 [kHz]], with just the first upper modes visible. In practice, the remaining higher wave modes are insignificant. The introduction of higher modes significantly impedes adequate comprehension of the structure's dynamic response.

The phase velocity range vp does not change. As can be shown, the fiber orientation angle has no effect on the phase velocity of the fundamental antisymmetric mode A0. At = 60°, however, the phase velocity Vp of the fundamental horizontal shear mode SH0 initially increases from Vp = 3.05 [km/s] to Vp = 4.4 [km/s]. As the fiber orientation angle rises, the phase velocity of the fundamental mode S0 decreases monotonically.

It should be noted that the frequency of coupling at which the initial horizontal shear mode SH0 occurs does not change appreciably across all parameter values studied, and the phase velocity is largely indifferent to the frequency f. As can be seen, changes in the orientation of the fiber have no influence on the speed of phase of the fundamental mode A0. The fundamental mode has the greatest influence on the phase velocity Vp; when the parameter value increases, the phase velocity increases monotonically. As previously stated, the effect of frequency fluctuation is very small up to a specific

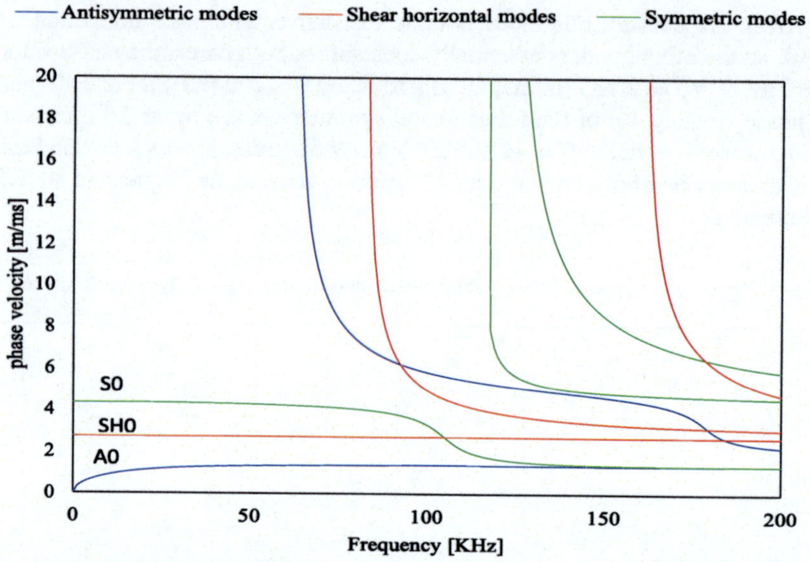

Fig. 3. Dispersion curves for the investigated CARALL composite "fiber orientation angle θ = 90°"

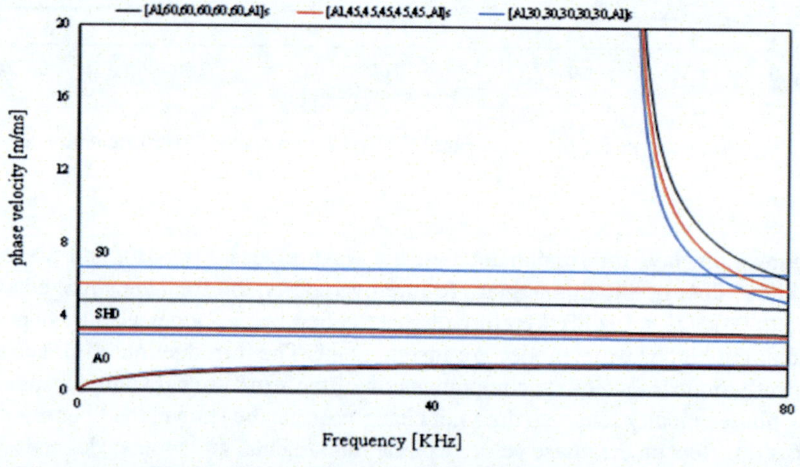

Fig. 4. Dispersion curves (fundamental modes A0, SH0, S0) and fiber orientation angles θ = 30°, 45°, and 60° were produced for the examined CARALL composite.

frequency; nevertheless, the larger the value of the fiber orientation angle, the greater the influence on frequency.

In the case of laminated composites, the different material layers have different mechanical properties, such as modulus of elasticity, and elastic wave propagation velocity. The dispersion curves of laminated composites depend on the stacking sequence because the stacking sequence affects the elastic properties of the composite, which in turn affects the dispersion curves. This can lead to differences in the travel time of elastic waves through the different layers, which can affect the dispersion curve. Therefore, the stacking sequence is an important factor to consider when analyzing the dispersion curves of laminated composites.

5 Conclusions

The current study looks at how the fiber orientation angle affects the dispersion relationships of the CARALL composite. The examined composite material has 14 layers, four of which are aluminum alloy and the remaining is prepreg Carbon/epoxy. The total thickness of the considered composite material is constant and equal to $t = 10$ [mm]. The angle is the parameter with values ranging from [0°, 90°] and the typical configuration of the CARALL composite is [Al, $+\theta$, Al]s. The gathered data may be summarized as follows. The existence of the upper horizontal shear mode SH1 limits the frequency's useful range. The basic antisymmetric mode A0 is almost insensitive to angle change in this region. The phase velocity of the basic modes SH0 and S0, on the other hand, is parameter-dependent. The unambiguous maximum is found at $\theta = 60°$ in the first example, and the phase velocity falls monotonically with the rise and decrease of 60°. In the second situation (S0), the phase velocity is maximum at 0° and falls monotonically as the parameter value increases.

References

Bellini, C., Di Cocco, V., Iacoviello, F., Sorrentino, L.: Performance evaluation of CFRP/Al fibre metal laminates with different structural characteristics. Compos. Struct. **225**, 111117 (2019)

Hora, P., Červená, O.: Determination of lamb wave dispersion curves by means of Fourier transform. Appl. Comput. Mech. **6**(1) (2012)

Harb, M.S., Yuan, F.G.: A rapid, fully non-contact, hybrid system for generating lamb wave dispersion curves. Ultrasonics **61**, 62–70 (2015)

Honarvar, F., Enjilela, E., Sinclair, A.N.: An alternative method for plotting dispersion curves. Ultrasonics **49**(1), 15–18 (2009)

Haskell, N.A.: The dispersion of surface waves on multilayered media. Bull. Seismol. Soc. Am. **43**(1), 17–34 (1953)

Hosten, B., Castaings, M.: Transfer matrix of multilayered absorbing and anisotropic media. Measurements and simulations of ultrasonic wave propagation through composite materials. J. Acoust. Soc. Am. **94**(3), 1488–1495 (1993)

Karpfinger, F., Gurevich, B., Bakulin, A.: Modeling of wave dispersion along cylindrical structures using the spectral method. J. Acoust. Soc. Am. **124**(2), 859–865 (2008)

Knopoff, L.: A matrix method for elastic wave problems. Bull. Seismol. Soc. Am. **54**(1), 431–438 (1964)

Lowe, M.J.: Matrix techniques for modeling ultrasonic waves in multilayered media. IEEE Trans. Ultrason. Ferroelectr. Freq. Control **42**(4), 525–542 (1995)

Mal, A.K.: Guided waves in layered solids with interface zones. Int. J. Eng. Sci. **26**(8), 873–881 (1988)

Muc, A., Barski, M., Stawiarski, A., Chwał, M., Augustyn, M.: Dispersion curves and identification of elastic wave modes for fiber metal laminates. Compos. Struct. **255**, 112930 (2021)

Nayfeh, A.H.: The general problem of elastic wave propagation in multilayered anisotropic media. J. Acoust. Soc. Am. **89**(4), 1521–1531 (1991)

Nayfeh, A.H., Chimenti, D.E.: Propagation of guided waves in fluid-coupled plates of fiber-reinforced composite. J. Acoust. Soc. Am. **83**(5), 1736–1743 (1988)

Pavlakovic, B., Lowe, M., Alleyne, D., Cawley, P.: Disperse: a general purpose program for creating dispersion curves. In: Thompson, D.O., Chimenti, D.E. (eds.) Review of Progress in Quantitative Nondestructive Evaluation, vol. 16, pp. 185–192. Springer, Boston (1997). https://doi.org/10.1007/978-1-4615-5947-4_24

Packo, P., Uhl, T., Staszewski, W.J.: Generalized semi-analytical finite difference method for dispersion curves calculation and numerical dispersion analysis for Lamb waves. J. Acoust. Soc. Am. **136**(3), 993–1002 (2014)

Press, F., Harkrider, D., Seafeldt, C.A.: A fast, convenient program for computation of surface-wave dispersion curves in multilayered media. Bull. Seismol. Soc. Am. **51**(4), 495–502 (1961)

Press, W.H., Flannery, B.P., Teukolsky, S.A., Vetterling, W.T., Chipperfield, J.R.: Numerical Recipes: The Art of Scientific Computing. Cambridge University Press, Cambridge (1987)

Sasso, M., Mancini, E., Dhaliwal, G.S., Newaz, G.M., Amodio, D.: Investigation of the mechanical behavior of CARALL FML at high strain rate. Compos. Struct. **222**, 110922 (2019)

Schöpfer, F., et al.: Accurate determination of dispersion curves of guided waves in plates by applying the matrix pencil method to laser vibrometer measurement data. CEAS Aeronaut. J. **4**(1), 61–68 (2013)

Schwab, F., Knopoff, L.: Surface-wave dispersion computations. Bull. Seismol. Soc. Am. **60**(2), 321–344 (1970)

Sorohan, Ş, Constantin, N., Găvan, M., Anghel, V.: Extraction of dispersion curves for waves propagating in free complex waveguides by standard finite element codes. Ultrasonics **51**(4), 503–515 (2011)

Rokhlin, S.I., Chimenti, D.E., Nagy, P.B.: Physical Ultrasonics of Composites. Oxford University Press, Oxford (2011)

Towfighi, S., Kundu, T., Ehsani, M.: Elastic wave propagation in circumferential direction in anisotropic cylindrical curved plates. J. Appl. Mech. **69**(3), 283–291 (2002)

Callister, W.D.: Material Science and Engineering An Introduction. John Wiley, Hoboken (2002)

Uncertainties Propagation Through Robust Reduced Non-linear Dynamic Model in Large Displacements

Mohamed Guedri[1,2,3(✉)] and Noureddine Bouhaddi[4]

[1] College of Technology at Makkah and Jeddah, Technical and Vocational Training Corporation, Riyadh, Kingdom of Saudi Arabia
mguedri@tvtc.gov.sa
[2] Laboratory/LR18ES45 - Mathematical Physics, Quantum Modeling and Mechanical Design, Nabeul Preparatory Engineering Institute, Nabeul, Tunisia
mohamed.guedri@isetn.rnu.tn
[3] National Higher Engineering School of Tunis, University of Tunis, Tunis, Tunisia
[4] FEMTO-ST Institute UMR 6174 – Applied Mechanics Department, University of Franche-Comté, 24 Chemin de l'Epitaphe, 25000 Besançon, France
noureddine.bouhaddi@univ-fcomte.fr

Abstract. This paper proposes a reduction order method (ROM) adapted to geometrical and large displacement nonlinearities and large size dynamic models with uncertainties. The nonlinear behavior used to evaluate the nonlinear force at each step, leads to high computation times that are not compatible with optimization and robustness analyses. In this context, model reduction using projection bases enriched with respect to geometrical nonlinearities and uncertainties is one of the ways to reduce the computational cost. The presented reduction order method (Combined Approximations Stochastic Enriched CASE) is based on the combined approximations method (CA method) introduced by Kirsch. The main contribution concerns the determination of an optimal robust reduction basis using a singular value decomposition procedure. This method has been successfully applied to large displacement nonlinear analysis with uncertainty, resulting in more accurate response predictions. Two numerical examples are presented to illustrate the performance of the proposed method.

Keywords: Uncertainties · Geometrical nonlinearities · Large displacements · Structural reanalysis · Reduction order method (ROM) · Combined approximations stochastic enriched (CASE) · Robustness

1 Introduction

In nonlinear dynamic analysis, time iteration algorithms are used to solve the differential equations governing the system's motion (Zienkiewicz and Taylor 2014). The computation of the revised tangent stiffness matrix typically incurs the greatest computational expense. Particularly when dealing with large scale structures with uncertainties, the effort necessary to solve the related linear equation sets at each iteration cycle might very quickly become prohibitive.

A. Akrout et al. (Eds.): ICAV 2022, ACM 22, pp. 189–203, 2023.
https://doi.org/10.1007/978-3-031-34190-8_22

Approximated reanalysis techniques might therefore be seen as a compelling alternative (Balmès 1996, Guedri et al. 2006). These are designed to efficiently analyze the modified system without having to solve the entire set of modified equations; all that is required is knowledge of the initial state and the modification data. The literature has demonstrated that a Ritz basis reduction method can produce accurate results in the case of localized nonlinearities when enriched by static residual vectors caused by modification forces (Balmès 1996, Guedri et al. 2006, Bouazizi et al. 2006, Rabhi et al. 2011, Chikhaoui et al. 2017).

However, these techniques continue to be unable to foretell the structure's nonlinear behavior when dealing with generalized nonlinearities. A nonlinear analysis form of Kirsch's (Kirsch 2003) mixed approximations technique has been developed to overcome this problem (Guedri et al. 2010, Gerges et al. 2012).

This research examines the structure's behavior under large displacements. Its responses are generated using a traditional iterative approach in the time domain. Two numerical examples are used to demonstrate the proposed method's accuracy.

In this paper, we propose a reduction order method (ROM) for a nonlinear large displacement stochastic model. In Sect. 2, a review of the total Lagrangian formulation-based theory of finite elements for geometrically non-linear elastic structures is presented. In Sect. 3, the stochastic nonlinear dynamical system with uncertainties is represented. Model reduction by a variant of the combined approximations method is presented in Sect. 4 to reduce the computation time and examine the robustness of the model. The proposed methodology is then illustrated by numerical examples in Sect. 5.

2 Large Displacements Nonlinear Formulation

This section reviews the total Lagrangian formulation-based theory of finite elements for geometrically non-linear elastic structures (GNS) (Zienkiewicz and Taylor 2014, Imai and Frangopol 2000).

2.1 Beam Structures

The total lagrangian formulation can be characterized as follows, according to (Zienkiewicz and Taylor 2014, Bathe and Bolourchi 1979, Crisfield 1991, Kanchi 1993, Felippa 1996). At first, the displacements u of structures are given by the product:

$$u = N \cdot U \tag{1}$$

where N is the vector of shape functions and U the vector of nodal displacements. Because of large displacements and rotations, Green's strain is adopted for the nonlinear relationships between strains and displacements. The Green's strain ε_G includes both linear and nonlinear terms, respectively B_L and $B_{NL}(U)$, , relating strain and nonlinear strain to the nodal displacements (Imai and Frangopol 2000):

$$\varepsilon_G = (B_L + B_{NL}(U)) \cdot U \tag{2}$$

The virtual work δW is given by the virtual displacement principle as follows (Imai and Frangopol 2000):

$$\delta W = \int_{V_0} (\sigma_S \delta \varepsilon_G) dV_0 - q_{ext} \delta U \tag{3}$$

where σ_S is the second Piola-Kirchhoff stress, $\delta\varepsilon_G$ the incremental form of the strain-displacement relationship, V_0 the volume of initial configuration and q_{ext} the vector of external loads. The stress can be determined by Hook's law because Green's strain is predicated on a small strain (Imai and Frangopol 2000):

$$\sigma_S = E\varepsilon_G = E(B_L + B_{NL}(U)) \cdot U \tag{4}$$

where E is the modulus of elasticity. Substituting $\delta\varepsilon_G$ from Eq. (2) into Eq. (3) results in,

$$\delta W = \left[\int_{V_0} (\sigma_S (B_L + B_{NL}(U))) dV_0 - q_{ext} \right] \delta U \tag{5}$$

Since δU is arbitrary, the vector of internal forces q_{int} is,

$$q_{int} = \int_{V_0} (\sigma_S (B_L + B_{NL}(U))) dV_0 \tag{6}$$

Taking the derivative of q_{int} with respect to the nodal displacements U gives the tangent stiffness matrix K_T:

$$K_T = \frac{\partial q_{int}}{\partial U} = \int_{V_0} \left(\frac{\partial \sigma_S}{\partial U} (B_L + B_{NL}(U)) + \sigma_S \frac{\partial B_{NL}(U)}{\partial U} \right) dV_0 \tag{7}$$

By substituting Eq. (4) into Eq. (7) K_T can be rewritten,

$$\begin{gathered} K_T = \int_{V_0} E B_L B_L dV_0 \\ + \int_{V_0} \sigma_S \frac{\partial B_{NL}(U)}{\partial U} dV_0 \\ + \int_{V_0} (E(B_L B_{NL}(U) + B_{NL}(U) B_L + B_{NL}(U) B_{NL}(U))) dV_0 \end{gathered} \tag{8}$$

It can be noticed that the three terms in Eq. (8) respectively stand for: the elastic stiffness matrix K_E, the geometric stiffness matrix K_G, and the initial displacement stiffness matrix K_U (Zienkiewicz and Taylor 2014, Crisfield 1991).

2.2 Plate Structures

A reminder of the thin plate problem formulation in nonlinear dynamics will be presented in this section. In various publications, you can get detailed formulations, such as (Zienkiewicz and Taylor 2014).

Consider a thin plate with a middle surface Γ_m and a thickness h in the reference $\mathfrak{R}(x.y.z)$. Consideration is given to the Love-Kirchoff model for thin plates' plane stress theory. As a result, the displacements are determined by understanding the middle surface motion, represented by the symbol $\mathfrak{R}(x.y.z)$:

$$u(x,y,z,t) = \begin{Bmatrix} u(x,y,z,t) \\ v(x,y,z,t) \\ w(x,y,z,t) \end{Bmatrix} - z \begin{Bmatrix} w_{,x}(x,y,t) \\ w_{,y}(x,y,t) \\ 0 \end{Bmatrix} \tag{9}$$

u and v are the in-plane displacements and w is the transverse displacement. Small deformations and moderate rotations hypotheses are used to write the 2nd order Green-Lagrange strain tensor in a vector form as follows:

$$E = \frac{1}{2}\left(\nabla u + \nabla u^t + \nabla u \nabla u^t\right) = \begin{Bmatrix} u_{,x} \\ v_{,y} \\ u_{,y} + v_{,x} \end{Bmatrix} + \begin{Bmatrix} \frac{1}{2}(w_{,x})^2 \\ \frac{1}{2}(w_{,y})^2 \\ w_{,x}w_{,y} \end{Bmatrix} - Z \begin{Bmatrix} w_{,xx} \\ w_{,yy} \\ w_{,xy} \end{Bmatrix} \tag{10}$$

Equation (10) can be rewritten,

$$E = E^m - ZK^b$$

With:

$$E^m = \begin{Bmatrix} u_{,x} \\ v_{,y} \\ u_{,y} + v_{,x} \end{Bmatrix} + \begin{Bmatrix} \frac{1}{2}(w_{,x})^2 \\ \frac{1}{2}(w_{,y})^2 \\ w_{,x}w_{,y} \end{Bmatrix} \text{ and } K^b = \begin{Bmatrix} w_{,xx} \\ w_{,yy} \\ w_{,xy} \end{Bmatrix} \tag{11}$$

where E^m represents the in-plane deformation and $E^b = ZK^b$ the transverse deformations due to the bending effect.

The weak formulation of the equation of motion is written as:

$$\int_{\Gamma_m} (\delta E^m)^T D^m E^m dS + \int_{\Gamma_m} (\delta K^b)^T D^b K^b dS + \delta\Pi_{inertia} - \delta\Pi_{ext} = 0 \tag{12}$$

where D^m and D^b represents the material stiffness matrices respectively dedicated to the membrane and bending effects.

$$D^m = \frac{Eh}{1-\nu^2}\begin{bmatrix} 1 & \nu & 0 \\ \nu & 1 & 0 \\ 0 & 0 & \frac{1-\nu}{2} \end{bmatrix}; \quad D^b = \frac{h^2}{12}D^m \tag{13}$$

The development of virtual work of deformation leads to:

$$\delta\Pi_{def} = \delta U^T \int_{\Gamma_m} \begin{bmatrix} B^{m^T} D^m B^m & 0 \\ 0 & B^{b^T} D^b B^b \end{bmatrix} U dS + \ldots$$

$$\delta U^T \int_{\Gamma_m} \begin{bmatrix} 0 & \frac{1}{2} B^{m^T} D^m B^{nl}(w) \\ B^{nl^T}(w) D^m B^m & \frac{1}{2} B^{nl^T}(w) D^m B^{nl}(w) \end{bmatrix} U dS \tag{14}$$

with:

$$K_0 = \int_{\Gamma_m} \begin{bmatrix} B^{m^T} D^m B^m & 0 \\ 0 & B^{b^T} D^b B^b \end{bmatrix}$$
$$\Delta K = \int_{\Gamma_m} \begin{bmatrix} 0 & \frac{1}{2} B^{m^T} D^m B^{nl}(w) \\ B^{nl^T}(w) D^m B^m & \frac{1}{2} B^{nl^T}(w) D^m B^{nl}(w) \end{bmatrix} \tag{15}$$

In this paper, the nonlinear dynamic behavior of structures in large displacements with uncertainties is studied. The responses are computed in the time domain by an implicit method using Newmark nonlinear time integration scheme (Newmark 1959, Gérardin and Rixen 1997), (see Sect. 3).

3 Stochastic Nonlinear Time Analysis

In this study, the Latin Hypercube Sampling (LHS) method is used to approximate the solution of nonlinear mechanical systems which can generally be represented in the time domain by the differential equation:

$$\begin{cases} M\ddot{U} + B\dot{U} + f_{int}(U) = f_{ext} \\ U(t_0) = U_0 \\ \dot{U}(t_0) = \dot{U}_0 \end{cases} \tag{16}$$

where the internal forces vector is of the form:

$$f_{int}(U) = \left(K + f_{NL}\left(U.\dot{U}\right)\right)U \tag{17}$$

where M, K and B stand for the mass, stiffness and damping stochastic matrices of the system, respectively, while f_{ext} the exciting force vector.

The time solution of Eq. (16) can be approximated using the Newmark nonlinear time integration scheme (Newmark 1959, Gérardin and Rixen 1997, Krenk 2009, Lulf et al. 2013, Wenneker and Tiso 2014). At time t_{n+1}, this equation is expressed as follows:

$$M\ddot{U}_{n+1} + B\dot{U}_{n+1} + f_{int}(U_{n+1}) - f_{ext}(t_{n+1}) = r_{n+1} \tag{18}$$

where r_{n+1} is the generalized residual force vector which must be minimized using an iterative Newton – Raphson algorithm. For the iteration i, the incremental solution ΔU^i_{n+1} is calculated by:

$$\Delta U^i_{n+1} = -\left(\tilde{K}^i\right)^{-1} r^i_{n+1} \tag{19}$$

where $\tilde{K}^i$ is the instantaneous stiffness matrix (Jacobian of the system) defined by:

$$\tilde{K}^i = K^i + \frac{\gamma}{\beta h} B + \frac{1}{\beta h^2} M \tag{20}$$

function of the tangent stiffness matrix:

$$K^i = \left[\frac{\partial f_{int}(U)}{\partial U} \right]_{U^i_{n+1}} \tag{21}$$

It should be noted that, the uncertainties are introduced on the parameters through the following approximation expression:

$$P(\theta) = P_a(1 + \delta_P \xi(\theta)) \tag{22}$$

where P_a is the average parameter and δ_P represents the dispersion factor of parameter.

4 Model Reduction by a Variant of the Combined Approximations Method

The terms of the local binomial series expansion are used to compute the vectors of a global reduced basis using Kirsch's Combined Approximations (CA) approach (Kirsch 2003). For each parametric modification of the initial structure, a new eigenproblem must be solved.

The normal mode v of the modified structure verifies the equilibrium relation given by:

$$\left(K^l + \Delta K^{nl}(U)\right) T^v = \lambda^v (M_0 + \Delta M) T^v \tag{23}$$

Multiplying the Eq. (22) by $\left(K^l\right)^{-1}$ leads to:

$$(1 + B) T^v = r_0^v \tag{24}$$

with:

$$\begin{cases} B = \left(K^l\right)^{-1} \Delta K^{nl}(U) \\ r_0^v = \lambda^v \left(K^l\right)^{-1} (M_0 + \Delta M) T^v \approx \lambda^v \left(K^l\right)^{-1} (M_0 + \Delta M) T_0^v \end{cases} \tag{25}$$

The reduction basis is then constructed using the following recurrence relation:

$$\begin{cases} r_1^v = \left(K^l\right)^{-1} (M_0 + \Delta M) r_0^v \\ r_i^v = -B r_{i-1}^v \ (i = 2; 3; \ldots; s) \end{cases} \tag{26}$$

and can be finally be written:

$$r_B^{(v)} = \left[r_1^{(v)} \ r_2^{(v)} \ \cdots \ r_S^{(v)} \right] \tag{27}$$

Applying the CA method to complex structures has revealed severe limitations and convergence problems of the binomial series expansion. This results in less predictive approximate solutions, particularly of large modification and nonlinear structures that account uncertainties. The variant of the method proposed in this paper consists in retaining only the most relevant part of the information contained in each considered mode's reduction basis. It amounts to extracting the subspace that best spans the solution space (Guedri et al. 2010, Gerges et al. 2012).

At first, a global transformation matrix is obtained by concatenation of the reduction basis $r_B^{(v)}$ associated to m studied normal modes,

$$r_B^* = \left[r_B^1 \, r_B^2 \cdots r_B^{(v)} \, r_B^{(m)} \right] \tag{28}$$

Singular value decomposition is then performed:

$$r_B^* = U \sum V^T = U_1 \sum\nolimits_1 V_1^T + U_2 \sum\nolimits_2 V_2^T \tag{29}$$

The robust Ritz basis, with regard to the parametric modifications, is finally constructed with the column vectors of matrix U_1,

$$r_B = U_1 \tag{30}$$

To perform structural reanalysis on large-scale finite element models, the proposed CA version has successfully been extended to linear dynamic substructuring, as must be noted (Guedri et al. 2010, Gerges et al. 2012).

The corresponding program diagram of CASE method is shown in Fig. 1.

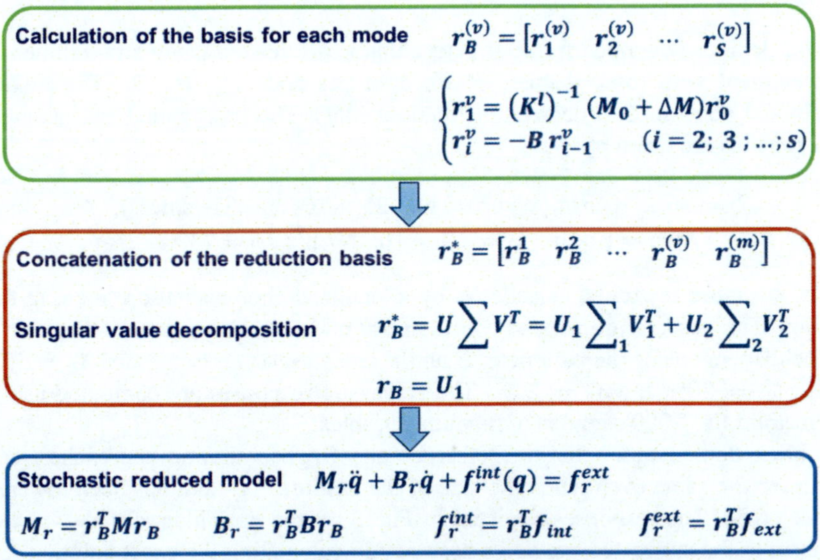

Fig. 1. CASE method

To illustrate the main features of the proposed method and to evaluate their robustness against uncertainties, two numerical examples are presented in Sect. 5.

5 Numerical Simulations

5.1 Evaluation Criteria: Temporal Moments

Temporal moments are used in transient responses in order to compare between different models. These temporal moments $M_i(t_s)$ have been proposed for transitory analysis (Smallwood 1994, Hemez and Doebling 2003). They are determined as balanced summations of the quadratic temporal signal and are similar to static moments:

$$M_i(t_s) = \int_{-\infty}^{+\infty} (t - t_s)^i (u(t))^2 dt \tag{31}$$

where t_s corresponds to a temporal shift and the index i represents the order of the moment. For more simplicity, the temporal moments M_i are defined for $t_s = 0$. The central moments are thus defined as follows:

$$\begin{array}{ll} E = M_0 & \text{energy}(\text{m}^2\text{s}) \\ T = \frac{M_1}{M_0} & \text{central time "Centroid"(s)} \\ D^2 = \frac{M_2}{M_0} - \left(\frac{M_1}{M_0}\right)^2 & \text{rms duration(s)} \end{array} \tag{32}$$

5.2 Example 1

Figure 2 depicts a clamped frame structure that is discretized using a two-dimensional beam element with three degrees of freedom per node U_x, U_y, θ_z. The finite element model contains 510 degrees-of-freedom (dof). The mechanical and geometrical characteristics are given by:

$$\text{b} = 36 \times 10^{-3}\text{m (Width)};\quad \text{h} = 25 \times 10^{-3}\text{m (Thickness)};$$
$$\text{Area} = \text{b} \times \text{h};\quad \text{E} = 2.1 \times 10^{11}\text{N/m}^2;\quad \rho = 7800\ \text{kg/m}^3.$$

The structure is excited at node N_f by a localized choc excitation force, in the U_x direction. The observation point is considered at node N_o, also in the U_x direction.

The dispersions of the parameter E and ρ are: Case 1: $\delta_E = 5\%$ and $\delta_\rho = 5\%$ and Case 2: $\delta_E = 20\%$ and $\delta_\rho = 20\%$. The stochastic responses are computed using the LHS method for 1,000 samples of random variables.

Table 1 demonstrates that the reference and CASE variation approaches' energy criteria are the same even when the size of the nonlinear system has been lowered by a factor of 50. The responses depicted in Fig. 3 on the time interval $[0 - 4s]$ clearly demonstrate the suggested method's accuracy in calculating the exact solution for the first 5 modes using 2 basis vectors.

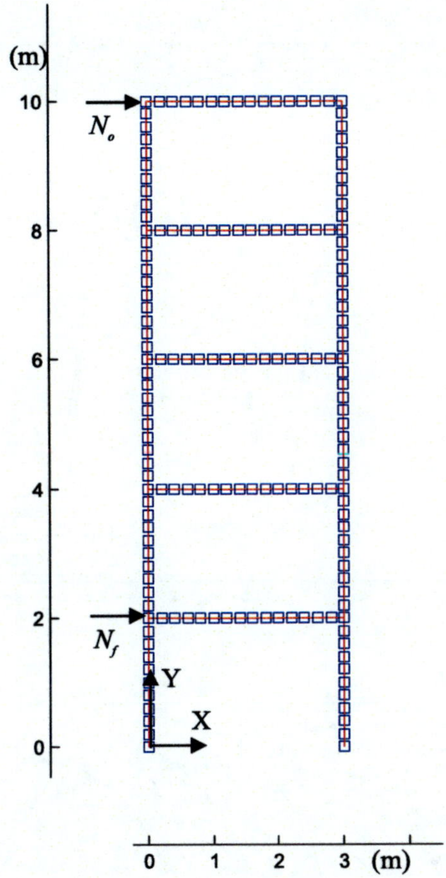

Fig. 2. Finite element model of the frame

Table 1. Reduced basis size and related energy criteria

		Size (dofs)	E (m^2s) × 10^{-6}	T (s)	D^2 (s)	CPU (%)
	Exact - Deterministic	510	2.4197	2.1484	1.1748	—
Case 1	Exact - Stochastic	510	2.4464	2.1421	1.1696	100
	CA – Stochastic (CASE)	10	2.4461	2.1421	1.1697	10
Case 2	Exact - Stochastic	510	2.6274	2.1169	1.1393	100
	CA – Stochastic (CASE)	10	2.6265	2.1167	1.1392	10

(a)

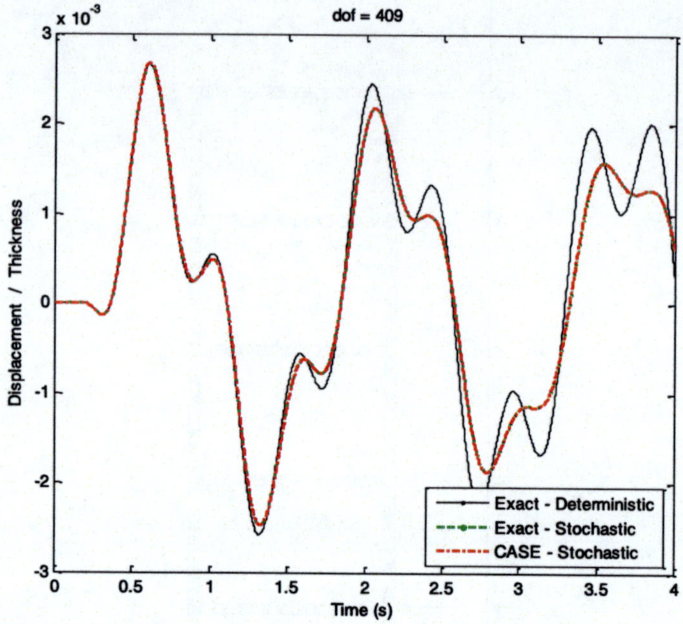

(b)

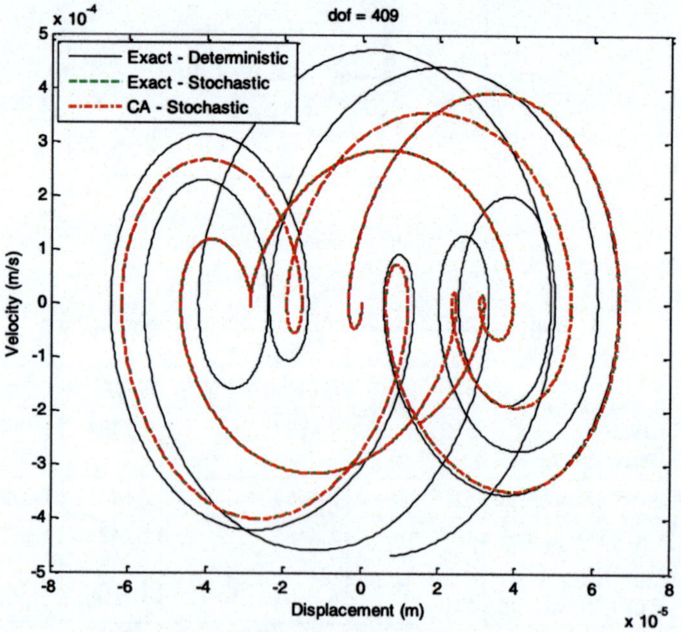

Fig. 3. Mean random nonlinear responses of the reference and condensed models at dof 409 on the time interval $[0 - 4s]$. **Case 1**: **(a)**, **(b)** and **Case 2**: **(c)**, **(d)**.

(c)

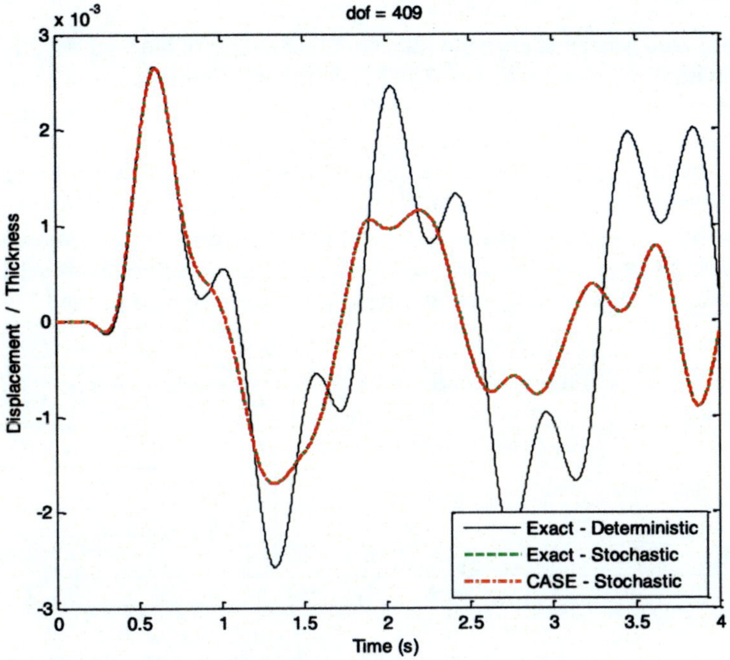

(d)

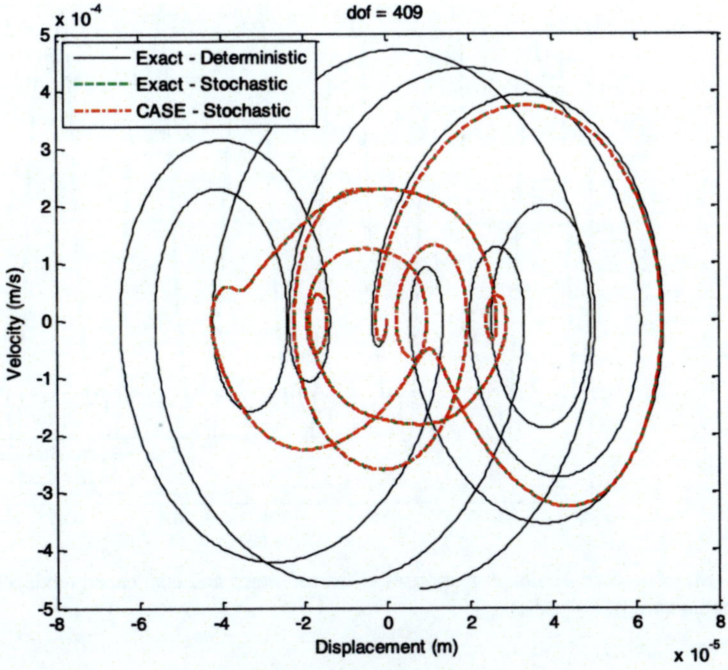

Fig. 3. *(continued)*

5.3 Example 2

The proposed academic example illustrates the interest of the method. It is a rectangular plate, with dimensions $0.654 \times 0.527 \times 0.002$ m^3. The finite element model contains 2645 dof, valid in the frequency band [0–300 Hz]. The structure is excited at the center by a harmonic force of pulsation $\omega = 135$ rad/s, of amplitude 100N. The mechanical characteristics are given by: $E = 70 \times 10^9$ N/m^2; $\rho = 2778$ kg/m^3. The parameter dispersions$E$, ρ and h are: $\delta_E = 20\%$; $\delta_\rho = 20\%$ and$\delta_h = 5\%$. The stochastic responses are computed using the LHS method for 1,000 samples of random variables.

Table 2 shows that the reference and the CASE variant methods are identical. The responses plotted in Fig. 4, state the accuracy of the proposed method.

Table 2. Reduced basis size and related energy criteria

	Size (dofs)	E (m^{2}s) $\times 10^{-3}$	T (s)	D^2 (s)) $\times 10^{-3}$	CPU (%)
Exact - Stochastic	2645	7.9945	0.35082	7.5145	100
CA – Stochastic (CASE)	72	7.9946	0.35081	7.5145	14

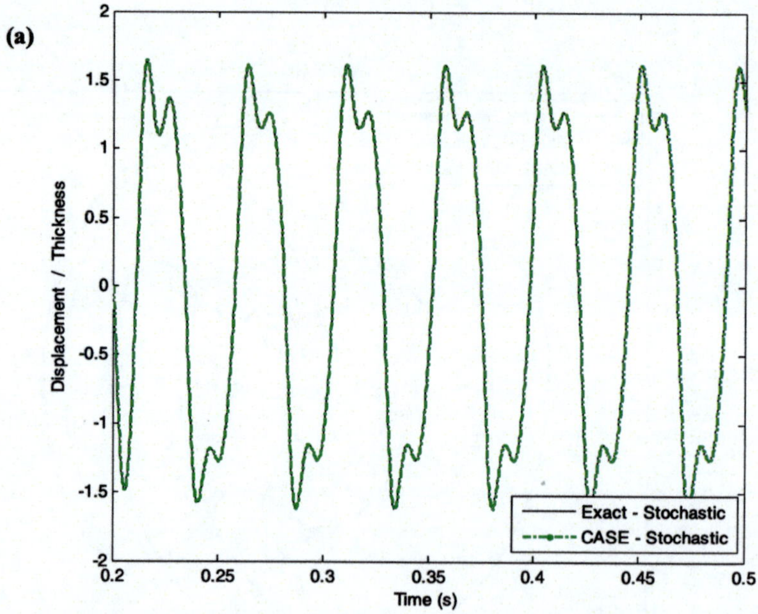

Fig. 4. Mean random nonlinear responses of the reference and condensed models at dof 1323 on the time interval [0.2 – 0.5s].

(b)

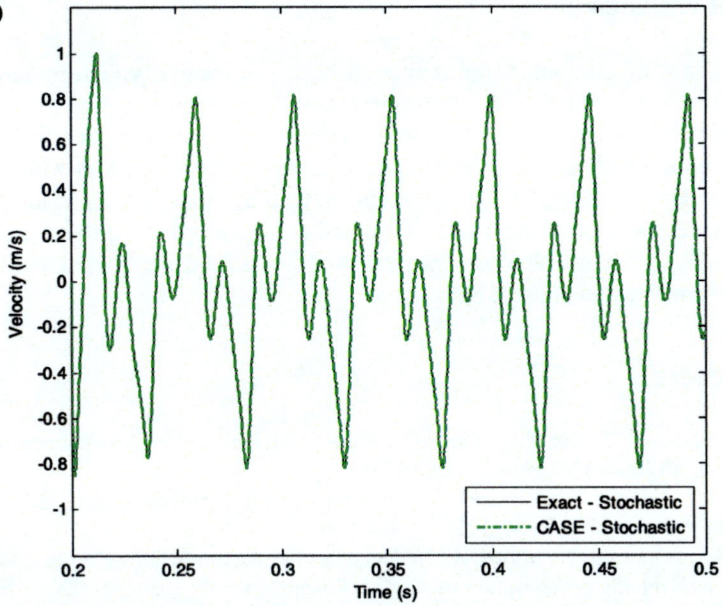

(c)

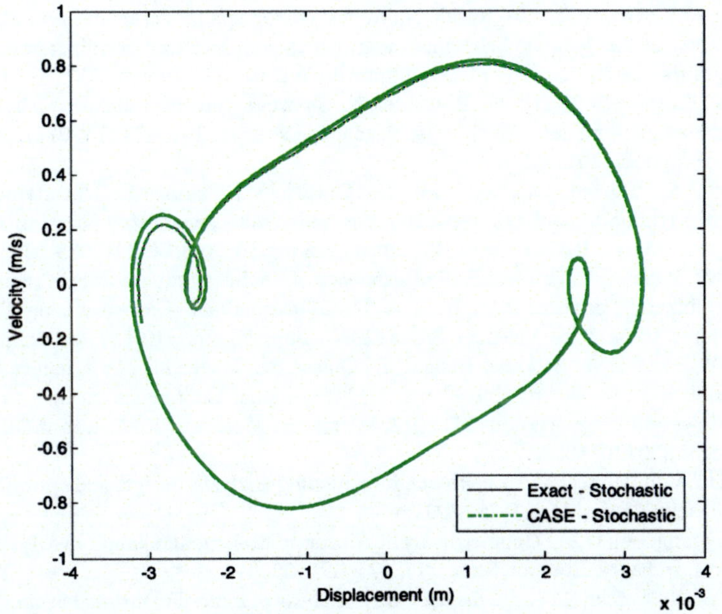

Fig. 4. *(continued)*

6 Conclusions

A variation of the combined approximation reanalysis method with uncertainties (CASE) is provided in this study. The main contribution focuses on selecting the best reduction foundation through the use of a singular value decomposition approach. The technique has proven to be effective for large displacement nonlinear analysis with uncertainties. When working with large scale models, in particular, it allows for a significant decrease in the tangent matrix, which must be updated at each stage of the iterative process. The simulations carried out show the robustness of the method with regard to different levels of nonlinearities and uncertainties.

References

Zienkiewicz, O.C., Taylor, R.L.: The Finite Element Method for Solid and Structural Mechanics, 7th edn. McGraw-Hill, New York (2014)

Balmès, E.: Optimal Ritz vectors for component mode synthesis using the singular value decomposition. AIAA J. **34**, 1256–1260 (1996)

Guedri, M., Bouhaddi, N., Majed, R.: Reduction of the stochastic finite element models by using a robust dynamic condensation method. J. Sound Vib. **297**, 123–145 (2006). https://doi.org/10.1016/j.jsv.2006.03.046

Bouazizi, M.L., Guedri, M., Bouhaddi, N.: Robust component modal synthesis method adapted to the survey of the dynamic behaviour of structures with localized non-linearities. Mech. Syst. Signal Process. **20**, 131–157 (2006). https://doi.org/10.1016/j.ymssp.2005.02.002

Rabhi, N., Guedri, M., Hassis, H., Bouhaddi, N.: Structure dynamic reliability: a hybrid approach and robust meta-models. Mech. Syst. Signal Process. **25**, 2313–2323 (2011). https://doi.org/10.1016/j.ymssp.2011.02.014

Chikhaoui, K., Kacem, N., Bouhaddi, N., Guedri, M., Soula, M.: Uncertainty quantification/propagation in nonlinear models: robust reduction - generalized polynomial chaos. Eng. Comput. J. **34**(4), 1082–1106 (2017). https://doi.org/10.1108/EC-11-2015-0363

Guedri, M., Weisser, T., Bouhaddi, N.: Reanalysis of nonlinear structures by a reduction method of combined approximations. In: The Tenth International Conference on Computational Structures Technology, Valencia, Spain (2010). https://doi.org/10.4203/ccp.93.312

Gerges, Y., Guedri, M., Sadoulet-Reboul, E., Ouisse, M., Bouhaddi, N.: A reduced order model for nonlinear vibroacoustic problems. In: International Conference on Structural Nonlinear Dynamics and Diagnosis CSNDD 2012, Marrakech, Morocco (2012). https://doi.org/10.1051/matecconf/20120110002

Kirsch, U.: A unified reanalysis approach for structural analysis, design, and optimization. Struct. Multidiscip. Optim. **25**, 67–85 (2003)

Imai, K., Frangopol, D.M.: Geometrically nonlinear finite element reliability analysis of structural systems. I: theory. Comput. Struct. **77**, 677–691 (2000)

Bathe, K.J., Bolourchi, S.: Large displacement analysis of three-dimensional beam structures. Int. J. Numer. Meth. Eng. **14**, 961–986 (1979)

Crisfield, M.A.: Non-linear finite element analysis of solid and structures. Wiley, Chichester (1991)

Kanchi, M.B.: Matrix methods of structural analysis. Wiley-Eastern, New Delhi (1993)

Felippa, C.A.: Lecture notes in nonlinear finite element methods, Center for Aerospace Structures, University of Colorado, Boulder, CO (1996)

Newmark, N.: A method of computation for structural dynamics. J. Eng. Mech. Div. ASCE **85**(7), 67–94 (1959)

Gérardin, M., Rixen, D.: Mechanical Vibrations: Theory and Application to Structural Dynamics, 2nd edn. Wiley, Hoboken (1997)

Krenk, S.: Non-linear Modeling and Analysis of Solids and Structures. Cambridge University Press, Cambridge (2009)

Lulf, F.A., Tran, D.M., Ohayon, R.: Reduced bases for nonlinear structural dynamic systems: a comparative study. J. Sound Vib. **332**(3), 897–921 (2013)

Wenneker, F., Tiso, P.: A substructuring method for geometrically nonlinear structures. Dyn. Coupled Struct. **1**, 157–165 (2014)

Smallwood, D.O.: Characterization and simulation of transient vibrations using band limited moments. Shock. Vib. **1**(6), 507–527 (1994)

Hemez, F.M., Doebling, S.W.: From shock response spectrum to temporal moments and vice-versa, International Modal Analysis Conference (IMAC-XXI), Kissirnmee, FL (2003)

Suspension of Heavy Trucks with Intelligent Control Using Artificial Neural Networks Particle Swarm Optimization (ANN-PSO)

Anis Hamza(✉), Issam Dridi, Kamel Bousnina, and Noureddine Ben Yahia

Mechanical, Production and Energy Laboratory (LMPE), National School of Engineering of Tunis (ENSIT), University of Tunis, Avenue Taha Hussein Montfleury, 1008 Tunis, Tunisia
anis7amza@gmail.com

Abstract. In this paper, we discuss the motivation to create a novel active suspension control method for trucks and Heavy Goods Vehicle (HGV) based on intelligent control using hybrid Artificial Neural Network Particles Swarm Optimization (ANN-PSO). There are three types of suspension systems: passive, semi-active and active. Ride comfort, suspension travel and handling are all factors in the performance required of an active suspension system. The model was created using the MATLAB toolkit, and it was efficient and validated using information gathered from experiments using a vehicle. Modeling a system can be done in different ways. One of them is to describe the system using the rules of physics and then determine the parameters of the system using experimental data or other information used using an intelligent algorithm ANN-PSO. The objective of this research is to create a reliable hybrid intelligent controller that can improve the functionality of heavy truck nonlinear active suspension system and its validations using motion and graphic output.

Keywords: Heavy Good Vehicle (HGV) · Active Suspension System · MATLAB · Artificial Neural Network (ANN) · Particle Swarm Optimization (PSO)

1 Introduction

Continuation of our previous research work (Hamza and Ben Yahia 2022, 2021, 2019) and within the framework of improvement and development of our intelligent controller for the active suspension of heavy trucks, we propose in this article to study another more sophisticated solution using a controller based on a hybrid model of Artificial Neural Networks Particle Swarm Optimization (ANN-PSO). Recently, machine learning methods have also been used to model the problem. However, the application of machine learning has been mostly limited to artificial neural networks (Hamza and Ben Yahia 2023). In addition, the ANN model turns out to be faster and more stable. Therefore, the contribution of this paper is to advance an ANN-based model to further improve the active suspension performance of heavy-duty trucks. To do this, the hybrid ANN-PSO method is proposed which provides an optimized neural network with a high level of adaptation, stability and generalization.

A. Akrout et al. (Eds.): ICAV 2022, ACM 22, pp. 204–213, 2023.
https://doi.org/10.1007/978-3-031-34190-8_23

Our case study uses hybrid optimization control based on control strategies. And in this framework, (Jiangtao et al. 2008) described the state of the art of inference systems for suspension control problems using fuzzy inference systems, neural networks, algorithms evolutionary and their combination in chronological order. Then, fuzzy logic control (FLC) and hybrid FLC were developed by (Mahbubur et al. 2011) to regulate the semi-active suspension system of the vehicle. Compared to the PID controller, the results indicated that fuzzy and hybrid logic controllers were perfectly adequate to significantly reduce road disturbances for semi-active suspension systems. Also, a nonlinear adaptive Neuro-Fuzzy Wavelet Network (NFWN) control methodology has been proposed by (Khan et al. 2012) to control eight-degree-of-freedom (8 DoF) car suspension systems. The results obtained from the simulations were compared to passive and semi-active suspension systems and show better performance in terms of ride comfort and vehicle stability when using the proposed controller.

Another application of the hybrid algorithms, (Ruey-Jing 2013) created the Enhanced Adaptive Self-Organizing Fuzzy Sliding Mode Controller (EASFSC) for active suspension systems. Instead of using output error and system error change, the EASFSC controller used a slip surface and its differential as input variables from the fuzzy logic control (FLC) to the self-organized fuzzy controller (SOFC) to provide fuzzy control input. Lyapunov's stability theorem was used to demonstrate the stability of the EASFC. To compare the performance of artificial neural networks (ANN) and adaptive neuro fuzzy inference systems (ANFIS) in terms of error rate minimization. Also, the authors (Vibha et al. 2014) used performance indicators such as the correlation coefficient (CORR), the normalized root mean square error (NRMSE) and the coefficient of determination. They concluded that the hybrid learning algorithm of ANFIS proves effective as a backward propagation of the ANN learning algorithm. The STFIS (Self-Tuning Fuzzy Inference System) controller has been studied by (Souilem et al. 2015) to show that the STFIS is more effective in obtaining driving comfort and handling qualities compared to the conventional PID controller and the passive suspension.

Two ways of controlling the suspension system have been presented by (Geweda et al. 2017). The first method is optimization using Genetic Algorithm (GA) to find the optimum values of spring stiffness and damping coefficient at different speeds and the other method is active control of the spring suspension system using the proportional integral (PI) controller. The results show a significant improvement in terms of sprung mass acceleration when using the presented controllers on the passive system. Next, (Khodadad and Ghadiri 2018) developed a fuzzy logic self-tuning PID controller based on fuzzy logic and $H\infty$ to improve the performance of the suspension system. Thanks to the proposed controller, the working space of the suspension (SWS) is minimized and the best driver comfort is achieved. And at the same time (Leon-Vargas et al. 2018) proposed a new adaptive control algorithm that combines a PID controller for suspension deflection with an external slip mode reference conditioning loop that uses acceleration vertical of the body as a source of additional control. The results show an improvement in driving comfort compared to the same PID controller without the external conditioning loop and the passive system. The hybrid ANFIS and PID controller (HANFISPID) was developed by (Devdutt 2018). The results demonstrate that the proposed HANFISPID control plane allows better driving comfort and better travel safety compared to the passive system and the ANFIS controller. And finally, a chatter-free sliding mode control strategy was proposed by (Yang et al. 2019). The numerical simulation results verify that this control method is effective for vibrations which have nonlinear characteristics of the car's suspension model and can improve driving comfort by reducing the vibration amplitude compared to the passive system of suspension.

In the following sections, the data and materials are presented where the details of the active suspension of the studied heavy trucks and the intelligent hybrid ANN-PSO method are described. Simulations are performed using MATLAB 2018a simulation software. Finally, the results of modeling and comparative analysis of ANN-PSO and ANN are presented.

2 Problematic Study

In this section, the tools and techniques used to gather data, develop a virtual controller based on a hybrid model called Artificial Neural Networks Particle Swarm Optimization (ANN-PSO), validate it, and test it are detailed. The goal is to lessen the acceleration of the suspended mass. The parameters of the heavy-duty truck under study and its mathematical model, which includes scenarios and descriptions of the software, are first provided. The ANN-PSO model utilized for the virtual controller, along with its structure and strategy for hyper parameter optimization, is then discussed.

3 Mathematical Model

In this section, the optimal control problem is presented and the control law is formulated. Active suspension systems, as shown in Fig. 1, add hydraulic actuators to the passive components of the suspension system. The advantage of this type of active suspension control system is that even if the active actuator or control system, the passive components come into action. This presents a very important safety factor, especially for heavy trucks.

Using Newton's second law of motion and free-body diagram concept, the following equations of motion are determined:

$$M_p\ddot{Z}_p = F_p - K_p(Z_p - Z - X_p\theta - Y_p\phi) - C_p(\dot{Z}_p - \dot{Z} - X_p\dot{\theta} - Y_p\dot{\phi}) \tag{1}$$

$$M_1\ddot{Z}_1 = K_1(Z - Z_1 - a\theta + W\phi) + C_1(\dot{Z} - \dot{Z}_1 - a\dot{\theta} + W\dot{\phi}) + K_t(Q_1 - Z_1) - F_1 \tag{2}$$

$$M_2\ddot{Z}_2 = K_2(Z - Z_2 + b\theta + W\phi) + C_2(\dot{Z} - \dot{Z}_2 + b\dot{\theta} + W\dot{\phi}) + K_t(Q_2 - Z_2) - F_2 \tag{3}$$

$$M_3\ddot{Z}_3 = K_3(Z - Z_3 - a\theta - W\phi) + C_3(\dot{Z} - \dot{Z}_3 - a\dot{\theta} - W\dot{\phi}) + K_t(Q_3 - Z_3) - F_3 \tag{4}$$

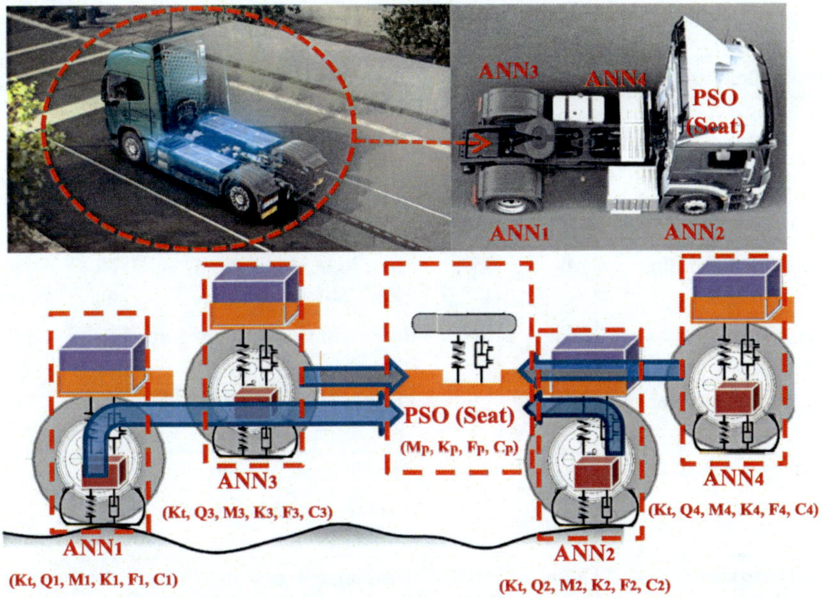

Fig. 1. Full car model of active suspension system for heavy truck (8 DoF)

$$M_4\ddot{Z}_4 = K_4(Z - Z_4 + b\theta - W\phi) + C_4(\dot{Z} - \dot{Z}_4 + b\dot{\theta} - W\dot{\phi}) + K_t(Q_4 - Z_4) - F_4 \tag{5}$$

Figure 1 shows an 8 DoF full vehicle model, it consists of the driver seat, sprung mass which refers to the part supported by active suspension and unsprung masses which refer to the front and rear wheels assembly. Table 1 describes the parameters used in the motion equations of the active suspension system of our truck under study.

The equations below present the vertical and angular movements of the truck:

$$\begin{aligned} M\ddot{Z} = {} & (F_1 + F_2 + F_3 + F_4 - F_p) + K_p(Z_p - Z - X_p\theta - Y_p\phi) + C_p(\dot{Z}_p - \dot{Z} - X_p\dot{\theta} - Y_p\dot{\phi}) \\ & - K_1(Z - Z_1 - a\theta + W\phi) - C_1(\dot{Z} - \dot{Z}_1 - a\dot{\theta} + W\dot{\phi}) - K_2(Z - Z_2 + b\theta + W\phi) \\ & - C_2(\dot{Z} - \dot{Z}_2 + b\dot{\theta} + W\dot{\phi}) - K_3(Z - Z_3 - a\theta - W\phi) - C_3(\dot{Z} - \dot{Z}_3 - a\dot{\theta} - W\dot{\phi}) \\ & - K_4(Z - Z_4 + b\theta - W\phi) - C_4(\dot{Z} - \dot{Z}_4 + b\dot{\theta} - W\dot{\phi}) \end{aligned} \tag{6}$$

$$\begin{aligned} \frac{I_x}{W}\ddot{\phi} = {} & (F_1 + F_2 - F_3 - F_4 + F_p) - K_1(Z - Z_1 - a\theta + W\phi) + C_1(\dot{Z} - \dot{Z}_1 - a\dot{\theta} + W\dot{\phi}) \\ & - K_2(Z - Z_2 + b\theta + W\phi) - C_2(\dot{Z} - \dot{Z}_2 + b\dot{\theta} + W\dot{\phi}) + K_3(Z - Z_3 - a\theta - W\phi) \\ & + C_3(\dot{Z} - \dot{Z}_3 - a\dot{\theta} - W\dot{\phi}) + K_4(Z - Z_4 + b\theta - W\phi) + C_4(\dot{Z} - \dot{Z}_4 + b\dot{\theta} - W\dot{\phi}) \\ & - \frac{Y_p}{W}K_p(Z_p - Z - X_p\theta - Y_p\phi) - \frac{Y_p}{W}C_p(\dot{Z}_p - \dot{Z} - X_p\dot{\theta} - Y_p\dot{\phi}) \end{aligned} \tag{7}$$

$$\frac{I_y\ddot{\theta}}{ab} = (\frac{F_2 + F_4}{a} - \frac{F_1 - F_3}{b} + \frac{F_p}{ab}) + \frac{K_1}{b}(Z - Z_1 - a\theta + W\phi) - \frac{C_1}{b}(\dot{Z} - \dot{Z}_1 - a\dot{\theta} + W\dot{\phi})$$

Table 1. Mathematical heavy truck parameters used in simulation.

Parameter Definition	Symbol	Unit	Parameter Definition	Symbol	Unit
Passenger seat mass	Mp	kg	Passenger seat actuator force	Fp	N
Sprung mass	M	kg	Front left side actuator force	F_1	N
Front left side unsprung mass	M_1	kg	Rear left side actuator force	F_2	N
Rear left side unsprung mass	M_2	kg	Front right side actuator force	F_3	N
Front right side unsprung mass	M_3	kg	Rear right side actuator force	F_4	N
Rear right side unsprung mass	M_4	kg	Road input at front left side	Q_1	m
Passenger Seat Stiffness	Kp	N/m	Road input at rear left side	Q_2	m
Front left side spring stiffness	K_1	N/m	Road input at front right side	Q_3	m
Rear left side spring stiffness	K_2	N/m	Road input at rear right side	Q_4	m
Front right side spring stiffness	K_3	N/m	Angular displacement	Ø, θ	
Rear right side spring stiffness	K_4	N/m	Vertical displacement	Z	
Tyre stiffness	Kt	N/m	Vertical displacement of wheel	Z_1,.. Z_4	
Passenger seat damping coefficient	Cp	Ns/m	Wheel track	2W	m
Front left side suspension damping coefficient	C_1	Ns/m	Mass moment of inertia for roll	Ix	kg.m^2
				Iy	kg.m^2
Rear left side suspension damping coefficient	C_2	Ns/m	Centre of Gravity location (C.G) from front axle	a	m
Front right side suspension damping coefficient	C_3	Ns/m	Centre of Gravity location (C.G) from rear axle	b	m
Rear right side suspension damping coefficient	C_4	Ns/m	Distance of seat position from Centre of Gravity (CG) of sprung mass	Xp	m
				Yp	m

$$
\begin{aligned}
&-\frac{K_2(}{a}Z - Z_2 + b\theta + W\phi) - \frac{C_2}{a}(\dot{Z} - \dot{Z}_2 + b\dot{\theta} + W\dot{\phi}) + \frac{K_3}{b}(Z - Z_3 - a\theta - W\phi) \\
&+ \frac{C_3}{b}(\dot{Z} - \dot{Z}_3 - a\dot{\theta} - W\dot{\phi}) - \frac{K_4}{a}(Z - Z_4 + b\theta - W\phi) - \frac{C_4}{a}(\dot{Z} - \dot{Z}_4 + b\dot{\theta} - W\dot{\phi}) \\
&- \frac{X_pK_p}{ab}(Z_p - Z - X_p\theta - Y_p\phi) - \frac{X_pC_p}{ab}(\dot{Z}_p - \dot{Z} - X_p\dot{\theta} - Y_p\dot{\phi}) \qquad (8)
\end{aligned}
$$

A state space model is a mathematical representation of a physical system as a set of input, output, and state variables related by first-order differential equations. State variables define the values of input and output variables.

Using following state space variables:

$$\begin{cases} X_1 = Zp \; X_2 = \dot{Z}p \;\; X_3 = Z \;\; X_4 = \dot{Z} \\ X_5 = Z_1 \;\; X_6 = \dot{Z}_1 \;\; X_7 = Z_2 \;\; X_8 = \dot{Z}_2 \\ X_9 = Z_3 \; X_{10} = \dot{Z}_3 \; X_{11} = Z_4 \; X_{12} = \dot{Z}_4 \\ X_{13} = \phi \;\; X_{14} = \dot{\phi} \;\; X_{15} = \theta \;\; X_{16} = \dot{\theta} \end{cases} \tag{9}$$

Substituting above variables in equations of motion and writing the equations in state space representation form:

$$\{\dot{X}\} = [A]\{X\} + [B]\{Q\} + [C]\{F\} \tag{10}$$

where, $[A]$ is the State space matrix, $\{X\}$ is the vector of State Space Model (SSM) variables, $[B]$ is the Input matrix, {Q} is the vector of Input of system (Road input), $[C]$ is the Output matrix and {F} is the vector of Output of system.

The optimization process entails identifying the controller input F that minimizes J, the necessary performance characteristic, and the controller input restrictions.

$$J = \int_0^\infty \left(X^T PX + F^T RF\right) dt \tag{11}$$

$$\begin{gathered} X = \{X_1 \; X_2 \; X_3 \; X_4 \; X_5 \; X_6 \; X_7 \; X_8 \; X_9 \; X_{10} \; X_{11} \; X_{12} \; X_{13} \; X_{14} \; X_{15} \; X_{16}\}^T \\ F = \{F_P \; F_1 \; F_2 \; F_3 \; F_4\}^T \end{gathered} \tag{12}$$

where, P and R represent the weighting matrices and are positive definite solution of the Riccati equation, X and F represent the vector of state space variables and the vector of actuator forces, respectively, and J represents the performance index characteristic requirement and the controller input constraints.

Particle Swarm Optimization (PSO) is a computational method to optimize a problem. It is a population-based optimization algorithm that is inspired by the social behavior of birds in a flock or fish in a bowl.

4 Simulations and Control Methods

The full vehicle model is created using MATLAB 2018a: a flow chart from the input (Road profile and speed) to the car body and finally the outputs (Seat and sprung mass vertical displacements) described in Fig. 2. Table 2 shows the different parameters of the heavy truck studied in our simulation.

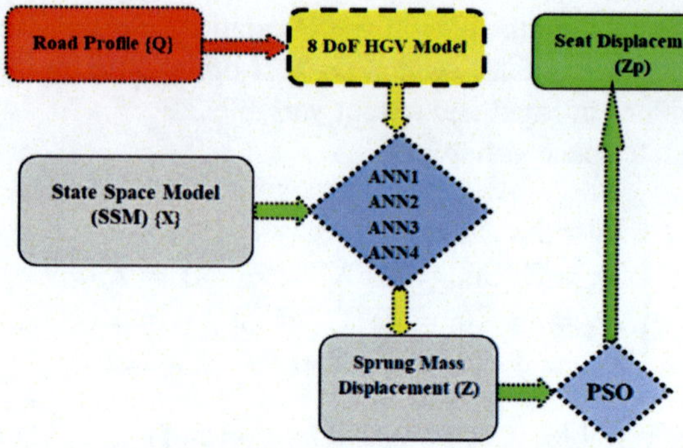

Fig. 2. MATLAB model flow chart.

Table 2. Nominal truck parameters values used in simulation.

Parameter Definition	Symbol	Value	Unit
Distance between the Front and Intermediate Axles	a	3.8	m
Distance between the Intermediate and Rear Axles	b	1.305	m
The effective bulk modulus of the system	B	15500	bar
Loaded total mass	M	25	tons
Pitch Moment of Inertia	I_x, I_y	62.10^3	kg.m^2
Front axle load capacity	M_1, M_3	3.5	tons
Rear axle load capacity	M_2, M_4	9	tons
Passenger seat mass	Mp	54.3	kg
Front axle elastic stiffness	K_1, K_3	58500	N/m
Rear axle elastic stiffness	K_2, K_4	117000	N/m
Passenger seat stiffness	Kp	19960	N/m
Elastic stiffness of the tire	Kt	175500	N/m
Passenger seat damping coefficient	Cp	260	N·s/m
Front axle suspension damping coefficient	C_1, C_3	1290	N.s/m
Rear axle suspension damping coefficient	C_2, C_4	1690	N.s/m

* Other input parameters are selected according to ISO 2631–5 (2018) and ISO 8608 standards.

5 Results and Discussion

After completing the training phase, a test period is analyzed while comparing the new desired RMS with the predicted RMS to check the reliability of the neural network. The suspension model is tested on several random road profiles according to the ISO 8608 standard, which classifies the road roughness from class A to class F based on the PSD (power spectral density) as shown in the Fig. 3. The simulated suspension system for a speed of 50 km/h and the hybrid model Artificial Neural Networks Particle Swarm

Optimization (ANN-PSO) is excited by several road profiles classified according to the ISO 8608 standard from class A to class F.

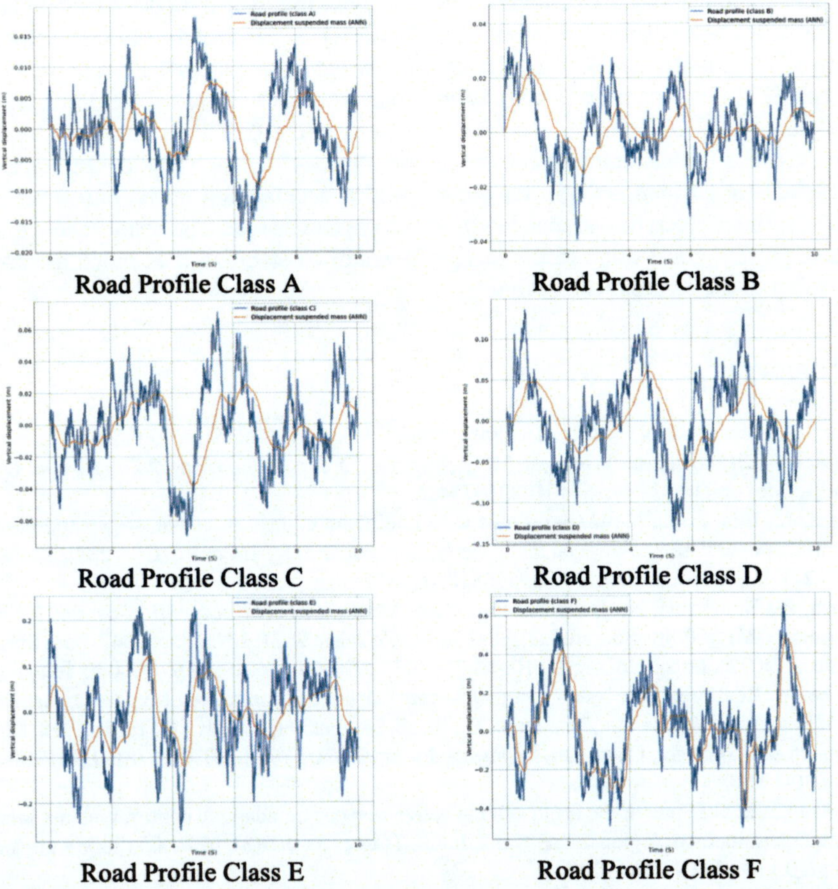

Fig. 3. Road profiles excitations from class A to class F according to ISO 8608 standards.

Root-mean-square error (RMSE) is a commonly used measure of the difference between predicted and actual values. It is a measure of the difference between a predicted value and an actual value, or between two sets of predicted and actual values. The ANN-PSO model has the best capability for all parameter outputs. It revealed the lowest train MSE for the RMS of the suspended acceleration. The same interpretation can be observed for the value of the correlation coefficient (R^2) which is 0.992 for the suspended acceleration of the four wheels (RMS Z_1, RMS Z_2, RMS Z_3 and RMS Z_4) and 0.987 for the acceleration suspended from the passenger seat (RMS Zp).

6 Conclusion

Considering the ability of neural networks to deal with unstructured problems, predict outcomes, and classify and analyses data, we chose the hybrid model Artificial Neural Networks Particle Swarm Optimization (ANN-PSO) as the tool to solve our active suspension. We initially developed a database extracted from a literature study is based on the ISO 2631-5 standard that studies driver comfort, with a sample of condition types on a different road (road profiles classified according to the ISO 8608 standard from class A to class F). The same interpretation can be observed for the value of the correlation coefficient (R^2) which is 0.992 for the suspended acceleration of the four wheels and 0.987 for the acceleration suspended from the passenger seat. The results show that the system gives a good value to the damper actuator to stabilize it as much as possible according to the desired RMS value.

References

Hamza, A., Ben Yahia, N.: Artificial neural networks controller of active suspension for ambulance based on ISO standards. Proc. Inst. Mech. Eng. Part D: J. Automob. Eng. **237**(1), 34–47 (2023). https://doi.org/10.1177/09544070221075456

Hamza, A., Ben Yahia, N.: Heavy trucks with intelligent control of active suspension based on artificial neural networks. Proc. Inst. Mech. Eng. Part I: J. Syst. Control Eng. **235**(6), 952–969 (2021). https://doi.org/10.1177/0959651820958516

Hamza, A., Ben Yahia, N.: Deep learning based intelligent active suspension control for heavy trucks (DMPSO). In: Bouraoui, T., et al. (eds.) CoTuMe 2021. LNME, pp. 347–354. Springer, Cham (2022). https://doi.org/10.1007/978-3-030-86446-0_46. ISBN 978-3-030-86445-3

Hamza, A., Ben Yahia, N.: Intelligent neural network control for active heavy truck suspension. In: Benamara, A., Haddar, M., Tarek, B., Salah, M., Fakher, C. (eds.) CoTuMe 2018. LNME, pp. 16–23. Springer, Cham (2019). https://doi.org/10.1007/978-3-030-19781-0_2. ISBN 978-3-030-19780-3

Jiangtao, C., et al.: State of the art in vehicle active suspension adaptive control systems based on intelligent methodologies. IEEE Trans. Intell. Transp. Syst. **9**(3), 392–405 (2008). https://doi.org/10.1109/TITS.2008.928244

Muhammad, M.R., et al.: Analysis and experimental study of magneto rheological-based damper for semi active suspension system using fuzzy hybrids. IEEE Trans. Ind. Appl. **47**(2), 1051–1059 (2011). https://doi.org/10.1109/TIA.2010.2103292

Khan, L., Qamar, S., Khan, M.U.: Neuro-fuzzy wavelets based network for full car active suspension system. In: IEEE International Conference on Engineering Technologies (ICET) (2012). https://doi.org/10.1109/ICET.2012.6375430

Ruey-Jing, L.: Enhanced adaptive self-organizing fuzzy sliding-mode controller for active suspension systems. IEEE Trans. Industr. Electron. **60**(3), 958–968 (2013). https://doi.org/10.1109/TIE.2012.2190372

Vibha, G., et al.: Comparative analysis of ANFIS and ANN for evaluating inter-agent dependency requirements. Int. J. Comput. Inf. Syst. Industr. Manage. Appl. **6**, 23–34 (2014)

Souilem, H., Mehjoub, S., Derbel, N.: Intelligent control for a half-car active suspension by self-tunable fuzzy inference system. Int. J. Fuzzy Syst. Adv. Appl. **18**(3), 9–15 (2015)

Geweda, A., El-Gohary, M., El-Nabawy, A., Awad, T.: Improvement of vehicle ride comfort using genetic algorithm optimization and PI controller. Alex. Eng. J. **56**(4), 405–414 (2017). https://doi.org/10.1016/j.aej.2017.05.014

Khodadadi, H., Ghadiri, H.: Self-tuning PID controller design using fuzzy logic for half car active suspension system. Int. J. Dyn. Control **6**(1), 224–232 (2018). https://doi.org/10.1007/s40435-016-0291-5

León-Vargas, F., Garelli, F., Zapateiro, M.: Limiting vertical acceleration for ride comfort in active suspension systems. Proc. Inst. Mech. Eng. Part I: J. Syst. Control Eng. **232**(3), 223–232 (2018). https://doi.org/10.1177/0959651817745469

Devdutt, S.: Passenger body vibration control in active quarter car model using hybrid ANFIS PID controller. Int. J. Intell. Syst. Appl. **10**(5), 51–60 (2018). https://doi.org/10.5815/ijisa.2018.05.06

Yang, Z., Liang, S., Zhou, Y., Zhao, D.: Sliding mode control for vibration comfort improvement of a 7-DOF nonlinear active vehicle suspension model. J. Robot. Mechatron. **31**(1), 95–103 (2019). https://doi.org/10.20965/jrm.2019.p0095

International Organization for Standardization (IOS), Mechanical vibration and shock – Evaluation of human exposure to whole-body vibration - Part 5, ISOi2631–5 (2018)

International Organization for Standardization (IOS), Mechanical Vibration Road Surface Profiles Reporting of Measured Data; ISOi8608 (2016)

British Standard Institution (BSI), Proposals for Generalised Road Inputs to Vehicles; BSI (MEE/158/3/1), London, BSI (72/34562) (1972)

Damping Behavior of Bio-Based Anti-trichiral Materials Made with Additive Manufacturing

Anis Hamrouni[1,2](✉), Jean Luc Rebiere[1], Abderrahim El Mahi[1], Moez Beyaoui[2], and Mohamed Haddar[2]

[1] Acoustics Laboratory of Le Mans University (LAUM), UMR CNRS 6613, Le Mans University, Av. O. Messiaen Le Mans, France
anis_hamrouni@yahoo.com, {jean-luc.rebiere, abderrahim.elmahi}@Univ-lemans.fr

[2] Laboratory of Mechanics Modeling and Production (LA2MP), National School of Engineers of Sfax, University of Sfax, Sfax, Tunisia
Mohamed.haddar@enis.rnu.tn

Abstract. Anti-chiral metastructures made up of ring nodes and elastic bending ligaments provide excellent flexibility in design for compliant architecture. Metastructures with auxetic behavior are known for their enhanced energy absorption. In this article, the Poisson's ratio and damping properties of the anti-trichiral architecture and sandwich composite are studied. The nucleus and the sandwiches are made from the same bio-based material which is polylactic acid PLA reinforced with flax fibers. Additive manufacturing technology is used to fabricate the specimens. Uniaxial tensile tests are performed on the anti-trichiral nuclei. Vibration tests were conducted on the auxetic nuclei, face sheets and sandwich structures with a clamp-free configuration. Those tests are carried out on specimens with different geometric parameters of the anti-trichiral unit cell in order to study their effect on the static and dynamic properties of this metastructure. In addition, a finite element model was used to estimate the Poisson's ratio values. The experimental and numerical results were in close agreement. The results show that the structural Poisson's ratio of the anti-trichiral nucleus largely depends on the radius (r) of the cylindrical nodes of the unit cell. The results also show that the damping properties of the auxetic structure depend on the radius (r).

Keywords: Vibration · Poisson's ratio · Auxetic · 3D Printing · Bio-sourced

1 Introduction

Structures and materials with negative Poisson's ratio (NPR) have particular mechanical behavior and expand or contract laterally under uniaxial tension or pressure, respectively. They present numerous advantages compared to conventional materials in terms of energy absorption capacity, compressive resistance, shear-bearing capacity and fracture resistance (Hou et al. 2016). Therefore, NPR materials present a wide industrial application in biomedical implants (Kuribayashi et al. 2006), aerospace (Bettini et al. 2010), artificial prostheses (Scarpa 2008), protection mats (Wang and Lakes 2002), sound

A. Akrout et al. (Eds.): ICAV 2022, ACM 22, pp. 214–223, 2023.
https://doi.org/10.1007/978-3-031-34190-8_24

insulations (Bertoldi et al. 2010), shock absorbers (Ma et al. 2013) and others. Lakes (1987) developed a polyurethane foam with a Poisson ratio of −0.7. More attention was paid to the studies of auxetic structures. The most known auxetic structures are the chiral and re-entrant honeycombs. Gibson et al (1982) first proposed the 2D re-entrant structure in 1982. Chiral honeycombs were suggested relatively late with respect to re-entrant structures. Prall and Lakes (1997) were the first authors to prove a two-dimensional chiral lattice with a Poisson's ratio toward -1. Alderson et al (2010) measured the in-plane Poisson's ratio and elastic modulus of hexachiral, trichiral, anti-trichiral tetrachiral and anti-tetrachiral honeycombs made with rapid prototyping.

Hu et al. (2019a; 2019b) studied the mechanical behavior of anti-trichiral and tetrachiral structures through both experimental and analytical analysis. Mousanezhad et al. (2016) studied the effects of two geometric refinement strategies, chirality and hierarchy, on the in-plane elastic response. Sandwich materials made with an auxetic core have shown their energy absorption capacity in a recent study by Hamrouni et al. (2021; 2022; 2023a). Scarpa et al. (2007) found that the topology of the material has considerable effects on the mechanical properties under tensile and bending tests as well as on the energy absorption of metamaterials. Ma et al. (2013) proposed an anti-tetrachiral design for a vibration damper sandwich structure, where the nodes are filled with rubber material. Static and vibration tests are carried out to evaluate the performances of the integrated rubber material damper structures. It is observed that the slightly integrated auxetic damping structure has higher loading capability and damping. Although distinct configurations of anti-trichiral and other structures have been studied, the effect of geometrical parameters on the dynamic behavior of a bio-based anti-trichiral structure obtained by additive printing technology has not been thoroughly investigated.

In this study, a two-dimensional anti-trichiral structure with auxetic behavior was suggested based on the node rotation mechanism of chiral honeycombs. The anti-trichiral nucleus and the sandwiches are made using a 3D printer. Four different radii of the ring nodes of the unit cell were studied. Experimental tensile tests are carried out to study the properties of this material. The Poisson's ratio of the specimens was compared to a numerical prediction. A good agreement was noted between the experimental and FE prediction. Then, damping properties of nuclei and sandwich structures are also measured by vibration tests.

2 Materials and Methods

2.1 Structure Design and Manufacturing

The proposed material is manufactured using polylactic acid (PLA) reinforced with short flax fibers with a density of 1000 kg.m^{-3} and the Young's modulus of 3400 MPa. The fiber volume fraction is less than 20%. The Flax/PLA filaments are provided by NANOVIA and have a diameter of 1.75 mm. This bio-composite metamaterial, developed for an additive manufacturing process, is a biodegradable, biosourced and recyclable. The metamaterials are made with a RAISE3D Pro2 Plus 3D printer. A reel pulls a filament from the supply spool and introduces it into the heating block, where it is heated to 220 °C before being placed on the printing platform (which is heated around 60°) using a printing with a diameter of 0.4 mm, a travel speed of 60 mm/s, a layer thickness of 0.2 mm and infill density of 100%. A CAD model of the materials is needed to create the desired shapes and the model is then converted into instructions that the printer can understand using dedicated Idea Maker software.

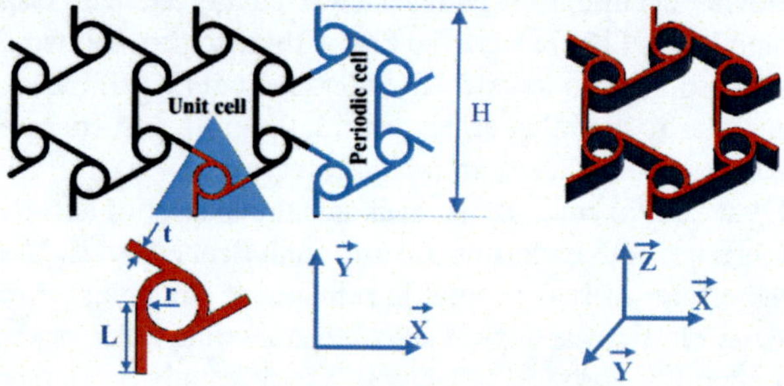

Fig. 1. Anti-trichiral honeycomb

Figure 1 shows the anti-trichiral honeycomb geometry employed in this work. The specimens are designed with Solidworks CAO cads and manufactured with different cell sizes (Table 1). The relative density of the core is determined using the expression (1) (Mousanezhad et al. 2016). All unit cells are inserted in an equilateral triangle. L is the ligament length, r is the radius of cylindrical components and t is the cell wall thickness. The plane directions of the cells are carried by (X, Y) directions and the thickness is along the Z direction. H and e are the width in the Y direction and the specimen thickness respectively.

$$\frac{\rho}{\rho_s} = \frac{1}{\sqrt{3}}\frac{t}{L}(1 + \frac{2\pi}{3}\frac{r}{L}) \tag{1}$$

Table 1. Anti-trichiral honeycombs parameters

r (mm)	ρ/ρs (%)	t (mm)	H (mm)	L (mm)	e (mm)
1.2	13.31	0.6	25	4.16	5
1.7	15.40				
2.2	17.48				
2.7	19.56				

2.2 Tensile Tests

The metamaterial properties are investigated through tensile tests performed on 3D printed dogbone according to the ASTM D638 standard test method (International 2014). The standard hydraulic machine INSTRON is used with a load cell of 1 kN and a rate of

1 mm/min. Figure 2 presents the tensile setup of the architectural nucleus. A transverse extensometer is used to estimate the transverse displacement of the specimens during tests. This architected material with different geometric sizes was tested in order to define its structural Poisson's ratio and Young's modulus. Specimens were printed with a length of 100 mm, a width of 25 mm and a thickness of 5 mm as shown in Fig. 2 (c). Two square bloc 25 mm long was printed with the specimens in order to avoid damaging the auxetic cells during clamping. Five samples are tested for each configuration during mechanical tests in order to take into account the variability of the experimental results.

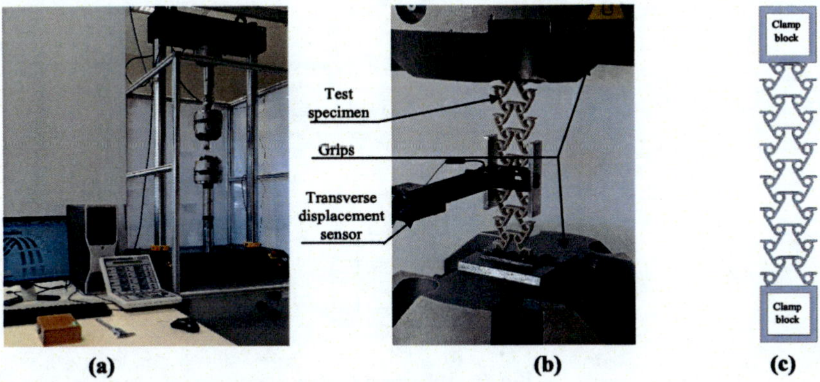

Fig. 2. (a) INSTRON Standard hydraulic machine and (b) Experimental tensile test equipment (c) Tensile specimen

2.3 Free Vibration Experimental Setup

To investigate the vibration behavior of composite and sandwich beams, free vibration tests were performed with the device shown in Fig. 3. The beams were evaluated using a clamped-free setup in accordance with ASTM E-756 (ASTM 1998). Experimentally, modal vibration analysis with impact excitation has the benefit of being very straightforward to implement. The specimen is excited at a point near the fixed end with a special impact hammer and the response is detected near the free end of the beam with a laser vibrometer. The excitation signals of the structure and those of the vibrometer output are then detected and processed by a device developed by SigLab whose function is to analyze the dynamic signals. This analyzer is essentially composed of an acquisition and processing card, coupled with software for checking and processing signals. The amplitude and frequency were measured during beam bending for each resonant peak of the frequency response of the beam. These values of the peak frequencies as well as the loss factor of each mode can be obtained using this approach. Modal damping factors may be calculated using the Half Power Bandwidth (HPB) approach. Equation (2) is used to calculate it. The damping factor η_n is calculated by dividing the bandwidth frequency at which the amplitude resonance decreases by 3 dB, by the resonance frequency (f_n). Different methodologies may be used to characterize the material's dynamic properties, such as the dynamic mechanical analysis methodology (Xu et al. 2019) and the experimental frequency response function (FRF) (Daoud et al. 2017) and (Essassi et al. 2019; Hamrouni et al. 2023b), for example. The FRF method is used in this study to determine typical flexural frequencies.

$$\eta_n = \frac{\Delta f_n}{f_n} = \frac{f_2 - f_1}{f_n} \tag{2}$$

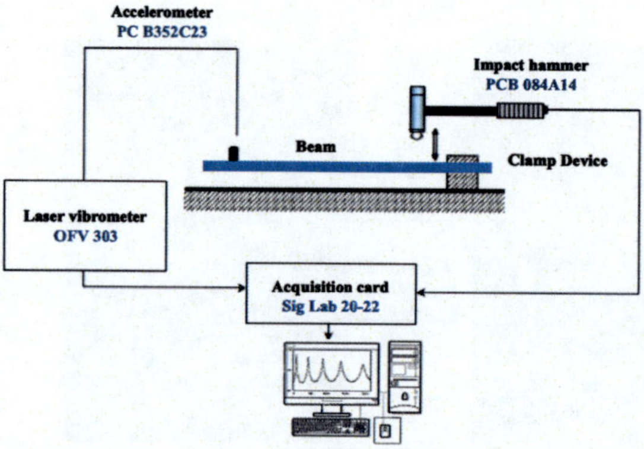

Fig. 3. Vibration device

3 Results

3.1 Anti-trichiral Core Properties

Tensile tests are carried out on 3D printed anti-trichiral cores of various cylindrical node radii of the unit cell. The structural Poisson's ratio can be determined using transverse/longitudinal strain curves in the linear-elastic strain domain. Extensometers are used to measure strains. Figure 4 shows the structural NPR of anti-trichiral honeycombs with different radii. The results show that the ratio is negative for this material. As a radius r increases, the NPR increases. Indeed, this metamaterial can grow when it is stretched under the effect of a tensile force thanks to the rotation of the cylindrical nodes. As r increases, in the anti-trichiral cell, the nucleus becomes auxetic, as the cylindrical components become the major source that composes the structure, while the straight components become increasingly negligible. In addition, a finite element model was used to estimate the Poisson's ratio values. The experimental and numerical results were in close agreement, which will allow us to exploit them to define the static properties of the cores. The nucleus material is assumed to have linear elastic behavior. The behavior law used in the computational software is that provided by the manufacturer Nanovia. Tetrahedral quadratic elements were used to mesh the nucleus. The number of elements used to mesh each specimens depends essentially on the geometrical parameters of its unit cell and varies between 15,145 for the nucleus with $r = 1.2$ mm and 17,720 for the nucleus with $r = 2.7$ mm.

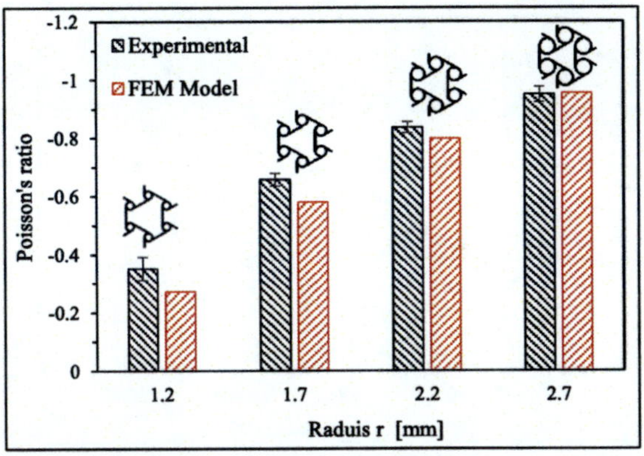

Fig. 4. Poisson's ratio as a function of the radius (r) of the anti-trichiral cores

3.2 Damping Performance of Anti-trichiral Materials

Free vibration tests were performed to study the damping properties of the anti-trichiral cores. Figure 5 shows the results of the vibration tests carried out on the anti-trichiral nuclei as a function of frequency. The curve shows the damping factor for two radii (r) of the cylindrical node. As the frequency increases, the structural damping factor increases; at low frequencies, the loss factor for the nuclei is between 1.55% and 1.80% for all sizes of the unit cells. Then the results diverge as the frequency increases. Indeed, an increase in frequency maximizes the relaxation of the architectural PLA Flax material after its deformation, which implies an increase in the structural loss factor. However, for auxetic nuclei with structural NPR, the results show that the damping properties improve as the node radius (r) increases.

Figure 6 shows the evolution of structural loss factors as a function of the node radius for different frequencies. An improvement of damping properties is noted for the higher radii. As r tends towards 1.2 the core loss factor decreases until it reaches its lowest value (1.7% at 100 Hz and 3.2% at 1500 Hz). The factor begins to increase gradually to reach its best state, which corresponds to r = 3.2 mm (4% at 1500 Hz). This behavior is observable for all the proposed frequency levels except for the lowest frequency. When the frequency F = 100 Hz, the damping factor remains almost constant between 1.65% and 1.8%, which represents the damping behavior of the base material PLA flax. The results of damping properties of the sandwich structures are presented in Fig. 7. For all the beams, a constant behavior of the loss factor is noticed except for high frequency where the damping factor is slightly higher when r toward 3.2 mm.

Several hypotheses can be proposed to explain certain damping properties of the anti-trichiral core. The architecture of the unit cell and its geometric parameters greatly influence damping properties. As shown in the previous paragraph, the variation of the radius causes the auxeticity of the nucleus, the auxetic behavior has a significant impact on the energy absorption and therefore on the beam loss factor. As the degree of negative Poisson's ratio increases, the damping factor of the auxetic core increases. In addition, the cylindrical geometries of the structure have a good energy absorption capacity, unlike the rectilinear forms. However, like the structural Poisson's ratio tends towards. To sum

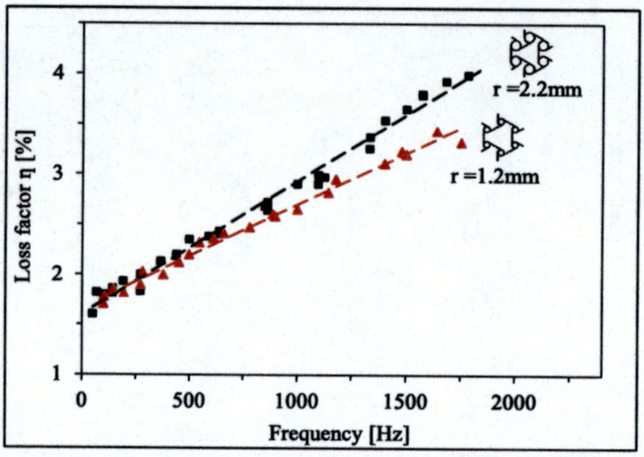

Fig. 5. Evolution of the loss factor of the anti-trichirale cores as a function frequency

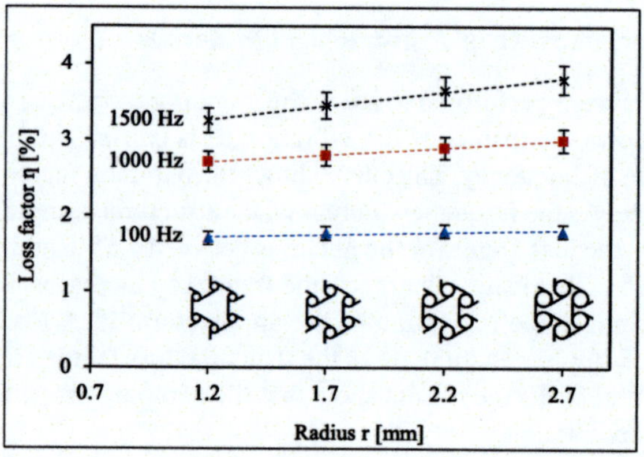

Fig. 6. Loss factor of anti-trichiral cores as a function of the radius (r)

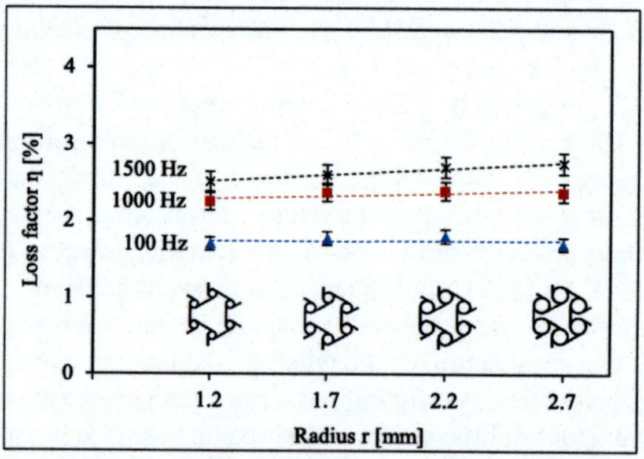

Fig. 7. Loss factor of sandwich structures as a function of the radius (r)

up, we can say that the damping properties of the architectural material depend on the value of the Poisson's ratio of the structure.

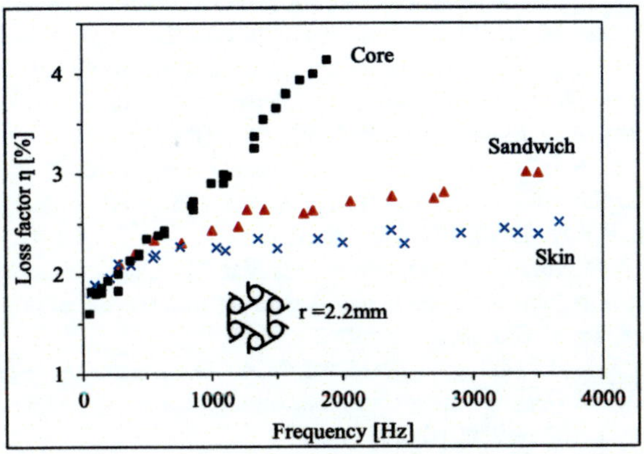

Fig. 8. Evolution of the loss factor with the frequency of the skin, auxetic core and sandwich beam

Figure 8 represents a comparison between the different components studied which are: a solid skin 7 mm thick, an anti-trichiral core 5 mm thick and a sandwich 7 mm thick consisting of a core and two thin face sheets. The loss factor variation in these components is shown for $r = 2.2$ mm. Clearly, at high frequencies, the damping factor of the sandwich beams is higher than that of the skins, but it is lower than that of the anti-trichiral core. This study defines the damping range of sandwiches and cores compared to skins, as well as the improvement in damping properties that can be provided when replacing a solid structure with an anti-trichiral structure.

4 Conclusion

Meta-structure with architectural nucleus manufacturing using a 3d printer is studied. Cores and anti-trichiral sandwiches are made from the same material which is polylactic acid (PLA) reinforced with flax fibers. Four radii of cylindrical components of the anti-trichiral unit cell are chosen to study their effect on the static and damping properties of the nucleus and sandwich composites. The anti-trichiral nucleus shows auxetic behavior in tensile tests. The results show that the Poisson's ratio varies from −0.25 to −1 depending on the radius (r). The degree of auxeticity increases as the radius of the cylindrical node increases. The results demonstrated that the experimental indications were in good agreement with the numerical predictions. In addition, vibration tests are carried out for the sandwich made with the same nuclei studied in the tensile tests. It is found that the specimens made with a high elementary cell radius of nucleus, exhibit high damping properties.

References

Alderson, A., et al.: Elastic constants of 3-, 4-and 6-connected chiral and anti-chiral honeycombs subject to uniaxial in-plane loading. Compos. Sci. Technol. **70**(7), 1042–1048 (2010)

Astm, E.: Standard Test Method for Measuring Vibration-Damping Properties of Materials, pp. 756–798. American Society for Testing and Materials Standards (1998)

Bertoldi, K., Reis, P.M., Willshaw, S., Mullin, T.: Negative Poisson's ratio behavior induced by an elastic instability. Adv. Mater. **22**(3), 361–366 (2010)

Bettini, P., Airoldi, A., Sala, G., Di Landro, L., Ruzzene, M., Spadoni, A.: Composite chiral structures for morphing airfoils: Numerical analyses and development of a manufacturing process. Compos. B Eng. **41**(2), 133–147 (2010)

Daoud, H., El Mahi, A., Rebiere, J.-L., Taktak, M., Haddar, M.: Characterization of the vibrational behaviour of flax fibre reinforced composites with an interleaved natural viscoelastic layer. Appl. Acoust. **128**, 23–31 (2017)

Essassi, K., Rebiere, J.-L., El Mahi, A., Ben Souf, M.A., Bouguecha, A., Haddar, M.: Dynamic characterization of a bio-based sandwich with auxetic core: experimental and numerical study. Int. J. Appl. Mech. **11**(02), 1950016 (2019)

Gibson, L.J., Ashby, M.F., Schajer, G.S., Robertson, C.I.: The mechanics of two-dimensional cellular materials. Proc. R. Soc. London A **382**(1782), 25–42 (1982)

Hamrouni, A., Rebiere, J.-L., El Mahi, A., Beyaoui, M., Haddar, M.: Experimental analysis of the Static and dynamic behavior of 3D printed bio-based conventional and auxetic architectural materials. Int. J. Appl. Mechanics **14**, 2250025 (2022). https://doi.org/10.1142/S1758825122500259

Hamrouni, A., Rebiere, J.-L., Mahi, A.E., Beyaoui, M., Haddar, M.: Dynamic behavior of sandwiches with an auxetic or a conventional core: experimental study. In: Walha, L., et al. (eds.) Design and Modeling of Mechanical Systems - V: Proceedings of the 9th Conference on Design and Modeling of Mechanical Systems, CMSM'2021, December 20-22, 2021, Hammamet, Tunisia, pp. 374–384. Springer International Publishing, Cham (2023a). https://doi.org/10.1007/978-3-031-14615-2_42

Hamrouni, A., Rebiere, J.L., Mahi, A.E., Beyaoui, M., Haddar, M.: Experimental analysis of the dynamic behavior of a sandwich with a bio-based auxetic core. In: Ben Amar, M., Bouguecha, A., Ghorbel, E., El Mahi, A., Chaari, F., Haddar, M. (eds.) A3M 2021. LNME, pp. 73–82. Springer, Cham (2021). https://doi.org/10.1007/978-3-030-84958-0_8

Hamrouni, A., Rebiere, J.-L., El Mahi, A., Beyaoui, M., Haddar, M.: Experimental and finite element analyses of a 3D printed sandwich with an auxetic or non-auxetic core. J. Sandwich Struct. Mater. **25**, 426–444 (2023b). https://doi.org/10.1177/10996362231151454

Hou, X., Deng, Z., Zhang, K.: Dynamic crushing strength analysis of auxetic honeycombs. Acta Mech. Solida Sin. **29**(5), 490–501 (2016)

Hu, L.L., Luo, Z.R., Zhang, Z.Y., Lian, M.K., Huang, L.S.: Mechanical property of re-entrant anti-trichiral honeycombs under large deformation. Compos. B Eng. **163**, 107–120 (2019)

Hu, L.L., Ye, W.K., Wu, Z.J.: Mechanical property of anti-trichiral honeycombs under large deformation along the x-direction. Thin-Walled Struct. **145**, 106415 (2019)

International A: Standard Test Method for Tensile Properties of Plastics. ASTM International (2014)

Kuribayashi, K., et al.: Self-deployable origami stent grafts as a biomedical application of Ni-rich TiNi shape memory alloy foil. Mater. Sci. Eng., A **419**(1–2), 131–137 (2006)

Lakes, R.: Foam structures with a negative Poisson's ratio. Science **235**(4792), 1038–1040 (1987)

Ma, Y., Scarpa, F., Zhang, D., Zhu, B., Chen, L., Hong, J.: A nonlinear auxetic structural vibration damper with metal rubber particles. Smart Mater. Struct. **22**(8), 084012 (2013)

Mousanezhad, D., Haghpanah, B., Ghosh, R., Hamouda, A.M., Nayeb-Hashemi, H., Vaziri, A.: Elastic properties of chiral, anti-chiral, and hierarchical honeycombs: a simple energy-based approach. Theor. Appl. Mech. Lett. **6**(2), 81–96 (2016)
Prall, D., Lakes, R.S.: Properties of a chiral honeycomb with a Poisson's ratio of—1. Int. J. Mech. Sci. **39**(3), 305–314 (1997)
Scarpa, F.: Auxetic materials for bioprostheses [In the Spotlight]. IEEE Signal Process. Mag. **25**(5), 126–128 (2008)
Scarpa, F., Blain, S., Lew, T., Perrott, D., Ruzzene, M., Yates, J.R.: Elastic buckling of hexagonal chiral cell honeycombs. Compos. A Appl. Sci. Manuf. **38**(2), 280–289 (2007)
Wang, Y.-C., Lakes, R.: Analytical parametric analysis of the contact problem of human buttocks and negative Poisson's ratio foam cushions. Int. J. Solids Struct. **39**(18), 4825–4838 (2002)
Xu, X., Koomson, C., Doddamani, M., Behera, R.K., Gupta, N.: Extracting elastic modulus at different strain rates and temperatures from dynamic mechanical analysis data: a study on nanocomposites. Compos. B Eng. **159**, 346–354 (2019)

Static Study of Bio-Based Architectural Materials Made with 3D Printing Technology

Anis Hamrouni[1,2](✉), Jean-Luc Rebiere[1], Abderrahim El Mahi[1], Moez Beyaoui[2], and Mohamed Haddar[2]

[1] Acoustics Laboratory of Le Mans University (LAUM) UMR CNRS 6613, Le Mans University, Av. O. Messiaen, Le Mans, France
anis_hamrouni@yahoo.com, {jean-luc.rebiere, abderrahim.elmahi}@Univ-lemans.fr

[2] Laboratory of Mechanics Modeling and Production (LA2MP), National School of Engineers of Sfax, University of Sfax, Sfax, Tunisia
moez.beyaoui@yahoo.fr, Mohamed.haddar@enis.rnu.tn

Abstract. In several industrial sectors, there are continuous requirements to develop lighter materials for various loading situations. The special process for achieving it is mainly the use of composite sandwiches where two skins are glued or printed on an architectural core. In this article, the static behavior and the failure mechanisms in sandwich composite with anti-trichiral architecture are studied. The core and the sandwiches are made from the same bio-based material which is polylactic acid PLA with short flax fibers. A Raise3D Pro2 Plus printer is used to elaborate the tested specimens. Several tensile tests are carried out on the anti-trichiral core while bending tests are performed on the sandwich structures. Those tests are conducted on specimens of the metamaterial for different node radius of the anti-trichiral unit cell in order to study their effect on the static properties of this architectured material and the damage characteristics of the sandwiches. Acoustic emission technology (AE) is used to monitor and quantify the failure mechanisms of sandwich materials. The results show that the structural Poisson's ratio and the Young's modulus of the anti-trichiral cores largely depends on the radius (r) of the round nodes of the unit cell.

Keywords: Poisson's ratio · Anti-trichiral · 3D Printing · Acoustic emission · Bending

1 Introduction

Sandwich structures have been widely used in the automotive, sports and aerospace industries due to their strength-to-weight ratio and flexural rigidity (Schaedler and Carter 2016; Toubia and Elmushyakhi 2017). Sandwich panels consist of a core thick confined by two thin and rigid skins. The core is usually wood, foam or alveolar materials like architectural material. The mechanical behavior of the architectured structure depends on the design of the core and the base material used. Scarpa et al. (2000; 2007) found that the topology of the material has considerable effects on the mechanical properties

A. Akrout et al. (Eds.): ICAV 2022, ACM 22, pp. 224–233, 2023.
https://doi.org/10.1007/978-3-031-34190-8_25

under tensile and bending tests as well as on the energy absorption of meta-materials. Auxetic honeycombs with a negative Poisson's ratio (NPR) are the structures used in some sandwich panels. Sandwich materials made with an auxetic core have shown their energy absorption capacity in a recent study by Hamrouni et al. (2021; 2022; 2023a; 2023b) and show reduced deflection upon application of bending (Romanoff and Varsta 2007; Hou et al. 2013; Xia et al. 2022). Mousanezhad et al. (2015a; 2015b; 2016) studied the effects geometric refinement strategies, chirality and hierarchy, on the in-plane elastic response. Additionally, the mechanical properties of sandwich panels with auxetic and conventional cores have been studied and discussed (Hou et al. 2018). They found that the auxetic core has a higher strength and durability, but a lower energy absorption capacity than a non-auxetic core. Recently, additive manufacturing technology has seen a good development which allows to easily supervise the intricate details of architectural materials (Yu et al. 2019; Wang et al. 2020). Today's environmental issues require the use of bio-based rather than synthetic products for their biodegradability, recyclability and sometimes low cost (Baghaei et al. 2013; Hamrouni et al. 2021).

In this work, a bio-based material that combines excellent sandwich panel bending performance with an architectural core is developed. The anti-trichiral core and the sandwich are made using a 3D printer. Four different radii of the cylindrical components of the unit cell are studied. Experimental tensile tests are carried out to study the properties of this material. The Young's modulus of the anti-trichiral nucleus is compared to a theoretical prediction. The static properties of sandwich panels are also determined by three-point bending tests. Then, the detection of the initiation and the monitoring of the propagation of damage is carried out using the acoustic emission method (AE). The results are discussed in terms of maximum load capacity combined with the appearance of AE data points.

2 Materials and Methods

2.1 Structural Design and Manufacturing

In The proposed material is manufactured using polylactic acid (PLA) with short flax fibers with a density of 1000 kg.m^{-3} and the Young's modulus of 3400 MPa. The fiber volume fraction is less than 20%. The Flax/PLA filaments are provided by NANOVIA and have a diameter of 1.75 mm. This bio-composite metamaterial, developed for an additive manufacturing process, is a biodegradable, biosourced and recyclable. The metamaterials are made with a RAISE3D Pro2 Plus 3D printer. A reel pulls a filament from the supply spool and introduces it into the heating block, where it is heated to 220 °C before being placed on the printing platform (which is heated around 60 °C) using a printing nozzle with a diameter of 0.4 mm, a travel speed of 60 mm/s, a layer thickness of 0.2 mm and infill density of 100%. A CAD model of the metamaterials is needed to create the desired shapes and the model is then converted into instructions that the printer can understand using dedicated Idea Maker software.

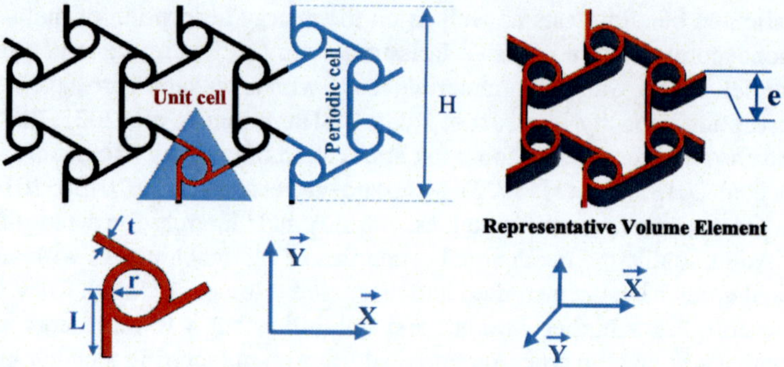

Fig. 1. Anti-trichiral honeycomb

Figure 1 shows the anti-trichiral honeycomb geometry employed in this work. The specimens are designed with Solsworks CAO cads and manufactured with different cell sizes (Table 1). The relative density of the core is determined using the expression (1) (Mousanezhad et al. 2016) where ρ and ρs are the density of the core and the Flax/PLA material respectively. All unit cells are inserted in an equilateral triangle. L is the ligament length, r is the radius of cylindrical components and t is the cell wall thickness. The plane directions of the cells are carried by (X, Y) directions, e is the specimen thickness along the Z direction and H is the specimen width in the Y direction.

$$\frac{\rho}{\rho_s} = \frac{1}{\sqrt{3}}\frac{t}{L}(1 + \frac{2\pi}{3}\frac{r}{L}) \quad (1)$$

Table 1. Anti-trichiral honeycombs parameters

r (mm)	ρ/ρs (%)	t (mm)	H (mm)	L (mm)	e (mm)
1.2	13.31	0.6	25	4.16	5
1.7	15.40				
2.2	17.48				
2.7	19.56				

2.2 Tensile Tests

The metamaterial properties are investigated through tensile tests performed on 3D printed dogbone according to the ASTM D638 standard test method (International 2014). The standard hydraulic machine INSTRON is used with a load cell of 1 kN and a rate of 1 mm/min. Figure 2 presents the tensile setup of the architectural core. A transverse extensometer is used to estimate the transverse displacement of the specimens during tests. This architected material with different radius values was tested in order to define its structural Poisson's ratio and Young's modulus. Specimens were printed with a length of 100 mm, a width of 25 mm and a thickness of 5 mm as shown in Fig. 2 (c). Each tensile specimen consists of one periodic cell in width and seven cells in length. Two square bloc 25 mm long was printed with the specimens in order to avoid damaging the auxetic cells during clamping. Five samples are tested for each configuration during mechanical tests in order to take into account the variability of the experimental results.

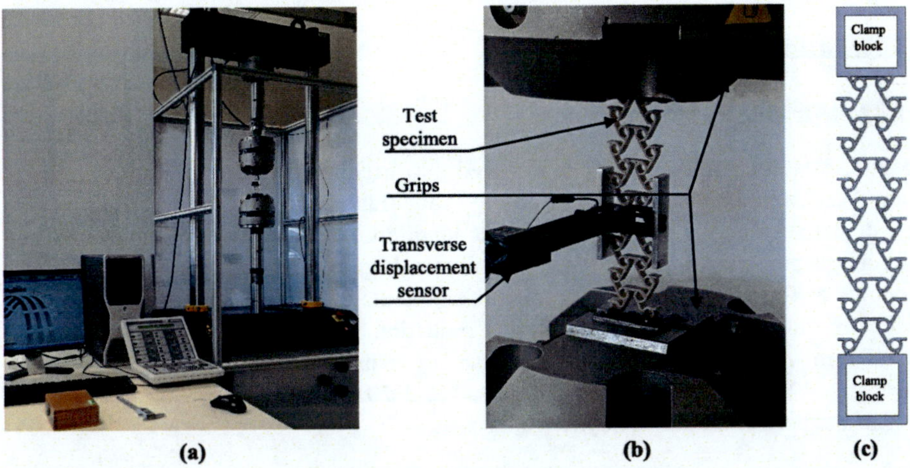

Fig. 2. (a) INSTRON Standard hydraulic machine, (b) Experimental tensile test equipment and (c) Tensile specimen

2.3 Three-Point Bending Test

Three-point bending tests are performed on the sandwich beams with different cell sizes according to ASTM C393 (International 2016) standard test methods as shown in Fig. 3(a). The quasi-static load is applied at a displacement speed of 3 mm/min. The tests are carried out with the hydraulic machine INSTRON equipped with a 1 kN load cell. The sandwiches shown in Fig. 3(b) are designed to have a total length of 130 mm, a width of 25 mm and a thickness of 7 mm (1 mm for each skin and 5 mm for the core). All specimens are tested to failure with a span between supports of 110 mm in order to analyze the properties of the sandwich at failure. To take into account the variability of the results due to the experimental conditions, five specimens for each configuration are tested.

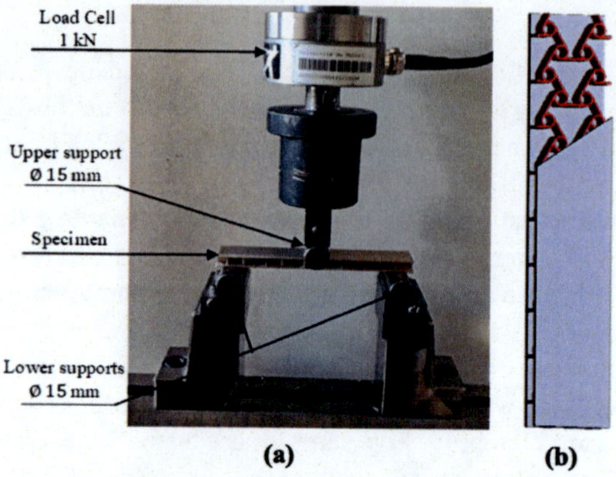

Fig. 3. (a) Experimental 3 points bending test equipment (b) Specimen bending test

3 Results

3.1 Anti-trichiral Core Properties

Tensile tests are carried out on 3D printed anti-trichiral cores of various cylindrical component radii of the unit cell. The Young's modulus as a function of the radius (r) may be calculated using the experimental stress-longitudinal strain curves, then normalized by Young's modulus of the core with r = 1.2 mm. Figure 4 shows the structural Young's modulus of the different anti-trichiral cores. When r increases, the stiffness decreases gradually until reaching approximately a reduction of 50% for the core with r equal to 2.7 mm. An analytical approach defined by expression 2 is used to estimate the Normalized Young's modulus (Mousanezhad et al. 2016). Results are in close agreement between experimental and analytical approaches.

$$\frac{E}{E_0} = \frac{1}{2\sqrt{3}}\left(\frac{t}{L}\right)^3\left[\frac{1}{1+6(r/L)^2}\right] \tag{2}$$

Additionally, the structural Poisson's ratio of this metamaterial can be determined using transverse/longitudinal strain curves in the linear-elastic strain domain. Extensometers are used to measure strains. Figure 5 shows the structural negative Poisson's ratio (NPR) of anti-trichiral honeycombs with different radii. The results show that the ratio is negative for this material. As a radius r increases, the NPR increases. Indeed, this meta-material can grow when it is stretched under the effect of a tensile force thanks to the rotation of the cylindrical components of the cells around their axes. As r increases, in the anti-trichiral cell, the nucleus becomes auxetic, as the cylindrical components become the major source that composes the structure, while the straight components become increasingly negligible.

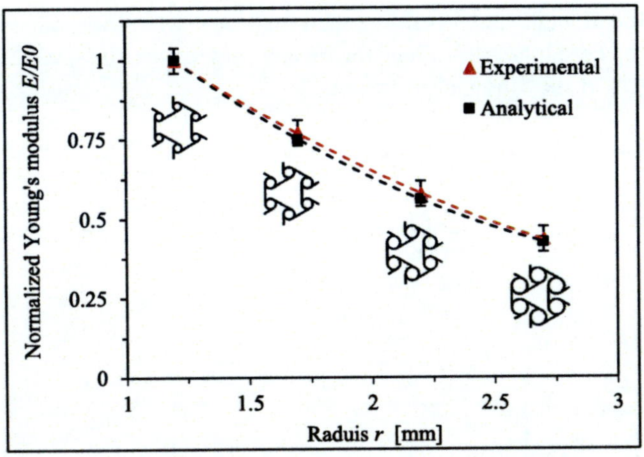

Fig. 4. Normalized Young's modulus as a function of the radius (r) of the anti-trichiral core

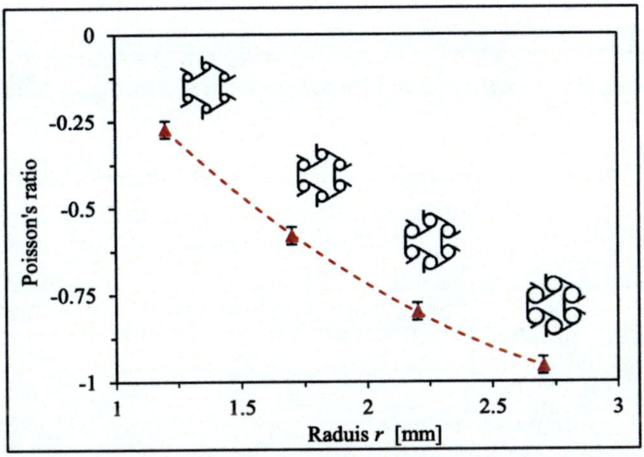

Fig. 5. Poisson's ratio as a function of the radius (r) of the anti-trichiral core

3.2 Bending Performance of Sandwiches

The sandwich structures can be exposed to bending loads in use. Experimental tests are used to determine the flex properties of the sandwich with an anti-trichiral core. The tests are carried out for 4 radii (r) of cylindrical components. All specimens are tested in three-point bending with a total length of 130 mm and a span of 110 mm between supports. Figure 6 shows an illustration of the static bending properties of sandwiches with a radius (r) equal to 2.7 mm. The curve can be divided into 3 parts representing linear behavior, then non-linear behavior and finally the failure of the sandwich. Table 2 shows the properties of the sandwich materials at failure. The results confirm that sandwiches with a core consisting of a high radius of cylindrical components have the highest bending strain and loading forces. There are many explanations for this behavior. As r increases, the number of cell walls under bending load increases. Moreover, the contact between the skin and the core increases. Thus, the density of the core rises, which improves the flexural strength of the sandwich structure.

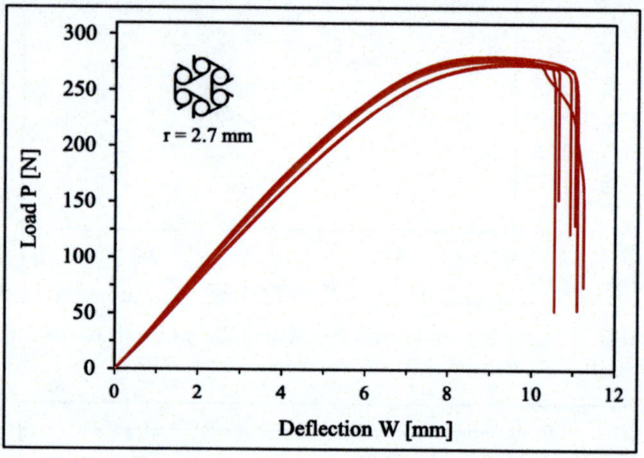

Fig. 6. Bending characteristic of sandwich with cylindrical components radius r = 2.7 mm

Table 2. Bending properties of the cores for different r values

r(mm)	Deflection *W* (mm)	Displacement at failure d_{rup} [mm]	Peak load P_{rup} [N]
1.2	$9.0^{\pm 0.5}$	6	$230^{\pm 5}$
1.7	$9.8^{\pm 0.5}$	6	$245^{\pm 5}$
2.2	$10.5^{\pm 0.5}$	6	$265^{\pm 5}$
2.7	$11.0^{\pm 0.5}$	6	$270^{\pm 5}$

3.3 Failure Mechanisms of Sandwiches

Figure 7 presents the results curve of the average loading time with the acoustic emission counts (AE) for two sandwiches made with an anti-trichiral core with r equal to 1.2 mm and 2.7 mm. It can be seen that the two sandwiches have almost the same shape of the loading time curve as well as the appearance of the AE data during the bending tests. A small difference is noted at the level of the high load supported by the architectural material. Obviously, as r increases, the peak load also increases. It is observed that the majority of AE data points are higher at the end of the test, which predicts cell failure.

It is also observed that the starting failure of the anti-trichiral core happens with a constant highest load. The acoustic amplitude range between 35 dB and 45 dB characterizes the damage initiation which corresponds to a crack in the bottom skin. Then, when the amplitudes vary between 45 dB and 65 dB, the damage propagation is already transferred to the cell wall of the core combined with a core-skin separation.

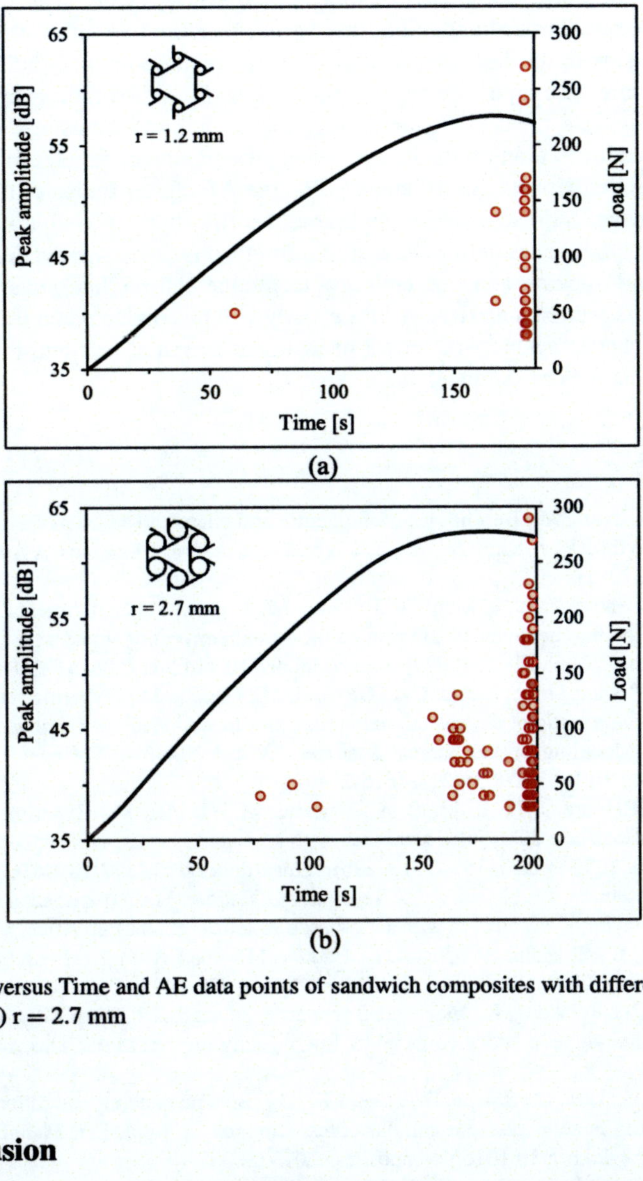

Fig. 7. Load versus Time and AE data points of sandwich composites with different radii (r): (a) r = 1.2 mm (b) r = 2.7 mm

4 Conclusion

Bio sourced sandwiches with architectural materials produced using a 3d printer is studied. Metamaterial cores and anti-trichiral sandwiches are made from the same material which is polylactic acid (PLA) with short flax fibers. Four radii of cylindrical components of the anti-trichiral unit cell are chosen to study their effect on the mechanical properties of core and sandwich composites. The anti-trichiral core shows auxetic behavior in tensile tests. The results show that the Poisson's ratio varies from -0.25 to -1 depending on the radius (r). The degree of auxeticity increases and the stiffness decreases as the radius of cylindrical components increases. The structural Young's modulus is consistent with the theoretical prediction. In addition, three-point bending tests are carried out for the sandwich made with the same cores studied in the tensile tests. In bending, specimens made with a core with high elementary cell radius exhibit the biggest static properties. During bending tests, the damage monitoring approach is applied using the AE acoustic emission technique to monitor the initiation and propagation of damage in the sandwich. At low amplitude (between 35 dB and 45 dB), the AE cluster indicates the appearance of minor damage. Indeed, at high amplitude, the EA shows a visible propagation of damage in the sandwiches until failure. Accordingly, it is demonstrated that this damage monitoring and tracking acoustic emission technique can predict damage and failure processes in 3D printed materials. A future study will be conducted on the static behavior of sandwiches to evaluate the effect of ligament length of unit anti-trichiral cell on the overall properties of sandwich materials.

References

Baghaei, B., Skrifvars, M., Berglin, L.: Manufacture and characterisation of thermoplastic composites made from PLA/hemp co-wrapped hybrid yarn prepregs. Compos. A Appl. Sci. Manuf. **50**, 93–101 (2013)

Hamrouni, A., Rebiere, J.-L., El Mahi, A., Beyaoui, M., Haddar, M.: Experimental analysis of the Static and dynamic behavior of 3D printed bio-based conventional and auxetic architectural materials. Int. J. Appl. Mech. (2022). https://doi.org/10.1142/S1758825122500259

Hamrouni, A., Rebiere, J.-L., El Mahi, A., Beyaoui, M., Haddar, M.: Dynamic behavior of sandwiches with an auxetic or a conventional core: experimental study. In: Walha, L., et al. (eds.) Design and Modeling of Mechanical Systems – V, pp. 374–384. Springer, Cham (2023a). https://doi.org/10.1007/978-3-031-14615-2_42

Hamrouni, A., Rebiere, J.-L., El Mahi, A., Beyaoui, M., Haddar, M.: Experimental and finite element analyses of a 3D printed sandwich with an auxetic or non-auxetic core. J. Sandwich Struct. Mater. 10996362231151454 (2023b). https://doi.org/10.1177/10996362231151454

Hamrouni, A., Rebiere, J.L., Mahi, A.E., Beyaoui, M., Haddar, M.: Experimental analysis of the dynamic behavior of a sandwich with a bio-based auxetic core. In: Ben Amar, M., Bouguecha, A., Ghorbel, E., El Mahi, A., Chaari, F., Haddar, M. (eds.) A3M 2021. LNME, pp. 73–82. Springer, Cham (2021). https://doi.org/10.1007/978-3-030-84958-0_8

Hou, S., Li, T., Jia, Z., Wang, L.: Mechanical properties of sandwich composites with 3d-printed auxetic and non-auxetic lattice cores under low velocity impact. Mater. Des. **160**, 1305–1321 (2018)

Hou, Y., Tai, Y.H., Lira, C., Scarpa, F., Yates, J.R., Gu, B.: The bending and failure of sandwich structures with auxetic gradient cellular cores. Compos. A Appl. Sci. Manuf. **49**, 119–131 (2013). https://doi.org/10.1016/j.compositesa.2013.02.007

ASTM International: Standard test method for tensile properties of plastics. ASTM International (2014)

ASTM International: Standard test method for core shear properties of sandwich constructions by beam flexure. ASTM International (2016)

Mousanezhad, D., et al.: Hierarchical honeycomb auxetic metamaterials. Sci. Rep. **5**(1), 18306 (2015a). https://doi.org/10.1038/srep18306

Mousanezhad, D., et al.: Spiderweb honeycombs. Int. J. Solids Struct. **66**, 218–227 (2015b). https://doi.org/10.1016/j.ijsolstr.2015.03.036

Mousanezhad, D., Haghpanah, B., Ghosh, R., Hamouda, A.M., Nayeb-Hashemi, H., Vaziri, A.: Elastic properties of chiral, anti-chiral, and hierarchical honeycombs: a simple energy-based approach. Theor. Appl. Mech. Lett. **6**(2), 81–96 (2016)

Romanoff, J., Varsta, P.: Bending response of web-core sandwich plates. Compos. Struct. **81**(2), 292–302 (2007). https://doi.org/10.1016/j.compstruct.2006.08.021

Scarpa, F., Blain, S., Lew, T., Perrott, D., Ruzzene, M., Yates, J.R.: Elastic buckling of hexagonal chiral cell honeycombs. Compos. A Appl. Sci. Manuf. **38**(2), 280–289 (2007)

Scarpa, F., Panayiotou, P., Tomlinson, G.: Numerical and experimental uniaxial loading on in-plane auxetic honeycombs. J. Strain Anal. Eng. Des. **35**(5), 383–388 (2000)

Schaedler, T.A., Carter, W.B.: Architected cellular materials. Annu. Rev. Mater. Res. **46**, 187–210 (2016)

Toubia, E.A., Elmushyakhi, A.: Influence of core joints in sandwich composites under in-plane static and fatigue loads. Mater. Des. **131**, 102–111 (2017)

Wang, Q., Yang, Z., Lu, Z., Li, X.: Mechanical responses of 3D cross-chiral auxetic materials under uniaxial compression. Mater. Des. **186**, 108226 (2020)

Xia, F., Durandet, Y., Tan, P.J., Ruan, D.: Three-point bending performance of sandwich panels with various types of cores. Thin-Walled Struct. **179**, 109723 (2022). https://doi.org/10.1016/j.tws.2022.109723

Yu, S., Hwang, Y.H., Hwang, J.Y., Hong, S.H.: Analytical study on the 3D-printed structure and mechanical properties of basalt fiber-reinforced PLA composites using X-ray microscopy. Compos. Sci. Technol. **175**, 18–27 (2019)

Detailed Specification for an Intelligent Mobile Sensor for Air Quality Monitoring Based on a Risk and Functional Analysis

Mohamed Abdessamia Chakchouk[1,2](✉), Pierre Richard Dahoo[3], Abdelkhalak El Hami[4], Azzedine Lakhlifi[5], Wajih Gafsi[2], and Mohamed Haddar[2]

[1] LMN, INSA de Rouen, 76800 Saint Etienne de Rouvray, France
mohamed.chakchouk@insarouen.fr
[2] LA2MP, Ecole Nationale des Ingénieurs de Sfax, Sfax, Tunisia
wajih.gafsi@enis.tn, mohamed.haddar@enis.rnu.tn
[3] LATMOS, Université Paris Saclay; UVSQ, CNRS, 78290 Guyancourt, France
pierre.dahoo@uvsq.fr
[4] 76800 Saint Etienne de Rouvray, France
abdelkhalak.elhami@insarouen.fr
[5] LATMOS, Université de Franche-Comté, UFC, CNRS, UTINAM, 25000 Besançon, France
azzedine.lakhlifi@obsbesancon.fr

Abstract. The knowledge of an atmosphere chemical composition is very important for environment protection purposes, risk analysis or new areas discovery. In this context different technologies, including optical and spectroscopic sensors, are employed to identify accurately the air molecules. However, the use of such technologies is limited due to complexity and sometimes severity of the earth's atmosphere. The originality of this study is an elaboration of a technical specification for a new miniaturized optical sensor mounted on a quadcopter drone, via a risk and a functional analysis study.

Methods used in this paper include a criticity level study for each risk and a design functional analysis based on the pieuvre diagram for the detection layer and the processing layer. The evaluation of the intended flexibility criterion for the primary and constraint service functions determines the numeric values for the technical specification. The developed approach for a mobile infra-red spectroscopy sensor prompted the investment of a vibration isolation system and an innovative cooling solution suited for use on drones.

Keywords: Mechatronic device · Spectroscopy · Atmosphere specification · Risk analysis · Functional analysis

1 Introduction

For the determination of the chemical composition of a sample, spectrometric analysis is preferred for several factors including the speed of analysis, and the high selectivity of these approaches. These advantages qualify them to handle several problems involving

A. Akrout et al. (Eds.): ICAV 2022, ACM 22, pp. 234–243, 2023.
https://doi.org/10.1007/978-3-031-34190-8_26

various analytes. The link between the analyte concentration in a substance or sample and the recorded signal is quite complex. In addition, the accuracy of a chemical analysis is highly dependent on how the examined sample is produced, as the technique of preparation is unique to chemical measurements [1, 2].

The use of IR spectroscopy technique is mainly related to laboratories due to the sensitivity of the optical sensors [3]. For this reason, IR spectroscopy research has been concentrated on laboratory applications, as the primary application of IR spectroscopy outside of laboratories is planet discovery.

This research makes the scientific contribution of studying the implementation of laboratory-developed sensors on quadcopter drones, to promote more dynamic applications of IR spectrometers. The desired solution is based on an IR sensor flown by a drone between N designated points to measure the air quality index. However, examining prior employment of these optical sensors justifies a thorough examination of the proposed system. In fact, the measurement process in spectroscopy, the tested object and measuring equipment cannot be separated [4, 5].

In the earth's atmosphere, the interaction of these variables takes place as a complicated phenomenon. Interactions between temperature, airflow direction, chemical makeup of the air, and air wind speed favor measurement uncertainty. To correctly interpret a measurement result, it is necessary to consider all factors that might influence the outcome [6, 7].

The aim of this research is to provide technical specifications for a mobile IR spectroscopic instrument used for air quality monitoring. For this purpose, a risk analysis, a pieuvre diagram, and an examination of the service function will be conducted.

The remainder of this article is structured as follows: Section 2 presents the risk analysis of the on-board sensor, examining the influence of the risk variables and offering some technological solutions; Section 3 examines the functional analysis of the detection layer and the processing layer; and Section 4 summarizes the work done in the previous sections and provides specifics for the air and ground compartments of the system.

2 Risk Analysis for the On-Board Sensor

The Table 1 below demonstrates the effect of each parameter on an IR spectrometer measurement which have an illustration in Chart 1.

Table 1. Risks for the implementation of a sensor on a drone

Description	seriousness	Occurrence	Criticity
	1–4	1–4	
very low gas concentrations (very restricted pollutant chamber)	3	3	9
the fragile nature of the fibers and the feedback from the laser	3	2	6

(continued)

Table 1. (*continued*)

Description	seriousness	Occurrence	Criticity
	1–4	1–4	
which reduces the signal to noise ratio			
Angular divergence contributes to noise in the spectra	2	3	6
the presence of water vapor in the medium to be studied	2	2	4
higher temperatures lower the accuracy levels of species Concentrations	4	1	4
the accuracy of the measurement	2	2	4

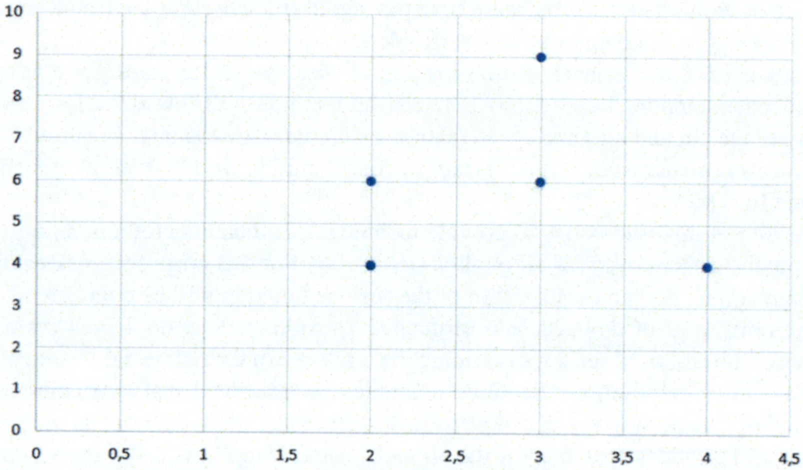

Chart 1. Criticity presentation

Fluctuating loads and uncertainties in analysis models contribute to the optimal design performing differently than expected [8, 9]. A discussion of the steps that must be taken to address the previously mentioned scientific problematics are discussed in Table 2.

Table 2. threats and repairing solutions for the IR sensor.

Prevention	repairing solutions
low concentration of gas molecules	Optimize the volume of the pollutant chamber and use of neural network for interferogram pattern recognition
Exposure to high Temperatures	Low loss hollow core waveguides (HCW) for mid-IR beam transmission
laser beam diffusion and bad beam uniformity	use of adaptative source sizing
water vapor interfering with the measurement	Use of the wavelength modulation spectroscopy technique to differentiate the spectral characteristics of H2O by completely suppressing the interferences

To ensure effective and safe functioning of the drone device, it is vital to master the data fusion component. In fact, the drone must have an overall state of knowledge on its status as well as the location of the other members of the team. All these variables must be considered. The risk analysis process revealed the number of factors that may damage our system, emphasizing the need of establishing an efficient strategy for project execution. To address this need, a functional analysis in the next two parts will be presented, concentrating on the most crucial layers, the detecting layer, and the processing layer [10].

3 Functional Analyses of the Different Project Compartments

Functional analysis is the process of discovering and defining the functions provided by a product when it is put in a system to meet the demands of its user. There are two kinds of needs: objective and subjective.

3.1 Functional Analysis of the Different Project Compartments

A system is a collection of pieces that together create an organized whole that meets many cohering requirements. It is a large group that often includes all sorts of merchandise (equipment, process, service). The term "product" is used in its broadest definition. It may be a physical, piece of equipment, administrative or technical procedure, a service, or software [11]. SADT (Structured Analysis and Design Technique) is the first tool that will be used for the functional analysis, it's a graphical tool for modular and hierarchical top-down analysis. It permits the representation of a model of the actual system as shown in Fig. 1.

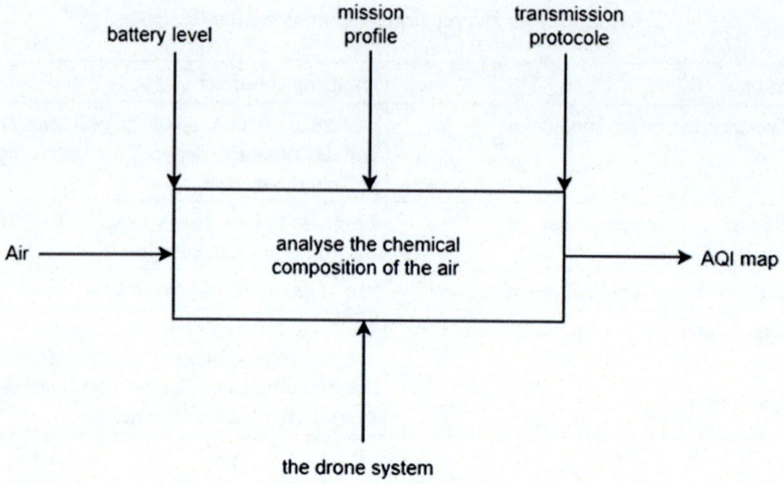

Fig. 1. SADT for the imbedded sensor

3.2 Detection Layer

The meant by detection layer is the FTIR sensors the pieuvre diagram for the sensors are mentioned in Fig. 2. The service functions are detailed in Table 3. The feasibility will be discussed in Table 4 when studying the criteria needed for each service function.

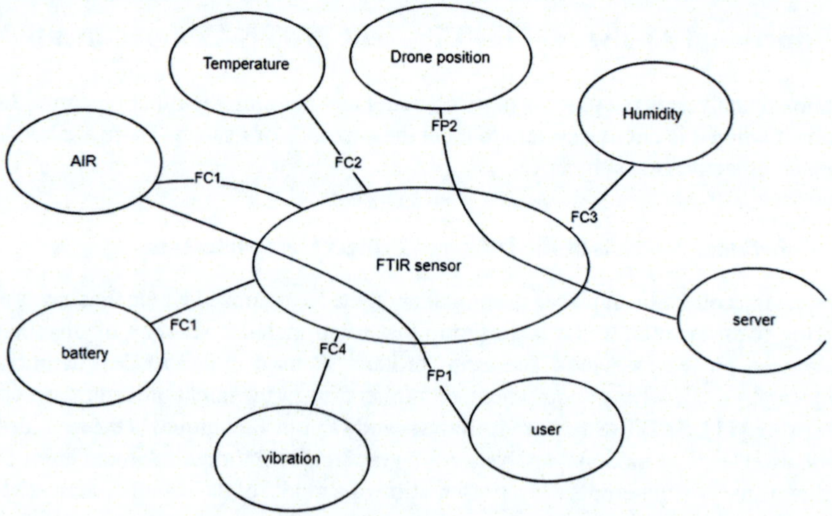

Fig. 2. The Pieuvre diagram for the embedded FTIR engine

Table 3. Service functions for the detection layer

service function	FP	FC
Measure the air quality index for the user	FP1	
provide the data from the drone position needed for the server for the measurement	FP2	
limit the measurement to the battery discharge level considered a safety margin		FC1
must be able to conduct measurement in high temperatures		FC2
must be made from materials that can deal with humidity		FC3
must have a system for hybrid vibration absorption		FC4

Table 4. Service functions analysis level and flexibility

service function	Criteria	Level	flexibility
Measure the air quality index for the user	Measure several points in one mission profile Measurement in real time	10 points 2 s	20% F1
provide the data from the drone position needed for the server for the measurement	Able to stay in one position for an adequate amount of time Equipped with tools to provide exact location information	10s 0.1 m	F2 F3
limit the measurement to the battery discharge level considered a safety margin	Maintain a sufficient battery level to ensure the drone returns to its base	15%	F0
must be able to conduct measurement in high temperatures	Capable of resisting the high temperatures both in external protection and internal cables	50 °C	F2
must be made from materials that can deal with humidity	Designed mainly from Fiber Carbone alloys	80%	F2
must have a system for hybrid vibration absorption	Removable with minimal weight Removable with minimal volume Utilizing materials that provide thermal insulation	0.25 kg 10 cm^3 Class F	F1 10% F1

3.3 Data Analysis Layer

The data analysis layer or the ground base will be mentioned as the server for simplification the pieuvre diagram for the server is shown in Fig. 3.the service functions are detailed in Table 5. The feasibility will be discussed in Table 6 when studying the criteria needed for each service function.

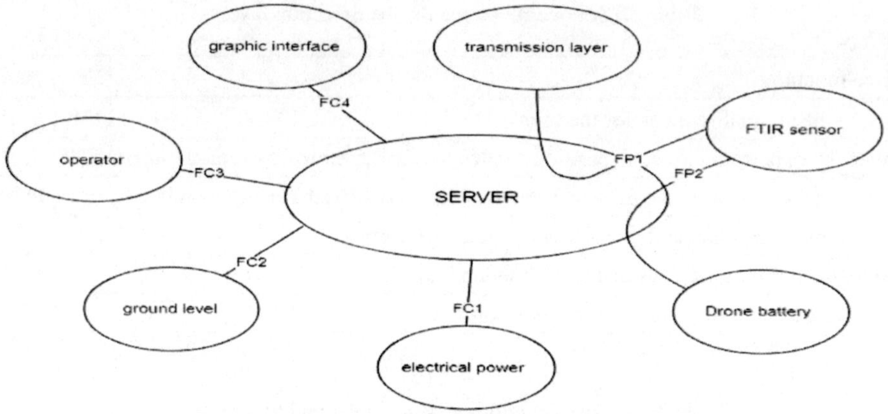

Fig. 3. The Pieuvre diagram for the embedded FTIR engine

Table 5. Service functions for the server

service function	FP	FC
must abort the mission when the UAV's battery level reaches a certain level	FP1	
treat the data the FTIR and send instruction to the drone based on it	FP2	
must be powered by a public power supply and have the capacity to deal with fluctuation of the electric current		FC1
must have the means to communicate with other components of the ground level		FC2
must provide the tools for the operator to intervene when needed on the mission profile		FC3
must be capable to generate the graphical interface and be able to modify It when needed		FC4

4 Specification of the System

The drone system is a group of drones with a small number of members that fly autonomously in vertical take-off (VTOL) mode.

It has at least four (4) electric motors that power the fins, guaranteeing optimum air drainage.

When parked at the measurement spot, the drone must retain its position and height.

To provide this characteristic, the fin supports must have the greatest amount of flexibility possible. It is outfitted with spectroscopy-specific infrared sensor payloads.

The following components make up the entire drone system:

Table 6. Service functions analysis for the server

service function	Criteria	Level	flexibility
must abort the mission when the UAV's battery level reaches a certain level	Able to send a notification for the operator in short time	5 s	F1
treat the data the FTIR and send instruction to the drone based on it	Quick feedback from the data base	2 s	F0
must be powered by a public power supply and have the capacity to deal with fluctuation of the electric current	Use the power supply to Make the computers functional	220 V	F0
must have the means to communicate with other components of the ground level	Multiple antennas should be mounted	6	±2
must provide the tools for the operator to intervene when needed on the mission profile	Have an emergency backup system	One system	F0
must be capable to generate the graphical interface and be able to modify It when needed	A normal user interface Expert user interface	Google maps users Spectroscopy experts	F0 F2

4.1 Air Vector Part

- A drone fleet composed of at least 2 drones.
- Each drone must have 4 motors that power its fins.
- Its power supply (2 sets of batteries).
- A removable IR sensor payload (Table 4).
- A cooling system of the sensor devices and the air that will be analyzed (Table 4).
- A communication protocol between drone team members.
- A data transmission protocol to the ground base (Table 6).

4.2 Ground Station Part

- A ground-based receiving and processing station for calibrating received measurements, selecting the ideal cycle between N sites, and constructing the air quality index map from a selected measuring companion.
- The Fourier transform and calibration algorithms, which will analyze the sensor input to generate a spectrograph (eliminate the water vapor imprint, isolate noise due to sudden overheating) (Table 6).

- A graphical interface for displaying the results as an air quality index map.
- Transmission equipment capable of ensuring secure communication with the fleet (Table 6)

4.3 Spare Parts, Tools, and Documentation

- Rechargeable batteries (500 cycles)
- Tools for assembly and disassembly of the entire drone.
- spare parts for RC antennas.

4.4 Means of Packaging for Storage and Transport

- Robust and thermally insulated storage cases.
- Storage means that can protect the sensor devices in harsh environments.

5 Conclusion

Although infrared spectroscopy is a powerful technique for assessing chemical composition, it is notoriously difficult to use due to the sensitivity of the optical sensor. The intended mobile sensor design will be more challenging to implement, as mounting the sensor on a drone would present additional constraints. Therefore, it is advantageous to invest more in the design study process.

To the best of our knowledge, this paper represents the first design specification approach of an FTIR sensor mounted on a drone. Interpreting the risk analysis section and the service function feasibility considering the pieuvre diagrams of the detection layer and the processing layer. Based on the system's technical specifications, the feasibility of the desired design is decided by the battery lifespan optimization, data fusion, and temperature and vibration isolation.

However, the design process revealed that technical solutions are limited due to the drone design requirements that impose limitations on mass and clutter. As a result, having measurement uncertainties remains.

This research pave the way to an efficient design tool and it will continue to work on the remaining technical solutions in one hand and enhance the analytical power of the instrument, employing the reliability-based design optimization (RBDO) technique for example in the other hand.

Acknowledgement. This work was done in a collaboration between the LATMOS laboratory of the University of Versailles St-Quentin Paris-Saclay and the LMN laboratory of INSA Rouen Normandie in support by LA2MP laboratory of Sfax and we thank all these institutions for support. Fruitful discussions with Doctor Hichame Maaname are greatly acknowledged.

References

1. Haas, J., Mizaikoff, B.: Advances in mid-infrared spectroscopy for chemical analysis. Ann. Rev. Anal. Chem. **9**, 45–68 (2016)
2. Beć, K.B., Huck, C.W.: Breakthrough potential in near-infrared spectroscopy: spectra simulation. A review of recent developments. Front. Chem. **7**, 48 (2019)
3. De Bruyne, S., Speeckaert, M.M., Delanghe, J.R.: Applications of mid-infrared spectroscopy in the clinical laboratory setting. Crit. Rev. Clin. Lab. Sci. **55**(1), 1–20 (2018)
4. Korablev, O.: SPICAM IR acousto-optic spectrometer experiment on Mars Express. J. Geo-Phys. Res. Planets **111**(9) (2006)
5. Thirupathaiah, P., Haider, S.A., Masoom, J.: Simulation of photoelectron flux, electron density, emission rate and limb intensity of CO(a3π) Cameron bands in the Martian thermo-sphere: comparisons with (1) SPICAM and IUVS observations and (2) other model calculations. J. Earth Syst. Sci. **131**(2) (2022)
6. Dahoo, P., Lakhlifi, A.: Infrared Spectroscopy of Triatomics for Space Observation 2 (2018)
7. Lanin, A.A., Voronin, A.A., Fedotov, A.B., Zheltikov, A.M.: Time-domain spectroscopy in the mid-infrared. Sci. Rep. **4**, 1–8 (2014)
8. Yang, Y., Zheng, Z., Bian, K., Song, L., Han, Z.: Real-time profiling of fine-grained air quality index distribution using UAV sensing. IEEE Internet Things J. **5**(1), 186–198 (2018)
9. Chakchouk, M.A., El Hami, A., Haddar, M.: La fiabilité appliqué à la conception du dispositifs spectromètre mobile I; optimisation des paramètres de vols (2022)
10. Yang, Y., Zheng, Z., Bian, K., et al. Real-time profiling of fine-grained air quality index distribution using UAV sensing. IEEE Internet Things J. **5**(1), 186–198 (2017)
11. Villacís, S.A., Burneo, P.S.: UAVs' efficient assembly: Lean Manufacturing implementation in an UAVs' Assembly Company. Int. J. Ind. Eng. Manage. **11**(4), 237–252 (2020)

Porous Functionally Graded Cylindrical Shells' Buckling Study

Jamel Mars[1], Hanen Jrad[1,2(✉)], Mondher Wali[1,2], and Fakhreddine Dammak[3]

[1] Laboratory of Electromechanical Systems (LASEM), National Engineering School of Sfax, University of Sfax, Route de Soukra Km 4, 3038 Sfax, Tunisia
hanen.jrad@essths.u-sousse.tn
[2] Higher School of Sciences and Technology of Hammam Sousse, University of Sousse, Rue Lamine Abassi, 4011Hammam, Sousse, Tunisia
[3] Laboratory of Electrochemistry and Environment (LEE), National Engineering School of Sfax, ENIS, University of Sfax, Sfax, Tunisia
Fakhreddine.dammak@enis.tn

Abstract. Traditionally, shell finite elements are best suited for the numerical simulation of thin structure applications, while solid elements are regularly used for bulk structures. However, engineering structures often combine thin components with thick/bulk geometries in the same assembly. Thus, the finite element modeling of such applications would be considerably simplified if the same type of finite element could be successfully used in both zones. Further, functionally graded material (FGM) have been broadly applied to numerous engineering applications such as aerospace, defense, automotive, nuclear power, bio-engineering and other areas. However, the occurrences of porosities are inevitable during the process of manufacturing FGMs. Knowing the mechanical behavior of composite FGM structures taking into account porosities is essential in the design and optimization process of engineering projects. In this paper, a geometrically nonlinear analysis of porous FGM cylindrical shell is investigated. A user defined subroutine (UMAT) is developed and implemented in Abaqus/Standard to study the FG shells in large displacements and rotations. The effects of functionally graded power index, porosity coefficient, porosity arrangements and geometrical design parameters are examined.

Keywords: Buckling analysis · FGM · porosity distribution

1 Introduction

Shell structures are present in various shapes in both nature and technology. Indeed, shells are actually employed in several engineering applications such as automotive, aerospace industry and in civil engineering due to their high ratio of load-carrying capacity to self-weight. The need of a useful finite element model to capture the mechanical response of structural components like shells is related to the fast growth and increase of computational power contemporary computers and the development of commercial purpose FE programs that are accessible for users, over the years, numerical methods are widely used

A. Akrout et al. (Eds.): ICAV 2022, ACM 22, pp. 244–250, 2023.
https://doi.org/10.1007/978-3-031-34190-8_27

to provide a way to approximate solutions to problems in many engineering disciplines such as aerospace, mechanical, and electrical, quickly and easily compared to analytic solutions and when analytical solutions aren't possible. In addition, finite element analysis (FEA) allows simulations to predict and understand the structure behavior under several physical conditions, Viola et al. (2013); Belabed et al. (2014) and Avey et al. (2021). It is a numerical approach to finding the real results and providing an approximate solution to different problems in mechanical engineering. A design optimization of dental implant geometrical characteristics enhancing primary stability is proposed by (Elleuch et al. 2021) using FEA of stress distribution around dental prosthesis. A numerical methodology to characterize aluminum behavior considering non-associated plasticity model coupled with isotropic ductile damage was carried out by (Bouhamed et al. 2021, 2022) in order to identify ductile damage constitutive equations for thin sheet metal applications. Numerical results closely matched experimental data.

Further, functionally graded materials (FGM) have been attracting attention between conventional composites. Formed by a gradual transition of composition between materials from ceramic to metal surface, FGMs have been designed to eliminate stress concentration and problems of discontinuities which are generally met with conventional laminated composites, (Jrad et al., 2018; Bagheri, et al., 2018; Li & Han, 2018; Alibeigloo & Nouri, 2010; Tornabene & Viola, 2013 and Xu et al., 2022). Meanwhile, due to technical issues, the occurrences of porosities during the process of fabricating FGMs cannot be avoided, which may cause weakness in the strength of the materials. Thus, amply understanding the mechanical behavior of porous FGMs is a major concern, (Wattanasakulpong et al., 2012; Chen et al., 2015; Wang & Wu, 2017; Gao et al., 2018; Li et al., 2019; Liu et al., 2022).

In this context, this study proposes a numerical analysis of the static buckling of Porous FGM cylindrical shells under uniform compressive axial loading. The material properties are supposed to vary continuously along the thickness referring to power law distribution including even and uneven porosity distributions. The material properties according to the coordinates of the integration points are defined using both UMAT subroutines, in ABAQUS software.

2 Basic Formulation of Cylindrical FGM Shell with Porosities

While solid elements are frequently utilized for bulk structures, traditionally, shell finite elements are best suited for numerical modeling of thin structure applications. Yet, engineering structures frequently assemble thin components and thick/bulk geometries together. Therefore, if the same type of finite element could be employed effectively in both zones, the finite element modeling of such applications would be greatly eased. Solid–shell elements combine a shell-like response with three-dimensional element geometry, thus naturally matching solid elements in the same mesh, (Sze & Yao, 2000; Hauptmann et al., 2001; Kim et al., 2005).

The basic formulation and numerical implementation of cylindrical FGM shell with porosities is briefly reviewed in this section. Let's consider a functionally graded cylindrical shell made of metal and ceramic components in which porosity is dispersed following two types of disposals, namely even and uneven arrangements as shown in Fig. 1. The upper surface of the structure is ceramic-rich while the bottom surface is metal-rich. The effective material properties such as the Young's modulus E can be expressed, according to (Wattanasakulpong and Ungbhakorn, 2014; Wattanasakulpong & Chaikittiratana, 2015), as function of the z variable and porosity parameter (α):

$$E(z,\alpha)=\begin{cases}(E_c-E_m)V_c+E_m-\frac{\alpha}{2}(E_c+E_m) & \text{(Even porosity)}\\ (E_c-E_m)V_c+E_m-\frac{\alpha}{2}(E_c+E_m)\left(1-\frac{2|z|}{h}\right) & \text{(Uneven porosity)}\end{cases} \quad (1)$$

where $0 \prec \alpha \prec 1$ is the porosity coefficient denoting the ratio between the vacant volume and the entire volume. For ($\alpha = 0$), the FGM shell is assumed perfect, as illustrated in Fig. 1. E_c and E_m are the Young's modulus of ceramic and metal portions and V_c is the volume fraction of the ceramic phase which is given by, (Mellouli et al., 2019):

$$V_c=\left(\frac{1}{2}+\frac{z}{h}\right)^n,\ V_m=1-V_c,\quad z\in\left[-\frac{h}{2},\frac{h}{2}\right] \quad (2)$$

The exponent n is the material power-law index representing material profile gradation. h denotes the structure thickness and z is the coordinate measured along the thickness direction.

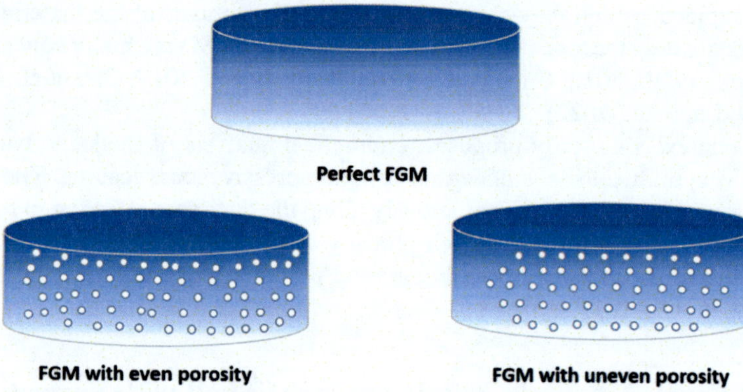

Fig. 1. Representations of perfect FGM and FGM with even and uneven porosity distributions

3 Numerical Results

Numerical simulation is performed using the commercial software ABAQUS to study the static buckling shell response. To avoid stress discontinuity at the interfaces, two interfaces are implemented into ABAQUS in order to define the material properties according to the coordinates of the integration points along the thickness. The FGM cylinder shell is subjected initially to uniform compressive axial loading, along z axis. The material properties are gathered in Fig. 2.

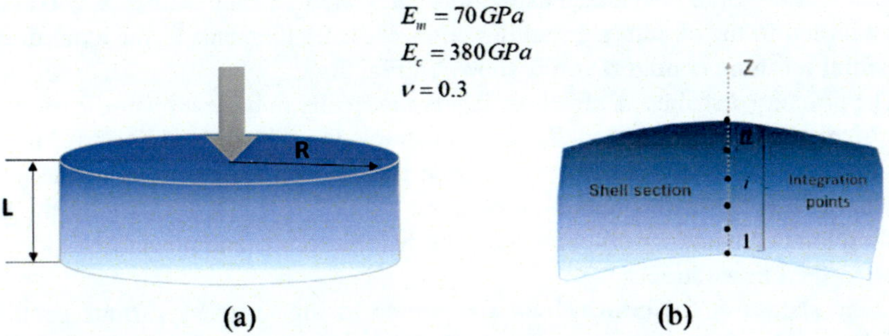

Fig. 2. (a) Axially compressed FGM cylinder, (b) Integration through the thickness of the FGM shell

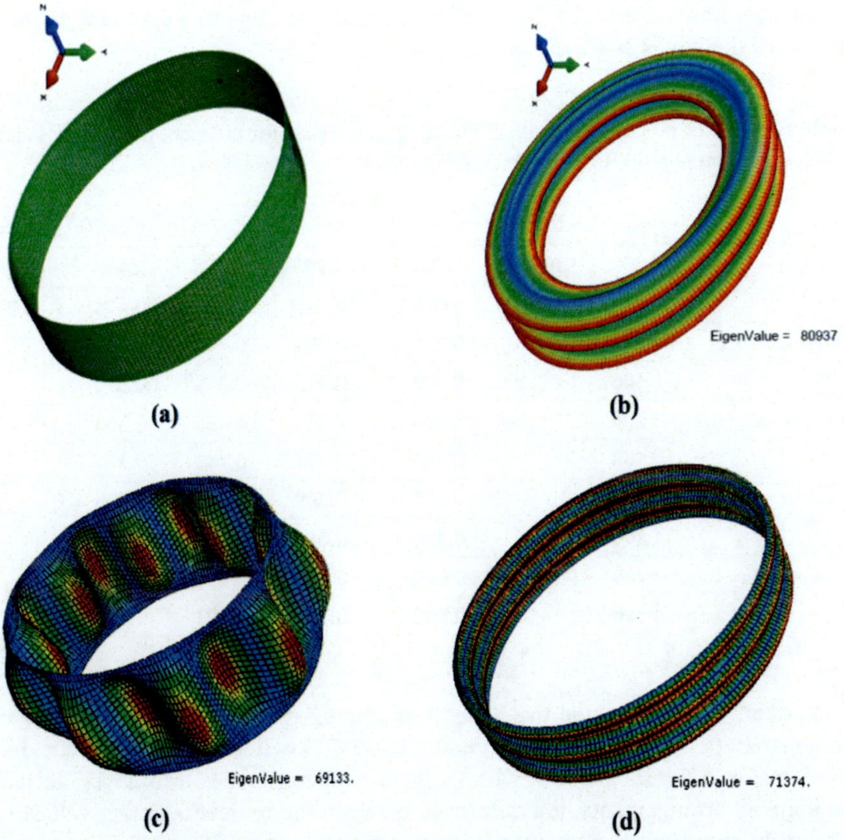

Fig. 3. Initial meshing configuration of the FGM cylindrical shell (a), Mode shapes at the critical load h = 0.01, L = 2 (b) perfect FGM, n = 1, R/h = 500; (c) even FGM, n = 1, R/h = 200, α = 0.1, (d) even FGM, n = 1, R/h = 500, α = 0.1

The cylinder shell is modeled using fully integrated S4 shell elements, 240 elements were adopted in the circumferential direction while 21 elements in the axial direction. The initial meshing configuration is shown in Fig. 3(a).

The porosity distribution effect on the geometrically nonlinear response of the FGM cylindrical shell is investigated. Table 1 illustrates the critical buckling load considering different power law index *n*. Both even and uneven distributions of the porosity are examined with various radius-to-thickness ratio R/h. Table 1 illustrates the critical buckling load considering different porosity distributions α and power law index *n* and radius-to-thickness ratio.

Mode shapes at the critical load are plotted in Fig. 3 under various geometrical conditions and results shown are visually in good agreement with the classical behavior of cylindrical shells presented by Simo et al. (1990).

From Table 1 and Fig. 3, it is noticed that an increase in the index power is accompanied by an increase in the critical buckling load. Indeed, critical buckling load for a pure ceramic cylinder has the maximum value and the buckling load diminishes as the power law index *n* raises due to the increase in the metal constituent in the FGM cylinder with the rise of the power law index *n*.

Table 1 Effect of porosity on Critical buckling load (108) for different power law index n and radius-to-thickness ratio R/h (h = 0:01, L/h = 200)

	R/h	n = 0	n = 0.5	n = 1	n = 5	n = 10	n = ∞
Perfect	**200**	1.458	1.007	0.795	0.475	0.403	0.268
	500	1.484	1.019	0.809	0.481	0.414	0.273
Even α = 0.1	**200**		0.915	0.691	0.373	0.309	
	500		0.928	0.713	0.379	0.316	
Uneven α = 0.1	**200**		0.974	0.761	0.438	0.366	
	500		0.985	0.772	0.442	0.377	
Even α = 0.2	**200**		0.823	0.599	0.263	0.208	
	500		0.836	0.616	0.276	0.210	
Uneven α = 0.2	**200**		0.942	0.725	0.397	0.327	
	500		0.950	0.735	0.401	0.338	

Further, it is well noted that when introducing the porosity effects and changing from perfect (α = 0) to imperfect (α = 0.1;0.2) FGM cylinder, the critical buckling load decreases, in a significant way, with the high buckling modes. Moreover, according to the porosity arrangements, the difference between the critical buckling values is more noticeable with uneven porosity pattern compared to even pattern. In fact, the way of porosity is distributed through the entire structure and the concentration of pores in the mid-surface of the cylinder effects essentially the buckling behavior more than even distributed pores, which is in accordance with references in literature.

4 Conclusion

Static buckling analysis of cylindrical FGM porous shell using FSDT were investigated and discussed. By using the UMAT subroutine, implemented into ABAQUS software, the material properties according to the coordinates of the integration points are defined so that stress discontinuity at the interfaces are eliminated. The effects of functionally graded power index, porosity coefficient, porosity arrangements and geometrical design parameters are examined. It was shown that porosity distribution significantly affects the rigidity of the structure and hence the critical buckling load.

References

Alibeigloo, A., Nouri, V.: Static analysis of functionally graded cylindrical shell with piezoelectric layers using differential quadrature method. Compos. Struct. **92**(8), 1775–1785 (2010)

Avey, M., Fantuzzi, N., Sofiyev, A.H., Kuruoglu, N.: Nonlinear vibration of multilayer shell-type structural elements with double curvature consisting of CNT patterned layers within different theories. Compos. Struct. **275**, 114401 (2021)

Bagheri, H., Kiani, Y., Eslami, M.R.: Asymmetric thermal buckling of temperature dependent annular FGM plates on a partial elastic foundation. Comput. Math. Appl. **75**(5), 1566–1581 (2018)

Belabed, Z., Houari, M.S.A., Tounsi, A., Mahmoud, S.R., Bég, O.A.: An efficient and simple higher order shear and normal deformation theory for functionally graded material (FGM) plates. Compos. Part B: Eng. **60**, 274–283 (2014)

Bouhamed, A., et al.: Identification of fully coupled non-associated-ductile damage constitutive equations for thin sheet metal applications: numerical feasibility and experimental validation. Thin-Walled Struct. **176**, 109365 (2022)

Bouhamed, A., Mars, J., Jrad, H., Wali, M., Dammak, F.: Experimental and numerical methodology to characterize 5083-aluminium behavior considering non-associated plasticity model coupled with isotropic ductile damage. Int. J. Solids Struct. **229**, 111139 (2021)

Chen, D., Yang, J., Kitipornchai, S.: Elastic buckling and static bending of shear deformable functionally graded porous beam. Compos. Struct. **133**, 54–61 (2015)

Elleuch, S., Jrad, H., Kessentini, A., Wali, M., Dammak, F.: Design optimization of implant geometrical characteristics enhancing primary stability using FEA of stress distribution around dental prosthesis. Comput. Methods Biomech. Biomed. Eng. **24**(9), 1035–1051 (2021)

Gao, K., Gao, W., Wu, B., Wu, D., Song, C.: Nonlinear primary resonance of functionally graded porous cylindrical shells using the method of multiple scales. Thin-Walled Struct. **125**, 281–293 (2018)

Hauptmann, R., Doll, S., Harnau, M., Schweizerhof, K.: Solid-shell'elements with linear and quadratic shape functions at large deformations with nearly incompressible materials. Comput. Struct. **79**(18), 1671–1685 (2001)

Jrad, H., Mars, J., Wali, M., Dammak, F.: An extended finite element meth-od for modeling elastoplastic FGM plate-shell type structures. Struct. Eng. Mech.: Int. J. **68**(3), 299–312 (2018)

Kim, K.D., Liu, G.Z., Han, S.: A resultant 8-node solid-shell element for geometrically nonlinear analysis. Comput. Mech. **35**(5), 315–331 (2005)

Li, H., Pang, F., Chen, H., Du, Y.: Vibration analysis of functionally graded porous cylindrical shell with arbitrary boundary restraints by using a semi analytical method. Compos. Part B: Eng. **164**, 249–264 (2019)

Li, W., Han, B.: Research and application of functionally gradient materials. In: IOP Conference Series: Materials Science and Engineering, vol. 394, No. 2, p. 022065. IOP Publishing (2018)

Liu, Y., Qin, Z., Chu, F.: Analytical study of the impact response of shear deformable sandwich cylindrical shell with a functionally graded porous core. Mech. Adv. Mater. Struct. **29**(9), 1338–1347 (2022)

Mellouli, H., Jrad, H., Wali, M., Dammak, F.: Geometrically nonlinear meshfree analysis of 3D-shell structures based on the double directors shell theory with finite rotations. Steel Compos. Struct.: Int. J. **31**(4), 397–408 (2019)

Simo, J.C., Fox, D.D., Rifai, M.S.: On a stress resultant geometrically exact shell model. Part III: computational aspects of the nonlinear theory. Comput. Methods Appl. Mech. Eng. **79**(1), 21–70 (1990)

Sze, K.Y., Yao, L.: A hybrid stress ANS solid-shell element and its generalization for smart structure modelling. Part I—solid-shell element formulation. Int. J. Numer. Methods Eng. **48**(4), 545–564 (2000)

Tornabene, F., Viola, E.: Static analysis of functionally graded doubly-curved shells and panels of revolution. Meccanica **48**(4), 901–930 (2013)

Viola, E., Tornabene, F., Fantuzzi, N.: General higher-order shear deformation theories for the free vibration analysis of completely doubly-curved laminated shells and panels. Compos. Struct. **95**, 639–666 (2013)

Wang, Y., Wu, D.: Free vibration of functionally graded porous cylindrical shell using a sinusoidal shear deformation theory. Aerosp. Sci. Technol. **66**, 83–91 (2017)

Wattanasakulpong, N., Chaikittiratana, A.: Flexural vibration of imperfect functionally graded beams based on timoshenko beam theory: Chebyshev collocation method. Meccanica **50**(5), 1331–1342 (2015)

Wattanasakulpong, N., Ungbhakorn, V.: Linear and nonlinear vibration analysis of elastically restrained ends FGM beams with porosities. Aerosp. Sci. Technol. **32**(1), 111–120 (2014)

Wattanasakulpong, N., Prusty, B.G., Kelly, D.W., Hoffman, M.: Free vibration analysis of layered functionally graded beams with experimental validation. Mater. Des. **1980–2015**(36), 182–190 (2012)

Xu, W.Z., Fu, Z.J., Xi, Q.: A novel localized collocation solver based on a radial Trefftz basis for thermal conduction analysis in FGMs with exponential variations. Comput. Math. Appl. **117**, 24–38 (2022)

Acoustic Velocity Estimation in the Presence of Steady Flow Using Particle Image Velocimetry

Simon Rampnoux[1(✉)], Islam Ramadan[2], Solène Moreau[1], and Mabrouk Ben Tahar[1]

[1] Roberval (mechanics, energy and electricity), Centre de recherche Royallieu, Université de Technologie de Compiègne, CS 60319-60203, Compiègne Cedex, France
simon.rampnoux@utc.fr

[2] Institut PPrime, CNRS - Université de Poitiers - ENSMA, Département Fluides-Thermique-Combustion, ENSIP, 6 rue Marcel Doré Bât. B17-BP 633, 86022 Poitiers Cedex, France

Abstract. Phase-locked PIV measurement technique is used to estimate the acoustic velocity in the presence of turbulent flow inside a rectangular duct. The acoustic field is generated by compression drivers mounted on the sides of a rectangular duct. The PIV measurements are synchronized with the compression drivers. The measured velocity fields can be decomposed to mean flow velocity, turbulent fluctuations and acoustic velocity. A post-processing technique is proposed to extract the acoustic velocity field from the total measured velocity field. Also, the uncertainty of the estimated acoustic velocity is calculated using two different techniques, namely peak ratio method and particle disparity method. The estimated acoustic velocity agrees with the acoustic velocity estimated from classical microphone's measurement technique. The estimated uncertainty of the acoustic velocity is found to be 18%.

Keywords: PIV · uncertainty · acoustic velocity

1 Introduction

The Particle Image Velocimetry (PIV) system was developed in 1980 [1]. Short time after, the digital PIV has been developed due to the advancement in the computers and cameras capacities [2]. The utilization of the PIV system in measuring acoustic velocity was first tried in the late 1980s [3]. In the late 1990s, Hann and Greated [4] developed a correlation method to estimate the acoustic velocity amplitude of a pure oscillating flow. Also, Nabavi et al. [5] developed a method to measure the acoustic and streaming velocities inside a standing-wave resonator. More recently, the time resolved PIV system along with microphones are used to estimate the acoustic velocity inside a duct in the presence of turbulent flow [6].

The low-frequency PIV systems have been used extensively to measure the acoustic velocity in duct using the phase-locking technique. In this technique,

A. Akrout et al. (Eds.): ICAV 2022, ACM 22, pp. 251–259, 2023.
https://doi.org/10.1007/978-3-031-34190-8_28

the PIV system is synchronized with the acoustic source. To our knowledge, this technique has never been used to measure the acoustic velocity in the presence of turbulent flow. Although the measurement uncertainty of the PIV is very low in ordinary flow fields measurements (e.g. steady and turbulent flows). The value of the uncertainty could be significant when measuring the acoustic velocity in the presence of turbulent flows.

The current study focuses on the utilization of the phase-locked PIV technique to estimate the acoustic velocity in the presence of turbulent flow inside a rectangular duct. In addition, the uncertainty in the acoustic velocity estimation will be provided. Section 2 provides a theoretical background of the principle of the PIV and the methods for estimating the uncertainties. The experimental setup is presented in Sect. 3. Results are discussed and conclusions are drawn respectively in Sects. 4 and 5.

2 Theoretical Background

2.1 PIV Measurement Technique

PIV system is a non-intrusive measurement technique that allows measuring a velocity field of a seeded flow. The system consists of a pulsed laser light source synchronized with a camera. The laser source illuminates the seeded flow and the camera registers two successive photos of seeding particles separated by time delay Δt. These photos are divided into small areas, known as interrogation area. The particles' displacement is determined in each interrogation area by applying a cross-correlation technique between the two images. Hence, the velocity field can be estimated from the displacement flow map.

In the current study, the measured instantaneous flow velocity can be decomposed as follows:

$$v(t) = \bar{v} + v'(t) + v_{ac}(t), \tag{1}$$

where t is the time, $\bar{v}$ is the mean flow velocity, v' is the turbulent velocity and v_{ac} is the acoustic velocity [8]. The focus of the current study is to estimate the acoustic flow velocity. It is worthwhile to mention that the acoustic velocity is three order of magnitude lower than the mean flow velocity. Hence, it is important to estimate the measurement uncertainty and compare it with the estimated acoustic velocity.

2.2 Measurement Uncertainties

Despite the emergence of PIV in the 1980s, the topic of uncertainty was addressed in the late 1990s [9,10]. Later, the works of both Sciacchitano et al. [11] and Neal et al. [12] presented a comparative assessment of several approaches proposed in the literature. In the current study, two different methods are selected to estimate the uncertainty of the measured velocity. These two methods are described in the following paragraphs.

The first method is called **peak ratio method (Charonko et Vlachos [13])** and is based on the assumption that, in the correlation plane, the measured displacement error is directly related to the peak ratio of the cross-correlation function. This peak ratio is defined as the ratio of the largest detectable peak representing particle displacement to the second highest peak due to parasitic particle image pairing (see Fig. 1a) [14]. The method gives an empirical relationship between the uncertainty of the measured particle displacement and the peak ratio. The estimated uncertainty is provided with respect to the magnitude of the velocity vector. Hence, the uncertainty for each velocity component cannot be determined.

The second method is referred to as **particle disparity method (Sciacchitano et al. [15])** and uses the calculated velocity field as a predictor to match the particle images of the records via an image processing algorithm. In each interrogation area, particle pairs are detected. For accurate measurement, the particle images of the two captures must match perfectly. In a real experiment, the paired particle images do not match exactly and a positional disparity appears. The latter is evaluated as the distance between the centers of gravity of the images of particles with an accuracy less than a pixel. Finally, the measurement uncertainty is calculated for each interrogation area from the mean value and the statistical dispersion of the disparity vector (see Fig. 1b). This method estimates the uncertainty of each velocity component x and y.

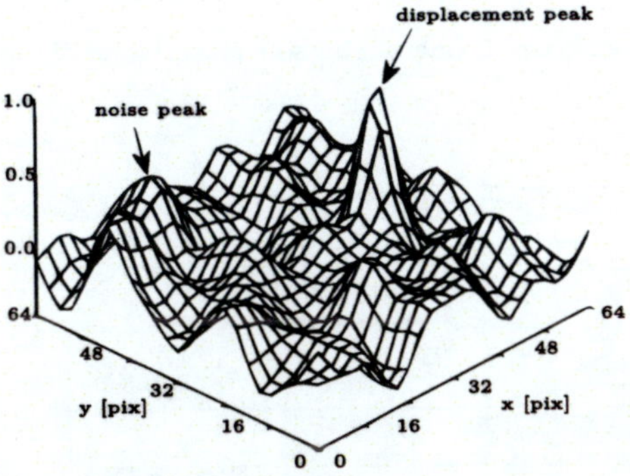

(a) Cross-correlation function representation of a PIV pair images.

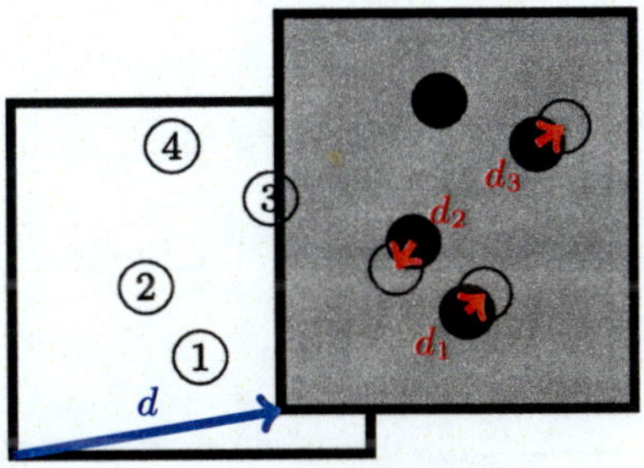

(b) Particle Disparity representation between two interrogation areas.

Fig. 1. Graphical representation of the uncertainties' estimation methods.

3 Experimental Setup and Measurement Techniques

3.1 Experimental Setup

As shown in Fig. 2, the experimental setup consists of a rectangular duct (100 mm × 200 mm). The central section of the rectangular duct includes 4 plexiglass windows to grant optical access for the PIV measurements. Hereinafter, this section is referred to as the measurement section. As shown in Fig. 3, the rectangular duct is composed of several sections: two anechoic terminations, two source sections and one measurement section. The anechoic terminations are used to eliminate the sound reflections. Two compression drivers (Model: BMS 4591) are mounted in the source section to generate the acoustic wave. The air flow is generated by a centrifugal variable speed fan, which is connected to the rectangular duct via a network of silencers and stabilization tank. The fan can provide a maximum flow rate of 800 kg/h. A flow meter is used to measure the flow rate through the duct.

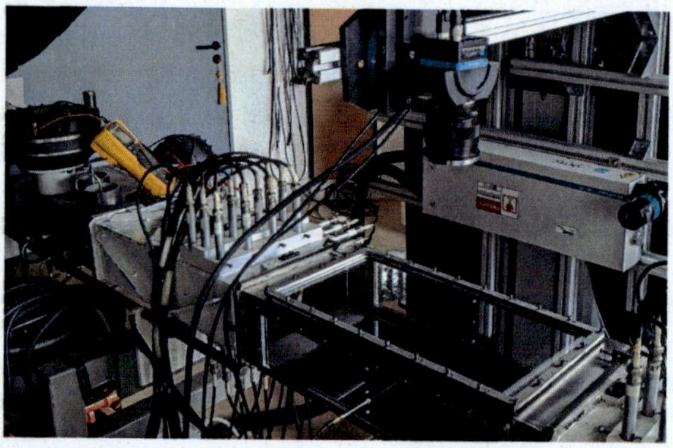

Fig. 2. Photography of the measurement section and one of the source sections of the experimental setup.

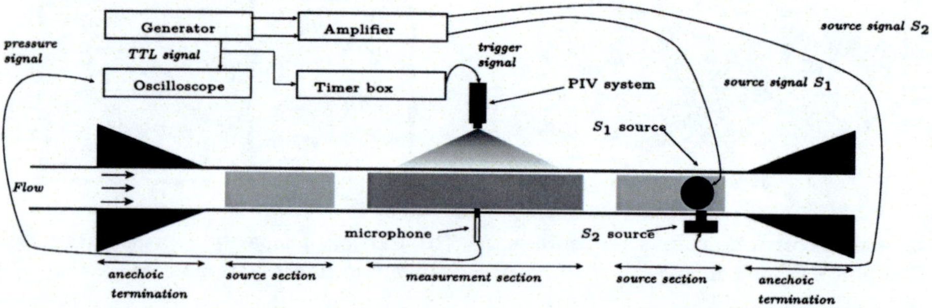

Fig. 3. Experimental setup diagram for PIV acoustic velocity measurement.

The PIV system in the lab is branded DANTEC DYNAMICS and has a sampling frequency of 7.5 Hz. The two pulsed lasers emit two pulses of green light at a wavelength of 532 nm. The Charge Coupled Device (CCD) camera (Model: FlowSense EO 4M, 2048 $pix \times 2048$ pix) is used to capture light scattered by particles. A seed generator (Model: Safex Fog2010) is used to produce liquid droplets with a mean particle diameter of about 1 μm. The field of view of the camera is 85 mm × 85 mm, and hence the size of each interrogation area is 2.65 mm × 2.65 mm.

3.2 Acoustic Velocity Measurement

The PIV system is synchronized with the sound pressure signal measured by the microphone at the center of the measurement section (see Fig. 3). This synchronization takes place via a Transistor Transistor Logic (TTL) signal. This TTL signal is generated by a function generator. This TTL signal is sent to both the PIV sytem (as a trigger signal) and to the power amplifier that drives the compression drivers. A trigger delay can be adjusted in the PIV software, which allows starting the measurements at a certain phase over the acoustic cycle. The complete acoustic cycle can be divided into a number of phases. In the current study, 20 phases, each separated by 18° are selected (see Fig. 4).

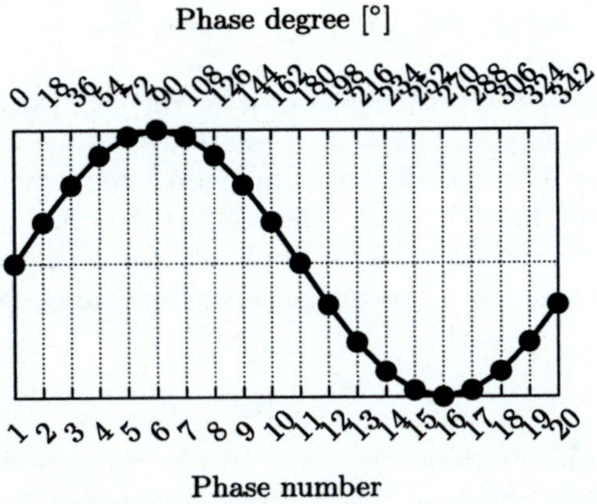

Fig. 4. Representation of the 20 phases of the acoustic period.

Time between two pulses Δt is adjusted to allow the seeding particles to move a distance of about half the size of the interrogations area (IA) [7]. In the current study, the mean flow velocity is 2.5 m/s, hence the time between pulses is set to 532 µs. It is worthwhile to mention that a multi-pass post processing technique is used to calculate the adaptive correlation starting from IA size of 128 pixels to final IA size of 64 pixels.

As explained in the Subsect. 2.1, the v particle velocity is the sum of the mean flow velocity $\bar{v}$, the turbulent velocity v', and the acoustic velocity v_{ac}. For this study, the measurement is carried out at the center of the duct so the average turbulent velocity $\overline{v'}$ is considered zero. The acoustic velocity for each phase point is obtained by the formula

$$v_{ac} = v - \bar{v} \tag{2}$$

where $\bar{v}$ is the ensemble phase-average velocity. The acoustic velocity is calculated for each phase of the acoustic cycle. In this study, the acoustic signal emitted is a sinus. Hence, the acoustic velocity can be represented as follows

$$v_{ac}(t) = |v_{ac}| \sin(2\pi f_{ac} t + \phi_{ac}) \tag{3}$$

where $|v_{ac}|$, f_{ac} and ϕ_{ac} are respectively the amplitude, frequency and phase of the acoustic velocity.

4 Results and Discussion

4.1 Velocity Measurement

In order to validate the method presented in the previous subsection for the estimation of the acoustic velocity, PIV measurements have been performed and analyzed. The air mass flow rate is set to (200 kg/h). The voltage signal fed to the compression drivers has a sinusoidal form with a frequency 100 Hz and voltage of 5 Vrms.

As shown in Fig. 5, the estimated acoustic velocity from the PIV measurements at the different phases is plotted. These data points are interpolated using a sinusoidal function. The best fit is obtained at a frequency of 100.8 Hz (a difference of 0.8% from the input signal). The estimated acoustic velocity amplitude is compared with the acoustic velocity amplitude deduced from an acoustic pressure measurement using the following acoustic impedance relationship:

$$Z_{ac} = \frac{p_{ac}}{v_{ac}}, \tag{4}$$

where p_{ac} is the sound pressure measured by the microphone at PIV measurement point. Also, it is observed that there is a phase shift of 30° between the two signals. In the previous equation, it is assumed that the acoustic impedance is real (pure travelling acoustic wave). This assumption is not valid because the reflection coefficient of the anechoic termination is not zero at a frequency 100 Hz. Hence, this leads to the aforementioned phase shift.

The PIV measurements are conducted at 20 phases over the acoustic cycle (see Fig. 4). 1000 velocity fields are measured at each phase. To validate that the data are stationary, a conversion study is performed, in which evolution of the relative error with the number of velocity fields is investigated. The relative error is defined as:

$$\text{Relative error} = \left|\frac{\text{value at } n \text{ averages} - \text{value at 1000 averages}}{\text{value at 1000 averages}}\right| \tag{5}$$

As shown in (Fig. 6), the relative error drops to 0.4% at 600 velocity fields.

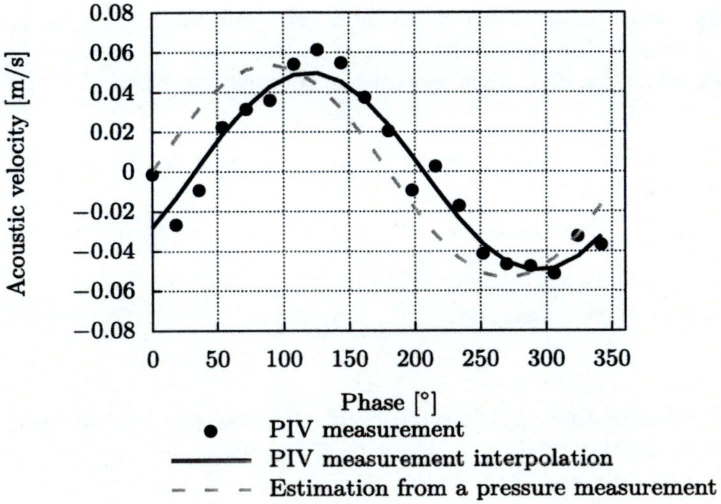

Fig. 5. Acoustic velocity over an acoustic period.

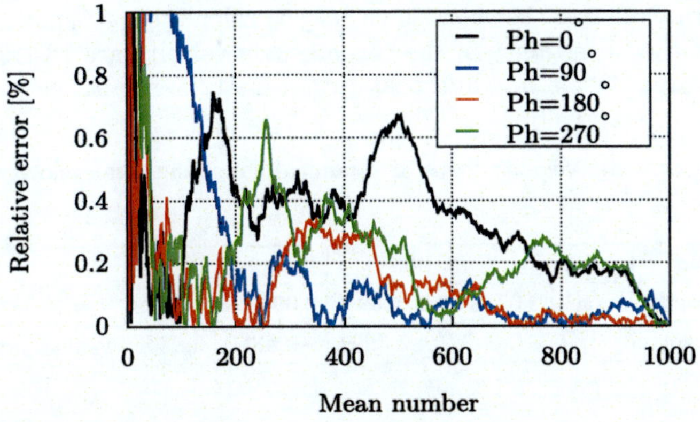

Fig. 6. Evolution of the relative error on the total velocity for 4 different phases as a function of the number of averaged velocity fields.

4.2 Uncertainty Calculation

The uncertainty of the measured total velocity is calculated using the two previously mentioned methods (i.e. Peak Ratio method and Particle Disparity method). For each velocity field, an uncertainty field is calculated. To compare the results of the two methods, each uncertainty field is spatially averaged. Also, for each phase, the spatially-averaged uncertainty values of the 1000 velocity fields are averaged. Table 1 summarizes the spatial-temporal-average of the uncertainty at each acoustic phase for the two methods. It is observed that the uncertainty values using the two methods are comparable for all the phases. Also, the values of the uncertainty are comparable to the value of the relative error obtained in the convergence study at 600 velocity fields.

Table 1. Spatial-temporal-average uncertainty estimated at different phases by two different methods.

Phase points [°]	0	90	180	270
Particle Disparity method [%]	0.3914	0.3849	0.3852	0.3910
Peak Ratio method [%]	0.3675	0.3736	0.3702	0.3919

As the focus of the current study is to estimate the acoustic velocity, it is important to estimate the uncertainty of the acoustic velocity. In the previous subsection, the estimated uncertainty of the total velocity is found to be about 0.38%. As presented in Table 2, the value of the acoustic velocity represents a very small percentage of the total velocity. Consequently, the estimated uncertainty of the acoustic velocity will be high. As the acoustic velocity varies with phases, the rms value of the acoustic velocity is calculated. As shown in Table 3, the values of the uncertainty of the rms acoustic velocity are 18.7% and 18.1% for particle disparity method and peak ration method, respectively.

Table 2. The acoustic velocity value as a percentage of the total velocity at different phases.

Phase point [°]	0	90	180	270	
Total velocity [m/s]	2.0723	2.0259	2.0643	2.1349	$\bar{\mathbf{v}} = \mathbf{2.0744\,m/s}$
Acoustic velocity [%]	−0.0997	−2.3921	−0.4859	2.8366	$\mathbf{v_{ac-rms}} = \mathbf{2.07\%}$

Table 3. The uncertainty of the rms acoustic velocity estimated by different methods.

Particle disparity method	18.7%
Peak ratio method	18.1%

5 Conclusion

A method to extract the acoustic velocity from PIV measurement of a turbulent flow inside a duct is presented. The method is based on the utilization of the phase-locked PIV technique.

A case study is investigated. The acoustic velocity is extracted from the velocity fields measured by the PIV system. The extracted acoustic velocity amplitude is in a good agreement with the estimated acoustic velocity amplitude from a pressure measurement using a microphone. The measurement uncertainty of the measured flow velocity is calculated based on two different methods. The uncertainty of the flow velocity is found to be 0.38%. As the acoustic velocity represents a very small percentage of the total flow velocity, the estimated uncertainty of the acoustic velocity is found to be 18%.

Improvements on the proposed method in estimating the acoustic velocity are the subject of our ongoing research project.

References

1. Adrian, R.: Laser velocimetry. Fluid Mech. Measur. **43**, 155–244 (1983)
2. Westerweel, J.: Fundamentals of digital particle image velocimetry. Measur. Sci. Technol. **8**, 1379–1392 (1997). https://doi.org/10.1088/0957-0233/8/12/002
3. Sharpe, J., Greated, C., Gray, C., Campbell, D.: The measurement of acoustic streaming using particle image velocimetry. Acta Acustica United Acustica **68**, 168–172 (1989)
4. Hann, D., Greated, C.: The measurement of flow velocity and acoustic particle velocity using particle-image velocimetry. Measur. Sci. Technol. **8**, 1517–1522 (1997). https://doi.org/10.1088/0957-0233/8/12/014
5. Nabavi, M., Siddiqui, M., Dargahi, J.: Simultaneous measurement of acoustic and streaming velocities using synchronized PIV technique. Measur. Sci. Technol. **18**, 1811–1817 (2007). https://doi.org/10.1088/0957-0233/18/7/003
6. Léon, O., Piot, E., Sebbane, D., Simon, F.: Measurement of acoustic velocity components in a turbulent flow using LDV and high-repetition rate PIV. Exp. Fluids **58**(6), 1–19 (2017). https://doi.org/10.1007/s00348-017-2348-4
7. Prasad, A.: Particle image velocimetry. Curr. Sci. Assoc. Indian Acad. Sci. **79**, 51–60 (2000)
8. Fischer, A., Sauvage, E., Röhle, I.: Acoustic PIV: measurements of the acoustic particle velocity using synchronized PIV-technique. In: 14th International Symposium on Applications of Laser Techniques to Fluid Mechanics, January 2008
9. Huang, H., Dabiri, D., Gharib, M.: On errors of digital particle image velocimetry. Measur. Sci. Technol. **8**, 1427–1440 (1997). https://doi.org/10.1088/0957-0233/8/12/007
10. Forliti, D., Strykowski, P., Debatin, K.: Bias and precision errors of digital particle image velocimetry. Exp. Fluids **28**, 436–447 (2000). https://doi.org/10.1007/s003480050403
11. Sciacchitano, A., et al.: Collaborative framework for PIV uncertainty quantification: comparative assessment of methods. Measur. Sci. Technol. **26**, 074004 (2015). https://iopscience.iop.org/article/10.1088/0957-0233/26/7/074004
12. Neal, D., Sciacchitano, A., Smith, B., Scarano, F.: Collaborative framework for PIV uncertainty quantification: the experimental database. Measur. Sci. Technol. **26**, 074003 (2015). https://iopscience.iop.org/article/10.1088/0957-0233/26/7/074003
13. Charonko, J., Vlachos, P.: Estimation of uncertainty bounds for individual particle image velocimetry measurements from cross-correlation peak ratio. Measur. Sci. Technol. **24**(6), 17 (2013)
14. Keane, R., Adrian, R.: Optimization of particle image velocimeters. I. Double pulsed systems. Measur. Sci. Technol. **1**, 1202–1215 (1990). https://doi.org/10.1088/0957-0233/1/11/013
15. Sciacchitano, A., Wieneke, B., Scarano, F.: PIV uncertainty quantification by image matching. Measur. Sci. Technol. **24**(4), 17 (2013)

Comparative Study of Particle Representations on the Performance of MOPSO Algorithm in Solving Capacitated Lot-Sizing Problem

Hanen Ben Ammar[1](✉), Wafa Ben Yahia[2], Omar Ayadi[1], and Faouzi Masmoudi[1]

[1] Mechanics, Modelling and Production Research Laboratory, Mechanical Department, National School of Engineering of Sfax, University of Sfax, Sfax, Tunisia
{Hanen.benammar,faouzi.masmoudi}@enis.tn

[2] Mechanics, Modelling and Production Research Laboratory, Mechanical Department, National School of Engineering of Gabes, University of Gabes, Gabes, Tunisia

Abstract. Capacitated lot-sizing problem (CLSP), which is NP-Hard problem, is an important and challenging optimization problem in production planning. Therefore, this problem is often solved using different metaheuristic methods, like multi-objective particle swarm optimization (MOPSO) algorithm. To use this method, the first step that should be performed is to design a particle representation, which will be used for representing the solutions of the addressed problem. This paper seeks to investigate the effect of using different particle representations while developing a MOPSO algorithm to solve the CLSP. Two kinds of particle representations are considered. The first one is binary particle representation and the second one is real-binary particle representation. The treated problem was designed to minimize simultaneously the total cost and the total inventory level. Three performance evaluation metrics, namely number of non-dominated solution, quality of obtained Pareto fronts and runtime, were examined for comparing the particle representations effect on the performance of MOPSO algorithms. Numerical experiments indicate that the MOPSO algorithm with binary representation of the particle outperforms the one with real-binary representation of the particle in solving CLSP. This research highlights the importance of particle representation in the development of MOPSO algorithm.

Keywords: Multi-objective optimization · Particle Swarm Optimization · Binary Particle representation · Real-Binary representation · lot-sizing

Nomenclature

Symbol		Description
Indexes	p	Item index, p = *1,…,P*
	t	Planning period index, $t = 1,\dots,T$
Set indexes	T	Set of planning period
	P	Set of products
Parameters	D_{pt}	Demand of product p at planning period t

(continued)

A. Akrout et al. (Eds.): ICAV 2022, ACM 22, pp. 260–270, 2023.
https://doi.org/10.1007/978-3-031-34190-8_29

(*continued*)

Symbol		Description
	s_p	Unit setup cost of product p
	c_p	Unit production cost of product p
	st_p	Necessary time to prepare the machine for producing product p
	v_p	Necessary time for producing one unit of product p
	C_t	Available capacity in planning period t
	M_{pt}	Production upper bound of product p at planning period t
Decision variables	Z_{pt}	Setup Variable, $Z_{pt}\begin{cases} = 1, \textit{if the machine is prepared for} \\ \textit{producing} \text{ product } p \text{ in planning } \textit{period } t, \\ = 0, \textit{otherwise} \end{cases}$
	X_{pt}	Quantities of product p to be produced in planning period t
	I_{pt}	Production upper bound of product p at planning period t

1 Introduction

Meta-heuristic algorithms have been commonly used to solve complex space-search problems, as lot sizing problem. Such kind of algorithms basically follow an evolutionary paradigm ((Ben Yahia et al., 2017) and (Wang et al., 2022)). Although the diversity of evolutionary algorithms proposed in the literature, the Particle Swarm Optimization (PSO), which was originally developed by (Kennedy & Eberhart, 1995), represents one of the widely and most popular used algorithms ((Ning et al., 2019) and (Chachan & Hassan Ali, 2021)).

These nature-inspired algorithms always necessitate reflection on the definition of the suitable solution representation of the problem to be solved (Mlakar et al., 2020). In fact, the choice of the solution representations is one of the key factors for the success to find optimal solutions.

(Chen et al., 2013) studied the effect of adopting two different solution (chromosome) representations while Tabu-search algorithm to solve a scheduling problem. As well as, (Wu et al., 2017) proposed a new chromosome representation for Genetic Algorithm (GA) to solve a scheduling problem. They investigated the effect of using different solution representations by comparing four variants of GA chromosome representations. In addition, (Vlašić et al., 2020) compared seven solution encoding for GA, regarding different criteria, to solve a specific scheduling problem. They demonstrated that the choice of the solution representation has a significant impact on the optimization results. (Mlakar et al., 2020) compared the performance of two variants of solution representations, which are binary-coded and real-coded, when applying four algorithms: Differential Evolution, Artificial Bee Colony, PSO, and GA, to solve feature selection problem. The obtained results show that the binary-coded algorithms variant is preferred for the addressed problem.

To solve a multi-item capacitated lot-sizing problem (CLSP) with setup times, a multi-objective real-binary PSO (MORBPSO) algorithm was proposed by (Ben Ammar et al., 2020) as well as a multi-objective binary PSO (MOBPSO) (Ben Ammar et al., 2021). Since, the solution representation is a key issue in designing a PSO algorithm for solving the problem at hand, this study investigates the impact of particle representations on the performance of the multi-objective particle swarm optimization (MOPSO) algorithms in for solving the CLSP. According to the knowledge of the authors, no paper has investigated the impact of the particle representations in PSO algorithm when solving CLSP.

The remainder of the paper is structured as follows. Section 2 presents the problem formulation. In Sect. 3 both variant of particle representation are described. Section 4 is reserved for reporting the experimental results are detailed. Finally, the paper is concluded in Sect. 5.

2 Problem Modeling

The addressed capacitated lot-sizing problem seeks to determine the optimal production plan of a set of products over a planning horizon while satisfying customer demands with minimal total cost and total inventory level.

The following assumptions are considered for modeling the target problem:

- Demand is assumed to be deterministic,
- Inventory level is equal to zero at the beginning and the end of the planning horizon,
- Inventory level is determined at the end of the planning period.

A list of indexes, set indexes, parameters, and decision variables is given in the Nomenclature section.

Objective Functions:

$$Min\ \ Total\_Cost = \sum_{t=1}^{T}\sum_{p=1}^{P}\left[(c_p X_{pt}) + (s_p Z_{pt})\right] \tag{1}$$

$$Min\ \ Total\_Inventory\_Level = \sum_{t=1}^{T}\sum_{p=1}^{P} I_{pt} \tag{2}$$

Subjects to:

$$X_{pt} + I_{p,t-1} = D_{pt} + I_{pt} \quad \forall t = 1 \ldots T;\ \ p = 1 \ldots P \tag{3}$$

$$X_{pt} \le M_{pt} Z_{pt} \quad \forall t = 1 \ldots T;\ \ p = 1 \ldots P \tag{4}$$

$$\sum_{p=1}^{P} (v_p X_{pt} + st_p Z_{pt}) \le C_t \qquad \forall\, t = 1 \ldots T \tag{5}$$

$$M_{pt} = \min\left\{\sum_{t'=t}^{T} D_{pt'}, \frac{(C_{t'} - st_p)}{v_p}\right\} \quad \forall p = 1...P \tag{6}$$

$$X_{pt}, I_{pt} \geq 0 \qquad \forall t = 1\dots T;\ \forall p = 1\dots P \tag{7}$$

$$Z_{pt} \in \{0, 1\} \qquad \forall t = 1\dots T;\ \forall p = 1\dots P \tag{8}$$

The first objective function seeks to minimize the total cost that is represented by the sum of production and setup costs and defined by Eq. (1). Equation (2) defines the second objective function consists of minimizing the total inventory level over the planning horizon. Constraints (3) specify the material balance between customer demand, production quantities and inventory levels. Constraints (4) provide the relation between production quantities and setup variables. The capacity restrictions are expressed by constraints (5). The upper bound of production quantities is defined by formula (6). Constraints (7) and (8) specify the domain of the decisions variables.

3 Particle Representations

The performance of an evolutionary algorithm greatly depends on the solution representation (Chen et al., 2013). Thus, this study compares two different particle representations of MOPSO. The first one is binary particle representation, which is proposed by (Ben Ammar et al., 2021). The second particle representation is a real-binary one that is introduced by (Ben Ammar et al., 2020). In the following, the both representations are illustrated by referring to CLSP, with P products and T planning periods.

3.1 Binary Particle Representation

In this representation, the setup variable is selected to represent a particle, which is a potential solution, to design the MOPSO algorithm for solving the CLSP effectively. Thus, each particle i at iteration j is depicted by a binary matrix $Z_{i,j}(P \times T)$ (Ben Ammar et al., 2021).

For all $p = 1...P$ and $t = 1...T$, $Z_{i,j}(p, t)$ is equal to 1 if the product p will be produced at the planning period t, 0 otherwise.

$$Z_{i,j} = \begin{bmatrix} Z_{i,j}(1,1) & Z_{i,j}(1,2) & \dots & Z_{i,j}(1,T) \\ Z_{i,j}(2,1) & Z_{i,j}(2,2) & \dots & Z_{i,j}(2,T) \\ \dots & \dots & Z_{i,j}(p,t) & \vdots \\ Z_{i,j}(P,1) & Z_{i,j}(P,2) & \dots & Z_{i,j}(P,T) \end{bmatrix}$$

3.2 Real-Binary Particle Representation

In this kind of representation, the particle is presented by two main parts in order to design the MOPSO algorithm for solving the target CLSP. The first one encodes the setup variable and it is illustrated by a binary matrix $Z_{i,j}(P \times T)$, as the representation described below. The second part represents the real particle encoding, which denotes the anticipated percentage of the required production quantities to satisfy subsequent customer demand. It is represented by a real matrix $Q_{i,j}(P \times T)$ with value comprised between 0 and 1 (Ben Ammar et al., 2020).

$$Q_{i,j} = \begin{bmatrix} Q_{i,j}(1,1) & Q_{i,j}(1,2) & \ldots & Q_{i,j}(1,T) \\ Q_{i,j}(2,1) & Q_{i,j}(2,2) & \ldots & Q_{i,j}(2,T) \\ \ldots & \ldots & Q_{i,j}(p,t) & \vdots \\ Q_{i,j}(P,1) & Q_{i,j}(P,2) & \ldots & Q_{i,j}(P,T) \end{bmatrix}$$

4 Computational Experiments

In this section, the computational experiments that are conducted are provided. Firstly, the tested instances are described in Sect. 4.1. Secondly, Sect. 4.2 details the considered performance evaluation metrics. Finally, Sect. 4.3 provides the computational results and analysis.

4.1 Description of Tested Instances

To investigate the particle representation on the performance of MOPSO algorithm in solving CLSP, three instances are considered. The tested instances are classified into three main class, namely, small, medium and large size instances as shown in Table 1. The main features of these instances are shown in Table 1. The comparison tests of the considered representations are performed on the same problem instances and under the same conditions.

Table 1. Problem Instances

Class	Instances	Number of products	Number of periods
Small	P1	4	6
Medium	P 2	6	6
Large	P 3	12	10

The MOBPSO and MORBPSO algorithms are coded using MATLAB R2018a software. The parameters of both algorithms are presented in Table 2.

Table 2. Parameters of MORBPSO and MOBPSO algorithms

Parameters	Value
Maximum number of iterations	1000
Population size	50
Archive size	100
Inertia Weight	1
Social coefficient	1
Cognitive coefficient	1

4.2 Performance Evaluation Metrics

To evaluate the effect of the particle representation on the performance of the MOPSO algorithm in solving CLSP, obtained results are compared using three performance metrics. The considered metrics that are commonly used to evaluate the performance of multi-objective metaheuristics are described below:

- ***Number of Pareto Solutions:*** this metric represents the number of non-dominated solutions, which constitutes the Pareto Front generated by the algorithm. The higher value of this metric is preferred.
- ***Quality of Pareto front:*** shows the percentage of non-dominated solutions obtained by each algorithm, which appertain to the Final Pareto Front (FPF). The FPF is obtained by selecting all the non-dominated solutions after combining all Pareto solutions obtained with all the considered algorithms.
- ***Run Time:*** presents the time consumed by an algorithm to generate the Optimal Pareto Front.

4.3 Results and Discussion

Figures 1, 2, and 3 illustrate the comparison results of the performance of MOBPSO and MORBPSO algorithms in terms of number of non-dominated solutions provided, the quality of the obtained Pareto front and the computational time consumed. Moreover, an illustration of the obtained Pareto Front with both algorithms for the different tested instances is given in Fig. 4.

The results shown in Figs. 1 and 4 reveal that the binary version MOBPSO algorithm outperforms the real-binary version MORBPSO algorithm in all tested instances. In fact, the MOBPSO algorithm generates more non-dominated solutions than the MORBPSO algorithm. Thus, the decision maker has more flexibility in choosing the suitable production plan according to his preferences.

In addition to the higher number of the provided non-dominated solutions by MOBPSO algorithm, the quality of the generated Pareto fronts is better than the one generated by MORBPSO algorithm in all tested instances, as illustrated in Figs. 2 and 4. In fact, for example, for the tested instance P1, 87% of the solutions in the final Pareto Front are generated by the MOBPSO algorithm and the rest of the solutions (13%) are

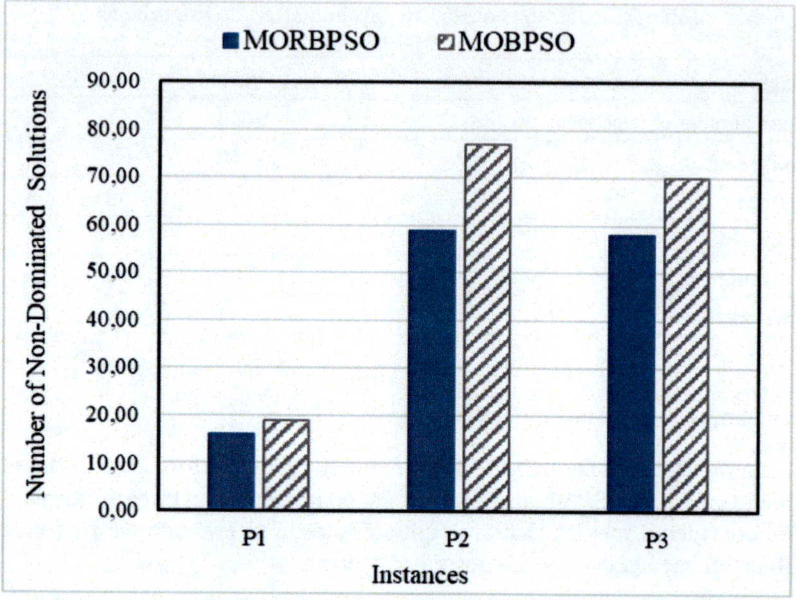

Fig. 1. Experimental results in terms of number of non-dominated solutions

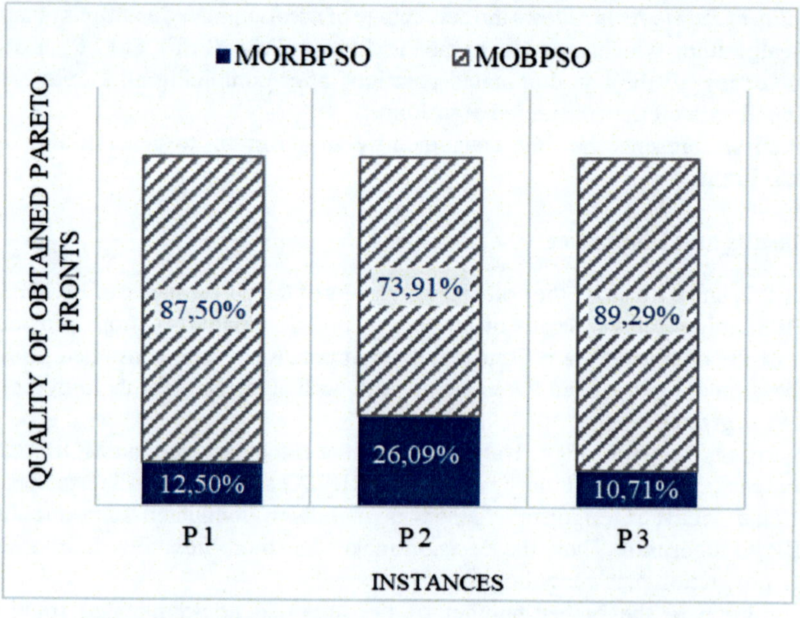

Fig. 2. Experimental results in terms of quality of Pareto fronts

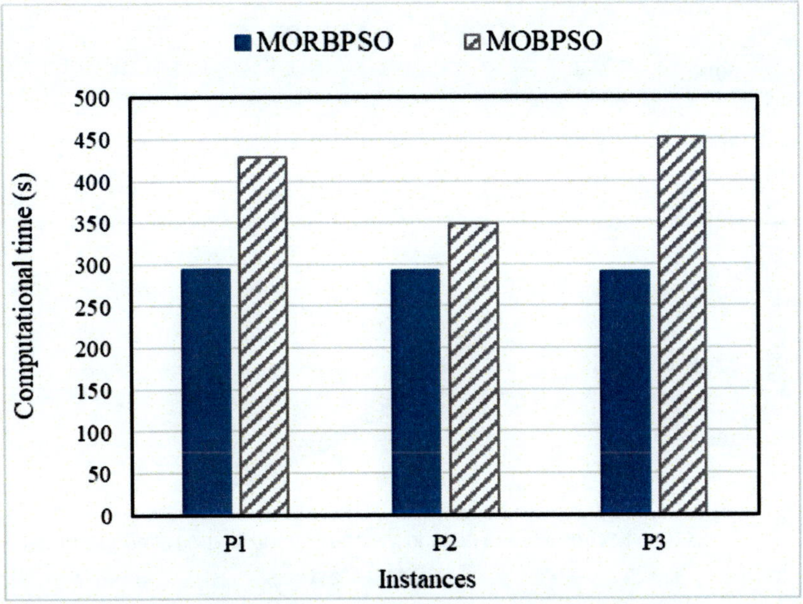

Fig. 3. Experimental results in terms of computational time

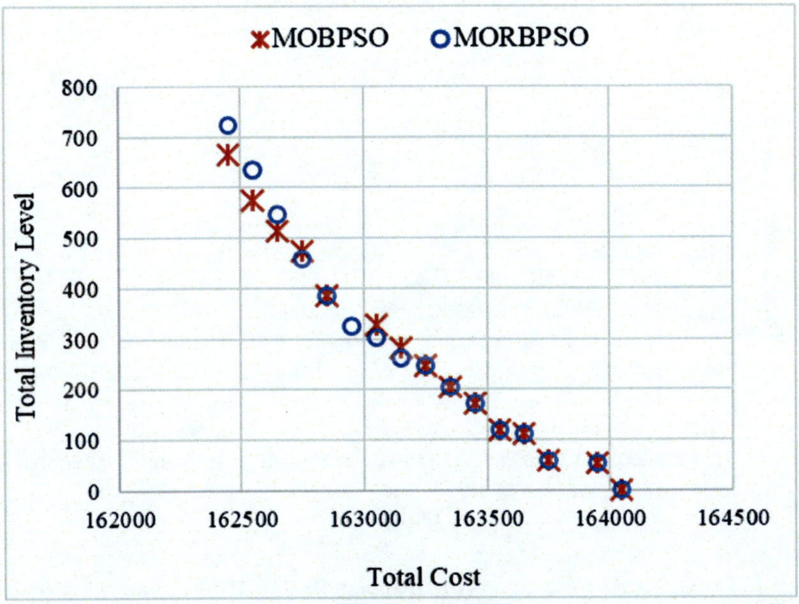

(a) Instance P1

Fig. 4. Obtained Pareto Fronts for considered Instances

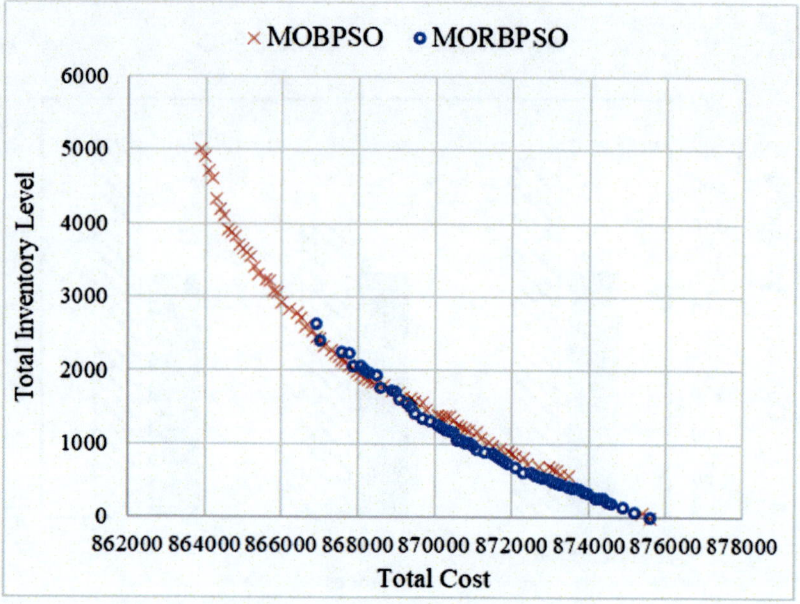

(b) Instance P2

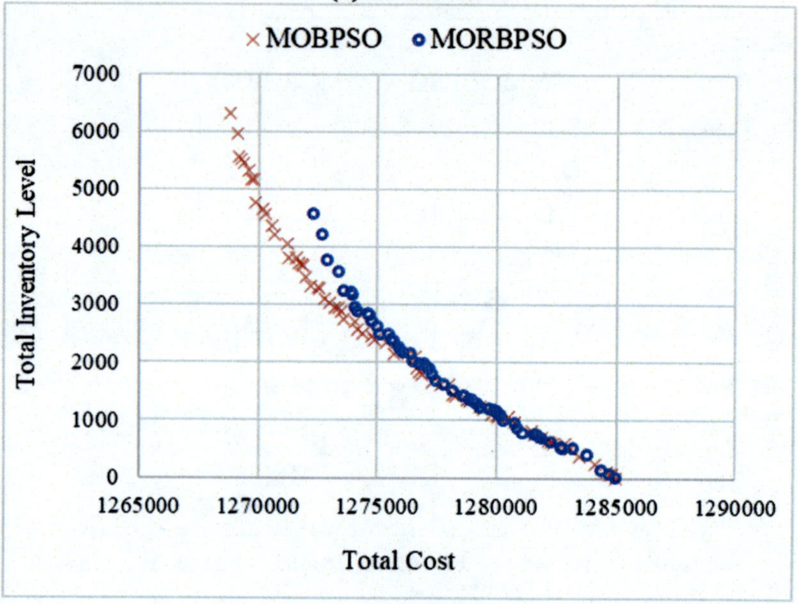

(c) Instance P3

Fig. 4. (*continued*)

generated by the MORBPSO algorithm as shown in Fig. 2. Therefore, the comparison results indicate that the MOBPSO algorithm has more exploration capability of the research space than the MORBPO algorithm.

Regarding to the run time metric, both variants of the PSO algorithm present a reasonable computational time that doesn't exceed eight minutes, as demonstrated in Fig. 3.

The experimental results demonstrate the importance of particle representation in enhancing the performance of MOPSO algorithm when solving the CLSP.

5 Conclusion

In this study, an investigation of the effect of the particle representation on the performance of the MOPSO algorithm in solving the CLSP is presented. Two variants of MOPSO with different particle representations are compared. The first one is the real-binary representation of the particle and the second one is the binary particle representation. The comparison tests of the considered representations are performed on the same problem instances and under the same conditions. The obtained results are evaluated using three performance metrics, namely number of non-dominated solutions, quality of obtained Pareto fronts and runtime.

Experimental results reveal that the performance of MOPSO algorithm with binary representation of the particle is better than the one with real-binary representation of the particle, with practically acceptable computational efforts. In fact, the MOBPSO algorithm outperforms the MORBPSO algorithm regarding the number of non-dominated solutions as well as the quality of the generated Pareto fronts. In addition, both variants of the MOPSO algorithm consume a reasonable computational time. Therefore, both kind of solution representations are suitable and efficient to adopt MOPSO for solving the CLSP. Moreover, with the binary representation of the solution, the decision maker has more flexibility in selecting the convenient production plan based on his preferences. Finally, the experimental results highlight the importance of the choice of the appropriate particle representation that enhance the performance of MOPSO algorithm in solving CLSP.

References

Ben Ammar, H., Ayadi, O., Masmoudi, F.: An effective multi-objective particle swarm optimization for the multi-item capacitated lot-sizing problem with set-up times and backlogging. Eng. Optim. **52**(7), 1198–1224 (2020). https://doi.org/10.1080/0305215X.2019.1636978

Ben Ammar, H., Ben Yahia, W., Ayadi, O., Masmoudi, F.: Design of efficient multiobjective binary PSO algorithms for solving multi-item capacitated lot-sizing problem. Int. J. Intell. Syst. **37**(2), 1723–1750 (2021). https://doi.org/10.1002/int.22693

Ben Yahia, W., Ayadi, O., Masmoudi, F.: A fuzzy-based negotiation approach for collaborative planning in manufacturing supply chains. J. Intell. Manuf. **28**(8), 1987–2006 (2015). https://doi.org/10.1007/s10845-015-1085-x

Chachan, H.A., Hassan Ali, F.: Using non-dominated sorting particle swarm optimization algorithm II for bi-objective flow shop scheduling problems. Iraqi J. Sci. **62**(1), 275–288 (2021). https://doi.org/10.24996/ijs.2021.62.1.26

Chen, C.F., Wu, M.C., Lin, K.H.: Effect of solution representations on Tabu search in scheduling applications. Comput. Oper. Res. **40**(12), 2817–2825 (2013). https://doi.org/10.1016/j.cor.2013.06.003

Kennedy, J., Eberhart, R.: Particle swarm optimization. In: IEEE International Conference of Neural Networking, vol. 4, pp. 1942–1948 (1995). https://doi.org/10.1109/ICNN.1995.488968

Mlakar, U., Fister, I., Fister, I.: Impact of solution representation in nature-inspired algorithms for feature selection. IEEE Access **8**, 134728–134742 (2020). https://doi.org/10.1109/ACCESS.2020.3011153

Ning, Y., Peng, Z., Dai, Y., Bi, D., Wang, J.: Enhanced particle swarm optimization with multi-swarm and multi-velocity for optimizing high-dimensional problems. Appl. Intell. **49**(2), 335–351 (2018). https://doi.org/10.1007/s10489-018-1258-3

Vlašić, I., Đurasević, M., Jakobović, D.: A comparative study of solution representations for the unrelated machines environment. Comput. Oper. Res. **123**, 105005 (2020). https://doi.org/10.1016/j.cor.2020.105005

Wang, S., Hui, J., Zhu, B., Liu, Y.: Adaptive genetic algorithm based on fuzzy reasoning for the multilevel capacitated lot-sizing problem with energy consumption in synchronizer production. Sustain. (Switz.) **14**(9) (2022). https://doi.org/10.3390/su14095072

Wu, M.C., Lin, C.S., Lin, C.H., Chen, C.F.: Effects of different chromosome representations in developing genetic algorithms to solve DFJS scheduling problems. Comput. Oper. Res. **80**, 101–112 (2017). https://doi.org/10.1016/j.cor.2016.11.021

Experimental and Numerical Investigation of Viscoelastic Layer Effect on Bio-Composite Dynamic Behaviour

Firas Meddeb[1,2](✉), Abderrahim El Mahi[1], Jean-Luc Rebiere[1], Hajer Daoud[2], Mohamed Amine Ben Souf[2], and Mohamed Haddar[2]

[1] Acoustic Laboratory of Le Mans University (LAUM), UMR CNRS 6613, AV. O. Messiaen, 72085 Le Mans Cedex 9, France
firas.meddeb@enis.tn

[2] Laboratory of Mechanics, Modelling and Production. (LA2MP), National School of Engineers of Sfax, University of Sfax, BPN °1173-3038 Sfax, Tunisia

Abstract. Biocomposite materials have the potential to address some of the health and environmental issues we face today. These materials are made from renewable resources. In addition to their environmental benefits, biocomposites also offer advantages over traditional synthetic composites in terms of their mechanical properties and durability. For example, some biocomposites have been shown to have high strength-to-weight ratios and excellent resistance to fatigue and impact This work shows the results of a numerical analysis of the damping characteristics of a bio-composite with and without a Viscoelastic Layer (VL). The insertion of this layer has a significant influence on the vibration behavior, bending stiffness and damping factor. The inserted viscoelastic layer in the middle plan of the composite is natural rubber and the external layers are flax fibers in PLA matrix composite. The composite is produced by the additive manufacturing technology. The aim of this article is the comparison of the results of numerical and the experimental vibration behavior of composite material: Flax/PLA composite and viscoelastic composite. This works has guiding significance for the calculation and application of elastic viscoelastic elastic sandwich structure.

Keywords: Bio composite · 3D Printing · Viscoelastic layer · Damping · Numerical simulation

1 Introduction

Bio-composite materials are used in the manufacturing of structural components such as automotive parts, due to their good balance of strength, rigidity and lightness. They are also environmentally friendly and durable [1]. These materials are used in a variety of applications including packaging, construction and consumer products [2]. Bio-composites are appreciated for their high energy dissipation capabilities and are used to reduce vibration [3, 4] and noise [5] in applications such as aircraft cabins and civil engineering structures improving durability and safety. Research shows that noise and

A. Akrout et al. (Eds.): ICAV 2022, ACM 22, pp. 271–278, 2023.
https://doi.org/10.1007/978-3-031-34190-8_30

vibration in aircraft cabins can negatively affect the health and well-being of passengers and crew members, leading to impacts on hearing and cardiovascular health, stress and sleep quality [6]. The effects of whole-body vibration on health in everyday conditions have been studied, however results should be interpreted with caution as they are based on assumptions and approximations. Research has also shown that railway vibrations and noise can affect passenger sleep quality [7]. Hamrouni et al. [8] characterize in static and vibration a biobased composite with short flax fibers and PLA. This composite was manufactured by additive manufacturing technology (3D printing). Daoud et al.[9] characterized a composite made from long flax fibers/epoxy matrix, followed by an integration of a viscoelastic layer and a comparison between their (with and without viscoelastic layer) vibratory behavior results. They have found that bio-based composites have an ability to dampen vibrations and the integration of a viscoelastic layer reinforces and multiplies this ability.

This study involves the use of numerical simulation to investigate the dynamic properties of short flax fiber/PLA sandwich composites and viscoelastic layers to reduce noise and vibration. Studies are carried out to determine the frequency responses and the damping properties of the structure as well as the influence of the addition of a natural rubber viscoelastic layer on its dynamic behavior. The objective of this work is to explore ways to improve the mechanical and dynamic properties of 3D printed composites while reducing vibrations.

2 Experimental Analyses

2.1 Sample Preparation

3D printing technology, also known as additive manufacturing, involves creating a physical object by building it layer by layer from a digital model. The Raise 3D 2 PLUS is a specific brand of 3D printer that uses material extrusion technology, where a material, such as plastic or composite, is melted and extruded through a nozzle to create the layers of the test tube. Solidworks CAD software is used to design the digital model of the specimen saved in stl format. Then this file is converted using specific software into instructions for printing it, which are stored on an SD card and read by the printing machine.

2.2 Vibration Experimental Setup

In this study, the clamped-free boundary condition is considered. The specimens are placed horizontally with one end fixed and the other end is free. To avoid damage to the composite material during tightening, the tightening was applied near the end of the specimens on a uniform block (see Fig. 1). The total length of the vibration specimens is 230 mm.

In this series of tests, an impact hammer was used to simulate the bending vibrations of the specimen near the clamped end. The frequency response to this excitation was measured using a laser vibrometer near the free end of the specimen. The signals recorded during excitation and response on the composites were processed and digitized by a dynamic signal analyzer. This equipment was used with a personal computer to acquire and analyze the collected data using MATLAB software.

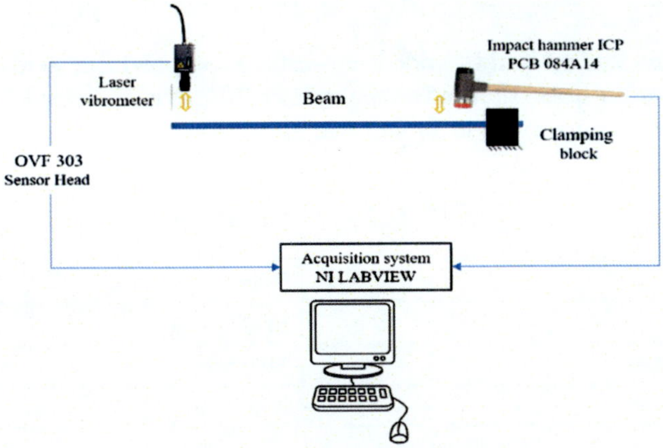

Fig. 1. Experimental setup.

3 Viscoelastic Layer Modelling

3.1 Geometric Modelling

Two types of specimens are tested, the first printed in Flax/PLA only and the other with a viscoelastic layer (Rubber) printed at mid-height inserted between two layers of Flax/PLA (see Fig. 2).

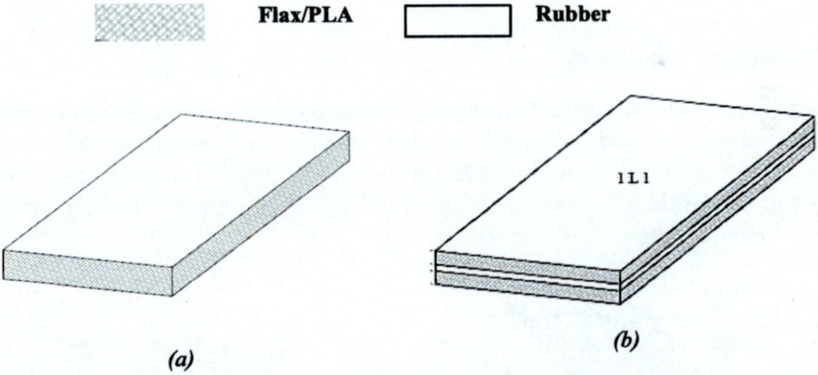

Fig. 2. Specimen (a) Flax/PLA elastic composite (b) Viscoelastic composite (with 1 mm of rubber)

The sandwich sizes are Length = 230 mm, width = 25 mm and thickness e_{tot} = 5 mm.

3.2 Materials

The composites studied in this work were made using polylactic acid with flax fiber (Flax/PLA) as a bio-based skin and natural rubber (NR) as a viscoelastic sandwich core. Their properties are presented in Table 1 and 2.

Table 1. Flax/PLA Property

Property	Value
Density	1070 kg/m3
Young's modulus	3400 MPa
Poisson ratio	0.29

Table 2. Rubber property

Property	Value
Density	1190 kg/m3
Young's modulus	13 MPa
Poisson ratio	0.39
Damping ratio	0.3

3.3 Boundary Conditions

The specimen is meshed with the elements mentioned below for the vibration analysis. The constrain set is made by removing all degrees of freedom at the fixed end (blue surface in Fig. 3) and the loading conditions is 1 N load applied at a point on the surface of the top face of the plate to model for modelling the effect of the loading carried out with the impact hammer (red arrow).

(a) (b)

Fig. 3. (a) Mesh model (b) Constrain boundary and loading conditions.

3.4 Finite Element Modelling

The 4-node tetrahedra (in 3D) have geometries of cubic reference. Their basic function is affine by relative to each local coordinate and is defined as the product of 3 1D affine functions, each applied to a different coordinate.

First, a convergence study (Fig. 4) is performed and only the meshes giving the converged results are used for the modal analysis.

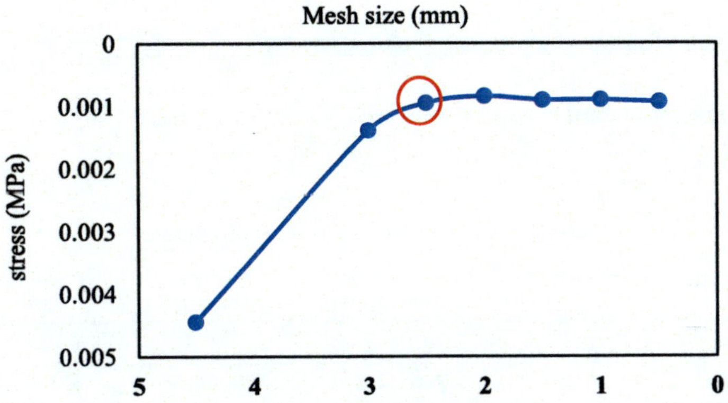

Fig. 4. Convergence study.

This study was with quadratic element order or we choose size 2.5 mm that it gives us 5278 nodes and 23761 elements.

4 Results and Discussions

4.1 Modal analysis

In a modal analysis, the only load that needs to be considered is the zero-displacement stress, so the only boundary conditions that must be defined in the finite element model are those related to this load. The numerical simulation is conducted, and the first four modes (Table 3) were depicted in Fig. 5.

4.2 Experimental and numerical results

Figure 6 shows the experimental results of the vibration tests for the studied composite with and without viscoelastic layer. This result illustrates the effect of the insertion of the viscoelastic layer. It can be noted that the amplitude decreases by 16% between 0 and 600 Hz and from 600 Hz the viscoelastic composite has greater damping.

To understand these following results, we performed a numerical validation for each bending mode on the studied composite specimen. We therefore combined the results of experimental tests and modelling predictions to create FRF curves showing the natural frequencies of the previously studied composite beam. These curves are shown in Fig. 7.

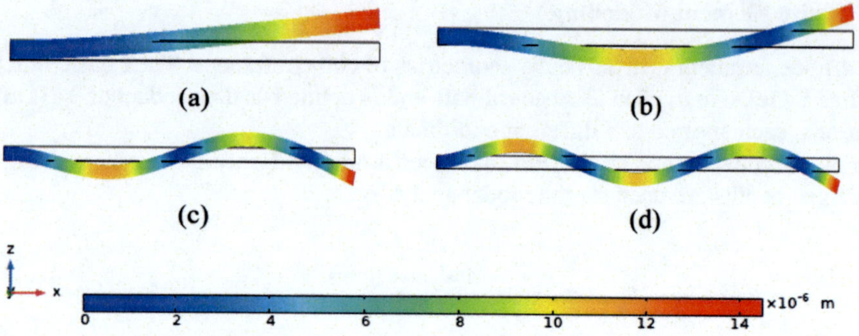

Fig. 5. The first four modal shapes of composite Flax/PLA: (a) mode 1 (b) mode 2 (c) mode 3 (d) mode 4

Table 3. Inherent frequency (Hz) of sandwich

Mode	Inherent frequency (Hz)
1	20
2	140
3	400
4	780

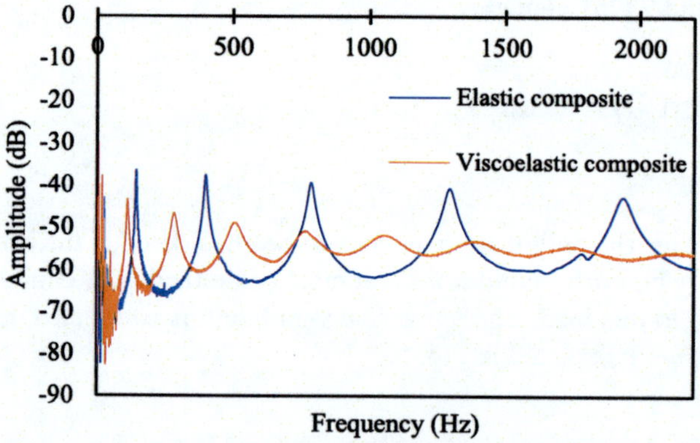

Fig. 6. Experimental results of vibration tests.

For all composite specimens, with and without viscoelastic layer (VL). We observed a good numerical validation of our experimental measurements.

The natural frequencies obtained from experimental tests and numerical simulation have a deviation of 10% to 20% (at 2200 Hz). This difference can be attributed on one hand to the variability in quality introduced by the use of 3D printing. On the other hand, the assumption of quasi-homogeneous material properties is made in the finite element analysis. The experimental tests were carried out under imprecise boundary conditions which could have an effect on the frequency response: The clamping conditions of the printed specimen which holds the sample, have an influence on the measured damping of each specimen composite.

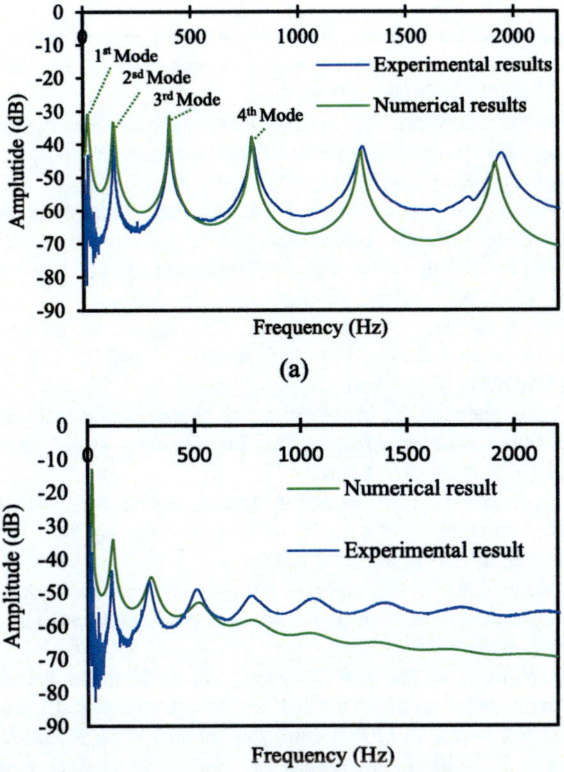

Fig. 7. FRF Numeric validation (a) elastic composite (b) Viscoelastic composite

5 Conclusions

The simulation of harmonic displacement response has been extended to the case of an elastic composite (Flax/PLA) including a viscoelastic layer. In addition, experimental tests were carried out on two types of composite specimens. Then, the effect of the inserted rubber layer is highlighted by a considerable damping in the FRF curves compared to the Flax/PLA composite. Also, one can see that experimental and numerical results are in good agreement. Finally, this work can be extended by doing a parametric study regarding the viscoelastic layer thickness effect and by considering different stacking sequences. These first results allow us to continue this work with different viscoelastic layer thicknesses and different stacking sequences.

Acknowledgements. The authors gratefully acknowledge the assistance and the financial support of the project 20PEJC 06–06 by the Tunisian Ministry of Higher Education and Scientific Research.

References

1. Yildizhan, Ş., Çalik, A., Özcanli, M., Serin, H.: Bio-composite materials: a short review of recent trends, mechanical and chemical properties, and applications. Eur. Mech. Sci. **2**(3), 3 (2018). https://doi.org/10.26701/ems.369005
2. Attaran, S.A., Hassan, A., Wahit, M.U.: Materials for food packaging applications based on bio-based polymer nanocomposites: a review. J. Thermoplast. Compos. Mater. **30**(2), 143–173 (2017). https://doi.org/10.1177/0892705715588801
3. Daoud, H., El Mahi, A., Rebiere, J., Taktak, M., Haddar, M.: Characterization of the vibrational behaviour of flax fibre reinforced composites with an interleaved natural viscoelastic layer. Appl. Acoust. **128** (2016). https://doi.org/10.1016/j.apacoust.2016.12.005
4. Kesentini, Z., El Mahi, A., Rebiere, J., Guerjouma, R., Beyaoui, M., Haddar, M.: Effect of hydric aging on the static and vibration behavior of 3D printed bio-based flax fiber reinforced poly-lactic acid composites. Polym. Polym. Compos. **30**, 096739112210818 (2022). https://doi.org/10.1177/09673911221081826
5. Bharath, K.N., Basavarajappa, S.: Applications of biocomposite materials based on natural fibers from renewable resources: a review. Sci. Eng. Compos. Mater. **23**(2), 123–133 (2016). https://doi.org/10.1515/secm-2014-0088
6. Mellert, V., Baumann, I., Freese, N., Weber, R.: Impact of sound and vibration on health, travel comfort and performance of flight attendants and pilots. Aerosp. Sci. Technol. **12**(1), 18–25 (2008). https://doi.org/10.1016/j.ast.2007.10.009
7. Mansfield, N.J., Aggarwal, G.: Whole-body vibration experienced by pilots, passengers and crew in fixed-wing aircraft: a state-of-the-science review. Vibration **5**(1), 1 (2022). https://doi.org/10.3390/vibration5010007
8. Hamrouni, A., Rebiere, J., El Mahi, A., Beyaoui, M., Haddar, M.: Experimental analysis of the static and dynamic behavior of 3D printed bio-based conventional and auxetic architectural materials. Int. J. Appl. Mech. **14** (2022). https://doi.org/10.1142/S1758825122500259
9. Daoud, H., Rebiere, J., El Mahi, A., Taktak, M., Haddar, M.: Effect of an interleave natural viscoelastic layer on the dynamic behaviour of a bio-based composite. Adv. Compos. Lett. **26** (2018). https://doi.org/10.1177/096369351702600601

Effect of Nanoparticles on the Heat Transfer of the Phase Change Materials

Maissa Bouguila[1,2](✉), Ahmed Samet[1], Mohamed Amine Ben Souf[1], Abdelkhalak El Hami[2], and Mohamed Haddar[1]

[1] Laboratory of Mechanics, Modeling and Manufacturing (La2MP), Mechanical Engineering Department, National School of Engineers of Sfax, University of Sfax, Sfax, Tunisia
{maissa.bouguila,mohamed-amine.bensouf,mohamed.haddar}@enis.tn
[2] Laboratory of Mechanics of Normandy (LMN), INSA Rouen Normandy, University of Rouen, 76801 St-Etienne de Rouvray, France
abdelkhalak.elhami@insa-rouen.fr

Abstract. This present study proposed a heat sink filled with NANO-enhanced Phase Change Materials (NANOPCM). The Computational Fluid Dynamics software (CFD) ANSYS Fluent 2021R1 is adopted to resolve the governing equations. The equations model the heat transfer and the fluid flow inside the PCM cavity. The used PCM is the n-eicosane with a melting temperature of 35 °C. The PCM-based heat sink filled is selected as a reference to reveal the effect of the nanoparticles. Three volume fraction: 1%vol, 2%vol, and 3%vol of Graphene are adopted to analyze the effect of the nanoparticle concentration over the heat transfer inside the heat sink. The results demonstrated that the graphene dispersion improve the heat sink performances. The base temperature reduce and the melting time decrease. Besides, the low percentage of the graphene 1 %vol volume fraction is more performed than the high nanoparticle concentration (2 %vol and 3 %vol volume fraction).

Keywords: NANO-enhanced Phase Change Materials (NANOPCM) · Thermal Management · Heat sink · Latent Heat Storage

1 Introduction

The high flux dissipated by the high-functionality electronic devices could cause several deteriorations. In the recent years, the Thermal management system based on Phase Change Materials (PCM) showed an efficient performances (Qureshi et al. 2018, Yang et al. 2021). However, the low thermal conductivity of the PCM is an important limitations of the use of this high capacity heat storage material.

Furthermore, the nanoparticle dispersion with high thermal conductivity present a relative solution to enhance the heat transfer inside the PCM. The Graphene is one of the recommended nanoparticles as it characterizes by a relative important thermal conductivity (Xiong et al. 2020). A numerical study (Bayat et al. 2018) is established to investigate the effect of two high thermal conductivity nanoparticles copper oxide

A. Akrout et al. (Eds.): ICAV 2022, ACM 22, pp. 279–286, 2023.
https://doi.org/10.1007/978-3-031-34190-8_31

and aluminum oxide over the PCM-based heat sink. Three nanoparticle concentration: 2%, 4%, and 6%, was adopted. The results demonstrated that the low percentage 2% present more efficient performances than the high concentration for both copper oxide and aluminum oxide. The same found is deducted by (Kothari et al. 2021). They proved that the heat sink with low nanoparticles concentration present the high enhancement ratio for SPT of 60 °C. (Bouguila et al., 2023) established a numerical study of finned heat sink filled with paraffin with nanoparticles: Graphene (GNP) and Multi-Walled Carbon Nanotubes (MWCNT). The same weight fraction 4% is adopted for both nanoparticles. The results showed that the GNP reduce the temperature and improve the efficiency of the heat sink while the MWCNT decline it.

In this study, the effects of the graphene concentration over the performances of the heat sink are detailed. The base temperature and the liquid fraction are analyzed. The first section is a presentation of the computational model and the second specifies the numerical simulation configuration and the third and the last part is for result discussion and conclusion.

2 Computational Model

The physical model is a heat sink filled with PCM and NANOPCM with dimensions 114 × 114 × 25 mm. The PCM cavity with dimensions 100 × 100 × 20 mm, as presented in Fig. 1. 2 mm-thick plate is placed under the heat sink and models the electronic device. A heat flux is applied at the bottom surface of the plate heater. The Thermo-physical properties of the based PCM: n-eicosane and the NANOPCMs: n-eicosane + 1 %vol GNP, n-eicosane + 2 %vol GNP, n-eicosane + 3 %vol GNP (Das et al. 2016).

Table 1. Thermo-physical properties of PCM and NANOPCMS (Mahdi et al. 2017)

Proprieties	n-eicosane	n-eicosane + 1%GNP	n-eicosane + 2%GNP	n-eicosane + 3%GNP
Density (Kg/m^3)	815(S) 780(L)	828.85 (S) 794.2(L)	842.7 (S) 808.4 (L)	856.5(S) 822.6(L)
Thermal conductivity (W/m.k)	0.412(S) 0.160(L)	0.705 (S) 0.25 (L)	0.995(S) 0.34(L)	1.26 (S) 0.44(L)
Specific heat (J/Kg.k)	1920(S) 2130(L)	1907 (S) 2092 (L)	1894(S) 2054(L)	1882 (S) 2017 (L)
Thermal expansion	$850 * 10^{-6}$	$827 * 10^{-6}$	$806 * 10^{-6}$	$784 * 10^{-6}$

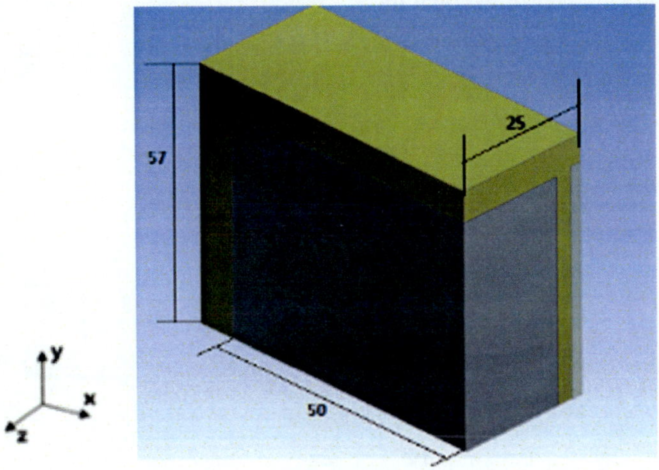

Fig. 1. Computational model

3 Numerical Model

3.1 Governing Equations

The governing equations are used to model the fluid flow during the charging phase (Yang and Wang, 2012). The ANSYS FLUENT 2021 is used to resolve of this equations and the following hypotheses are adopted:

- Uncompressible and laminar Fluid Flow (the pressure-based solver is used).
- The Thermo-physical properties are unchanged in the liquid and solid phases.
- No volume change during the melting process.
- No Radiation and no Natural heat convection

*Continuity Conservation Equation

$$\frac{\partial\rho}{\partial t}+\frac{\partial\rho u}{\partial x}+\frac{\partial\rho v}{\partial y}+\frac{\partial\rho w}{\partial z}=0 \tag{1}$$

With ρ the density of the PCM and u, y and w the velocity vector components along x, y and z.

*Energy Conservation Equation

$$\begin{gathered}\frac{\partial\rho c_pT}{\partial t}+u\frac{\partial\rho c_pT}{\partial x}+v\frac{\partial\rho c_pT}{\partial y}+w\frac{\partial\rho c_pT}{\partial z}= \\ \frac{\partial}{\partial x}(\lambda\frac{\partial T}{\partial x})+\frac{\partial}{\partial y}(\lambda\frac{\partial T}{\partial y})+\frac{\partial}{\partial z}(\lambda\frac{\partial T}{\partial z})+S_h\end{gathered} \tag{2}$$

Where C_p is specific heat, K is the thermal conductivity.

*Momentum conservation equation

$$
\begin{aligned}
&\frac{\partial \rho u}{\partial t}+\frac{\partial \rho u u}{\partial x}+\frac{\partial \rho u v}{\partial y}+\frac{\partial \rho u w}{\partial z}= \\
&\frac{\partial}{\partial x}(\mu\frac{\partial u}{\partial x})+\frac{\partial}{\partial y}(\mu\frac{\partial u}{\partial y})+\frac{\partial}{\partial z}(\mu\frac{\partial u}{\partial z})-\frac{\partial p}{\partial x}+Au \\
&\frac{\partial \rho v}{\partial t}+\frac{\partial \rho u v}{\partial x}+\frac{\partial \rho v v}{\partial y}+\frac{\partial \rho w v}{\partial z}= \\
&\frac{\partial}{\partial x}(\mu\frac{\partial v}{\partial x})+\frac{\partial}{\partial y}(\mu\frac{\partial v}{\partial y})+\frac{\partial}{\partial z}(\mu\frac{\partial v}{\partial z})-\frac{\partial p}{\partial x}+Av
\end{aligned} \tag{3}
$$

Effect of nanoparticles on the heat transfer of the phase change materials

$$
\begin{aligned}
&\frac{\partial \rho w}{\partial t}+\frac{\partial \rho u w}{\partial x}+\frac{\partial \rho w v}{\partial y}+\frac{\partial \rho w w}{\partial z}= \\
&\frac{\partial}{\partial x}(\mu\frac{\partial w}{\partial x})+\frac{\partial}{\partial y}(\mu\frac{\partial w}{\partial y})++\frac{\partial}{\partial z}(\mu\frac{\partial w}{\partial z})-\frac{\partial p}{\partial x}+Aw+S_b
\end{aligned}
$$

The solidification/melting model is used to describe the heat transfer and the fluid flow and the enthalpy-porosity formulation is adopted. The fusion front is modelled as porous zone with the porosity is equal to liquid fraction. It varies between 0 (solid) and 1 (liquid) and in the mushy zone is defined as:

$$
\begin{aligned}
&f=\frac{T-T_{solidus}}{T_{liquidus}-T_{solidus}} && T_{solidus}<T<T_{liquidus} \\
&f=0 && T<T_{solidus} \\
&f=1 && T>T_{liquidus}
\end{aligned} \tag{4}
$$

The two source term S_h, S_b and Au, Av are inserted to the energy and the momentum equations to ensure the convergence.

$$
S_b=\frac{\rho_{ref}\, g\beta(h-h_{ref})}{C_p} \quad S_h=\frac{\partial \rho \Delta \mathrm{H}}{\partial t}+\frac{\partial \rho \Delta \mathrm{H}}{\partial x}+\frac{\partial \rho \Delta \mathrm{H}}{\partial y}+\frac{\partial \rho \Delta \mathrm{H}}{\partial z} \quad A=C\frac{(1-s^2)}{s^3+b} \tag{5}
$$

With C and b are constants: C = C = 10^5 and b = 0.01 and $\Delta H = fL$: Latent heat.

3.2 Boundary Conditions

The PCM is totally solidified and the temperature is set at the ambient temperature 18 °C (290 K) at the initial time t = 0 s (Eq. 6). The heat sink is insulted (Eq. 7). At the bottom surface of the heater, a heat flux Q equal to 2000 W/m^2 is applied (Eq. 8) and K_{AL} is the thermal conductivity of the aluminum block.

$$T_0 = 290K,\ u = 0,\ v = 0,\ w = 0 \tag{6}$$

$$\left.\frac{\partial T}{\partial x}\right|_{\substack{x=0,x=114\\ y=0-114\\ z=0-30}} = 0,\quad \left.\frac{\partial T}{\partial y}\right|_{\substack{y=0,y=114\\ x=0-114\\ z=0-30}} = 0,\quad \left.\frac{\partial T}{\partial z}\right|_{\substack{y=0-114\\ x=0-114\\ z=30}} = 0 \tag{7}$$

$$\left.\frac{\partial T}{\partial z}\right|_{\substack{z=0\\ x=0-114\\ y=0-114}} = \frac{Q}{K_{AL}} \tag{8}$$

3.3 CFD Methodology

The computational fluid dynamics CFD software ANSYS Fluent 2021, solves the conservation equations defined in Subsect. 1.1. The resolution is based on a finite volume method scheme. The solidification-melting model is adopted. The enthalpy-porosity models the mushy zone (fusion zone), which relies the enthalpy (energy equation) to the value of the liquid fraction. The high-order Quadratic Upstream Interpolation for the Convective Kinematics is adopted for the energy and momentum equations. The PREssure STaggering Option (PRESTO) spatial interpolation is used for the pressure interpolation. The under-relaxation factors are 0.3, 0.7, 0.9 and 1.0 for pressure, velocity, liquid fraction, and energy respectively and the time step is chosen 0.5 s and the solution converge after 120 iterations. The choice of the time step is referenced to the convergence study carried out by (Bouguila et al., 2023).

4 Results and Discussions

The geometric model and the boundary conditions are taken referenced to the experimental study (Ali and Arshad, 2017). The numerical study results of the base temperature is compared with the experimental data, for the pure PCM: n-eicosane, as presented in Fig. 2. It can be seen that the numerical and the experimental results showed a good argument with 11% maximum discrepancy. The three phase: the sensible phase (pre-melting), the latent phase, and the sensible phase (post-melting), are presented. However, the experimental study showed that the latent phase start at melting temperature 42 °C but it theoretically 35 °C. It can be explained by the error in the experimental measurement and the heat loss.

Figure 3 present the base temperature of the heat sink filled with the based PCM: n-eicosane and the NANOPCMs with volume fraction (%vol): 1 %vol GNP, n- 2 %vol GNP, 3 %vol GNP. The results showed that the graphene (GNP) reduce the base temperature

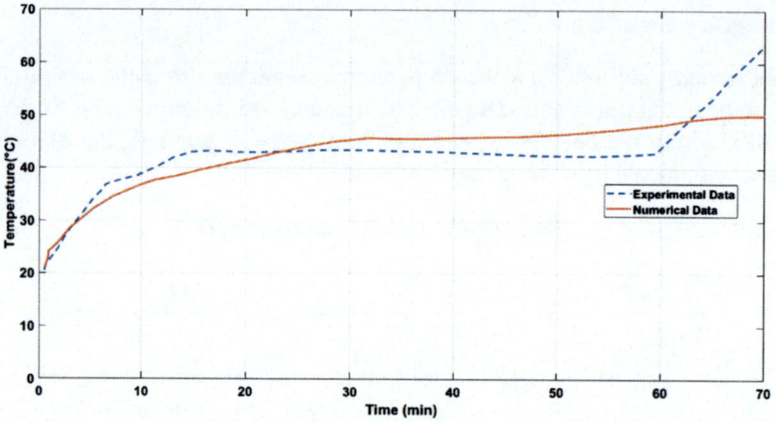

Fig. 2. Validation of the numerical study vs experimental data (Ali and Arshad, 2017)

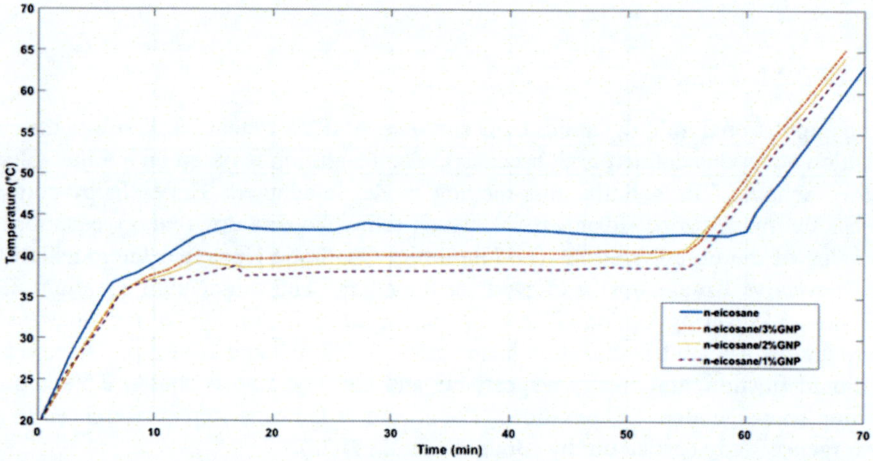

Fig. 3. Base temperature for PCM n-eicosane, NANOPCM n-eicosane: 1%GNP, 2%GNP, and 3%GNP

at the latent phase by 4 °C, 5 °C, and 6 °C for the n-eicosane + 3 %vol GNP, n-eicosane + 2 %vol GNP, and n-eicosane + 1%GNP.

The temperature variation of the based PCM: n-eicosane and the NANOPCM n-eicosane/1%GNP is presented in Fig. 4. The results demonstrated that the PCM/1%GNP reach the melting temperature 35 °C at 14.5 min while the based PCM reach it at 28.5. The melting time is 50 min and 59 min for NANOPCM and the the based PCM. The dispersion of the graphene,with high thermal conductivity, enhance the heat transfer inside the heat sink.

Figure 5 present the liquid fraction results in function of time for the PCM: n-eicosane and the NANOPCM: n-eicosane/1%GNP. It can be seen that the melting of both PCM

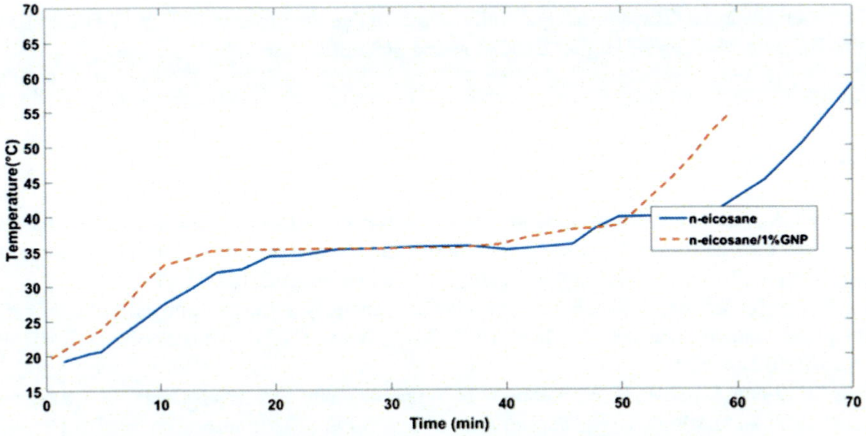

Fig. 4. Temperature variation for PCM and NANOPCM with 1%GNP

and NANOPCM start at 8 min. However, the melting ends early for the NANOPCM at 52 min and the PCM reach the completed melting at 62 min.

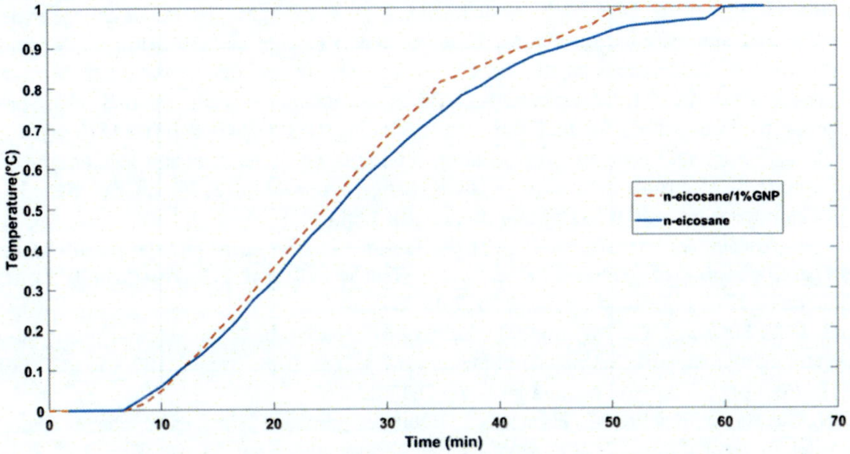

Fig. 5. Liquid fraction for PCM and NANOPCM with 1%GNP

5 Conclusion

The dispersion of the graphene improves the heat sink performance. In this study, the low percentage of the graphene 1% volume fraction is more performed than the high volume fraction 2% and 3%. The base temperature is reduced by 13.63% compared to the based PCM. Besides, the graphene (1 %vol GNP) accelerates the heat transfer inside the PCM and that caused the decrease of the melting time.

Acknowledgements. The authors gratefully acknowledge the Project "PRF 2019-D6P1" funded by the Tunisian Ministry of Higher Education and Scientific Research.

References

Qureshi, Z.A., Ali, H.M., Khushnood, S.: Recent advances on thermal conductivity enhancement of phase change materials for energy storage system: a review. Int. J. Heat Mass Transf. **127**, 838–856 (2018). https://doi.org/10.1016/j.ijheatmasstransfer.2018.08.049

Yang, L., Jin, X., Zhang, Y., Du, K.: Recent development on heat transfer and various applications of phase-change materials. J. Clean. Prod. **287**, 124432 (2021). https://doi.org/10.1016/j.jclepro.2020.124432

Xiong, T., Zheng, L., Shah, K.W.: Nano-enhanced phase change materials (NePCMs): a review of numerical simulations. Appl. Therm. Eng. **178**, 115492 (2020). https://doi.org/10.1016/j.applthermaleng.2020.115492

Bayat, M., Faridzadeh, M.R., Toghraie, D.: Investigation of finned heat sink performance with nano enhanced phase change material (NePCM). Therm. Sci. Eng. Prog. **5**, 50–59 (2018). https://doi.org/10.1016/j.tsep.2017.10.021

Kothari, R., Sahu, S.K., Kundalwal, S.I.: Investigation on thermal characteristics of nano enhanced phase change material based finned and unfinned heat sinks for thermal ma-nagement system. Chem. Eng. Process. **162**, 108328 (2021). https://doi.org/10.1016/j.cep.2021.108328

Bouguila, M., Samet, A., Souf, M.B., Abdelkhalak, E.H., Haddar, M.: The phase change materials for thermal applications. In: Walha, L., et al. (eds.) Design and Modeling of Mechanical Systems - V: Proceedings of the 9th Conference on Design and Modeling of Mechanical Systems, CMSM'2021, December 20–22, 2021, Hammamet, Tunisia, pp. 695–703. Springer International Publishing, Cham (2023). https://doi.org/10.1007/978-3-031-14615-2_78

Yang, Y.-T., Wang, Y.-H.: Numerical simulation of three-dimensional transient cooling application on a portable electronic device using phase change material. Int. J. Therm. Sci. **51**, 155–162 (2012). https://doi.org/10.1016/j.ijthermalsci.2011.08.011

Ali, H.M., Arshad, A.: Experimental investigation of n-eicosane based circular pin-fin heat sinks for passive cooling of electronic devices. Int. J. Heat Mass Transf. **112**, 649–661 (2017). https://doi.org/10.1016/j.ijheatmasstransfer.2017.05.004

Mahdi, J.M., Nsofor, E.C.: Melting enhancement in triplex-tube latent thermal energy storage system using nanoparticles-fins combination. Int. J. Heat Mass Transf. **109**, 417–427 (2017). https://doi.org/10.1016/j.ijheatmasstransfer.2017.02.016

Das, N., Takata, Y., Kohno, M., Harish, S.: Melting of graphene based phase change nanocomposites in vertical latent heat thermal energy storage unit. Appl. Therm. Eng. **107**, 101–113 (2016). https://doi.org/10.1016/j.applthermaleng.2016.06.166

Numerical and Experimental Study of the Lubricant Oil Leak Phenomenon on a Metro Traction Motor Gear Box

Mohamed Chadi Yakoubi[1,2](✉), Walid Najjar[2], and Hatem Mrad[1]

[1] School of Engineering, University of Quebec in Abitibi-Témiscamingue (UQAT), Rouyn-Noranda J9X 5E4, Canada
{MohamedChadi.Yakoubi,hatem.mrad}@uqat.ca

[2] LMPE-Ecole Nationale Supérieure d'Ingénieurs de Tunis, Université de Tunis, Tunis, Tunisia
walid.najar@ensit.u-tunis.tn

Abstract. One of the main problems of the urban metros is the leakage of quantities of lubricating oil into the traffic lanes. In addition to the fact that this leak has technical and financial repercussions, it is above all the subject of certain environmental and ecological concerns. It is then necessary to prevent this leak from occurring by improving the existing sealing system. In this work, a numerical and experimental analysis has been performed to identify the leak phenomena location and its physical origin. The numerical study was carried out in order to model and simulate the behavior of the oil in the traction motor. This study made it possible to determine the pressure distribution on a Polytetrafluoroethylene (PTFE) seal. An experimental study on an instrumented device was carried out to quantify the leak rate and location when the traction motor is subjected to pressurization cycles. Preliminary results showed that the leak rate is important and that its location is correlated with the high-pressure area of the seal.

Keywords: Traction motor · Oil leak · Sealing system · Simulation · Finite element

1 Introduction

Urban metros are equipped with motor and carrier bogies located under the car platforms. These bogies are cars whose job is to accelerate, brake, balance and steer the metro. They are equipped with a complex mobility system that allows the metro to follow the curved rail tracks. Few studies can be found in the literature on the behavior of fluids inside the mechanical structures (especially sloshing phenomena) as well as on the modeling of the pressurization of a fluid within the casing. (Al-Zughaibi et al. 2021; Ingle et al. 2016; Ganuga et al. 2014; Gómez-Goñi et al 2013; Akyildız and Unal 2006). No published material, however, can be found, in our knowledge, about the particular phenomena of the lubricant behavior within the differential gear box of metro traction motor and the induced oil leak phenomena. The study of this paper focuses on the traction motor of the bogie and aims mainly to figure out the correlation between the gearbox lubricant

A. Akrout et al. (Eds.): ICAV 2022, ACM 22, pp. 287–304, 2023.
https://doi.org/10.1007/978-3-031-34190-8_32

oil pressure during the movement and the oil leak phenomena. To achieve this goal a numerical study of the fluid behavior has been performed in order to predict the pressure on the sealing system. An experimental analysis has been therefore made to identify the approximate amount and especially the location of the leak.

2 Description of the System

As shown in Fig. 1 a bogie consists mainly of a frame structure, a pneumatic wheel, a braking cylinder, an axle shaft and a rubber tyred guide wheel.

Fig. 1. Composition of the motor bogie

The associated drive system consists of the traction motor and differential gear, which includes a gear train, screws, bearing and lip seal. The transmission elements are bathed in lubricating oil. The oil has the function of lubricating the gears and dissipating heat.

3 Numerical Modeling of Oil Behavior in the Differential Gear Box

The modeling process was made gradually by carrying out several simulations to obtain a convergent model with the optimal parameters (especially the convergent mesh). The used software for simulation is SolidWorks Flow (Matsson 2019). The aim of these simulations is to analyze the behavior of the fluid in the real operating conditions of the traction motor, and thus to identify the pressure applied by the oil to the seal area (Fig. 2).

3.1 Choice of Parameters

A set of parameters must be introduced into the software and several modeling assumptions must be made in order to set up our model.

- **Transient mode**
  As a hypothesis, we decided to highlight a physical time that is worth 53 s for the simulation, and to this time interval we applied a critical cycle of acceleration and deceleration, to cause a phenomenon of extreme sloshing to appear.

Fig. 2. Perspective view of traction motor differential gear box

Table 1. Characteristics of PluraSafe GL WI 220 oil

Parameters	Values
Density	1.006
Viscosity grade	220
Viscosity index	177
Kinematic viscosity at 40 °C	250 mm^2/s
Dynamic viscosity	0.2515 Pa.s

- **Oil characteristics**
  The oil characteristics has been configured with the properties of the PluraSafe GL WI 220 (BASF 2015) (Table 1):

It should be noted that the empty area of the differential box was also filled with air. The properties of this fluid were introduced automatically by SolidWorks using its standard fluid library.

3.2 Starting and Braking Cycle

To apply a strong perturbation to the oil behavior in the traction motor, a real acceleration and deceleration cycle representing an extreme operating scenario has been used. It should be mentioned that the maximum speed of the metro is around 72 km/h. Therefore, a fast start cycle has been defined, allowing us to reach this speed in the 14 s (which represents an extreme case). And then a sudden braking (deceleration) was also provided in the loading cycle, which corresponds to an acceleration of -1.43 m/s^2 (passing from a speed of 72 km/h to 0 in 14 s). This loading is presented by the curve of the Fig. 3.

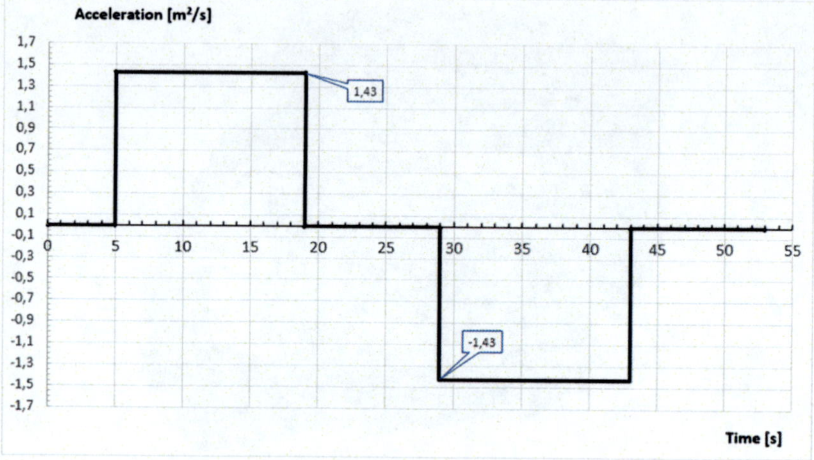

Fig. 3. Proposed acceleration and braking mode

It should also be noted that a cyclic loading as well as a constant gravity along the appropriate axis were applied as boundary conditions on the whole structure of the traction motor for a physical duration of 53 s (the dynamic simulation relates to 53 s). This choice was adopted in order to reveal the possible phenomena related to the sloshing of the system.

The volume of oil was designed in its idle state as a solid body, all its fluid characteristics were then applied, and thus SolidWorks quantifies it as a real fluid and models it by the finite volume method (Matsson 2019). SolidWorks Flow Simulation is a parametric flow simulation tool that uses the finite volume method to calculate product performance through design scenarios that allow optimization. The solver finished the calculation with the following balance sheet (Table 2):

Table 2. Calculation report for the fluid-structure simulation

Parameters	Value
Total number of cells	82 149
Fluid cells in contact with the structure	29,840
Number of iterations	350
Calculation time	04 h 21 m 54 s
Executed on	UROU1-D221-02
Number of bodies	8

3.3 Behavior of the Oil in the Traction Motor

The Fig. 4 shows the oil level in idle state.

Fig. 4. Oil in idle state

The Fig. 5 presents the oil level during metro acceleration, one can see that this level decreases in the seal side and slightly increases in the other side.

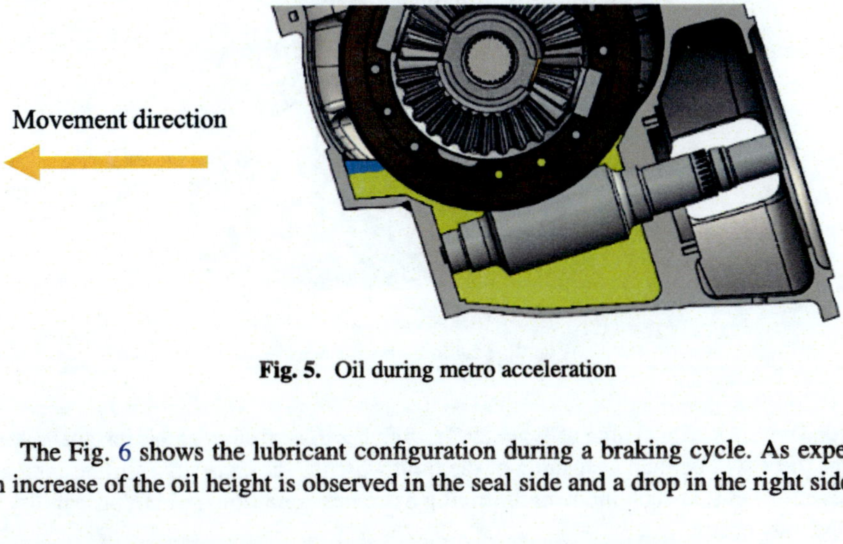

Fig. 5. Oil during metro acceleration

The Fig. 6 shows the lubricant configuration during a braking cycle. As expected, an increase of the oil height is observed in the seal side and a drop in the right side.

3.4 Evolution of the Maximum Pressure on the Seal Surface

Figure 7 shows the location of the seal.

Figure 8 shows the evolution of the maximum pressure at the surface of the lip seal as a function of the time. It can be seen that this evolution is correlated with the curve of Fig. 3 which describes the loading cycle (acceleration and braking). More precisely, when the metro accelerates, the fluid seems to press on the rear area of the gear box which will reduce the pressure on the seal between the instants of 5 s and 19 s.

On the other hand, when the metro decelerates, the oil will then swirl for a moment then press against the area which carries the seal, the pressure increases and locally reaches a maximum value for braking. This maximum pressure during the test has been

Fig. 6. Oil during metro braking

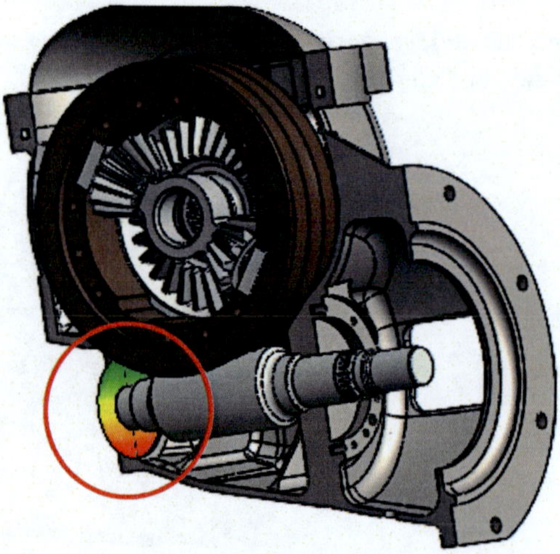

Fig. 7. Seal location surface

recorded within a critical time of $T_{cr} = 29.2$ s, which is logical since at this instant braking begins, there will be then a sloshing of the fluid which is the cause of a further increase in pressure. Thus, at such moment of driving extremes conditions, a high probability of leakage may occur.

3.5 Pressure Distribution on the Seal Location Surface

The mapping of the distribution of the pressure applied to the seal surface at time 29.2 s can be seen in Fig. 9.

From these figures one can observe that the pressure is higher in the lower zone of the seal. These pressure values can deform locally the seal and then could be one of the potential sources of the leak phenomena, which should logically appear in the lower area of the seal. To confirm this hypothesis an experimental analysis which will be presented in the following section has been then done.

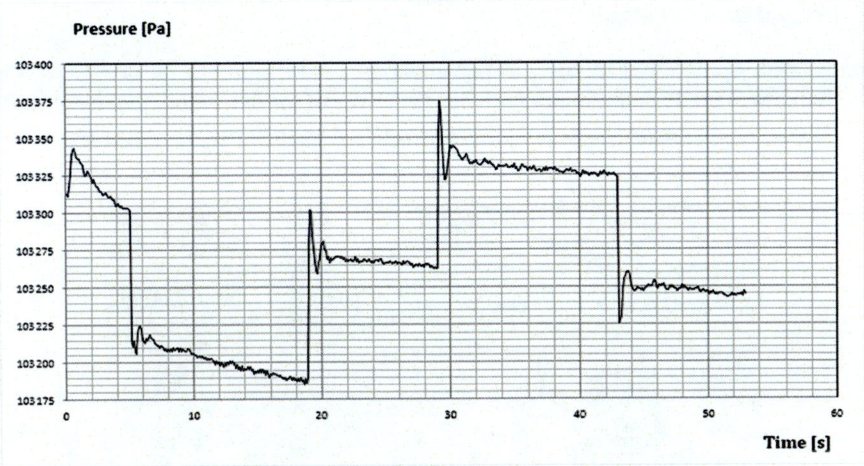

Fig. 8. The evolution of the maximum pressure at the surface of the joint

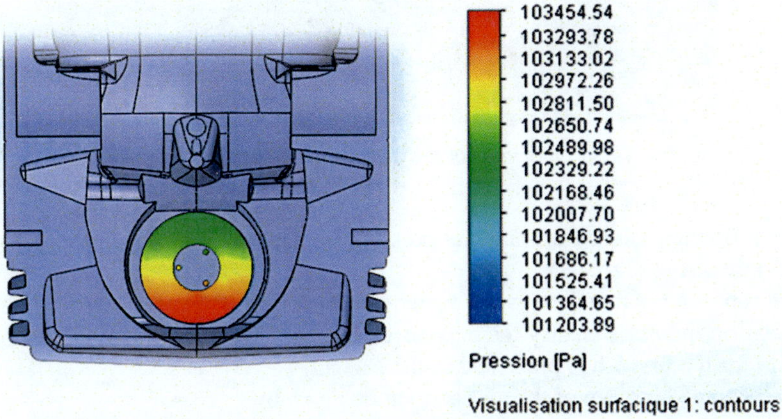

Fig. 9. Pressure distribution at $T_{cr} = 29.2$ s

4 Experimental Study

The aim of this analysis is to identify the oil leak rate and location in the closed differential gear box. Within the laboratory of the Ecole de genie -UQAT, a real test bench of the metro traction motor has been made available to carry out the experimental study in order to detect the leak location and approximate its amount (Fig. 10). The experimental test started by identifying all the areas that could be a potential site of a leak. This allowed us to apply a foaming solution to these areas which would clearly indicate the path of the leak. Subsequently, the test bench was put under pressure for a time interval of the order of 24 h.

4.1 Pressurization Circuit

The inside of the traction motor was put under P = 40 Psi = 275.79 kPa = 2.75 Bar. This choice was made, in harmony with the directives of the laboratory technical staff, in order to accelerate the process and to have tangible results within acceptable deadlines.

Fig. 10. Perspective view of the traction motor under instrumentation

The following table illustrates the pressure values taken during this experiment at the pressure gauge (Table 3):

One can observe that there was a variation (decrease) of the pressure during the test, this variation indicates clearly the existence of a leak. The hourly rate of this pressure variation for the first 9 h is presented in the Fig. 11.

The average leak rate of the air can then be given by:

$$\sum_{i=1}^{9} \frac{L_i}{9} = 3.064 \text{ Kpa/h} \tag{1}$$

Table 3. Internal pressure values taken during the experiment

Measure	Time [Hour]	Pressure [Kpa]	Pressure [psi]
1	0	275.79	40
2	1	272.342	39.5
3	2	270.274	39.2
4	3	266.827	38.7
5	4	2,62,000	38
6	5	259.932	37.7
7	6	256.484	37.2
8	7	255.106	37
9	8	251.658	36.5
10	9	248.211	36
11	23	197.879	28.7
12	24	194.432	28.2

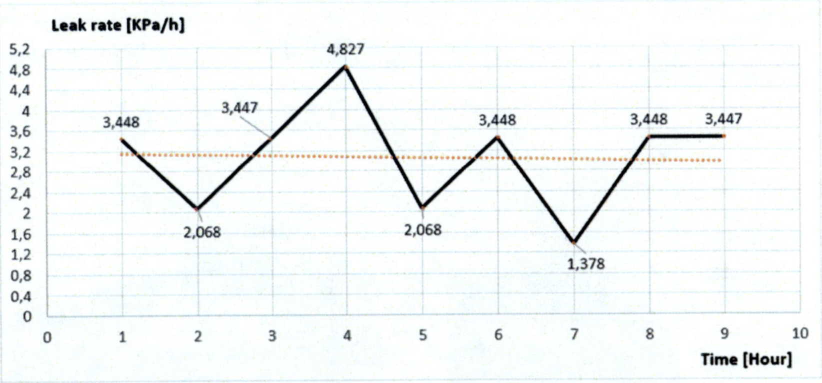

Fig. 11. Hourly leak rate

4.2 Leak Location

Significant agitation of the foaming solution at the lower area of the differential gear cover, precisely at the lower side of the small cover enclosing the bearing and the single lip seal, has been observed Fig. 12.

This leak area is in perfect agreement with the zone of high pressure (red zone) predicted by the simulation (Fig. 9). This demonstrates that the leakage is primarily caused by the deformation of the seal, which is induced by the pressure exerted on it during the metro's operation. The study that follows focuses on this seal deformation.

Fig. 12. Reaction of the foaming solution during the leak

5 Finite Element Simulation of the Gasket Structure

Firstly, a numerical structural study is conducted using Abaqus/Explicit to examine the behavior of the gasket, particularly at the gasket-shaft interface, and to predict potential areas of leakage (Fig. 13).

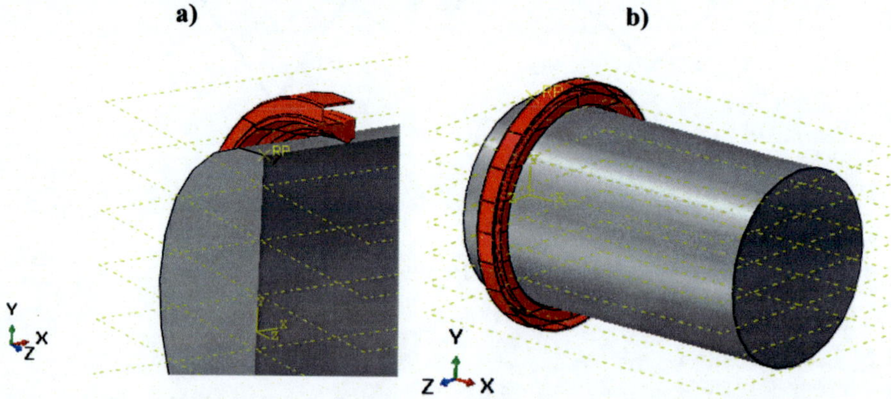

Fig. 13. Perspective view of the assembly Gasket-Shaft: a) Section view – b) Perspective view

5.1 Analysis of the Single Lip Seal Behavior

In this part, we simulate the behavior of the existing gasket using as loading the pressure identified previously Fig. 9. Figure 14 shows the actual gasket used as well as two partial section views of its 3D design:

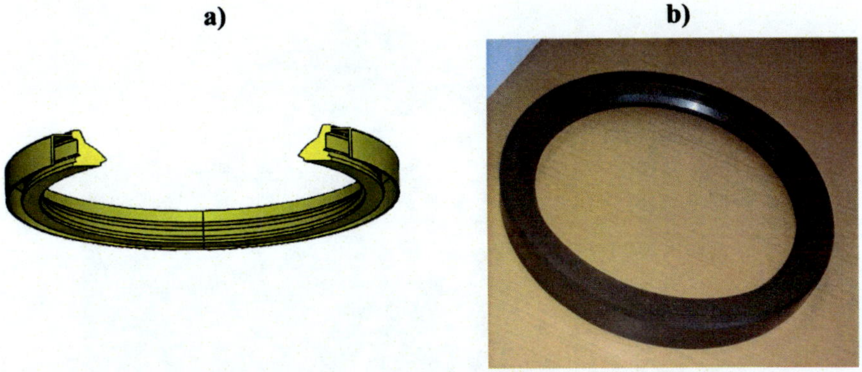

Fig. 14. a) Partial sectional view of the existing single lip gasket – b) Actual view of the existing single lip gasket

- ***Gasket material***

The used material model to describe the behavior of the gasket is the polynomial hyper-plastic law written in Mooney-Rivlin form. The strain energy density function is then expressed as follows.

$$W = C_{01}(\overline{I}_2 - 3) + C_{10}(\overline{I}_1 - 3) + D_1(J - 1)^2 \quad (2)$$

where:

C_{10}, C_{01}, D_1: Material constants. I_1, I_2: Invariants of strain tensor. J: The elastic volume ratio.

In our case (PTFE), the coefficients that characterize the existing gasket are shown in Table 4 (Ahmed and Alemdar 2021).

Table 4. Mooney-Rivlin coefficients

Coef.	Value
C_{10}	0.18186
C_{01}	0.036242
D_1	0.000645

- ***Boundaries conditions***

Since the radial surface of the gasket is completely supported by the bridge structure, we decided to embed this gasket zone against the structure body to represent a realistic boundary condition modelling (Fig. 15).

- ***Loading cycles***

The study considered a time cycle of 20 s, of which the first 15 s are devoted to braking, the left 5 s are to restore the oil to its initial state. Indeed, at instant 0 s, the metro is

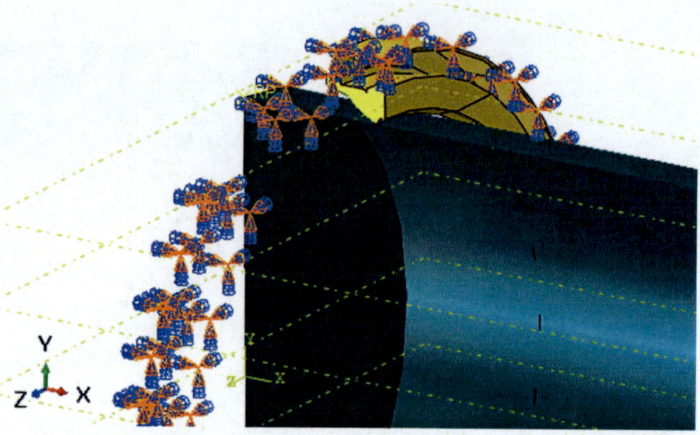

Fig. 15. Perspective view of the different mates on the gasket

in braking mode and it takes 15 s for it to come to a complete stop. On the other hand, from instant 15 s to instant 20 s, the metro is at the idle state and the oil returns to its defined state when the metro presents any movement. Subsequently, this time interval was iterated 1000 times in order to obtain 1000 braking cycles, the total physical time is then 20,000 s. This was implemented in Abaqus through a dynamic-explicit step.

- ***Potential area of acquisition of results***

The objective of this numerical study is to analyze the deformation of the gasket. In particular, we focus on the zone at the shaft/gasket interface which has the maximum radial displacement. This area may possibly be associated with shaft/gasket gap and then with leaks. Many preliminary simulations made it possible to identify the node location where the radial displacement according to Y is maximum. The identified critical node is presented in the Fig. 16.

- ***Displacement results***

The displacement U2 along Y at the critical node as a function of time, is presented in the figure below. The curve of this figure describes this displacement, it can be seen that from the beginning there was a slight disturbance which varies between −0.006 mm and +0.006 mm for a time interval of 1600 s. Then, between 1600 s and 6200 s the displacement is almost stable. At this stage, we can guess that there will be no leak (Fig. 17).

Moreover, from 6200 s, i.e. 310 braking cycles, we begin to see a significant decrease in the U2 curve going down to -0.1 mm at 9000 s (450 braking cycles). It can therefore be considered at this stage that a leak may appear because of the gap which appears between the shaft and gasket. The amplitude of this gap at the end of the process is presented in the Fig. 18.

- ***Discussion***

These results clearly show that the current configuration and concept of the joint (single lip seal) lead to excessive deformations at the shaft/joint interface after a few functional

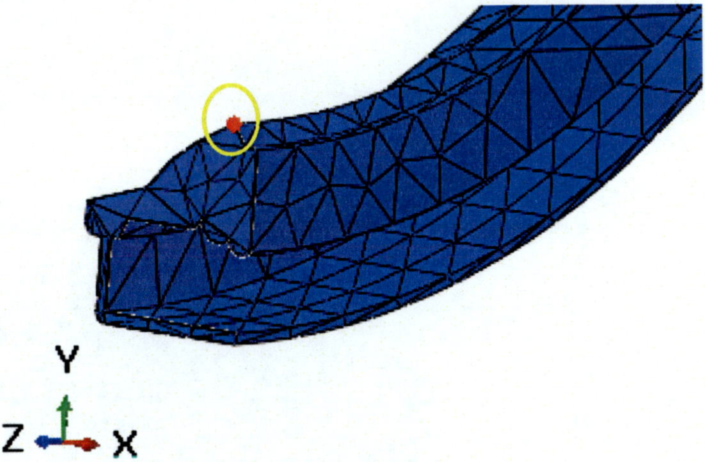

Fig. 16. Location of critical node for existing gasket

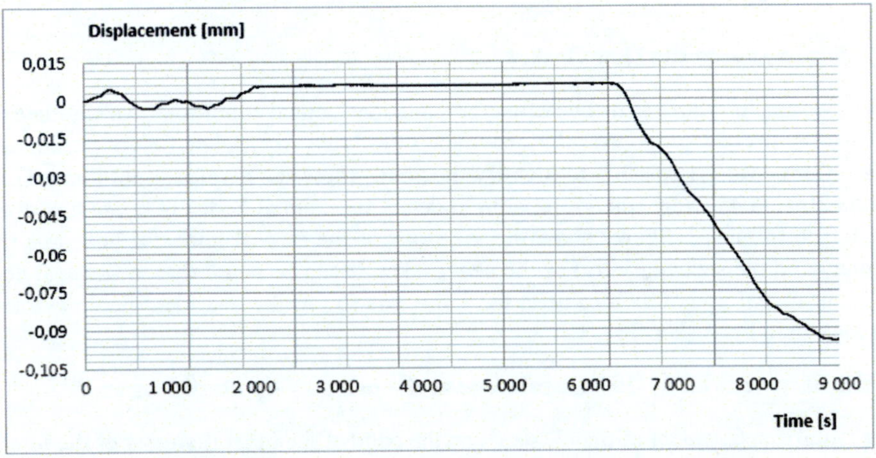

Fig. 17. Displacement U2 of critical node following Y for gasket 1

cycles and presumably causing leaks. In the following, we propose to remedy this problem through a technological solution, namely the use of a two-lip seal. Indeed, this solution seems interesting insofar as no changes in design or material will be implemented. The improvement will therefore be induced solely through the structural reinforcement provided by the second lip of the seal. The effectiveness of this solution is studied and tested in the next section.

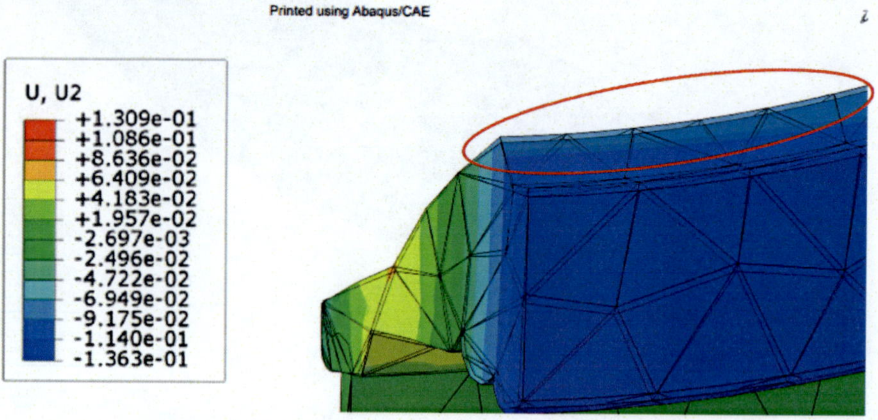

Fig. 18. Displacement map according to U2 for gasket 1.

5.2 Simulation of the New Gasket

With the same pressurization cycle and the same mechanical conditions, we simulate the behavior of the new proposed solution which is the double lip gasket. Sets of preliminary simulations made it possible to identify the zone where the displacement along U2 is maximal. And we were, therefore, able to detect the critical nodes in contact with the shaft. The following figures show the locations of the critical nodes of both lips (the primary and the secondary lip) of the new gasket. It can be noted that the critical node of the principal lip is found to be at the same axis (Y) as the one identified previously for the single lip gasket (Fig. 19).

- **Displacement U2 of the critical node of the primary lip according to Y**

The radial displacement of the elements on the edge of the inner diameter of the joint is presented in the Fig. 20 which shows a displacement along Y, of the order of 0.0135 mm at about 1250 s. This is much lower than the value found in the first solution (i.e. an improvement ratio of 7.8 times compared to the existing solution).

It can be seen in the Fig. 21 that there is nearly no visible gap at the critical node of the primary lip.

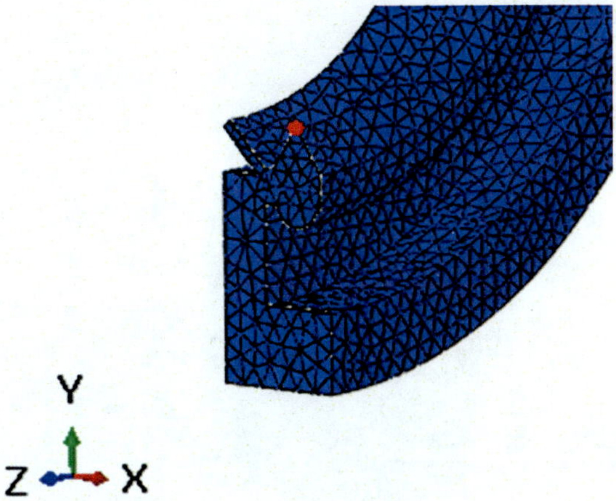

Fig. 19. Locations of critical nodes of the gasket 2 for primary lip and secondary lip.

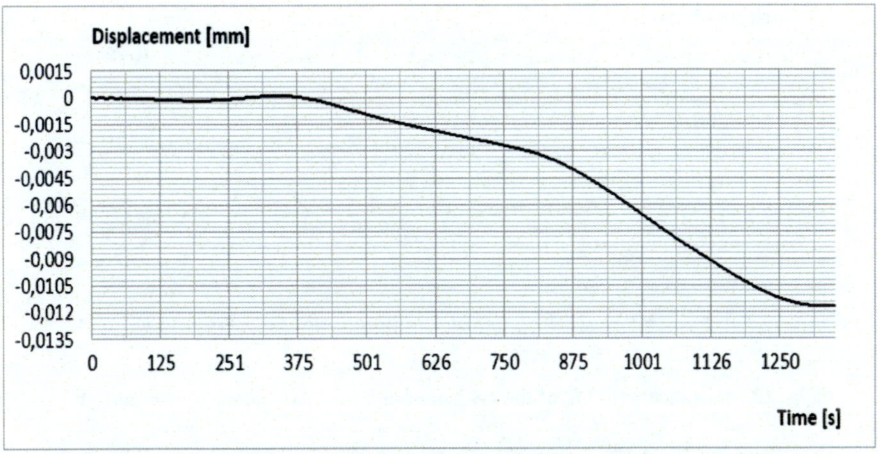

Fig. 20. Displacement U2 of the critical node of the primary lip for gasket 2

- **Displacement U2 of the critical node of the secondary lip according to Y**

A maximum displacement of the critical node of the secondary lip along Y of the order of 0.09 mm ≈ 0.1 mm has been found at 4500 s (Fig. 22).

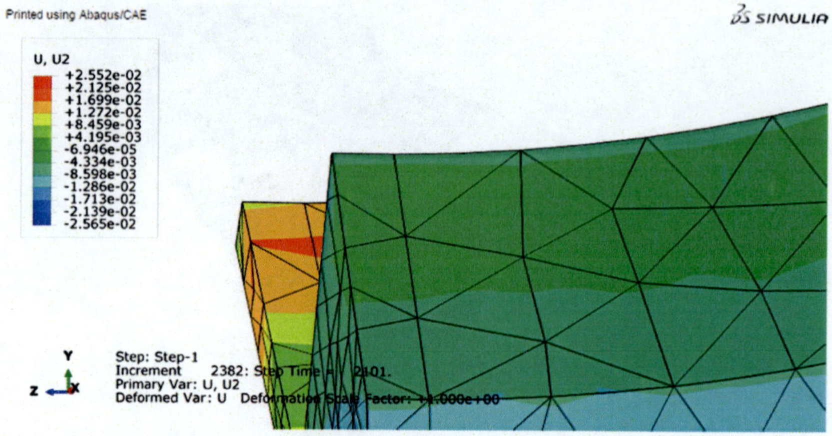

Fig. 21. Map of the displacement U2 of the primary lip for gasket 2

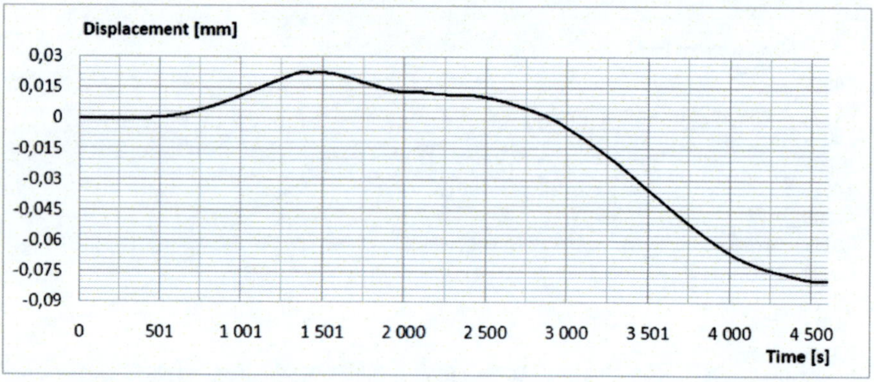

Fig. 22. Displacement U2 of the critical node of the secondary lip for gasket 2

The gap in the critical node of the secondary lip is presented in the Fig. 23.

- ***Discussion***

Based on the simulation results, the double lip gasket solution proves to be effective in addressing the leakage problem. Notably, a significant improvement in the radial displacement of the critical node at the primary lip can be observed, with a reduction of up to seven times. This can be attributed to the more effective distribution of the structural load over the two lips of the seal body.

In addition, this solution allows for the retention of the seal material, Polytetrafluoroethylene, which is notable for its excellent resistance and favorable properties for use in industrial environments.

Nonetheless, further experimental tests are required to validate the proposed solution. The next stage of this work will focus on conducting these tests.

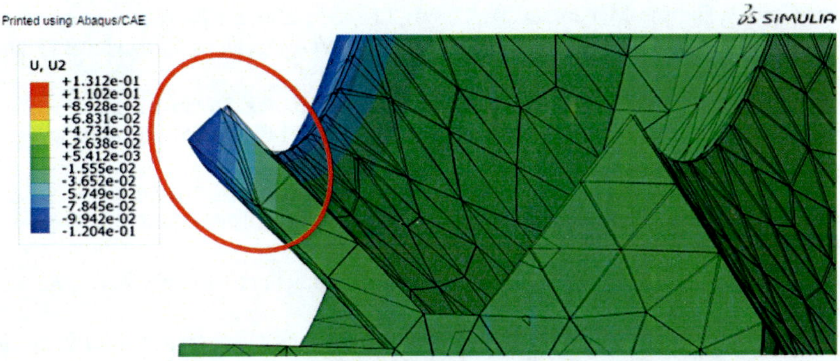

Fig. 23. Visualization spectrum of U2 of the secondary lip for gasket 2

6 Conclusion

In this paper a numerical and experimental analysis has been carried out to identify the location and the physical origin of the oil leak phenomena of a metro traction motor. The numerical study was carried out in order to model and simulate the behavior of the oil in the traction motor which allowed to predict the pressure distribution on the sealing lip seal. The experimental study on an instrumented device was carried out to quantify the leak rate and location. Preliminary results showed that the leak rate is important and that the leak zones are correlated with the high-pressure area of the seal predicted by the simulation. These results suggest that the leak is mainly caused by the deformation of the seal induced by the pressure applied to it. A solution consisting of the use of a double lip gasket joint, has been then numerically tested using Abaqus/Explicit. The simulation results show that this technological solution is potentially very effective for the leakage problem.

Acknowledgments. This research was supported by the University of Quebec in Abitibi - Témiscamingue – Canada and the Higher National School of Engineers of Tunis. (ENSIT) - Tunis-Tunisia.

This project is supported by Mitacs Research Organization.

References

Al-Zughaibi, A.I., Hussein, E.Q., Rashid, F.L.: Studying and analysis of nonlinear sloshing (vibrating) of interaction fluid structure in storage tanks. J. Mech. Eng. Res. Dev. **7**, 180–191 (2021)

Ingle, V.P., Shetty, M.V., Chopda, S.M.: Study of sloshing phenome-non in an automotive irregular shaped fuel tank using CFD. Int. J. Mod. Trends Eng. Res. **3**(4), 895–904 (2016)

Ganuga, R.S., Viswanathan, H., Sonar, S., Awasthi, A.: Fluid-structure interaction modelling of internal structures in a sloshing tank subjected to resonance. Int. J. Fluid Mech. Res. **2**, 145–168 (2014). https://doi.org/10.1615/InterJFluidMechRes.v41.i2.40

Gómez-Goñi, J., Garrido-Mendoza, C.A., Cercós, J.L., González, L.: Two phase analysis of sloshing in a rectangular container with Volume of Fluid (VOF) methods. Ocean Eng. **73**, 208–212 (2013). https://doi.org/10.1016/j.oceaneng.2013.07.005

Akyildız, H., Unal, N.E.: Sloshing in a three-dimensional rectangular tank: numerical simulation and experimental validation. Ocean Eng. **33**(16), 2135–2149 (2006). https://doi.org/10.1016/j.oceaneng.2005.11.001

Matsson, J.E.: An Introduction to SolidWorks Flow Simulation 2019 (2019). Sdcpublications. https://sdcpublications.com/Textbooks/Introduction-SOLIDWORKS-Flow-Simulation-2019/ISBN/978-1-63057-239-6/. Accessed 21 July 2022

BASF (2015) Plurasafe® GL WI 220. Setral. https://setral.net/products/068062/fr/. Accessed 09 Aug 2022

Ahmed, F., Alemdar, F.: Validation of an elastomeric bearing characterized with finite element hyperelastic models. Eur. J. Sci. Technol. **27**, 471–478 (2021). https://doi.org/10.31590/ejosat.930964s

An Adapted Formulation for the Locally Adaptive Weak Quadrature Element Method Using Gauss-Lobatto Points

Mohamed Ali Argoubi[1], Mohamed Trabelssi[1], and Molka Chiboub Hili[2(✉)]

[1] Applied Mechanics and Systems Research Laboratory, Tunisia Polytechnic School, University of Carthage, B.P. 743, 2078 La Marsa, Tunisia

[2] Laboratory of Mechanics, Modeling and Products, National School of Engineers of Sfax, University of Sfax, Soukra road km 4, 3038 Sfax, Tunisia

molka.chiboub@gmail.com

Abstract. In this manuscript, the Gauss-Lobatto-Legendre (GLL) points is proposed as an alternative to formulate elements of the Locally adaptive Weak Quadrature Element Method. This adapted form of the original formulation takes advantage of the reduced integration of the variational statement and produces a diagonal mass matrix. The effectiveness of this method is evaluated using the dynamic response of a strain gradient Euler-Bernoulli nanobeams for both linear and nonlinear case. Linear and nonlinear vibrational frequencies are calculated for a range of setups and boundary conditions. In few particular, cases, the computational cost of the proposed formulation can rival that of the Locally adaptive Differential Quadrature Method. It also achieved a similar convergence speed as the Weak Quadrature Element Methods, while providing much simpler implementation.

Keywords: Functionally graded nanobeam · Strain gradient theory · Locally adaptive Weak Quadrature Element Method (LaWQEM) · von Karman strain · Gauss-Lobatto quadrature points

Nomenclature

Physical variable	Symbol	Unit	Physical variable	Symbol	Unit
Upper material property	P_U	–	Coordinate from the geometry neutral surface	z_G	m
Lower material property	P_L	–	Power-law index	γ	Dimensionless
Nanobeam's width	b	m	Position of the physical neutral surface	c_0	m
Nanobeam's height	h	m	Coordinate from the physical neutral surface	z	m
Nanobeam's length	l	m	Longitudinal displacement of the neutral axis	u	m
Young's modulus	E	$N.m^{-2}$	Transverse displacement of the neutral axis	w	m
Material density	ρ	$kg.m^{-3}$	Magnitude of nonlinear vibration	A	Dimensionless
Dimension strain gradient length	l_s	m			

A. Akrout et al. (Eds.): ICAV 2022, ACM 22, pp. 305–316, 2023.
https://doi.org/10.1007/978-3-031-34190-8_33

1 Introduction

High order continuum mechanics, such as the Strain Gradient Theory (SGT), is typically viewed as an appealing substitute to molecular dynamic (MD) simulations used on nano and microstructure due to their relatively low computational cost. To account for the length-scale impacts of nanostructures, the SGT incorporates additional terms for strain energy gradients. Each variant of SGT is based on a specific strain gradient form [1–7] usually referred to as the SGT family. SGTs often provide boundary conditions that require extra derivative Degrees Of Freedom (DOFs) at the borders. In order to describe these types of systems, researchers combine SGT constitutive equations with widely used beam and plate theories including the Euler-Bernoulli beam theory (EBT), the Timoshenko beam theory (TBT), and the Reddy beam theory (RBT).

Both the Differential Quadrature Method (DQM) and the Locally Adaptive Differential Quadrature Method (LaDQM) [8] provide promising options for computing high-order solutions. In contrast to FEM, these techniques are not flexible geometrically and have particular limitations with regards to the concentrated loads and the application of boundary conditions (BCs). In general, most implementation of DQM matrices do not support the use of Hermite matrices required to model SGT EBT, as it requires two to three derivative DOFs at each boundaries depending on SGT implementation. This complicates the use of WQEM for high-order beams [9]. Recently Trabelsi and El-Borgi [10] presented a technique known as the Locally adaptive Weak Quadrature Element Method (LaWQEM) that employs both DQM differentiation matrices and quadrature integral. This weak-form formulation, is developed to avoid the explicit computation of the shape functions while providing a simple tool for computing various types of Hermit based WQEM formulations.

Integration of LaWQEM formulating was performed using Gauss quadrature points in combination to a transfer matrix-based approach. While such approach produces a fully integrated mass and stiffness matrices, it also generates a full mass matrix. Consequently, for large mesh systems, such a mass matrix can be computationally expensive. Hence, it's reasonable to estimate that a diagonal mass matrix generated using Gauss–Lobatto quadrature (GLL), can be computationally more efficient then [10] when dealing with a large mesh. GLL quadrature eliminates the need for interpolation between interior and boundary points, however, it may also require more element nodes to attain the same accuracy as Gauss quadrature. For this reason, it is always recommended that the efficiency vs accuracy outcomes of implementations be reevaluated. This is elaborated upon further in the current study.

The aim of this manuscript is to develop a GLL based LaWQEM for SGT EBT and evaluate its accuracy and convergence speed comparing to other WQEM based formulation from the literature. The following is the structure of this article. After this introductory section, Sect. 2 presents variational statements for a nonlinear SSG EBT. Then, Sect. 3 presents the applications of the conventional WQEM for SSG beam models, along with a detailed explanation of the proposed LaWQEM SSG EBT element formulations. The effectiveness of the suggested formulation is reported in Sect. 4 using data from the literature and the outcomes of well-established techniques. Concluding observations may be found in Sect. 5.

2 Integral Formulation for the Second Strain Gradient Euler-Bernoulli Nanobeam

An example of a graded nanobeam with a rectangular cross-section is shown in Fig. 1, in which, Latin letters b, h and l represent respectively, width, height and length. A functionally graded material properties are selected along the nanobeam thickness. Using the rule of mixtures, one can express the nanobeam's material characteristics P:

$$P(z_G) = (P_U - P_L)\left(\frac{z_G}{h} + \frac{1}{2}\right)^{\gamma} + P_L \tag{1}$$

where P can be either the Young's modulus E or the material density ρ. It is important to note that the geometric neutral surface of the FG nanobeam and the physical neutral surface are not identical. The coordinates of the physical neutral surface can be found by using the following formula [11,12]:

$$c_0 = \frac{\int_{-h/2}^{h/2} E(z_G)\, z_G dz_G}{\int_{-h/2}^{h/2} E(z_G)\, dz_G} = \frac{(E_U - E_L)\, h\gamma}{2(2+\gamma)(E_U + \gamma E_L)} \tag{2}$$

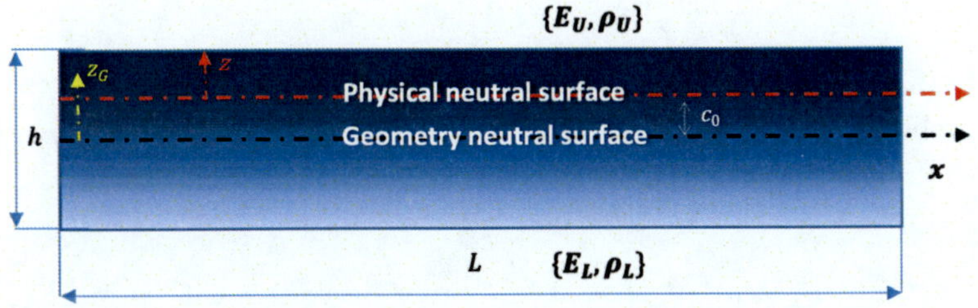

Fig. 1. Schematic illustration for FGM beams

For a rectangular FG nanobeam having a thickness coordinate z measured from the physical neutral surface as shown in Fig. 1, where

$$z = z_G - c_0 \tag{3}$$

The nanobeam's material characteristics are expressed in the new coordinate system by replacing Eq. (3) into Eq. (1). One has

$$P(z) = (P_U - P_L)\left(\frac{z + c_0}{h} + \frac{1}{2}\right)^{\gamma} + P_L \tag{4}$$

According to the Euler–Bernoulli beam theory, the displacements of the FG nanobeam are represented by

$$u_1(x,z,t) = u(x,t) - z\frac{\partial w(x,t)}{\partial x} \quad u_2(x,z,t) = 0, \quad u_3(x,z,t) = w(x,t). \tag{5}$$

The EBT nanobeam's Green-Lagrange strain tensor is calculated using von Karman's geometric relationship, which is given as

$$\varepsilon_{xx} = \underbrace{\frac{\partial u}{\partial x} + \frac{1}{2}\left(\frac{\partial w}{\partial x}\right)^2}_{\varepsilon_{xx}^{(0)}} - z\underbrace{\frac{\partial^2 w}{\partial x^2}}_{\varepsilon_{xx}^{(1)}} \qquad \gamma_{xz} = 0 \tag{6}$$

where u is the axial displacement of a point on the physical neutral surface and w is the transverse displacement. SGT is employed to take into consideration the small size scale effect in the nanobeam. The total strain energy of the second strain gradient FG nanobeam is calculated as [13]

$$U_{ssg} = \frac{1}{2}\int_V \left(\sigma_{ij}\varepsilon_{ij} + \sigma_{ijm}^{\langle s\rangle}\varepsilon_{ij,k}\right) dV \tag{7}$$

where

$$\sigma_{ij} = \lambda\varepsilon_{kk}\delta_{ij} + 2G\varepsilon_{ij} \tag{8a}$$

$$\sigma_{ijm}^{\langle s\rangle} = l_s^2(\lambda\varepsilon_{kk,m}\delta_{ij} + 2\mu\varepsilon_{ij,m}) = l_s^2\sigma_{ij,m} \tag{8b}$$

$$\varepsilon_{ij} = \frac{1}{2}(u_{i,j} + u_{j,i}) \tag{8c}$$

where l_s and ε_{ij} are the first non-classical elastic constant and the strain tensor respectively. This paper's notation convention for the derivative is given by

$$\begin{cases} \dfrac{\partial^2 u}{\partial t^2} = \partial_t^2 u = \ddot{u} & \text{time derivative} \\ \dfrac{\partial u}{\partial x} = \partial_x u = u_{,x} = u' & \text{spatial first order derivative with respect to } x \end{cases} \tag{9}$$

Using the Hamilton principle, the weak integral form for the nonlinear nanobeam may be derived

$$\int_{t_1}^{t_2} (\delta K - \delta U_{ssg})\, dt = 0 \tag{10}$$

where δK and δU_{ssg} represent the kinetic and the strain energy variations, respectively. The strain expression from Eq. (6) is substituted into Eq. (7) to yield

$$\delta U_{ssg} = \int_0^L \int_A \sigma_{xx}\delta\left(u_{,x} + \frac{1}{2}\left(w_{,x}\right)^2 - zw_{,xx}\right) dAdx + \int_0^L \int_A \sigma_{xxx}^{\langle s\rangle}\delta\left(u_{,xx} + w_{,x}w_{,xx} - zw_{,xx}\right) dAdx \tag{11}$$

$$\delta K_{ssg} = \int_0^L m_0\dot{u}\delta\dot{u}dx + \int_0^L m_0\dot{w}\delta\dot{w}dx + \int_0^L m_2\dot{w}_{,x}\delta\dot{w}_{,x}dx \tag{12}$$

By introducing the following stress resultant [10],

$$M_{ij}^{(m)} = \int_A z^m\sigma_{ij}(x,z)dA = b\int_{-\frac{h}{2}-c_0}^{\frac{h}{2}-c_0} z^m\sigma_{ij}(x,z)dz = b\int_{-\frac{h}{2}}^{\frac{h}{2}} \left(z_G - c_0\right)^m \sigma_{ij}(x_G, z_G)dz_G \tag{13}$$

and the following FGM constants

$$\tilde{E}_0 = \int_A E(z)dA \qquad \tilde{E}_2 = \int_A z^2E(z)dA \tag{14}$$

$$m_0 = \int_A \rho(z)\,dA \qquad m_2 = \int_A \rho(z)\,z^2dA \tag{15}$$

one can simplify the variational form of the strain energy Using Eqs. (11), (13) and taking into consideration simplification hypothesis in [10], the normalized variational statement is given by:

$$0 = \int_0^1 \left(\begin{array}{c}\underline{\ddot{w}}\delta\underline{w} + (\kappa_0\left[\int_0^1 \underline{w}_{,x}^2 d\xi\right]\underline{w}_{,x} + \underline{m}_2\underline{\ddot{w}}_{,x})\delta\underline{w}_{,x} \\ +\underline{w}_{,xx}\delta\underline{w}_{,xx} + \underline{l}_s^2\underline{w}_{,xxx}\delta\underline{w}_{,xxx}\end{array}\right) d\xi \tag{16}$$

where

$$\xi = \frac{x}{L},\ \underline{w} = \frac{w}{r},\ s^2 = \frac{\tilde{E}_2}{L^2\tilde{G}_0},\ \kappa_0 = \frac{r^2\tilde{E}_0}{2\tilde{E}_2} \tag{17a}$$

$$\tau = t\frac{1}{L^2}\sqrt{\frac{\tilde{E}_2}{m_0}},\ r = \sqrt{\frac{I}{A}},\ \underline{l}_s = \frac{l_s}{L},\ \underline{m}_2 = \frac{m_2}{L^2m_0} \tag{17b}$$

subjected to the following classical and non-classical boundary conditions for a Simply-Supported beam (SS) and a Clamped-Clamped beam (CC):

SS nanobeam	CC nanobeam	
$\underline{w}\left(\xi = \xi_b\right) = 0$	$\underline{w}\left(\xi = \xi_b\right) = 0$ and $\underline{w}_{,x}\left(\xi = \xi_b\right) = 0$	(18)
$\underline{w}_{,xx}\left(\xi = \xi_b\right) = 0$	$\underline{w}_{,xx}\left(\xi = \xi_b\right) = 0$	(19)

where $\xi_b \in \{0, 1\}$.

3 LaWQEM Formulation

The goal of this section is to present the LaWQEM formulation for SSG beams, which calls for multiple degrees of freedom at the boundaries. First, the WQEM fundamentals are covered, such as the number DOFs needed using Euler-Bernoulli beam theory, development of the variational statement as well as the integration quadrature in matrix form. Once the mapping between LaWQEM and LaDQM DOFs is established, a new formulation of LaWQEM based on transfer matrices is used. This mapping guarantees that the needed DOFs are compatible with the order of the polynomials used in the shape functions. In addition, this mapping permits the generation of multiple Hermite polynomials through linear combinations of Lagrange polynomials, and the use of standard WQEM tools with the degrees of freedom of Hermite-type.

3.1 Regular Weak Quadrature Element Method (WQEM)

The purpose of this subsection is to provide discretization of the variational integral statement (16). Let Y be the nodal displacement vector which can be expressed as

$$Y = \Bigg[\underbrace{\underline{w}_1''\ \underline{w}_1'}_{\text{Derivative degrees of freedom}} \quad \underbrace{\underline{w}_1 \cdots \underline{w}_n}_{\text{Internal degrees of freedom}} \quad \underbrace{\underline{w}_n'\ \underline{w}_n''}_{\text{Derivative degrees of freedom}} \Bigg]^T \tag{20}$$

The velocity and acceleration vectors are denoted by $\dot{Y}$ and $\ddot{Y}$, respectively. Similar to the finite element method (FEM), strain gradient Euler-Bernoulli WQEM needs continuity of second-order derivatives at the element boundaries due to the presence third order derivatives in the weak form of the SSG EBT. WQEM elements based on Lagrange interpolations cannot meet this condition. The current LaWQEM formulation, however, can provide such number of derivatives Degrees Of Freedom (DOFs) at the boundaries. Consequently, it can guarantee the SSG EBT system's continuity.

It is possible to write the discretized system in matrix form as to facilitates its implementation. The WQEM/LaWQEM matrix form is obtained by discretizing (16), and can be expressed as

$$\begin{aligned}([\omega_\xi]\,[M_0])^T\,.\left([M_0]\,.\ddot{Y} - \underline{m}_2\,[M_2]\,.\ddot{Y}\right) + ([\omega_\xi]\,[M_3])^T\,.\left(\underline{l}_s^2\,[M_3]\,.Y\right)\\ + ([\omega_\xi]\,[M_1])^T\,.\left(\kappa_0 i \mathcal{W}(Y)\,[M_1]\,.Y\right) + ([\omega_\xi]\,[M_2])^T\,.\left([M_2]\,.Y\right) = 0\end{aligned} \tag{21}$$

where the expression for the DQM k^{th} derivative matrix $[M_k]$ is given by [9] and where $[\omega_\xi]$ denote the integral quadrature weights arranged in a diagonal matrix. One can note here that the DQM matrices provided by [9] are not square, as they only compute the derivative at the internal nodes. Also note here that a diagonal mass matrix can be only achieved when the contribution of $\underline{m}_2$ is neglected. Finally, the discritization of the nonlinear term present in equation (21) is similar to the rest of the integrals as shown in [10]. The normalized variational statements was converted into WQEM matrix form using the following relation [10]

Fig. 2. An example of an LaDQM mesh with 11 internal nodes (black markers) and 4 external nodes (white markers)

$$
\begin{aligned}
\int_a^b f^{[k]}(x) g^{[l]}(x) dx &= \sum_{i=1}^{n} [\omega_\xi]_i f(\xi_i) g(\xi_i) \\
&= \left([\omega_\xi] \left[f^{[k]}(\xi) \right] \right)^T \cdot \left[g^{[l]}(\xi) \right] \\
&= \left([\omega_\xi] [M_k] \cdot [f(\xi)] \right)^T \cdot [M_l] \cdot [g(\xi)] \\
&= [f(\xi)]^T \cdot \left([\omega_\xi] [M_k] \right)^T \cdot [M_l] \cdot [g(\xi)]
\end{aligned} \tag{22}
$$

where $\left[f^{[k]}(\xi)\right]$is the k^{th}derivative of $f(x)$'s DOFs, $\xi_i, (i = 1, ..., n)$, is the coordinates of the quadrature point. It is important to keep in mind that $[\omega_\xi]$must be suitable for the chosen grid. For instance, for the present implementation, a Gauss-Lobatto-Legendre (GLL) grid with adequate integral quadrature weight coefficients is employed. In the previous formulation proposed in [10], a fully integrated system was adopted. For this end, Gauss quadrature integration was used. This was accomplished using another type of transfer matrices as Gauss points cannot be used for nodal displacement. This approach yields a full mass matrix which should be difficult to manipulate when a large number of nodes is used. For this reason, an adaptation of LaWQEM to use GLL nodes can be beneficial. Lets note that such formulation requires the use of the following mesh

$$
\text{Roots of } (1 - \xi^2) \partial_\xi \text{Legendre}_{n-2}(\xi). \;\; -1 \leq \xi \leq 1 \tag{23}
$$

This mesh can be adjusted to fit within $[0, 1]$ interval [14].

3.2 The LaWQEM Formulation with with GLL Reduced Integration

The LaWQEM formulation relies on transfer matrices, which help simplify the mapping between the LaWQEM and the LaDQM DOFs, guaranteeing that the required degrees of freedom are compatible with the order of the polynomial basis functions. It is then possible to construct a wide variety of Hermite polynomials by linearly combining Lagrange polynomials. This type of flexibility allow Second Strain Gradient nanobeams's extra DOFs to be handled using standard WQEM tools. A GLL mesh with two external nodes at each boundary, similar to the one used in LaDQM, is required for the proposed LaWQEM formulation as shown in Fig. 2. After discretizing the normalized displacement $\underline{w}$ on a LaDQM grid, Y is expressed as:

$$
Y = \left\{ \underbrace{\underline{w}_{-1} \; \underline{w}_0}_{\text{External degree of freedom}} \; \underbrace{\underline{w}_1 \cdots \underline{w}_n}_{\text{Internal degrees of freedom}} \; \underbrace{\underline{w}_{n+1} \; \underline{w}_{n+2}}_{\text{External degree of freedom}} \right\}^T \tag{24}
$$

where the external DOFs are associated with white external nodes in Fig. 2. In (24), the number of degrees of freedom of the LaDQM displacement vector is linked to the number of DOFs of the conventional WQEM displacement vector in (20) as follows:

$$\begin{bmatrix} \underline{w}_1'' \\ \underline{w}_1' \\ \underline{w}_1 \\ \vdots \\ \underline{w}_n \\ \underline{w}_n' \\ \underline{w}_n'' \end{bmatrix} = \underbrace{\begin{bmatrix} [\mathcal{M}_2]_{1,-1} & & \cdots & & [\mathcal{M}_2]_{1,n+2} \\ [\mathcal{M}_1]_{1,-1} & & \cdots & & [\mathcal{M}_1]_{1,n+2} \\ 0 & 0 & 1 \ \ 0 \cdots 0 & & 0 \\ \vdots & \vdots & \ddots & \vdots & \vdots \\ 0 & 0 \cdots & 0 \ \ 1 & 0 & 0 \\ [\mathcal{M}_1]_{n,-1} & & \cdots & & [\mathcal{M}_1]_{n,n+2} \\ [\mathcal{M}_2]_{n,-1} & & \cdots & & [\mathcal{M}_2]_{n,n+2} \end{bmatrix}}_{[T]} \cdot \begin{bmatrix} \underline{w}_{-1} \\ \vdots \\ \\ \vdots \\ \\ \vdots \\ \underline{w}_{n+2} \end{bmatrix} \tag{25}$$

$\mathcal{M}_i$ is the LaDQM i^{th} derivative matrix and $[T]$ is the LaWQEM mapping matrix. The LaWQEM Hermite derivation matrix can be obtained using the relation proposed in [15]

$$[M_k] = [T] \cdot [\mathcal{M}_k] \cdot [T]^{-1} \tag{26}$$

originally built for Strong form QEM which is the colocation version of QEM.

4 Results and Discussion

The purpose of this section is to validate the proposed LaWQEM approach. The method is evaluated using both analytical findings and well-established numerical methods, as well as data from the literature. The boundary conditions used in this study can be found in Table 1. For numerical convergence, high-order approaches often involve increasing the element order. This keeps the total number of elements minimal [14,16,17]. As a result, throughout this investigation, a discretization with a single element is used.

Table 1. Implemented Dirichlet boundary conditions for the SSG EBT

Reference	Classical boundary conditions	High-order boundary conditions
SS	$\underline{w}\,(\xi = \xi_b) = 0$	$\underline{w}''\,(\xi = \xi_b) = 0$
CC	$\underline{w}\,(\xi = \xi_b) = 0$ $\underline{w}'\,(\xi = \xi_b) = 0$	$\underline{w}''\,(\xi = \xi_b) = 0$

4.1 Linear Natural Frequencies

The novel LaWQEM formulation is first validated against the results of a research by Ishaquddin and Gopalakrishnan [16] on the free vibration problem using a simply supported BCs of a linear SSG EBT. The problem has been addressed both analytically and numerically in the literature. For the validation, simulation parameters have been defined in consistent units, where, $L = 1, \rho = 1$, $E = 3 \times 10^6$ and $\nu = 0.3$. The same BCs used by Ishaquddin and Gopalakrishnan (a simply supported case) have been employed for the linear SSG EBT. Figure 3 presents results of the analysis. The proposed formulation matches the classical WQEM and the regular LaWQEM and the analytical results. It also achieves a convergence speed a comparable to the classical WQEM across all the data points. The first eigenvalues actually converge at around the 11 nodes mark while higher modes requires up to 15 to17 nodes, depending on the value of $\underline{l}_s$. The proposed method requires a 2 to 4 more nodes match the accuracy of the fully integrated LaWQEM.

4.2 Nonlinear Natural Frequencies

This section focuses on the nonlinear vibration behavior of the SSG nanobeams using LaWQEM formulation. The proposed GLL-based LaWQEM data is compared to LaDQM and LaWQEM. For the purpose of determining eigenvalues. The nonlinear frequency can be obtained through several iterations [10,14,18–20] or using the harmonic balance approach [21]. In this section, for simply supported (SS) and clamped (CC) BCs, a combination including several material characteristics such as γ and $\underline{l}_s$ is employed. The magnitude of vibration A varies between 0 and 1. The value of $\frac{L}{h}$ is set to either 10 or 100. The data is presented in Table 2.

The presence of the transfer matrices $[T]$ don't appear to affect LaWQEM results. Using 11 nodes yields results as accurate as the LaDQM approach using 15 nodes. The GLL-based LaWQEM formulation delivers similar accuracy to

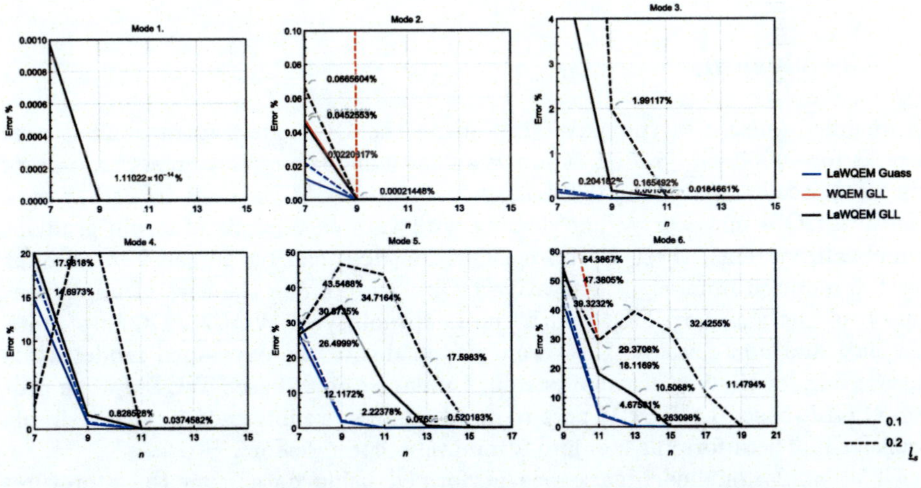

Fig. 3. Mesh convergence history for SS boundary conditions

Table 2. Nonlinear natural frequencies for an Euler-Bernoulli nanobeamin in comparison with regular DQM based methods.

	L/h			10			100			10			100		
	A	$\underline{l}_s$	γ	1	2	4	1	2	4	1	2	4	1	2	3
				SS nanobeam						CC nanobeam					
LaWQEM GLL (Current)	0.5	0.1		10.6132	10.5979	10.5723	10.6609	10.6468	10.6201	35.7964	35.7826	35.7750	36.0414	36.0346	36.0218
		0.2		11.8837	11.8697	11.8471	11.9371	11.9245	11.9007	55.8906	55.8746	55.8745	56.2867	56.2821	56.2736
		0.3		13.7426	13.7300	13.7108	13.8043	13.7934	13.7728	78.4952	78.4757	78.4814	79.0565	79.0530	79.0470
	1	0.1		11.5059	11.4527	11.3551	11.5575	11.5055	11.4064	36.2406	36.2069	36.1617	36.4884	36.4616	36.4109
		0.2		12.6873	12.6387	12.5506	12.7443	12.6971	12.6074	56.1892	56.1598	56.1342	56.5873	56.5691	56.535
		0.3		14.4431	14.4001	14.3231	14.508	14.4666	14.3879	78.7105	78.6813	78.6687	79.2732	79.26	79.2355
LaDQM	0.5	0.1		10.6132	10.5979	10.5723	10.6609	10.6468	10.6201	35.7964	35.7826	35.7750	36.0414	36.0346	36.0218
		0.2		11.8837	11.8697	11.8471	11.9371	11.9245	11.9007	55.8906	55.8746	55.8745	56.2867	56.2821	56.2736
		0.3		13.7426	13.7300	13.7108	13.8043	13.7934	13.7728	78.4952	78.4757	78.4814	79.0565	79.0530	79.0470
	1	0.1		11.5059	11.4527	11.3551	11.5575	11.5055	11.4064	36.2406	36.2069	36.1617	36.4883	36.4616	36.4109
		0.2		12.6873	12.6387	12.5506	12.7443	12.6971	12.6074	56.1892	56.1598	56.1342	56.5873	56.5691	56.5350
		0.3		14.4431	14.4001	14.3231	14.5080	14.4666	14.3879	78.7105	78.6813	78.6687	79.2732	79.2600	79.2355
LaWQEM Gauss	0.5	0.1		10.6132	10.5979	10.5723	10.6609	10.6468	10.6201	35.7964	35.7826	35.7750	36.0414	36.0346	36.0218
		0.2		11.8837	11.8697	11.8471	11.9371	11.9245	11.9007	55.8906	55.8746	55.8745	56.2867	56.2821	56.2736
		0.3		13.7426	13.7300	13.7108	13.8043	13.7934	13.7728	78.4952	78.4757	78.4814	79.0565	79.0530	79.0470
	1	0.1		11.5059	11.4527	11.3551	11.5575	11.5055	11.4064	36.2406	36.2069	36.1617	36.4884	36.4616	36.4109
		0.2		12.6873	12.6387	12.5506	12.7443	12.6971	12.6074	56.1892	56.1598	56.1342	56.5873	56.5691	56.5350
		0.3		14.4431	14.4001	14.3231	14.5080	14.4666	14.3879	78.7105	78.6813	78.6687	79.2732	79.2600	79.2355

the fully integrated LaWQEM SSG EBT formulation, despite the differences throughout the DQM matrix composition.

It should also be noted that the natural frequencies of the nanobeam increase slightly when normalized length scale $\underline{l}_s$ is increased but they are not too much affected by the variation of the Power-law index γ. These results are predictable and validated by several other studies on nanobeams. Obviously the non linear natural frequencies of a CC nanobeam are much higher than those of a SS nanobeam and these frequencies generally depend on the vibration amplitude.

5 Conclusion

A modified version of the LaWQEM using the GLL quadrature point is presented for nonlinear graded strain gradient nanobeams elements. As part of the proposed formulation, the Lagrange based DQM, is used to construct a Hermite DQM matrices. When dealing with various versions of strain gradient nanobeam systems, these Hermite DQM matrices can be adjusted to account for the required number of derivative DOFs. Combining the ease of implementation of the standard DQM with the adaptability of WQEM. DOFs of both classical and non-classical EBT were included into the developed model while accounting for the von Karman strain. Unlike Hermite-based WQEM, The proposed GLL-based LaWQEM uses reduced quadrature integration with a simple transfer matrices formulation and a minimum coding effort.

The validation benchmark was performed using data from the literature. Numerical results of the SSG EBT using standard Hermite WQEM element shows good correlation with the simulation results. In comparison to the SSG

EBT standard Hermite WQEM element, the LaWQEM formulation with GLL quadrature resulted in a good convergence rate and comparable accuracy. Multiple nonlinear response were then obtained for a wide set of material characteristics, vibration amplitudes, and boundary conditions. These results were compared to data from the literature.

Results from the computations led to the conclusion that the proposed LaWQEM approach may provide valid and reliable data for nonlinear strain gradient EBT systems.

References

1. Thai, H.T., Vo, T.P., Nguyen, T.K., Kim, S.E.: A review of continuum mechanics models for size-dependent analysis of beams and plates. Compos. Struct. **177**, 196–219 (2017)
2. Toupin, R.A.: Elastic materials with couple-stresses. Arch. Ration. Mech. Anal. **11**(1), 385–414 (1962)
3. Mindlin, R.D., Tiersten, H.F.: Effects of couple-stresses in linear elasticity. Arch. Ration. Mech. Anal. **11**(1), 415–448 (1962)
4. Koiter, W.T.: Couple-stress in the theory of elasticity. In: Proceedings of the Koninklijke Nederlandse Akademie van Wetenschappen, vol. 67, pp. 17–44. North Holland Publication (1964)
5. Yang, F., Chong, A.C.M., Lam, D.C.C., Tong, P.: Couple stress based strain gradient theory for elasticity. Int. J. Solids Struct. **39**(10), 2731–2743 (2002)
6. Mindlin, R.D.: Micro-structure in linear elasticity. Arch. Ration. Mech. Anal. **16**(1), 51–78 (1964)
7. Mindlin, R.D.: Second gradient of strain and surface-tension in linear elasticity. Int. J. Solids Struct. **1**(4), 417–438 (1965)
8. Ouakad, H.M., El-Borgi, S., Mousavi, S.M., Friswell, M.I.: Static and dynamic response of CNT nanobeam using nonlocal strain and velocity gradient theory. Appl. Math. Modell. **62**, 207–222 (2018)
9. Wang, X.: Novel differential quadrature element method for vibration analysis of hybrid nonlocal Euler-Bernoulli beams. Appl. Math. Lett. **77**, 94–100 (2018)
10. Trabelssi, M., El-Borgi, S.: A novel formulation for the weak quadrature element method for solving vibration of strain gradient graded nonlinear nanobeams. Acta Mech. **233**, 4685–4709 (2022)
11. Şimşek, M.: Nonlinear free vibration of a functionally graded nanobeam using nonlocal strain gradient theory and a novel hamiltonian approach. Int. J. Eng. Sci. **105**, 12–27 (2016)
12. El-Borgi, S., Rajendran, P., Trabelssi, M.: Nonlocal and surface effects on nonlinear vibration response of a graded Timoshenko nanobeam. Arch. Appl. Mech. **93**(1), 151–180 (2022)
13. Li, L., Tang, H., Yujin, H.: The effect of thickness on the mechanics of nanobeams. Int. J. Eng. Sci. **123**, 81–91 (2018)
14. Trabelssi, M., El-Borgi, S., Friswell, M.I.: A high-order FEM formulation for free and forced vibration analysis of a nonlocal nonlinear graded Timoshenko nanobeam based on the weak form quadrature element method. Arch. Appl. Mech. **90**, 2133–2156 (2020)

15. Trabelssi, M., El-Borgi, S.: Vibration of nonlocal strain gradientfunctionally graded nonlinearnanobeams using a novel locallyadaptive strong quadrature element-method. J. Nanomater. Nanoeng. Nanosyst. (2022)
16. Ishaquddin, Md., Gopalakrishnan, S.: A novel weak form quadrature element for gradient elastic beam theories. Appl. Math. Model. **77**, 1–16 (2020)
17. Wang, X.: Differential Quadrature and Differential Quadrature Based Element Methods: Theory and Applications. Butterworth-Heinemann (2015)
18. Ghorbanpour, A.A., Reza, K., Masoud, E.: Nonlinear vibration analysis of piezoelectric plates reinforced with carbon nanotubes using DQM. Smart Struct. Syst. **18**, 787–800 (2016)
19. Yang, J., Ke, L.L., Kitipornchai, S.: Nonlinear free vibration of single-walled carbon nanotubes using nonlocal Timoshenko beam theory. Phys. E **42**(5), 1727–1735 (2010)
20. Malekzadeh, P., Vosoughi, A.R.: DQM large amplitude vibration of composite beams on nonlinear elastic foundations with restrained edges. Commun. Nonlinear Sci. Numer. Simul. **14**(3), 906–915 (2009)
21. Krack, M., Gross, J.: Harmonic Balance for Nonlinear Vibration Problems. Springer, Heidelberg (2019). https://doi.org/10.1007/978-3-030-14023-6

Anomaly Detection in Ultrasonic Monitoring System Using Unsupervised Machine Learning

Yassine Kanoun[1,2](✉), Hatem Mrad[1], Bassem Zouari[2], and Tikou Belem[1]

[1] Université du Québec en Abitibi-Témiscamingue (UQAT), 445 Bd de l'Université, Rouyn-Noranda J9X 5E4, Canada
{yassine.kanoun,hatem.mrad,tikou.belem}@uqat.ca
[2] École Nationale d'Ingénieurs de Sfax, km4, Route de La Soukra, 3038 Sfax, Tunisia
Bassem.zouari@enis.tn

Abstract. The development of monitoring and control systems must be carried out to ensure the efficiency and safety of industrial systems. The detection of anomalies in the monitoring system is of great importance for the predictive maintenance of pressure vessels and pipelines. In this sense, it is crucial to analyze the sensor data and investigate the latest updates in machine learning (ML) technology to detect the various system anomalies immediately. Ultrasonic (UT) monitoring system is one of the traditional non-destructive testing (NDT) methods that is widely used to monitor the various industrial components. The traditional algorithms based on setting a threshold value is not able to detect outliers in the monitoring system dataset. This study investigates this problem by evaluating the performance of various unsupervised ML algorithms and applying them to detect anomalies in multivariate time series of UT wall thickness monitoring system. The following methods are evaluated: Interquartile range, distance-based method, density-based method, and isolation forest. The results provide important insights that the established algorithms can be effectively implemented for anomaly detection.

Keywords: Monitoring · maintenance · predictive · Anomaly · machine learning

1 Introduction

Pressure vessels and pipelines comprise a variety of sophisticated processes and operations related to the exploration and production of petrochemical products.

These industrial components represent the lifelines for the economic development of several countries. However, pressure vessels and pipelines can fail under a variety of circumstances, resulting in catastrophic repercussions, financial losses, and environmental damage. According to the Conservation of Clean Air and Water in Europe (CONCAWE) (Davis et al. 2010), damage to pipelines is caused by mechanical failure, operational malfunctions, natural hazards, corrosion/erosion, and foreign interference. As a result, several companies have implemented a number of intrusive techniques in order to ensure safety, improve system performance, prevent losses, and monitor the

A. Akrout et al. (Eds.): ICAV 2022, ACM 22, pp. 317–327, 2023.
https://doi.org/10.1007/978-3-031-34190-8_34

system closely. Monitoring is a challenging task in the industry 4.0 paradigm and is increasingly being investigated not only in the pipeline industry, buildings, etc., but also in vehicles, aircraft, and biomedical devices. Monitoring helps to anticipate how the various actors involved in the industrial chain will behave at different levels of operation and also to predict failures.

In this context, the ultrasonic monitoring system (UT) has been studied as one of the non-destructive methods with the aim of on-line inspection. This method is widely used to detect and/or predict failures in piping systems. CONCAWE data indicates that 30% of failures are due to corrosion. This complex phenomenon requires continuous monitoring of the wall thickness of the pipelines. However, the accuracy of the manual ultrasonic method depends on several factors (operator skills, environmental factors, etc.). Therefore, a consistent thickness measurement technique is needed without shutting down the system. For the UT monitoring system, the thickness measurement is calculated by computing the transit time in a time domain between successive echoes and knowing the value of the ultrasonic velocity along the transit time. False echoes in thickness measurement by UT can be affected by the sensitivity of the sensors to various factors: Temperature, environmental factors, transducer performance, and ultrasonic coupling (Lebowitz and Brown 1993). These errors can lead to an unexpected deviation in the recorded data, resulting in some aberrant values in the data set. As mentioned earlier, an outlier can be defined as a data point in a time series that is significantly different from the others. Outliers are unusual observations that interfere with data analysis and should be treated with caution in order to obtain a clean data set and make good predictions. These anomalies have led to the development of a variety of different anomaly detection algorithms. In recent years, much research has been conducted to develop new approaches to detect reliably and accurately the different types of anomalies using advances in ML technology. Anomaly detection algorithms can be applied either in online mode, known as "data streams," or in offline mode, known as "batch processing".

Several ML methods are tailored to anomaly detection, these methods can be supervised or unsupervised, depending on whether the dataset has labels in the training dataset or not. However, supervised algorithms are more restrictive than unsupervised methods because they require a labeled dataset. This requirement represents a significant cost if the labeling must be done by users. Managing a highly skewed class distribution, which is an intrinsic issue in anomaly detection, can affect the performance of supervised algorithms. The principle of supervised outlier detection approaches, such as the support vector machine (Schölkopf et al. 1999) and the decision tree (Song and Lu 2015), is achieved by training classification algorithms on annotated datasets. These models can then be used to predict whether the input data is anomalous or not. Unsupervised anomaly detection algorithms use unlabeled data to assign a score to each sample. The principle of this method is to assemble data with similar characteristics so that anomalous data can be weeded out. Some of these algorithms include density-based clustering, distance-based clustering (Chen, Wang and van Zuylen 2010), and isolation forest (Liu, Ting and Zhou 2008).

The present work deals with various unsupervised machine learning methods aimed at identifying and detecting anomalies in time series data for the thickness monitoring system UT. Different unsupervised ML techniques, based on the density method, the distance method, and the isolation forest, were implemented to identify, and analyze the anomalies.

2 Materials and Experimental Setup

2.1 Materials

The data set in this project is based on real data from a thickness monitoring system in which sensors are installed in order to monitor the thinning of pipes caused by sulfuric acid. The UT sensors utilized in this experiment was applied to measure a range of thickness between 1 mm and 150 mm and it was used for temperature range of [−30 °C – 132 °C]. The sensors were calibrated in the presence of an inspector using a ductile iron calibration gauge. A coupling agent is used for the connection between the sensor and the pipe surface. Dual-element contact transducer was used, with 5 MHz frequency. The sensors are placed at different locations of a ductile iron pipe system as shown in Fig. 1. A Smart PIMS Datalogger with integrated battery and memory, capable of storing up to 3000 thickness readings, was used for data storage. It records measurements at each time interval specified by the user.

2.2 Experimental Setup

In this experiment, two 90° ductile iron bends were monitored by installing 4 sensors at different locations, as shown in Fig. 1. The locations of monitoring points depend on factors such as the complexity of pipe system. 90° Elbow is an important part of the piping system, sudden changes in flow velocity and flow rate can lead to significant differences in corrosion behavior at the different locations. The time series of the acquisition system are aggregated to hourly data. In this work, the studied period is 20 days of May 2022. The obtained dataset contains 5 features: Date, Thickness (mm), Fluid Temperature (°C), Fluid Concentration (%), Fluid Flow Rate (m^3/h). The non-control of these parameters can result in unexpected deterioration of the piping system components such as brittle fracture, corrosion, and leakage in piping equipments (flange, valves, gasket, etc.). The process parameters are obtained from the historical data series of the connected measuring devices. A total of 468 values were recorded for each parameter and each UT probe.

a- 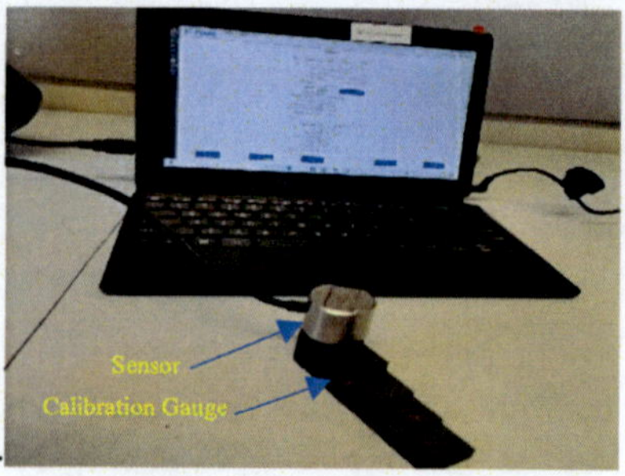

b- 

Fig. 1. Material and Experimental setup for thickness Ultrasonic monitor system. a) Calibration phase, b) Sensors Installation

3 Anomaly Detection Models

Various techniques and methods implemented in this work are discussed in this section. The unsupervised ML is perfectly well adapted to identify anomalies since the large datasets do not need to be labeled. In particular, we highlight the following algorithms: Interquartile Range, K-Nearest Neighbors (K-NN), Local Outlier Factor (LOF), and Isolation Forest (IF).

3.1 Interquartile Range Method

A common way to identify outliers is to analyze the statistical dispersion of the data set. The method presented in this section concern the Inter quartile range (IQR), this method reflects the spread of all data points about the median. IQR is a measure of statistical dispersion and is calculated as the difference between the first quartile Q1 and the third quartile Q3 of the dataset, it is represented by the formula IQR = Q3–Q1. To determine the outliers using IQR, is needed also to calculate the lower fence (F_L = Q1–1.5*IQR) and the upper fence (F_U = Q1 + 1.5*IQR), the 1.5 represents the sensitivity of the dataset. Data points that exceed the minimum and maximum range are considered outliers as illustrated in Fig. 5.

3.2 Based Distance Clustering

The based-Distance method is based on the nearest neighbor approach and uses a defined metric distance to detect outlier. In this work, the K-nearest neighbor approach is proposed for detecting outliers, which exploits the relationship between neighborhoods, where the distance between an object and its K-nearest neighbor provides information on whether or not the object is considered an outlier in the data points, where it presupposes that the outliers are sparse or far away from their neighbors. Dang, Ngan and Liu (2015) presents the basic step of this method: 1) preparing data, 2) defining K parameter (number of cluster), 3) obtaining K-distance for all data points, 4) Find the range of K-distance, 5) Perform Outlier detection method. Elbow method is used to choose the number of clusters.

3.3 Density Based Clustering

Since handling data points with different densities can be challenging for distance-based approaches, the density-based approach could be used to handle these type of data points. An outlier is recognized by density-based approaches if its local density is significantly different from that of its neighbors. More precisely, the locality is determined by the K-nearest neighbors, whose distance is used to evaluate the local density. This density-based method aims to quantify the degree to which each data sample is an outlier, or the "Local Outlier Factor" (LOF) (Breunig et al. 2000). The LOF defines outliers as those samples whose density is significantly lower than that of their neighbors.

3.4 Isolation Forest

The Isolation-Forest algorithm (IF) was introduced by (Liu, Ting and Zhou 2008) and is based on decision trees, focusing on segregating anomalies from the rest of the data points. For each data point, an anomaly score is determined by calculating the length of the average path from the root of the tree to the node enclosing the point. The Isolation Forest has the same classification concept as the Random Forest. If a point dips deeper into the tree, it is not expected to be an outlier. In contrast, if a point is in shorter branches, it is more likely to be an outlier because the assigned values deviate greatly from the norm.

4 Result and Discussion

Thickness monitoring in this project will use all active thickness measurements to perform corrosion analysis calculations. Otherwise, the monitoring system UT may exhibit various forms of anomalies. To ensure a reliable and accurate comparison between the anomaly detection algorithms used, tests were performed to investigate the effects of the process parameters and their correlation effect on the UT measurement. Figure 2 shows the evolution of the thickness data for the 4 sensors. As shown, the measured thicknesses

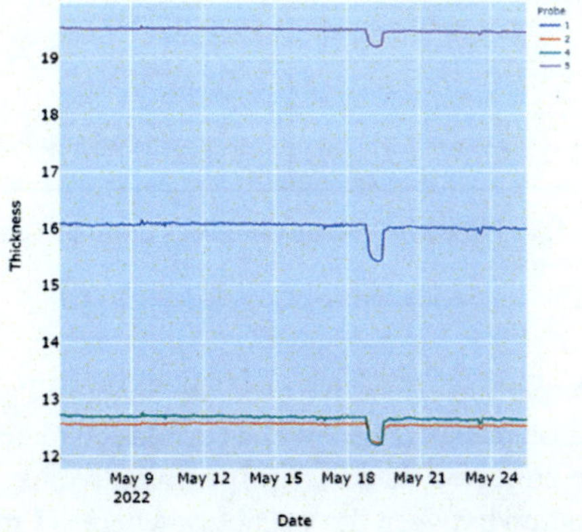

Fig. 2. Thickness measurement data of the installed sensors

are from 12 mm to 20 mm where the critical thickness is defined as $t_{cr} = 7.9$ mm, this confirms that the system should be closely monitored to avoid failure, especially when the thickness seems to be close to the critical value. On the other hand, it is clear that the trend of the thickness measurement coming from the different probes shows a similar behavior with respect to the evolution of the time series. For the rest, the probe number 5 is selected.

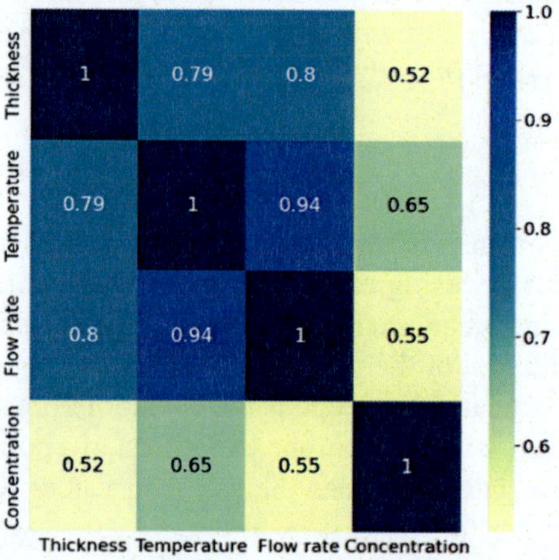

Fig. 3. Correlation matrix

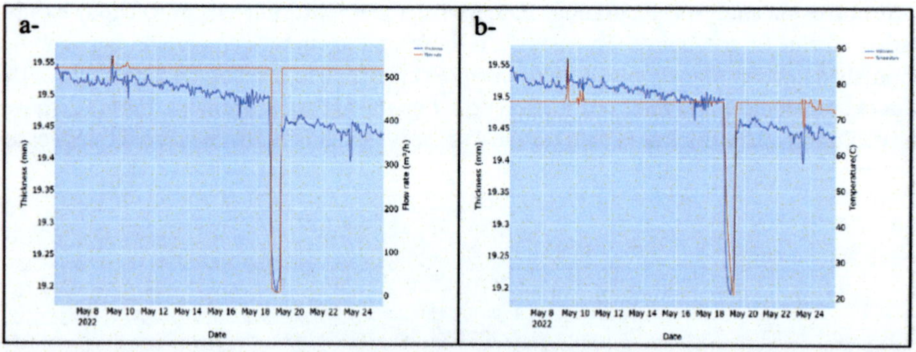

Fig. 4. Parameter evolution. a) Flow rate versus thickness data, b) Temperature versus thickness data (Color figure online)

To investigate the correlation effect, we used the Seaborn Python library to generate and visualize the correlation matrix (Fig. 3). The correlation matrix shows the correlation coefficients between the collected data using Spearman's rank correlation. The matrix calculates a linear correlation between the variables, where −1 means that the correlated variables have a strong negative correlation and 1 indicates a strong positive correlation.

The diagonal values denote the dependence of a variable on itself (also called autocorrelation). Figure 3 shows that both temperature and flow rate are strongly correlated in thickness measurements, while fluid concentration is less so. Thus, tests were conducted to analyze the measured thickness and fluid temperature as well as fluid flow rate, as shown in Fig. 4 (The blue line shows the hourly thickness data measured in real time).

Twenty-one anomalies were counted, spread over the May 18–19 shutdown period, where 17 anomalies occurred, and the remainder spread over the period from May 9 (1 anomaly) to May 23 (3 anomalies). Using Fig. 4, the thickness measurement error was confirmed as a function of temperature and flow rate variation in term of system sensitivity and accuracy, that's mean that a variation in fluid temperature, thermal expansion ($\Delta T = 54$), can affect the echoes of UT sensors. For the thickness degradation during shutdown, Fig. 4 shows that increasing the flow rate from 0 (m^3/h) to 520 (m^3/h) and increasing the temperature from 20 °C to 75 °C contribute to the reduction in wall thickness. Moreover, During the system shutdown, the sulfuric acid is diluted due to the decrease in acid concentration. In addition, the drop in temperature, the percentage of impurities in acid and the variations in relative humidity inside the pipeline and then it can promote a drastic thinning of the pipe material (Louie 2005). Due to those perps, thickness monitoring is needed in petrochemical industries to determine how process changes affect corrosion behavior.

4.1 Anomaly Detection Algorithms Performance

To investigate and explore the different approaches described in Sect. 3, we used the Python library for outlier detection (PyOD) and the Anomaly Detection Toolkit (ATDK), two libraries that contain several unsupervised tools for detecting anomalies in time series. Matplotlib is used to visualize the scatter of the thickness dataset using boxplots Fig. 5. This method helps to identify the range of outlier values.

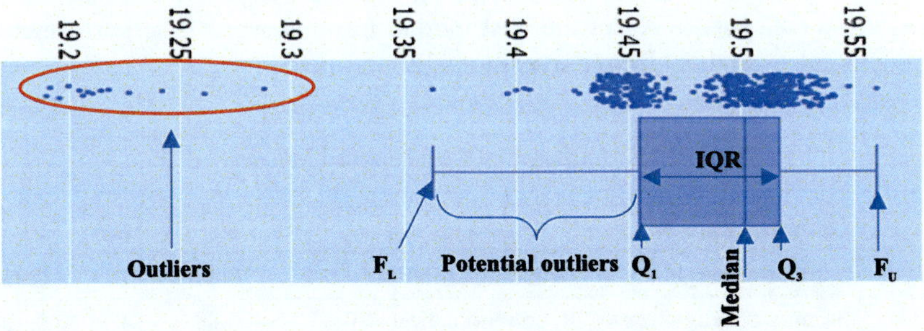

Fig. 5. Outlier detection by boxplot

Before the ability to predict outliers in time series signals can be tested, a set of essential parameters for the proposed method must be defined. For K-NN and LOF, the number of neighbors and the degree of contamination of the data set are fundamental parameters. For simplicity, contamination is understood as the expected proportion of outliers in the data. For each parameter, several values are tested in this work to investigate their effects, and then optimal values are determined. Some other standard parameters are used; standard Euclidean distance is used as the metric parameter. For IF, the number of estimators is a basic parameter related to the number of trees used to build the forest. IF has the same parameter of contamination as other methods. For IQR, only one parameter needs to be defined, c = 1.5.

The above methods aim to filter out the outliers and eliminate the normal points. In other words, the predicted anomalous point is marked with a value of 1. These results are useful to build the confusion matrix (Karimi 2021) as shown in Table 1, TP the true positive anomalies, FP the false positive anomalies and FN the false negative anomalies. To better demonstrate the excellent performance of the proposed algorithms, each model is evaluated by calculating the following indicators, Precision (P), Recall (R) and F1 score, $F1 = \frac{2*P*R}{P+R}$, $P = \frac{TP}{TP+FP}$ and $R = \frac{TP}{TP+FN}$.

Table 1. General confusion matrix

	Predicted (Abnormal)	Predicted (normal)
Actual (Abnormal)	TP	FN
Actual (normal)	FP	TN

The implementation of the optimal parameters for each model is studied to evaluate their performance, as shown in Table 2. The outcomes of each method are presented in Fig. 6. It can be noticed that IQR failed to detect all abnormal data with a total of 9 false anomalies. K-NN and LOF detected the same points, they managed to detect the correct anomalies (21 true positive anomalies), while 3 false anomalies were identified. Isolation Forest also detected the true positive anomalies. In addition, when each model was executed, the results showed that these models were sensitive to noisy cues and required more data sets to select the best model. In other words, the performance to distinguish a false anomaly from a true anomaly.

Table 2. Performance of Unsupervised ML method

Model	Optimal parameter	P	R	F1
K-NN	number of neighbors = 2, contamination = 0,05	0,88	1	0,93
LOF	number of neighbors = 20, contamination = 0,05	0,88	1	0,93
IF	number of estimators = 20, contamination = 0,05	0,88	1	0,93

For the next step, IF is used to clean the outliers on this period. The choice of the IF is based on his high performance and simplicity because it uses a sequence of tree and each tree tries to correct the errors of the one before in term of prediction. Figure 7. The thickness data (green points), at different times are plotted after the cleaning phase. As

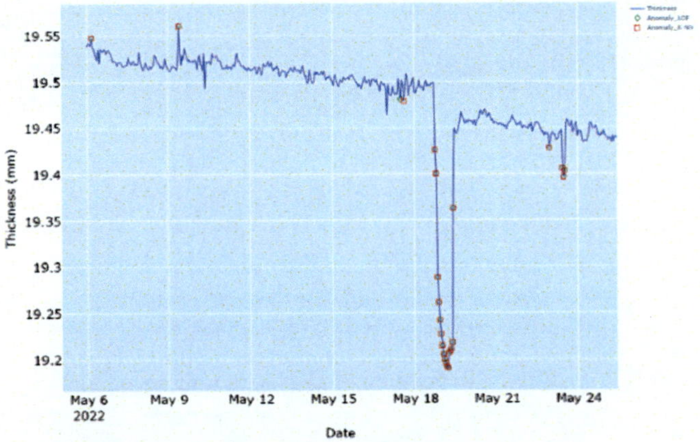

a) Anomaly detection from LOF and K-NN

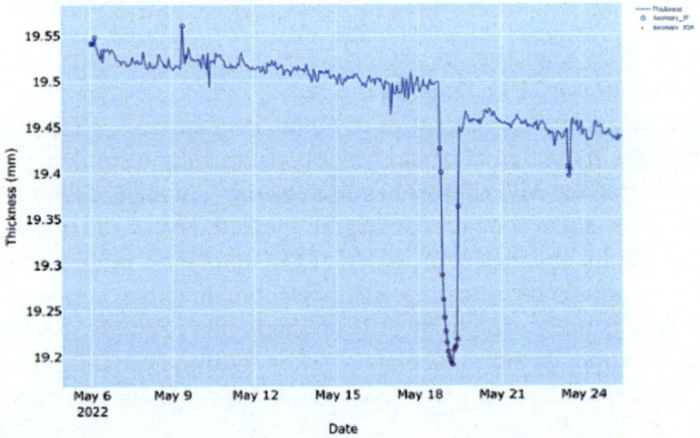

b) Anomaly detection from IF and IQR

Fig. 6. Anomaly Detection using Unsupervised ML models

mentioned above, the main intended application of the new on-line ultrasonic thickness monitoring system is to control the wall thinning measurements in corrosion processes. Therefore, 3 different polynomial features models are implemented (liner, quadratic, and cubic) to predict the thickness variation on this period as a function of time, and then corrosion rate can be estimated. Based on the computed R-Squared function, Fig. 7 shows that the applied model can perform to predict the thickness measurements with $R^2 \geq 0.87$. The estimated corrosion rate on this period is 60 mpy before and after shutdowns which an important degradation compared to the normal rate ≈6 mpy (Louie 2005). This important rate is due to the contribution of the several factors discussed above (shutdowns, thermal expansion…).

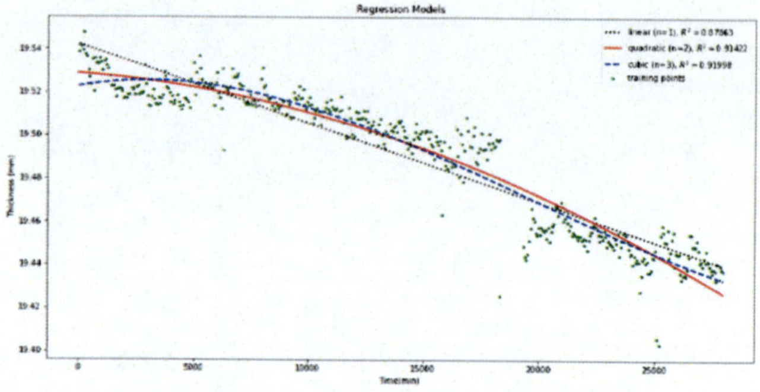

Fig. 7. Regression Models (Color figure online)

5 Conclusions and Future Work

Timely and accurate detection of anomalies is of great importance for the reliability and economy of industrial systems. In this work, several unsupervised ML algorithms were implemented to detect anomalies in monitoring systems. Correlation coefficients between thickness measurements and process parameters were determined to analyze the effectiveness of the ML approaches in detecting anomalies. The advantage of the proposed methods is that they do not require labeled data of the anomalous events. In summary, the results obtained with K-NN, LOF and IF on small and continuous data sets make them suitable for actual anomaly detection in a thickness monitoring system. In future work, it is interesting to investigate intelligent anomaly mitigation techniques for Big Data to extract the normal behavior when multiple variables are simultaneously anomalous in order to develop predictive models for controlling industrial process.

Acknowledgments. The authors would like to thank Joël Fortin and François Grégoire of Norda Stelo in Montreal for their support and help during this project.

References

Breunig, M.M., Kriegel, H.P., Ng, R.T., Sander, J.: LOF: identifying density-based local outliers. ACM SIGMOD Rec. **29**(2), 93–104 (2000). https://doi.org/10.1145/335191.335388

Chen, S., Wang, W., van Zuylen, H.: A comparison of outlier detection algorithms for ITS data. Expert Syst. Appl. **37**(2), 1169–1178 (2010). https://doi.org/10.1016/j.eswa.2009.06.008

Dang, T.T., Ngan, H.Y.T., Liu, W.: Distance-based k-nearest neighbors outlier detection method in large-scale traffic data. In: 2015 IEEE International Conference on Digital Signal Processing (DSP), pp. 507–510 (2015). https://doi.org/10.1109/ICDSP.2015.7251924

Davis, P.M., et al.: Performance of European crosscountry oil pipelines - statistical summary of reported spillages in 2010 and since 1971 (2011)

Karimi, Z.: Confusion Matrix (2021)

Lebowitz, C.A., Brown, L.M.: Ultrasonic measurement of pipe thickness. In: Thompson, D.O., Chimenti, D.E. (eds.) Review of Progress in Quantitative Nondestructive Evaluation, pp. 1987–1994. Springer, Boston (1993). https://doi.org/10.1007/978-1-4615-2848-7_255

Liu, F.T., Ting, K.M., Zhou, Z.H.: Isolation forest. In: 2008 Eighth IEEE International Conference on Data Mining, pp. 413–422. IEEE (2008). https://doi.org/10.1109/ICDM.2008.17

Louie, D.K.: Handbook of Sulfuric Acid Manufacturing. DKL Engineering Inc., Thornhill (2005)

Schölkopf, B., Williamson, R.C., Smola, A., Shawe-Taylor, J., Platt, J.: Support vector method for novelty detection. Adv. Neural Inf. Process. Syst. **12** (1999). MIT Press. https://proceedings.neurips.cc/paper/1999/hash/8725fb777f25776ffa9076e44fcfd776-Abstract.html

Song, Y.Y., Ying, L.U.: Decision tree methods: applications for classification and prediction. Shanghai Arch. Psychiatry **27**(2), 130–135 (2015). https://doi.org/10.11919/j.issn.1002-0829.215044

Pyod and ATDK documentation. https://pyod.readthedocs.io/en/latest/index.html

Implementation of a New Approach Based on Harmonic Quadrature Method on the Study of a Misaligned Rotor Supported in Ball Bearings Squeeze Film System

Farouk Thaljaoui[1], Mohamed Trabelssi[2,3], and Molka Hili[2,4](✉)

[1] Tunisia Polytechnic School, University of Carthage, B.P. 743, 2078 La Marsa, Tunisia

[2] Department of Mechanical Engineering, Tunis Higher National Engineering School, University of Tunis, 1008 Tunis, Tunisia
molka.chiboub@gmail.com

[3] Applied Mechanics and Systems Research Laboratory, Tunisia Polytechnic School, University of Carthage, B.P. 743, 2078 La Marsa, Tunisia

[4] Laboratory of Mechanical Modeling and Production (LA2MP), National School of Engineers of Sfax (ENIS), University of Sfax, B.P. 1173, 3038 Sfax, Tunisia

Abstract. This manuscript aims to study a misaligned rotor using a new approach to solving nonlinear rotor equations of motion based on the Harmonic Quadrature Method (HQM). The present rotor system is supported by ball bearings with a squeeze film damper. The dynamic model representing this rotor yields a system of nonlinear equations of motion with time-dependent coefficients.

HQM is a periodic discretization technique that does not involve costly Galerkin integrals while providing high-order computational accuracy and efficiency. The different steps of its implementation were presented, and a validation study was carried out. A good agreement was found between the numerical results found in the literature and those obtained by the HQM method. Numerical simulations of the dynamic misaligned rotor behavior generated by the HQM method were also performed. These simulations show the non-synchronous and chaotic motions (bifurcated response) generated by the non-linear interaction between the rotor misalignment and the fluid film forces induced by the damper.

1 Introduction

Rotating systems including shaft, disc, and bearing assembly are usually used in various types of machinery, such as compressors, turbines, and aero-engines. The motion of such systems is usually governed by a system of nonlinear differential equations. The presence of such nonlinearities is essentially due to forces

A. Akrout et al. (Eds.): ICAV 2022, ACM 22, pp. 328–341, 2023.
https://doi.org/10.1007/978-3-031-34190-8_35

induced by the support elements (bearings). Rotors are often supported on ball bearings with squeeze film dampers (SFD) to muffle or quell their vibration. However, conventional SFDs are highly nonlinear elements and often result in nonsynchronous vibration of the rotor. The occurrence of nonsynchronous and chaotic motion in rotating machinery is undesirable and should be avoided, as they introduce cyclic stresses in the rotor. Such stresses can rapidly provoke fatigue failures, especially in the presence of defects such as rotor misalignment. Modeling this type of system generally yields a system of nonlinear equations of motion. Solving such equations can provide a better understanding of these particular vibratory phenomena and mitigate their adverse shortcoming. The literature offers a variety of numerical methods which aim to solve the differential equations of motion and the numerical analysis of linear and nonlinear problems of rotating systems. These methods are generally classified into two categories: the time domain and the frequency domain methods. The Runge Kutta method [1,2] and Newmark method [3,4] are ones of the most common methods used to solve nonlinear differential equations in the time domain. Time domain methods can be coupled with specific techniques to reduce the computation time of the transition response, such as the shooting method [5]. For instance, Adiletta et al. [2] derive the equations of motion of a rigid rotor on short fluid bearings and solve them numerically using the Runge Kutta method. They obtain bifurcation diagrams and harmonic, subharmonic, quasi-periodic, and chaotic responses. Li and Xu [6] study a JEFFCOTT-type rotor supported on infinitely long oil-film bearings to compute periodic orbits using the generalized shooting method (generalized shooting method). Bouaziz S et al. [7] investigated the dynamic linear response of a rigid misaligned rotor mounted in two identical AMBs. Results of this work show that angular misalignment is such that the 2x and 4x running speed components are predominant in spectra of vibration.

The Harmonic Balance Method (HBM) [8], the Spectral Method (SM) [9], and the Method of Multiple Scales (MMS) [10] are very efficient for the resolution of nonlinear equations in the frequency domain. For instance, Beley-Sayettat [11] studies the movement of a rotor whose support (rigid and flexible) is subjected to seismic excitations. A spectral method with complex modes was used to solve the equations of motion. Chu and Zhang [12] use the simplified HBM method to study the nonlinear dynamic behavior of a JEFFCOTT rotor system due to rotor-stator rubbing contact forces. Sinou [13] develops an HBM variant associated with the alternating frequency/time (AFT) algorithm with a continuation by arc length to analyze the dynamics of a flexible rotor modeled by the finite elements of Timoshenko beams and supported by ball bearings with non-linear contact. Inayat-Hussain [14] studied the nonlinear response (bifurcation) of a rigid rotor supported by levitated squeeze-film dampers. The numerical continuation method is used to trace a solution branch, detects bifurcation points and determines the stability of these points. Li and Yu [15] developed a nonlinear model of a rotor-blade system with nonlinear supports at both ends. The nonlinear vibration and stability of the system are studied by the multiple scales method.

The Harmonic Quadrature Method (HQM) is a periodic discretization technique based mainly on Harmonic Lagrange basis. This method comes with built-in periodic initial conditions and does not come with a pre-imposed mesh type like SM. However, like SM, HQM is a high-order method, i.e. it makes use of all grid points to interpolate the derivatives. For this reason, HQM benefits from the same enhanced convergence speed and accuracy as SM. In the field of mathematics, HQM belongs to the family of techniques known as differential quadrature (DQ). DQ is a high-order numerical approach that is efficient for solving linear and nonlinear partial differential equations. DQM theory seeks to achieve the most accurate interpolation of an arbitrary function's derivative for a given mesh. This makes HQM a suitable alternative to other frequency discretization techniques. To the best of the authors' knowledge, this method was not used before to model nonlinear rotor systems. To fill in this gap, HQM will be used in the present study along with other classical techniques to assess its accuracy.

In a previous study [16], the authors investigated the linear response of a rigid misaligned rotor mounted on different supports: elastic, hydrodynamic, and Active Magnetic Bearings. The motion of a misaligned rotor found is governed by ordinary linear differential equations with time-dependent coefficients. In this paper, the misaligned rotor is supported on ball bearings with a squeeze film damper, the forces generated by this type of support are highly nonlinear. The equations to be solved are therefore nonlinear with periodic coefficients. Implementing the traditional methods presented earlier like HBM is thus very complicated. This model seems, therefore, ideal to assess the performance and accuracy of the new HQM-based approach.

In this manuscript, a misaligned rotor supported by two identical ball bearings with a squeeze film damper will be presented. The equations of motion of the system will be derived by LAGRANGE formalism. Then, a complete description of the implementation steps of HQM to determine the nonlinear response of the studied system will be described. Validation of the developed method is carried on using results from the literature. Finally, numerical simulations of the nonlinear misaligned rotor response, obtained by HQM, will be presented.

2 Dynamic Model

The misaligned rotor model is presented in Fig. 1a. The system consists of: a motor shaft (1) and a rigid weighing receptor shaft (4) supported by two identical ball bearings with squeeze film damper (BBSFD) (3) and (5) are shown in Fig. 1b. The present system has three degrees of freedom: $\alpha(t)$ corresponds to the angular misalignment and represents the small variations of the receptor shaft position around an imposed value α_E. $y3(t)$ and $y5(t)$ are respectively the outer races radial displacements of each bearing (the coupling effect between the rotor, ball bearing, and SFD was considered).

The equation of motion of the receptor shaft was obtained by writing the kinetic, potential, and dissipative energy and applying Lagrange's equation [17]:

$$[M(t)]\,\ddot{\delta} + [C(t)]\,\dot{\delta} + [K(t)]\,\delta = \left\{F(t,\omega,\delta,\dot{\delta})\right\} \quad (1)$$

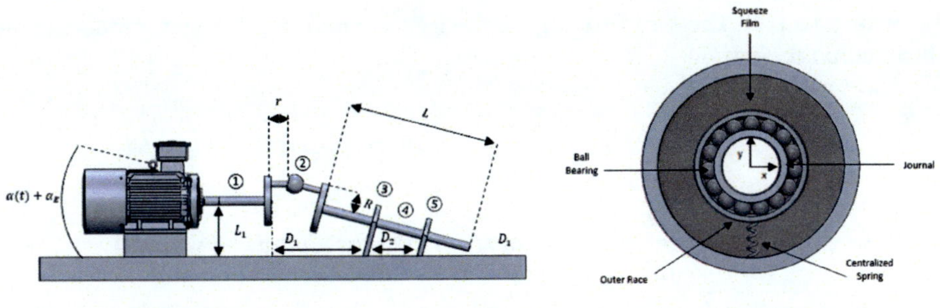

(a) Modeling of the misaligned rotor

(b) Nonlinear support

Fig. 1. Misaligned rotor elements

where the mass matrix $[M]$, the stiffness matrix $[K]$, the gyroscopic matrix $[G]$ and the damping matrix $[C]$ are given by:

$$[M] = \begin{bmatrix} M_s(t) & 0 & 0 \\ 0 & m_b & 0 \\ 0 & 0 & m_b \end{bmatrix} \quad [K] = \begin{bmatrix} K_s(t) & 0 & 0 \\ 0 & K_b & 0 \\ 0 & 0 & K_b \end{bmatrix} \quad [G] = \begin{bmatrix} G_s(t) & 0 & 0 \\ 0 & 0 & 0 \\ 0 & 0 & 0 \end{bmatrix} \quad [C] = \begin{bmatrix} C_s & & \\ & 0 & \\ & & 0 \end{bmatrix} \tag{2}$$

and where

$$\begin{aligned} M_s(t,\omega) &= m[L_{eff}^2 + R^2 cos^2(\omega t)] + I \\ K_s(t,\omega) &= mR^2 L_{eff} sin(\alpha_E) cos(\omega t) - mR^2\omega^2 cos(\alpha_E) cos(\omega t)^2 + K_\alpha \\ G_s(t) &= -2mR^2\omega sin(\omega t) cos(\omega t) \end{aligned} \tag{3}$$

ω is the rotational speed and C_s is the viscous damping of the shaft due to aerodynamic effect. The various constants are presented in Table 1. The displacement vector $\{\delta\}$ and $\left\{F(t,\omega,\delta,\dot{\delta})\right\}$ the external forces vector is given by:

$$\{\delta\} = \begin{Bmatrix} \alpha \\ y3 \\ y5 \end{Bmatrix} \qquad \{F\} = \begin{Bmatrix} F_\alpha(t,\omega) - a_3 F_{B_3}^{BBSF} - a_5 F_{B_5}^{BBSF} \\ -F_{B_3}^{BBSF} + f_{hyd_3} \\ -F_{B_5}^{BBSF} + f_{hyd_5} \end{Bmatrix} \tag{4}$$

where $F_\alpha(t,\omega)$, the effort induced by shaft misalignment, is given by:

$$F_\alpha(t,\omega) = -mR^2\omega^2(1 - cos(\alpha_E)) cos(\omega t) + mR^2\omega^2 sin(\alpha_E) cos(\omega t)^2 \tag{5}$$

a_3 and a_5 are giving by:

$$a_3 = -(Rcos(\omega t) + L1).(sin(\alpha_E) + \alpha cos(\alpha_E)) - (D_1 - r)(cos(\alpha_E) - \alpha sin(\alpha_E)) \tag{6a}$$

$$a_5 = -(Rcos(\omega t) + L1).(sin(\alpha_E) + \alpha cos(\alpha_E)) - (D_1 + D_2 - r)(cos(\alpha_E) - \alpha sin(\alpha_E)) \tag{6b}$$

According to [14], the ball bearing forces $F_{B_3}^{BBSF}$ and $F_{B_5}^{BBSF}$ generated can be obtained as follows:

$$F_{B_3}^{BBSFD} = \sum_{j=1}^{N_b} K_m (y_{B_3}^* sin\theta_j - r_0)^{\frac{3}{2}} H(y_{B_3}^* sin\theta_j - r_0) cos\theta_j$$
$$F_{B_5}^{BBSFD} = \sum_{j=1}^{N_b} K_m (y_{B_5}^* sin\theta_j - r_0)^{\frac{3}{2}} H(y_{B_5}^* sin\theta_j - r_0) cos\theta_j \quad (7)$$

$y_{B_3}^*$ and $y_{B_5}^*$ are the displacements of the journal associated with each of the BB-SFD 3 and 5.

$H(.)$ is the Heaviside function. The different constants used in the expression of the forces $F_{B_3}^{BBSF}$ and $F_{B_5}^{BBSF}$ are presented in Table 2.

The hydrodynamic forces f_{hyd_3} and f_{hyd_5} are given by [14]:

$$f_{hyd_3} = -B.\dot{y}_3 \left[\frac{\pi}{2} - \arctan\left(\frac{|y_3|}{\left(1 - \frac{y_3^2}{e}\right)^{\frac{1}{2}}} \right) \frac{1 + 2y_3^2}{(1 - y_3^2)^{\frac{5}{2}}} \right] \quad (8)$$

$$f_{hyd_5} = -B.\dot{y}_3 \left[\frac{\pi}{2} - \arctan\left(\frac{|y_5|}{\left(1 - \frac{y_5^2}{e}\right)^{\frac{1}{2}}} \right) \frac{1 + 2y_5^2}{(1 - y_5^2)^{\frac{5}{2}}} \right] \quad (9)$$

The parameter (B) is a measure of the amount of damping that a squeeze-film damper can provide, and it is given by:

$$B = \frac{\mu R l_c^3}{c^2} \quad (10)$$

The different constants used in the expression of the forces f_{hyd_3} and f_{hyd_5} are presented in Table 3.

It is clear that the equation to be solved is highly nonlinear with periodic coefficients. This model seems, therefore, ideal to assess the performance and accuracy of the new HQM-based approach.

Table 1. Geometric and mechanical features of the receptor shaft

Parameter	Feature	Given Value
m	Receptor shaft mass (Kg)	1.34
R	Disc radius (m)	0.08
r	Coupling radius (m)	0.002
I	moment of inertia	0.136
K_α	Receptor shaft stiffness (N/m)	10^5
L	Height of the bearing motor shaft (m)	0.55
L_1	geometric parameters (m)	0.05
D_1	Distances between the motor shaft and bearing (3) (m)	0.08
D_2	Distance between bearings (3) and (5) (m)	0.04
α_e	imposed angular misalignment (rad)	0.017

Table 2. Geometric and mechanical features of the ball bearings

Parameter	Feature	Given Value
R_c	outer race radius (m)	3.10^{-2}
r_c	inner race radius (m)	10^{-3}
r_0	balls radius (m)	10^{-4}
K_m	Hertzian contact stiffness (N/m)	$32.5.10^5$
m_b	Outer races mass (Kg)	0.5
K_b	Centralizing spring stiffness (N/m)	10^6
θ_j	angle location of the jth ball (rad)	$\frac{\omega r_c}{r_c+R_c} + \frac{2\pi}{N_b}(j-1)$

Table 3. Geometric and mechanical features of the SFD

Parameter	Feature	Given Value
l_c	damper length (m)	$8.3.10^{-3}$
e	damper radial clearance (m)	$2.3.10^{-4}$
μ	dynamic viscosity of lubricant (Ns/m^2)	0.1

3 Implementing of Harmonic Quadrature Method (HQM) for Misaligned Rotor Systems

3.1 Discretization of a Rotor Equation of Motion Using HQM

HQM is a member of the differential quadrature (DQ) method family. The goal of DQM theory is to achieve the most accurate interpolation of an arbitrary function's derivative for a particular mesh. While standard DQM is based on polynomial bases, HQM is formulated using a set of harmonic bases. The use of harmonic bases is helpful when modeling periodic phenomena. This section gives a detailed description of the implementation steps of HQM to model the forced vibration of a nonlinear rotor's differential equations. For this purpose, a simplified nonlinear rotor's differential equation is purposed

$$m(t)\ddot{q}(t) + c(t)\dot{q}(t) + k(t)q(t) = f^{ext}(\omega, t) + f^{nl}(\omega, t, q(t), \dot{q}(t)) \tag{11}$$

where $m(t)$, $c(t)$, $k(t)$, f^{ext} and f^{nl} are respectively the mass, viscosity, stiffness, external force and nonlinear force of the system. To simplify the implementation process, a new timescale is introduced

$$\tau = \frac{\omega}{2\pi}t. \tag{12}$$

Then, to discretize the equation of motion in the time domain, the first and the second time derivative are substituted by the HQM first and second derivative matrices $\left[D_t^{(1)}\right]$ and $\left[D_t^{(2)}\right]$ respectively [18,19]. The expression of $\left[D_t^{(1)}\right]$ and $\left[D_t^{(2)}\right]$ can be found in [19]. The time-dependent variables are substituted with diagonal matrices containing the values of the latter at each time step. Finally, the discretized equation of motion is written as follows:

$$\left[\left(\frac{\omega}{2\pi}\right)^2 [m(\tau)]\left[D_\tau^{(2)}\right] + \left(\frac{\omega}{2\pi}\right)[c(\tau)]\left[D_\tau^{(1)}\right] + [k(\tau)]\right] q(\tau) = {}_{nl}f(\tau, q) + {}_{ext}f(2\pi\tau) \tag{13}$$

where $[m(\tau)]$, $[c(\tau)]$, $[k(\tau)]$, ${}_{nl}f(\tau, q)$ and ${}_{ext}f(2\pi\tau)$ are giving by:

$$[m(\tau)] = \begin{bmatrix} m_{\tau=0} & \cdots & 0 \\ \vdots & \ddots & \vdots \\ 0 & \cdots & m_{\tau=1} \end{bmatrix}, [c(\tau)] = \begin{bmatrix} c_{\tau=0} & \cdots & 0 \\ \vdots & \ddots & \vdots \\ 0 & \cdots & c_{\tau=1} \end{bmatrix}, [k(\tau)] = \begin{bmatrix} k_{\tau=0} & \cdots & 0 \\ \vdots & \ddots & \vdots \\ 0 & \cdots & k_{\tau=1} \end{bmatrix}, \tag{14}$$

$${}_{nl}f(\tau, q) = \begin{Bmatrix} {}_{nl}f_{\tau=0} \\ \cdots \\ {}_{nl}f_{\tau=1} \end{Bmatrix}, {}_{nl}f(\tau) = \begin{Bmatrix} {}_{ext}f_{\tau=0} \\ \cdots \\ {}_{ext}f_{\tau=1} \end{Bmatrix} \tag{15}$$

A proper discretization of the time dimension is required for a numerical solution of the forced vibration system. The HQM may be used to achieve this goal since it implicitly incorporates periodic beginning conditions with the added precision of high-order techniques. As a result, the appropriate suitable mesh is used [19].

$$\hat{\tau}_i = \frac{i}{n_\tau},\ 0 \le \hat{\tau}_i < 1,\ 0 \le i \le n_\tau - 1 n_\tau \text{ is odd} \tag{16}$$

where n_τ designates the time increment number. Then the initial conditions are reformulated as:

$$\begin{cases} q|_{\tau=0} & = q|_{\tau=1} \\ \dot{q}|_{\tau=0} & = \dot{q}|_{\tau=1} \end{cases} \tag{17}$$

The residual vector is then written in the following form:

$$R = \left[\left(\frac{\omega}{2\pi}\right)^2 [m(\tau)]\left[D_\tau^{(2)}\right] + \left(\frac{\omega}{2\pi}\right)[c(\tau)]\left[D_\tau^{(1)}\right] + [k(\tau)]\right] q(\tau) - {}_{nl}f(\tau, q) - {}_{ext}f(2\pi\tau) = \{0\} \tag{18}$$

The response is evaluated at each time step size τ_i, by solving (18). The Fourier coefficients $\{\widehat{q}\}$ are obtained by the following transformation:

$$\{\widehat{q}\} = [\Gamma]^{-1} q(\tau) \tag{19}$$

where

$$[\Gamma] = \begin{bmatrix} 1 & cos(2\pi\tau_1) & sin(2\pi\tau_1) & ... & cos(2H\pi\tau_1) & sin(2H\pi\tau_1) \\ 1 & cos(2\pi\tau_2) & sin(2\pi\tau_2) & ... & cos(2H\pi\tau_2) & sin(2H\pi\tau_2) \\ ... & ... & ... & ... & ... & ... \\ 1 & cos(2\pi\tau_n) & sin(2\pi\tau_n) & ... & cos(2H\pi\tau_n) & sin(2H\pi\tau_n) \end{bmatrix} \tag{20}$$

It is clear from the present discretization of the equation of motion that the implementation of HQM is simpler than, for example, the Harmonic Balance Method. Given that HQM is a collocation method, each time increment requires less computational effort than HBM as there is no Galerkin integral to evaluate. Unlike other collocation methods, HQM comes with prebuilt periodic initial conditions and does not compute the transition regime. Furthermore, the fact, that HQM is built on the high-order DQ platform ensure fast convergence compared to other collocation methods.

Finally, it is important to mention that though HQM increment can be computed faster than HBM, being a collocation method, HQM usually requires a bit more harmonics than HBM to converge.

3.2 Arc Length Continuation

Obtaining the frequency response curve (FRC) for a nonlinear system is rather challenging, as this curve is not usually a simple function of the excitation frequency. In most cases, the FRC is generally a parametric curve rather. For this reason, the arc length continuation is used to follow all branches of the nonlinear response. The principle of this method is to construct a vector tangent to the curve, from the point of application of the previous solution and of unit norm. As soon as the normed tangential vector is evaluated, the approximate predictor solution is obtained from the previous converged solution. Corrections of this approximate solution are thus carried out in the direction orthogonal to the tangent vector to be brought back on the response curve.

4 Validation of the Proposed HQM Approach

To the best of the author's knowledge, HQM has not been used to plot the FRC of a complex, highly nonlinear rotor system. To this end, this section aims to test the accuracy of HQM for such systems. In this section, two numerical results are validated using the proposed HQM approach. A nonlinear two variables rotor system, proposed by Inayat Al-Hussein, is selected as the first validation example [14]. According to Inayat Al-Hussein, the equations of motion are given by [14]

$$x'' = -B[\frac{x}{\varepsilon}[I_1\varepsilon' + I_2\varepsilon\varphi'] - \frac{y}{\varepsilon}[I_2\varepsilon' + I_3\varepsilon\varphi']] - S^2x + Ucos(\tau) \quad (21)$$

$$y'' = -B[\frac{y}{\varepsilon}[I_1\varepsilon' + I_2\varepsilon\varphi'] + \frac{x}{\varepsilon}[I_2\varepsilon' + I_3\varepsilon\varphi']] - S^2y + Usin(\tau) - W \quad (22)$$

where x and y are respectively the non-dimensional displacement of the geometric center of the rotor in the X and Y directions and the different parameters of the equation are indicated in [14]:

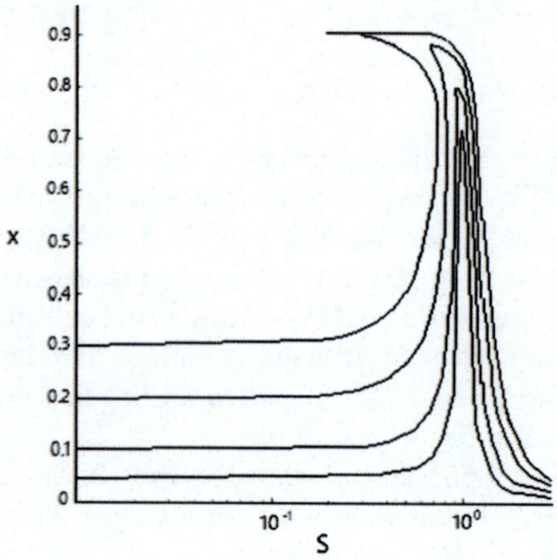

Fig. 2. Validation using Inayat Al-Hussein results

Figure 2 shows that the FRCs generated using HQM match the ones obtained by [14]: regardless of the value of the parameter U.

The second validation example presented in this manuscript is based on the rotor system proposed by Min Sun and Jianen Chen [20]. Min Sun and Jianen Chen studied the nonlinear response of a nonlinear primary oscillator with nonlinear energy sink under harmonic excitation. The equations of motion of selected oscillator are given by

$$\ddot{w} + \gamma_1\dot{w} + Kw + k_1w^3 + k_2(w - v)^3 + \gamma_2(\dot{w} - \dot{v}) = fcos(\Omega t) \quad (23)$$

$$\varepsilon\ddot{v} + k_2(v - w)^3 + \gamma_2(\dot{v} - \dot{w}) = 0 \quad (24)$$

The different parameters of the equation are indicated in [20]. Numerical results of I Min Sun and Jianen Chen [20] are consistent with HQM results shown in Fig. 3. Furthermore, according to Min Sun and Jianen Chen [20], nonlinear branch is not detectable since the use of a numerical method based on sequential continuation, however, this branch is detectable when the Arc length continuation coupled with HQM is implemented Fig. 3.

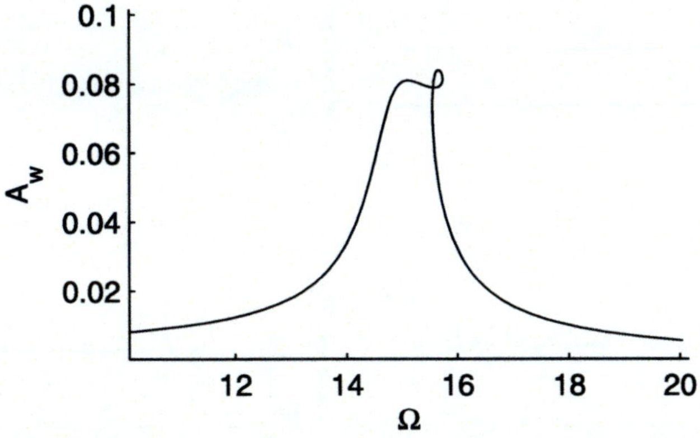

Fig. 3. Validation using Min Sun and Jianen Chen results

5 Numerical Results

After validating the proposed method, the equation of motion of the misaligned rotor can be now be solved using the HQM. This will provide further opportunities to assess the performance of the proposed approach to deal with such complicated dynamic nonlinear systems. First, the angular response was calculated by three different methods: Runge Kuta (time-domain method), HBM (frequency domain method) and HQM. The main geometrical parameters of the model are shown in Tables 1, 2, and 3. The results are presented in Fig. 4a. The response obtained using HQM agrees with those obtained using HBM. There are bistable responses near the first critical speed (ω_n = 649 rad/s). It is a softening type of nonlinearity because of the negative nonlinear terms. The response obtained by Runge Kutta Method agrees with the ones obtained by the frequency domain method. However, it is unable to follow the unstable branches of the curve. Such weakness is due to the absence of a continuation method like arc-length and a guessing algorithm like the shooting method.

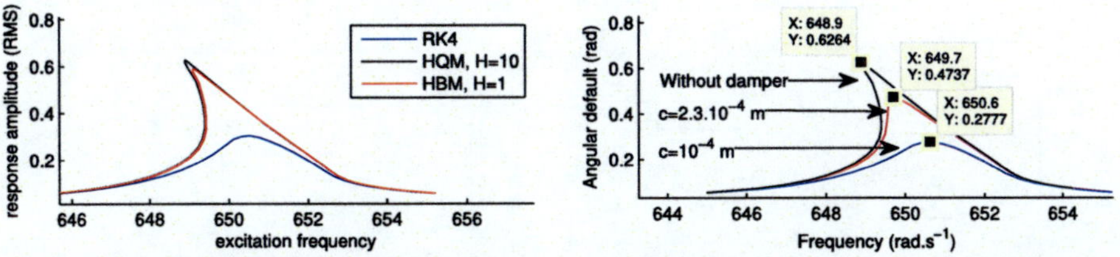

(a) Response obtained by HQM, HBM and RK4 method

(b) Influence of the radial clearence on the nonlinearity of the response

Fig. 4. Frequency response curve of misaligned rotor

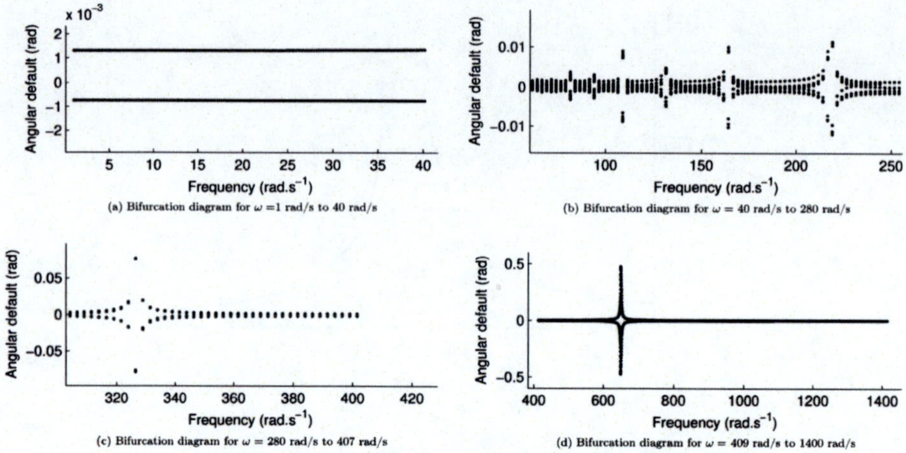

(a) Bifurcation diagram for ω =1 rad/s to 40 rad/s

(b) Bifurcation diagram for ω = 40 rad/s to 280 rad/s

(c) Bifurcation diagram for ω = 280 rad/s to 407 rad/s

(d) Bifurcation diagram for ω = 409 rad/s to 1400 rad/s

Fig. 5. Bifurcation diagram

The frequency response of the misaligned rotor supported by BBSFD is then compared to that of the misaligned rotor supported only by ball bearings (without squeeze film). Figure 4b shows that most angular displacement of the misaligned shaft supported by BBSFD is smaller than that of the misaligned shaft supported by BB. In addition, decreasing the radial clearance from $c = 2.3 \cdot 10^{-4}$ m to $c = 10^{-4}$ m, decreases the maximum angular displacement from $\alpha = 0.47$ rad to $\alpha = 0.27$ rad. This makes, the nonlinear branch disappeared, and the response became similar to a linear response. Therefore, BBSFD can restrain the bistable response, and have a better damping effect.

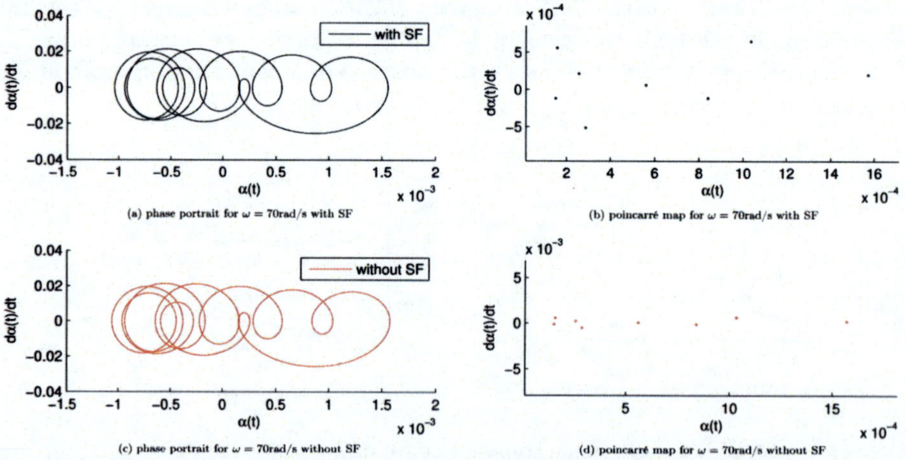

(a) phase portrait for ω = 70rad/s with SF

(b) poincarré map for ω = 70rad/s with SF

(c) phase portrait for ω = 70rad/s without SF

(d) poincarré map for ω = 70rad/s without SF

Fig. 6. Periodicity of the angular default response for the second range of speed

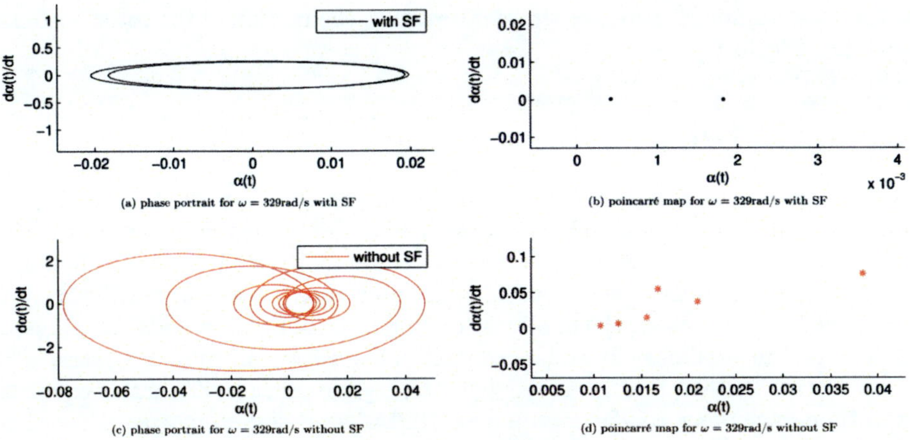

(a) phase portrait for $\omega = 329$rad/s with SF

(b) poincarré map for $\omega = 329$rad/s with SF

(c) phase portrait for $\omega = 329$rad/s without SF

(d) poincarré map for $\omega = 329$rad/s without SF

Fig. 7. Periodicity of the angular default response for the third range of speed

To study the performance of BBSFD, bifurcations diagrams are used to analyze the motion when the rotor speed is varied. Figure 5 shows bifurcation diagrams with different rotating speeds range (0~14000 rad/s). It can be seen from Fig. 5(a) that the motion of the rotor is periodic at the beginning until $\omega = 40$ rad/s, there is only one period branch of the movement. According to Fig. 5(b), it can be also seen that the speed range of the route to chaotic motion is from 40 rad/s to 280 rad/s, there are several branches of periods (a sequence of period-doubling bifurcations of the period-1 orbit). Such as $\omega = 70$ rad/s, the period-1 solution branch, is sequentially bifurcated to period-8 orbits, prior to the chaotic motion. The corresponding, portrait phases and Poincaré map for the two configurations studied (rotor supported by BBSFD and rotor supported by BB) are shown in Fig. 6. It is clear that in this speed range, the BBSFD does not manage to improve the dynamic behavior of the system. Indeed, this range corresponds to the different sub-harmonics of the critical speed ($\omega_n/3$, $\omega_n/4$, $\omega_n/5$...), indicating the presence of shaft misalignment defect and are due to modulation phenomenon because of the dependence of the mass and stiffness terms on time.

The motion of the rotor becomes bi periodic from 280 rad/s to 407 rad/s (Fig. 5(c)), there are two branches of periods. In order to be able to analyze the effect of SFD on the periodicity of the misaligned rotor system, the phase portrait and the Poincaré map were drawn at $\omega = 329$ rad/s which corresponds to $\frac{\omega_n}{2}$ (characteristic frequency of misalignment defect [16]) as shown in Fig. 7. SFD has a good damping crossover in the second-order critical speed range of the misaligned rotor system. Comparing the two conditions with SFD and without SFD, it was found that the angular response with SFD condition is basically in the two-period motion state (Fig. 7b). However, the periodicity of the angular response without SFD is basically in the chaotic motion state (Fig. 7d). Finally, as the rotating speed increases (from 407 rad/s), the motion of the rotor becomes periodic (Fig. 5(d)).

6 Conclusion

In this paper, a new approach based on Harmonic Quadrature Method (HQM) was adopted to investigate the vibrations of a misaligned rotor supported on ball bearings with squeeze film. A rotor dynamic model was developed in which the coupling effect between a misaligned rotor, ball bearings, and FSDs was considered. Furthermore, the nonlinear oil film force of FSDs was also considered, as well as the nonlinear Hertzian contact force between balls and races. The rotor motion is therefore governed by a highly nonlinear differential equation with time-dependent coefficients. A new HQM-based technique was adopted to solve this type of equation.

HQM was first validated using classical numerical methods and simulation data of a few nonlinear rotor systems from the literature. The validation results showed several interesting advantages such as reduced computational time, ease of obtaining the time domain solution, the detection of both stable and unstable solution branches thanks to arc length continuation, ease of raising the number of harmonics, and accuracy comparable to HBM.

Following the validation step, HQM was used to investigate the nonlinear dynamic response of the misaligned rotor supported on a ball bearing with FSDs. Numerical results showed that the nonlinearity of the system is of the softening type. Furthermore, it was noted that the increase in the radial clearance destabilizes the system and strengthens the effect of the nonlinearity. The bifurcation diagrams show the existence of a chaotic regime in the frequency range, including the sub harmonics of the natural frequency. The squeeze film damper did not prevent the occurrence of this motion, caused primarily by the interaction between shaft misalignment and nonlinear support forces.

References

1. Khonsari, M.M., Chang, Y.J.: Stability boundary of non-linear orbits within clearance circle of journal bearings. J. Vib. Acoust. **115**(3), 303–307 (1993)
2. Adiletta, G., Guido, A.R., Rossi, C.: Nonlinear dynamics of a rigid unbalanced rotor in journal bearings. Part I: Theoretical analysis. Nonlinear Dyn. **14**(1), 57–87 (1997)
3. Zheng, T., Hasebe, N.: An efficient analysis of high-order dynamical system with local nonlinearity. J. Vib. Acoust. **121**(3), 408–416 (1999)
4. Baguet, S., Jacquenot, G.: Nonlinear couplings in a gear-shaft-bearing system. Mech. Mach. Theory **45**(12), 1777–1796 (2010)
5. Filipov, S.M., Gospodinov, I.D., Faragó, I.: Shooting-projection method for two-point boundary value problems. Appl. Math. Lett. **72**, 10–15 (2017)
6. Li, D., Jianxue, X.: A method to determine the periodic solution of the non-linear dynamics system. J. Sound Vib. **275**(1–2), 1–16 (2004)

7. Bouaziz, S., Messaoud, N.B., Mataar, M., Fakhfakh, T., Haddar, M.: A theoretical model for analyzing the dynamic behavior of a misaligned rotor with active magnetic bearings. Mechatronics **21**(6), 899–907 (2011)
8. Hosen, Md.A., Chowdhury, M.S.H.: A new analytical technique based on harmonic balance method to determine approximate periods for duffing-harmonic oscillator. Alexandria Eng. J. **54**(2), 233–239 (2015)
9. Shirono, T., Kulatilake, P.H.S.W.: Accuracy of the spectral method in estimating fractal/spectral parameters for self-affine roughness profiles. Int. J. Rock Mech. Min. Sci. **34**(5), 789–804 (1997)
10. Nie, J.F., Zheng, M.L., Yu, G.B., Wen, J.M., Dai, B.: The method of multiple scales in solving nonlinear dynamic differential equations of gear systems. Appl. Mech. Mater. **274**, 324–327 (2013)
11. Beley-Sayettat, A.: Asymmetry and seismic effects in rotordynamics. Ph.D. thesis, Institut national des sciences appliquées de Lyon (Lyon). Organisme de soutenanceLMSt - Laboratoire de Mécanique des Structures, UMR 5514 (Lyon, INSA). Laboratoire associé á la thèseEcole Doctorale Mecanique, Energetique, Genie Civil, Acoustique (MEGA) (Villeurbanne) (1994)
12. Chu, F., Zhang, Z.: Bifurcation and chaos in a rub-impact Jeffcott rotor system. J. Sound Vib. **210**(1), 1–18 (1998)
13. Sinou, J.-J.: Non-linear dynamics and contacts of an unbalanced flexible rotor supported on ball bearings. Mech. Mach. Theory **44**(9), 1713–1732 (2009)
14. Inayat-Hussain, J.I., Kanki, H., Mureithi, N.W.: On the bifurcations of a rigid rotor response in squeeze-film dampers. J. Fluids Struct. **17**(3), 433–459 (2003)
15. Li, B., Ma, H., Xi, Yu., Zeng, J., Guo, X., Wen, B.: Nonlinear vibration and dynamic stability analysis of rotor-blade system with nonlinear supports. Arch. Appl. Mech. **89**(7), 1375–1402 (2019)
16. Hili, M.A., Fakhfakh, T., Hammami, L., Haddar, M.: Shaft misalignment effect on bearings dynamical behavior. Int. J. Adv. Manuf. Technol. **26**(5–6), 615–622 (2004)
17. Efendiev, B.I.: Lagrange formula for ordinary continual second-order differential equations. Differ. Equ. **53**(6), 736–744 (2017)
18. Chang, S.-Y.A., Lin, C.-S., Yau, H.-T.: Lectures on Partial Differential Equations. International Press (2003)
19. Trabelssi, M., El-Borgi, S., Friswell, M.I.: A high-order FEM formulation for free and forced vibration analysis of a nonlocal nonlinear graded Timoshenko nanobeam based on the weak form quadrature element method. Arch. Appl. Mech. **90**(10), 2133–2156 (2020)
20. Sun, M., Chen, J.: Dynamics of nonlinear primary oscillator with nonlinear energy sink under harmonic excitation: effects of nonlinear stiffness. Math. Probl. Eng. **2018**, 1–13 (2018)

Mechanical Properties and Fracture Toughness Behavior of Cold Worked AA 5754 Alloy

Wafa Taktak(✉) and Riadh Elleuch

Department of Mechanical Engineering, Laboratory of Systems Electro-Mechanical (LASEM), National School of Engineers of Sfax, University of Sfax, 3000, BP 599-3018 Sfax, Tunisia
taktak.2022.wafa@gmail.com, riadh.elleuch@gnet.tn

Abstract. In this present investigation, aluminum AA5754H111 alloy are cold-worked to different reduction percentage (30%, 50% and 80%) at ambient temperature to establish effect of cold working on mechanical properties and fracture toughness of AA5754H111. The mechanical properties are studied by tensile test. The fracture toughness behavior is investigated by fracture toughness tests according to ASTME 1820 standard utilizing CCP (central cracked panels) specimens. The fracture toughness tests were performed to define the crack-tip opening displacement at resistance curve (CTOD-R) of starting and cold worked AA5754H111 alloy. Experiment results proved that the cold working has a profound effect on the mechanical tensile properties and fracture toughness parameter (resistance at crack initiation $CTOD_{0.2}$) of AA5754H111 aluminum alloy. The accumulation of cold working enhances the work hardening impact, which is an excellent an agreement with the improvement of the tensile strength (ultimate tensile UTS and yield strength YS) and low the fracture toughness behavior (tenacity at crack initiation $CTOD_{0.2}$) and plasticity (ductility). This is attributed to the significantly enhance of the density of dislocations and the severely elongation of the equiaxed grains through the rolling direction. Increase in tensile strength and decrease of fracture toughness parameter modified the fracture from ductile to brittle.

Keywords: AA5754H111 alloy · Mechanical properties · Crack tip Opening Displacement · fracture toughness behavior

1 Introduction

Due to its excellent strength and toughness combination, low densities, good workability and excellent corrosion resistance, the non-heat treatable (5XXX series aluminum alloys) AA 5754 aluminum alloy has been extensively employed in automotive materials such as panels components (Ubertalli G. et al. 2020; Dhara S. et al. 2016; Sazali E.S. et al. 2015; Xia et al. 2014).

Cold worked aluminum alloy sheet is the conventional manufacturing procedure for automotive panels components (Ismail et al. 2016). It is well know that the cold working is a procedure of based deformation at room temperature to generate strain hardening.

A. Akrout et al. (Eds.): ICAV 2022, ACM 22, pp. 342–350, 2023.
https://doi.org/10.1007/978-3-031-34190-8_36

The plastic deformation created during the cold working leads to the modification in the mechanical behavior and fracture toughness of this alloys and appearing the cracks in the automotive panels (Cosham 2001; Madi Y. et al. 2020; Chang et al. 2017).

The cold working of non-heat treatable Al-Mg alloy can be performed not only flat sheets but also enhance the strength of these alloys because of the presence of the strain harding. On the other hand, the ductility of the cold worked alloy were decreased (Wang et al. 2015; Sarkar et al. 2001; Jin et al. 2006; Wowk et al. 2009). Some studies related to the cold working effect on the mechanical behavior of AA5XXX aluminum alloy are reported below.

Jin et al. 2006 studied the cold working effect on tensile properties of AA5754 alloy. They reported that the increasing of the cold working percentage leads to enhance of yield stress (YS) and ultimate tensile stress (UTS). However, the ductility (plasticity) is decreased. A similar effect was shown in the case of AA5052 alloy (Bora et al. 2020).

The cold working also influences the fracture toughness behavior. From the literature review, severely researches are concentrated on the cold working effect on the fracture toughness behavior of steel alloys (Hagiwara et al. 2001; Kim et al. 2018; Madi et al. 2020). They reported that the important reduction of the fracture toughness behavior of steel alloys when the cold working increase. But, the cold working effect of aluminum alloys has been few investigated. In the present study (Tajally et al. 2010), the fracture toughness behavior of cold rolled AA7075 alloy with different cold rolling reduction was studies by use of impact charpy test.

The objective of the present work is to study the cold working effect on tensile properties and fracture toughness behavior of AA5754H111 aluminum alloy by use of the crack resistance curve (CTOD-R curves) test method.

In this paper, we study this effect in further detail by comparing each of the cold worked alloy and non-cold worked alloy.

2 Material and Experimental Methods

The used aluminum alloy in this investigation is AA5754H111 sheet with a thickness of 3 mm. Table 1 illustrates the chemical composition of the investigated alloy.

Table 1. Chemical composition of AA5754-H111 (*wt%*).

Element	Composition (*wt%*)
Al	Balance
Mg	3.6
Mn	0.5
Fe	0.4
Si	0.4
Cr	0.3

(continued)

Table 1. (*continued*)

Element	Composition (*wt%*)
Zn	0.2
Ti	0.15
Cu	0.1

In the current investigation, rectangular plate, samples of the starting alloy were used. The sizes of these plates are (500 mm × 200 mm × 3 mm). The used plates were cold rolled in the rolling direction using a laboratory rolling mill at ambient temperature in the rolling direction. The rectangular plates were cold rolled until achieving the flowing thickness: 2,1 mm (30% reduction), 1,5 mm (50% reduction) and 0,6 mm (80% reduction).

According to NF EN 10002-1 standard depicted in Fig. 1, the plate tensile samples were used to investigate tensile strength and ductility of the starting and cold rolling materials. All the tensile specimens were charged along the rolling direction. All tests were conducted at ambient temperature using the 50 kN LLOYD universal testing machine with a constant cross lead speed of 10 mm/min. During this test, the displacement was measured by an extensometer (gauge length of 25 mm) mounted on the medium of tensile test sample. For each percentage of cold rolling, three samples by direction were used.

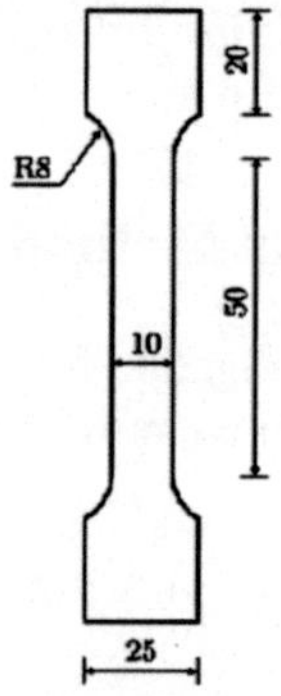

Fig. 1. Tensile test specimen (dimensions in mm).

The fracture toughness test of the starting and cold rolling alloys was carried out at ambient temperature using the 50 kN LLOYD universal testing machine at across lead speed of 1 mm/min. The central cracked panel (CCP) samples were invested to identify the parameters of toughness for starting and cold working alloy (see Fig. 2). The wire cutting machine was employed to processing this specimen in cold working direction. Using fatigue test, the CCP samples were pre-cracked to get initial half crack lengthiness «a» equal to 0,36 W (where W expresses the half width of sample). The resistance curve CTOD-R for the starting and cold worked samples were attained according to ASTME 1820 utilizing the single sample technique (Imad et al. 2003). Three samples were used for each percentage of cold working.

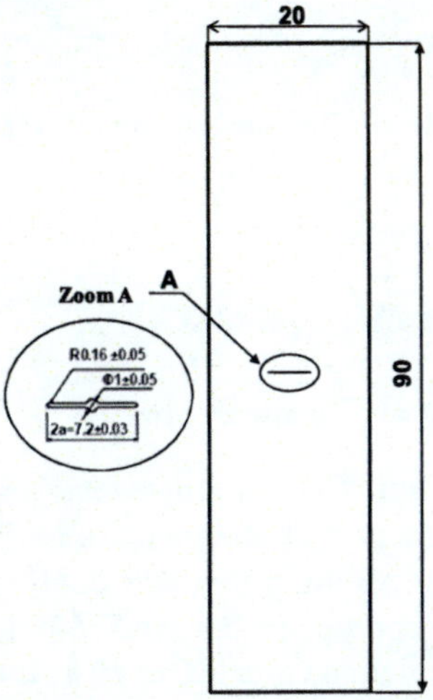

Fig. 2. Geometry of CCP specimen (in mm) used for J-integral measurements.

The Δa value ductile crack extension and the crack tip opining displacement CTOD can be determined through pictures of fracture surface of CCP samples acquired through the high resolution video camera situated in front of the specimen.

The elected sequence of high-resolution video camera pictures depicting the crack growth Δa in AA5754-H111 specimen are displayed in Fig. 3.

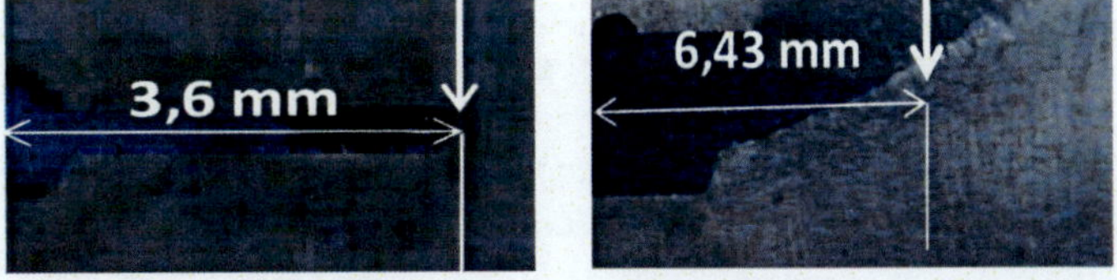

Fig. 3. Pictures exhibiting crack advance in starting α-AA5754H111 alloy.

The experimental value of CTOD corresponds to the crack opening estimated in the initial position of the crack tip (see Fig. 4).

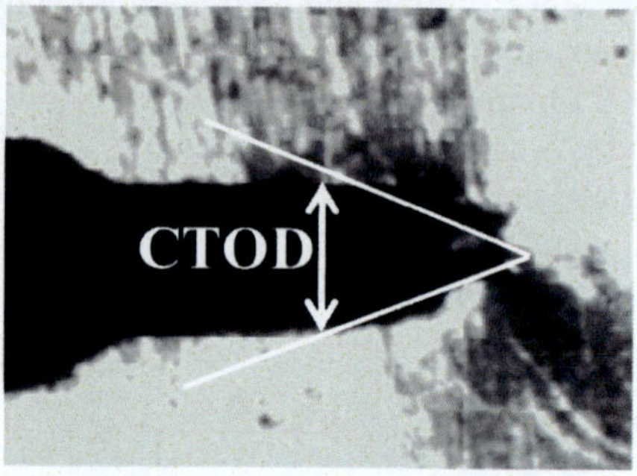

Fig. 4. Determination of COTD.

3 Results and Discussion

3.1 Effect of Cold Working on Tensile Properties

The evolution of the yield strength (YS), ultimate tensile strength (UTS) and elongation (A%) of AA5754H111 specimens with the increase in the cold working reduction are shown in Fig. 5. The present results prove that cold working leads to a significant modification of the tensile properties of the AA5754H111 aluminum alloy. With the increase in the cold rolling reduction from 0% to 80%, the ultimate tensile strength and yield strength of AA5754H111 have consistently improved, whereas the ductility has dropped.

After a cold working reduce of 80%, ultimate tensile strength increases from 206 MPa to 392 MPa and yield strength increases from 100 MPa to 265 MPa, which corresponds to 91% for UTS and 165% for YS, respectively. Al though, the elongation decreases from 16.5% to 1.95% after a cold rolling reduction of 80%, which indicates a considerable reduction of 88%.

The following data reveal that the behavior of AA5457H111 after cold working is marked by higher strength and lower ductility. Accordingly, cold working has a significant effect on the tensile properties of AA5754H111which has also been mentioned in various investigation (Jin et al. 2006; Bora et al. 2020; Kusmono et al. 2021; Mishra et al. 2018; Mhedhbi et al. 2017; Tajally et al. 2010; Kumar et al. 2017). These experimental results can be explained by the appearance of extremely elongated density dislocation and its augmentation created in the materials in the rolling direction during the cold working (Wang et al. 2015; Wowk et al. 2009; Howeyze et al. 2020).

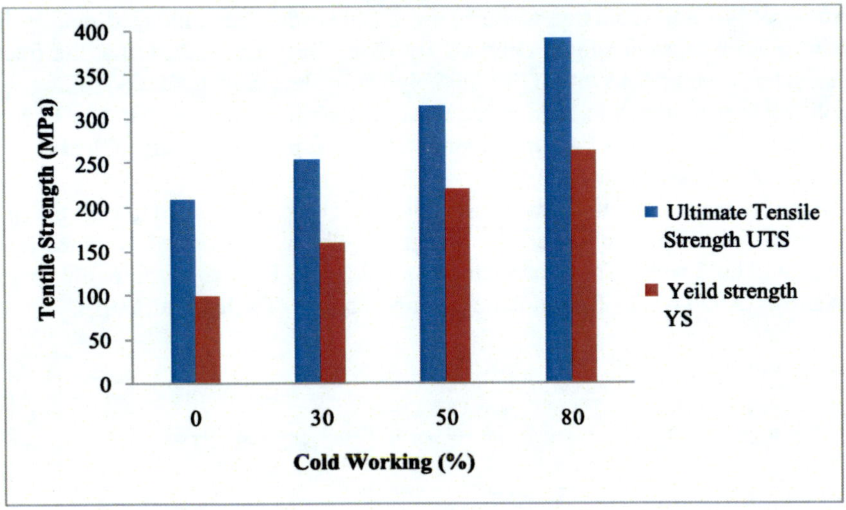

(a)

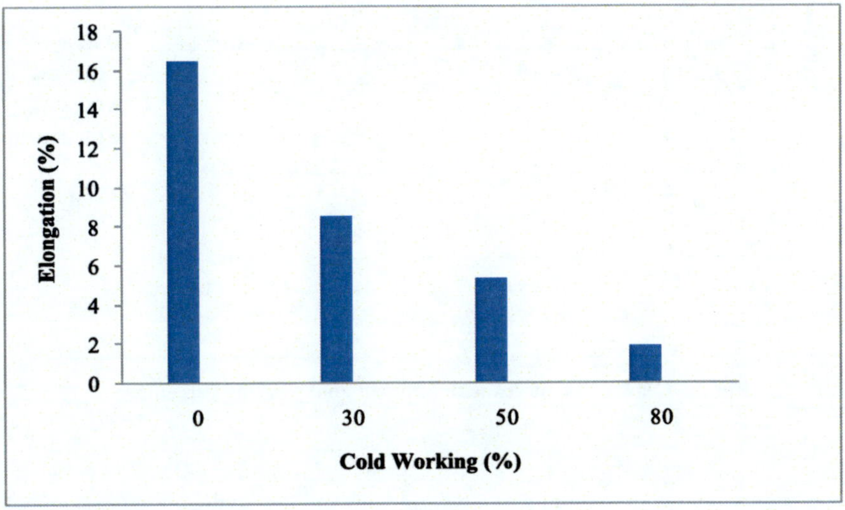

(b)

Fig. 5. The effect of cold working reduction on the (a) Ultimate tensile strength (UTS), and yield strength (YS) and (b) elongation of AA5754H111.

3.2 Effect of Cold Working on Fracture Toughness Behavior

Figure 6 illustrates the experimental CTOD-R curves of AA5754H111 aluminum alloy samples with different cold working reductions. The resistance at crack initiation $CTOD_{0.2}$ is estimated by the cross of CTOD-R curves and the 0.2 mm offset line corresponding to ASTEM 1820 (Frometa et al. 2020). It can be seen that the CTOD-R curves decreased with the rise in the amount of the cold working reduction.

After cold rolling reduction to 80%, the $CTOD_{0.2}$ has been reduced rapidly from 6.3 mm to 1.5 mm, indicating a decrease by 76%. This results show that the fracture toughness behavior of AA5754H111 is notably diminished through the cod rolling. Also, the cold rolling conducts to reduce the fracture toughness behavior in the evaluated AA5754H111. These results are in agreement with the previous results (Oh et al. 2007; Luu et al. 2006; Taktak et al. 2023).

These results reveal that, the weakest resistance curves are related to with the rise of the rate and the instability of the crack propagation which is performed by increasing the percentage of cold work. Thus, the increase in cold work leads to modifier the behavior of the alloy AA5754H111 from ductile tearing to brittle (Tajally et al. 2010).

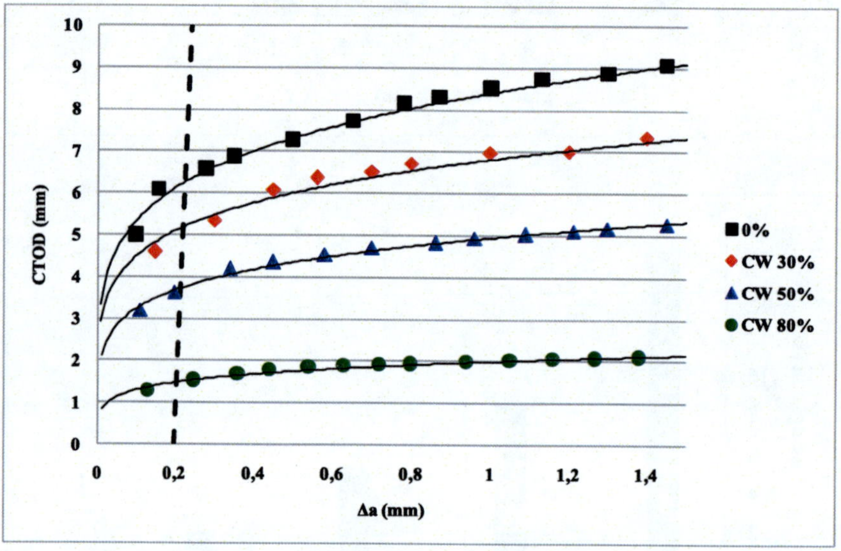

Fig. 6. CTOD- Δa curves with different cold working percentage.

4 Conclusion

The mechanical properties and the fracture toughness behavior of AA 5754H111 alloy under different cold working conditions were investigated and the following conclusions have been levied.

- The considerable in tensile ultimate strength and yield strength with the increase in the percentage of cold working reduction, can be explained by the occurrence of the work hardening phenomenon during cold working. This phenomenon leads to an increase of the dislocation density and a reduction of the mobility of these dislocations.
- The cold worked AA5754H111 alloy with 30% and 50% present a ductile tearing behavior however the cold worked alloy with 80% shows a brittle behavior.

- A notable reduction in values of the fracture toughness parameter (COTD0.2) of AA5754H1111 alloy was mentioned with the increase in the percentage of cold working. This show that the cold working causes the decrease fracture toughness behavior of AA5754H111.
- The cold working has a negative influence in the fracture toughness behavior of AA5754H111. The increase in the cold rolling reduction leads to the lowering of the fracture toughness resistance.
- The cold working has caused a loss of resistance failure and change in mechanism of rupture from ductile to brittle.

References

Ubertalli, G., et al.: High strain rate behavior of aluminum alloy for sheet metal forming processes. Metals **10**(2), 242 (2020). https://doi.org/10.3390/met9101129

Dhara, S., Basak, S., Panda, S.K., Hazra, S., Shollock, B., Dashwood, R.: Formability analysis of pre-strained AA5754-O sheet metal using Yld96 plasticity theory: role of amount and direction of uni-axial pre-strain. J. Manuf. Process. **24**, 270–282 (2016). https://doi.org/10.1016/j.jmapro.2016.09.014

Sazali, E.S., Sahar, M.R., Rohani, M.S.: Optical investigation of erbium doped lead tellurite glass: Judd-Ofelt analysis. Mater. Today Proc. **2**(10), 5241–5245 (2015). https://doi.org/10.1016/j.matpr.2015.05.027

Xia, S.L., Ma, M., Zhang, J.X., Wang, W.X., Liu, W.C.: Effect of heating rate on the microstructure, texture and tensile properties of continuous cast AA5083 aluminum alloy. Mat. Sci. Eng. A **609**, 168–176 (2014). https://doi.org/10.1016/j.msea.2014.05.002

Cosham, A.: A model of pre-strain effects on fracture toughness. J. Off. Mech. Arct. Eng. **123**(4), 182–190 (2001). https://doi.org/10.1115/1.1408613

Madi, Y., Shinohara, Y., Besson, J.: Effect of prestrain on ductility and toughness in a high-strength line pipe steel. Int. J. Fract. **224**(1), 15–29 (2020). https://doi.org/10.1007/s10704-020-00442-6

Chang, L., Zhou, C.Y., He, X.H.: The effects of prestrain and subsequent annealing on tensile properties of CP-Ti. Metals **7**(3), 99 (2017). https://doi.org/10.3390/met7030099

Wang, B., Chen, X.H., Pan, F.S., Mao, J.J., Fang, Y.: Effects of cold rolling and heat treatment on microstructure and mechanical properties of AA 5052 aluminum alloy. Trans. Nonferrous Met. Soc. China **25**(8), 2481–2489 (2015). https://doi.org/10.1016/S1003-6326(15)63866-3

Sarkar, J.K.T.R.G., Kutty, T.R.G., Conlon, K.T., Wilkinson, D.S., Embury, J.D., Lloyd, D.J.: Tensile and bending properties of AA5754 aluminum alloys. Mater. Sci. Eng. A **316**(1–2), 52–59 (2001). https://doi.org/10.1016/S0921-5093(01)01226-6

Wowk, D., Pilkey, K.: Effect of prestrain with a path change on the strain rate sensitivity of AA5754 sheet. Mater. Sci. Eng. A **520**(1–2), 174–178 (2009). https://doi.org/10.1016/j.msea.2009.05.019

Jin, Z., Mallick, P.Q.: Effect of cold work on the tensile and fatigue performance of aluminum alloy 5754. J. Mat. Eng. Perf. **5**, 540–548 (2006). https://doi.org/10.1361/105994906X136052

Bora, C., Salim, U.A.: Effects of cold rolling (CR) and annealing time on microstructure and mechanical properties of AA 5052 aluminum alloy. Metalurgija **59**(4), 485–488 (2020)

Hagiwara, N., Masuda, T., Oguchi, N.: Effects of prestrain on fracture toughness and fatigue-crack growth of line pipe steels. J. Pressure Vessel Technol. **123**(3), 355–361 (2001)

Kim, K., et al.: Improvement of strength and impact toughness for cold-worked austenitic stain-less steels using a surface-cracking technique. Metals **8**(11), 932 (2018)

Tajally, M., Zainul, H., Haji, H.M.: A comparative analysis of tensile and impact-toughness behavior of cold-worked and annealed 7075 aluminum alloy. Int. J. Imp. Eng. **37**(4), 425–432 (2010)

Imad, A., Wilsius, J., Abdelaziz, M.N., Mesmacque, G.: Experiments and numerical approaches to ductile tearing in an 2024-T351 aluminum alloy. Int. J. Mech. Sci. **45**, 1849–1861 (2003)

Kusmono, K., Bora, C., Salim, U.A.: Effects of cold rolling and annealing time on fatigue resistance of AA5052 aluminum alloy. Int. J. Eng. **34**(9), 2189–2197 (2021)

Mishra, V.D., Rao, B.C., Murthy, A.: Enhancement of mechanical properties by cold-rolling of Al6061. Mater. Today Proc. **5**(2), 8263–8270 (2018)

Mhedhbi, M., Khlif, M., Bradai, C.: Investigations of microstructural and mechanical properties evolution of AA1050 alloy sheets deformed by cold-rolling process and heat treatment annealing. J. Mater. Environ. Sci. **8**(23), 2967 (2017)

Kumar, C.N., Babu, R.N., Charyulu, T.N.A.: Comparative analysis of cold rolling on mechanical properties of aluminum. Int. J. Eng. Res. Technol. **6**(7), 353–359 (2017)

Frómeta, D.: Identification of fracture toughness parameters to understand the fracture resistance of advanced high strength sheet steels. Eng. Fract. Mech. **229**, 106949 (2020). https://doi.org/10.1016/j.engfracmech.2020.106949

Oh, C.K., Kim, Y.J., Baek, J.H., Kim, Y.P., Kim, W.: A phenomenological model of ductile fracture for API X65 steel. Int. J. Mech. Sci. **49**(12), 1399–1412 (2007)

Luu, T.: Déchirure Ductile Des Aciers A Haute Resistance Pour Gazoducs (X100). PhD thesis, Ecole nationale supérieure de Mines Paris (2006)

Taktak, W., Elleuch, R.: Cold working effect on the fracture toughness properties of AA1050H16 aluminum alloy. In: Walha, L., et al. (eds.) Design and Modeling of Mechanical Systems - V. LNME, pp. 894–900. Springer, Cham (2023). https://doi.org/10.1007/978-3-031-14615-2_100

Seismic Behavior of a Building Structure Reinforced with Composite Trusses

Sofiene Helaili[1,2](✉)

[1] ISTEUB, University of Carthage, Charguia 2, Carthage, Tunisia
sofiene.helaili@gmail.com
[2] LASMAP, Tunisia Polytechnic School, University of Carthage, BP 743, 2078 La Marsa, Tunisia

Abstract. Seismic design is a vital process of structural analysis while designing a building, which is subjected to earthquake ground motions such that the facility continues to function and serve its purpose even after an Earthquake. The seismic behavior of buildings is studied during the design and initial calculation of the building, even before its construction. The evolution of normative requirements and restrictions, the development of new construction materials, and the development of new calculation methods require additional studies on the structural behavior of the building. This paper studies the seismic behavior of a building in its initial state and after reinforcement with composite trusses. The building studied is a common reinforced concrete building composed of a structure of columns, beams, ribbed slabs, and solid slabs. The foundations of the building are assimilated to be fixed to the ground. The material of the reinforcements is varied and studied. The vibratory behavior of the building is improved by increasing the modal frequencies. Higher natural frequency response reduces the amplitude of the building's vibrations during a seismic event, making it more resistant to damage or collapse.

Keywords: Seismic behavior · reinforcement · composites · structures vibration

1 Introduction

The building structure design is based on standards, including specific criteria to be verified (American Concrete Institute, 2015; Foliente, 2000). Among these standards and criteria are seismic verifications (Sezen, 2002). Different seismic calculation methods exist (Datta, 2010): equivalent lateral forces analysis, modal and spectral analysis, and temporal dynamic analysis. Each method has its advantages and disadvantages. For example, in the Algerian seismic regulations (RPA) (Belazougui, 2017), two methods are recommended: the modal method coupled with equivalent static loads and the spectral method. The Eurocodes 08 (Lemaire, 2018) code suggests four methods: the equivalent lateral forces, the modal/spectral analysis, the direct temporal analysis, and the pushover method.

The verifications define a series of criteria to be satisfied by the constructions, such as the resistance of the elements and their connections, the ductility of sections and

A. Akrout et al. (Eds.): ICAV 2022, ACM 22, pp. 351–358, 2023.
https://doi.org/10.1007/978-3-031-34190-8_37

elements, the overall balance and stability (P-Λ effect) (Chian and Qin, 2020), the floor's resistance, the stability of the foundations, the strength and stability of secondary elements and non-structural elements and finally, the width of joints and relative floor deformations (Wang et al., 2022).

The seismic study of a structure depends on several factors. Seismic zones are constructed based on past seismic events (Di Ludovico et al., 2022). The importance of the work defines the degrees of security applied. The site of the work and its shape and regularity is also decisive. Several reinforcement techniques can be used at early design stages or for building rehabilitation. Steel structures, cables (Wang et al., 2022), composite plates (Aiello et al., 2009; Bakis et al., 2002), or composite structures can be used. Several research works studied the properties of synthetic composites or natural fiber-based composites (Arrakhiz et al., 2012; Helaili and Chafra, 2023) to prepare and facilitate the use of these properties in actual structures (Helaili et al., 2023a, 2023c, 2023b).

This paper studies the modal behavior of a real building in Tunisia. The building is not built on a seismic area but will serve as a case study. The building without any reinforcement is analyzed first. Then, a steel mesh reinforcement is added to the frame of the building. The natural modes and frequencies are analyzed. Finally, a reinforcement based on composites is applied. The three states are compared.

2 Materials and Methods

A multistage approach (see Fig. 1.) based on ANSYS finite element code is used in this study. The processing starts with a static analysis and ends with a prestressed and constrained modal analysis. The 3D finite elements ANSYS modal analysis results are compared to the Robot Structural Analysis code based on 1D and 2D elements.

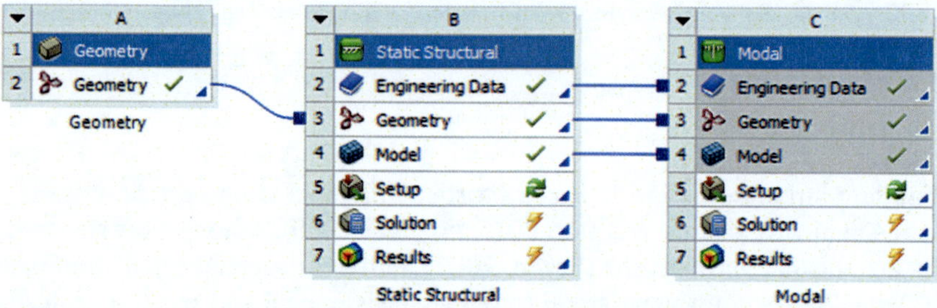

Fig. 1. Multistage ANSYS finite element process

The three used materials in models are concrete, steel, and composite. The property of each material is indicated in Table 1.

Table 1. Materials properties

Parameters	Concrete C30/37	Steel	Composite
Density (t/m3)	2.5	7.8	1.6
Young's Modulus (GPa)	34	210	150
Poisson's ratio	0.2	0.3	0.43
Damping ratio	0.04	0.04	0.4

The studied building uses a ground floor and four floors for residential use. The building is irregular in shape. The structure of the building is made up of columns, beams, floors, footings, and stringers. The geometry of the building is indicated in Fig. 2.

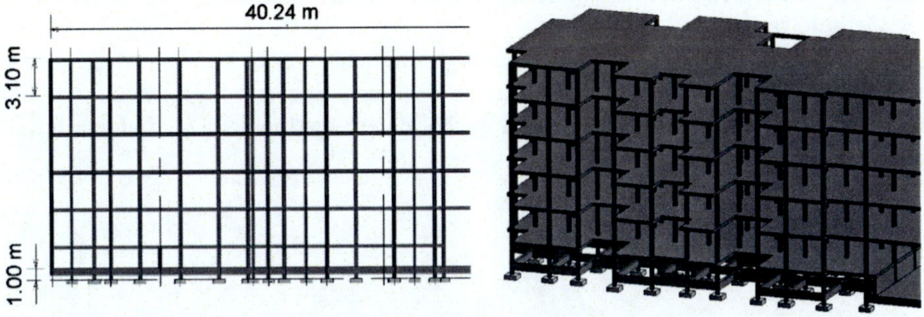

Fig. 2. Building geometry

The building is subject to its weight and operating loads. The truss structure used for reinforcement is presented in Fig. 3.

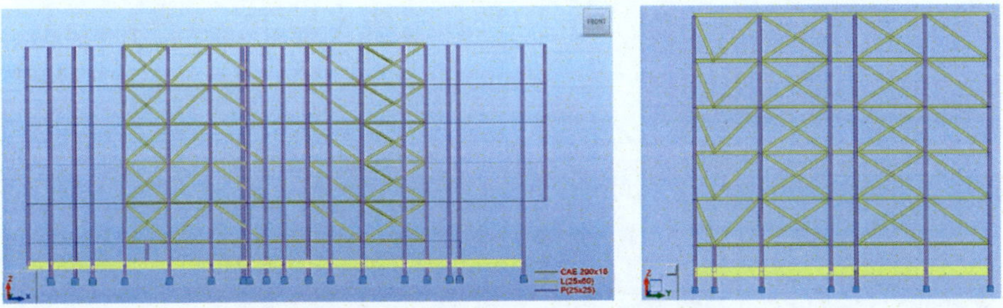

Fig. 3. Trusses structure reinforcement

3 Results

When a building is reinforced with trusses, it modifies the seismic response of the structure by decreasing its natural frequency due to the additional stiffness of the trusses. The seismic interpretation of this reduction in natural frequency is that it enables the building to resist low-frequency seismic vibrations that can cause excessive deformations. However, it is important to note that the decrease in natural frequency can also affect the building's response to high-frequency vibrations, which may require additional design adjustments.

There is no specific seismic frequency considered dangerous. The danger of a seismic event is determined by various factors, including the earthquake's magnitude, the depth at which it occurs, and the geology of the affected region. Earthquakes with magnitudes above 6.0 on the Richter scale are typically considered dangerous as they can cause significant damage to buildings, infrastructure, and people. However, lower-magnitude earthquakes can also be dangerous if they occur near densely populated areas or vulnerable structures.

It is important to note that the frequency of an earthquake does not determine its danger. A low-frequency earthquake can be as destructive as a high-frequency earthquake if it is powerful enough. The factors that determine the danger of an earthquake

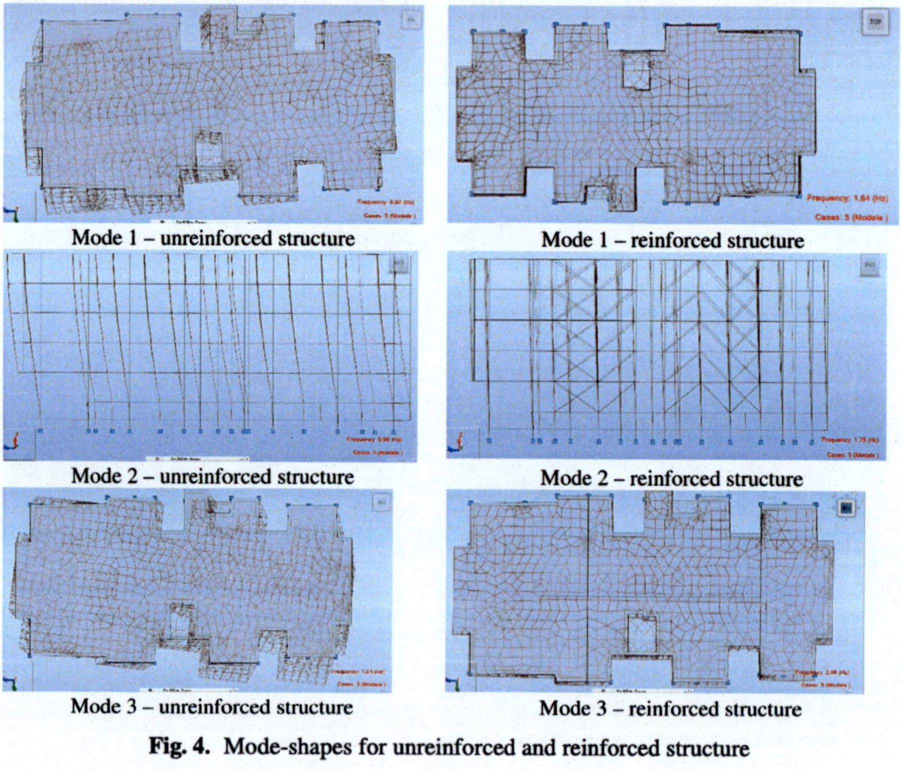

Fig. 4. Mode-shapes for unreinforced and reinforced structure

are its magnitude, depth, proximity to urban areas, and vulnerability of structures and infrastructure.

The mode shapes for unreinforced and reinforced structures resulting from the numerical simulations are shown in Fig. 4.

The frequencies of the unreinforced structure, the reinforced structure with steel trusses, and the reinforced structure with composite trusses are indicated in Table 2 and Table 3. The first twelve modes are extracted. These results are presented in Fig. 4.

Table 2. Modal analysis results for Unreinforced structure

	Unreinforced structure			
Mode	Frequency [Hz]	UX Mass [%]	UY Mass [%]	Total Mass [t]
1	0,92	1,65	40	2840,46
2	0,96	69,83	42,12	2840,46
3	1,01	70,36	70,16	2840,46
4	2,8	70,5	74,65	2840,46
5	2,92	78,37	74,87	2840,46
6	3,04	78,46	78,07	2840,46
7	4,72	78,51	79,69	2840,46
8	4,88	81,13	79,78	2840,46
9	5,09	81,17	80,9	2840,46
10	6,54	81,2	81,69	2840,46
11	6,68	82,41	81,74	2840,46
12	7,04	82,42	82,32	2840,46

When a building is strengthened, it can increase the structure's natural frequencies. This means the building is stiffer and responds more quickly to dynamic loads, such as earthquakes or wind. The effect is beneficial as the higher frequency response reduces the amplitude of the building's vibrations during a seismic event, making it less vulnerable to damage or collapse. This phenomenon is explained by the fact that the addition of reinforcement elements, such as trusses, braces, or shear walls (Berman, 2011), increases the stiffness of the building. This increase in stiffness leads to a shift in the structure's natural frequencies toward higher values (Federal Emergency Management Agency, U. S. Department of Homeland Security, 2013).

Table 3. Modal analysis results for reinforced structure

Steel reinforced structure					Composite reinforced structure			
Mode	Frequency [Hz]	UX Mass [%]	UY Mass [%]	Total Mass [t]	Frequency [Hz]	UX Mass [%]	UY Mass [%]	Total Mass [t]
1	1,2	15,04	9,88	2849,55	1,64	13,8	4,44	2851,3
2	1,26	72,01	13,27	2849,55	1,75	75	5	2851,3
3	1,47	72,05	72,66	2849,55	2,06	75,06	76,52	2851,3
4	3,62	73,55	73,81	2849,55	4,92	76,81	77,27	2851,3
5	3,8	80,74	74,15	2849,55	5,23	85,61	77,38	2851,3
6	4,4	80,75	81,95	2849,55	6,07	85,61	87,67	2851,3
7	5,77	80,75	81,96	2849,55	8,15	85,92	88,33	2851,3
8	5,77	80,75	81,96	2849,55	8,81	89,22	88,39	2851,3
9	5,77	80,75	81,96	2849,55	9,64	89,22	88,39	2851,3
10	5,77	80,75	81,96	2849,55	10,32	89,27	88,39	2851,3
11	5,77	80,75	81,96	2849,55	10,52	89,28	89,12	2851,3
12	6,01	81,26	82,38	2849,55	10,7	89,32	90,81	2851,3

4 Discussion

Several studies have shown that strengthening a building can significantly improve its seismic performance (Caliò and Marletta, 2005), especially for older buildings not originally designed to withstand seismic loads. The use of modern materials, such as fiber-reinforced composites, has also allowed for more efficient and cost-effective strengthening solutions. Most earthquakes in Tunisia are low to moderate magnitude (less than 5.5 on the Richter scale) and have relatively low frequencies, usually less than 5 Hz (Ben Ayed, 2020).

Based on the results in the previous section, the predominant modes that are determinants in seismic building behavior are the first three modes. The first three modes involve 70% of the total Mass of the building. The total Mass of the building involved in vibrations in X and Y directions have not significantly varied when using reinforcements, so we can conclude that the reinforcing strategies did not change the overall behavior of the building, and the weight of the reinforcing structures is transmitted properly to the foundations in the Z direction.

Concerning the modal frequencies, the sensitivity of the building to seismic vibrations has been improved by the reinforcements. The first mode for the unreinforced building is low and is equal to 0.92 Hz. The first mode is improved when reinforcing with steel and increased to 1.2 Hz. The best improvement is when composite trusses are used. The value of the first mode in the composite reinforcement case is equal to 1.64 Hz.

5 Conclusion

This paper studied the seismic behavior of a multistory building. The modal analysis is used. The reinforcing strategy using a trusses structure made of composites improves the first modal frequency from 0.92 Hz to 1.64 Hz. The vibrating Mass of the building in transversal directions remains unchanged. This type of reinforcement can be made for existing structures for their rehabilitation.

References

Aiello, M.A., Micelli, F., Valente, L.: FRP confinement of square masonry columns. J. Compos. Constr. **13**, 148–158 (2009). https://doi.org/10.1061/(ASCE)1090-0268(2009)13:2(148)

American Concrete Institute: Guide for the design and construction of structural concrete reinforced with fiber-reinforced polymer (FRP) bars, 1st printing. edn, ACI report. American Concrete Institute, Farmington Hills, MI (2015)

Arrakhiz, F.Z., Elachaby, M., Bouhfid, R., Vaudreuil, S., Essassi, M., Qaiss, A.: Mechanical and thermal properties of polypropylene reinforced with Alfa fiber under different chemical treatment. Mater. Des. **35**, 318–322 (2012). https://doi.org/10.1016/j.matdes.2011.09.023

Bakis, C.E., et al.: Fiber-reinforced polymer composites for construction—state-of-the-art review. J. Compos. Constr. **6**, 73–87 (2002). https://doi.org/10.1061/(ASCE)1090-0268(2002)6:2(73)

Belazougui, M.: Algerian seismic building code: main features of the new draft RPA 2015. In: 16th World Conference on Earthquake Engineering, 16WCEE. pp. 9–13 (2017)

Ben Ayed, N.: Les séismes en Tunisie. La Revue de l'énergie **71**, 619–624 (2020)

Berman, J.W.: Seismic behavior of code designed steel plate shear walls. Eng. Struct. **33**, 230–244 (2011). https://doi.org/10.1016/j.engstruct.2010.10.015

Caliò, I., Marletta, M.: Seismic performance of a reinforced concrete building not designed to withstand earthquake loading. In: Maugeri, M. (ed.) WIT Transactions on State of the Art in Science and Engineering. WIT Press, pp. 289–309 (2005). https://doi.org/10.2495/1-84564-004-7/16

Chian, S.C., Qin, C.: Seismic slope stability with discretization-based kinematic analysis. In: Geo-Congress 2020. Presented at the Geo-Congress 2020, pp. 284–294. American Society of Civil Engineers, Minneapolis, Minnesota (2020). https://doi.org/10.1061/9780784482810.031

Datta, T.K.: Seismic Analysis of Structures. John Wiley & Sons (2010)

Di Ludovico, M., et al.: Fragility curves of Italian school buildings: derivation from L'Aquila 2009 earthquake damage via observational and heuristic approaches. Bull. Earthquake Eng. **21**, 397–432 (2022). https://doi.org/10.1007/s10518-022-01535-4

Federal Emergency Management Agency, U. S. Department of Homeland Security: Risk Management Series Publication: Design Guide for Improving School Safety in Earthquakes, Floods, and High Winds (FEMA P-424/December 2010) (2013)

Foliente, G.C.: Developments in performance-based building codes and standards. For. Prod. J. **50**, 12 (2000)

Helaili, S., Bouajila, S., Kaddami, H., Chafra, M.: Turbine blade made of natural fiber composite structural and vibrational behavior. In: Walha, L., et al. (eds.) Design and Modeling of Mechanical Systems - V: Proceedings of the 9th Conference on Design and Modeling of Mechanical Systems, CMSM'2021, December 20-22, 2021, Hammamet, Tunisia, pp. 743–749. Springer International Publishing, Cham (2023). https://doi.org/10.1007/978-3-031-14615-2_83

Helaili, S., Chafra, M.: Identification of mechanical properties of a braided alfa stem (Stipa tenacissima L.) by an inverse method. J. Compos. Mater. **57**, 443–450 (2023). https://doi.org/10.1177/00219983221147685

Helaili, S., Guizani, A., Najar, F., Chafra, M.: Novel rebar made from epoxy and braided natural alfa fibers. In: Walha, L., et al. (eds.) Design and Modeling of Mechanical Systems - V: Proceedings of the 9th Conference on Design and Modeling of Mechanical Systems, CMSM'2021, December 20-22, 2021, Hammamet, Tunisia, pp. 188–194. Springer International Publishing, Cham (2023). https://doi.org/10.1007/978-3-031-14615-2_22

Helaili, S., Rezgui, T., Guizani, A., Najjar, F.: Reinforcement of load-bearing structural elements by curved composite plates made from braided natural fibers. In: Walha, L., et al. (eds.) Design and Modeling of Mechanical Systems - V: Proceedings of the 9th Conference on Design and Modeling of Mechanical Systems, CMSM'2021, December 20–22, 2021, Hammamet, Tunisia, pp. 171–178. Springer International Publishing, Cham (2023). https://doi.org/10.1007/978-3-031-14615-2_20

Lemaire, J.: Évaluation de la vulnérabilité sismique du bâti existant selon l'Eurocode : Essai méthodologique et application au cas de Mulhouse – Bâle (2018)

Sezen, H.: Seismic Behavior and Modeling of Reinforced Concrete Building Columns. University of California, Berkeley (2002)

Wang, X., Xie, C., Jia, Z., Vasdravellis, G.: Seismic behaviour of post-tensioned beam-to-column connection using slender energy-dissipating rectangles. Eng. Struct. **250**, 113444 (2022). https://doi.org/10.1016/j.engstruct.2021.113444

Multi-pass Optimization of a Face Milling Operation for Energy, Time, Cost and Surface Roughness Saving

Anoire Benjdidia(✉), Taissir Hentati, Mohamed Taoufik Khabou, and Mohamed Haddar

Laboratory Mechanics, Modeling and Production, National Engineering School of Sfax (ENIS), 1173-3038 Sfax, BP, Tunisia
benjdidia.anoire@gmail.com, mohamed.haddar@enis.rnu

Abstract. Reducing energy consumption during the cutting processes is the key to an efficient manufacturing process. The best way to achieve that aim is to select correctly the cutting parameters, which can enhance the quality and efficiency of production. This study proposes an optimization for four functions such as energy consumption, surface roughness, cutting time and cost production to achieve sustainability. Several studies are taken those problems but they don't take account the dynamic behavior of cutting force and naturally cutting energy. The innovation of this study consists on taking account of the variable chip thickness. The PSO algorithm is used and the study is performed for different pass numbers which can be beneficial in terms of time, energy and power for a machining range. In this paper four mono-objective optimization is performed taking account the consumed energy, the cutting time, the cost production and the surface roughness. After that, a mono-normalized objective function is elaborated for a single pass face milling operation. And finally the multi-pass case is considered. Results corresponding to each study are presented.

Keywords: optimization · multi-pass face milling · dynamic behavior · energy consumption

1 Introduction

The manufacturing sector seeks 30% of the global energy consumption, as proven by (Mushtaq et al. 2020). According to statistics data related to Chinese China energy (2018), the manufacturing industry consumed 55.65% of total energy consumption. Wang et al. (2018a) proved that optimizing the cutting tool path during face milling case is realized thanks to a compromise between cutting power, milling time, and surface integrity. Another work was performed by (Luan et al. 2018c) using GRA (Rey relational grade) and 3D surface plot methods to obtain the optimal values of radial depth of cut and cutting speed to optimize cutting energy, surface roughness and the tool wear. Wang et al. (2018) established a multi-objective formulation for face milling in order to reduce the production costs and the consumed energy. Furthermore, the researchers

A. Akrout et al. (Eds.): ICAV 2022, ACM 22, pp. 359–366, 2023.
https://doi.org/10.1007/978-3-031-34190-8_38

adopted an evolutionary strategy approach to identify optimal cutting conditions. The resolution of the optimization problem is illustrated by using the Genetic Algorithm (GA). Nguyen et al. (2020) optimized the machining inputs, such as feed, depth of cut, rotational speed, and tool radius. The reason to do so is to enhance the power factor and energy efficiency and improve the surface roughness for vertical milling. The adaptive simulated annealing algorithm was used to resolve the problem. Results show that the cutting parameters can, indeed, alter the mailing machines. The proposed model achieved in the latter study was used to reduce the surface roughness by 39.18%, while the improvements in power factor and energy efficiency are around 26.47% and 22.61%, respectively. Recently, the machining center processing system energy flow characteristics are analyzed by (Xiao et al. 2021). A multi-objective optimization model was proposed regarding the constraints machine tool performance constraints and tool life in the machining process. A weight coefficient is used to facilitate the solution to convert it into a single objective optimization model. The particle swarm optimization combined with NSGA-II are applied to solve the model.

It is well-known that when the tooth acts on the part, it causes dynamic chip thickness. If this dynamic nature is neglected during the removing process, the estimation of the consumed energy is not accurate, and consequently, the optimization is not performant. Thus, it is primordial to consider the dynamic nature of the cutting forces. This study aims to realize a reduction of the consumed energy considering the non-linearity of the cutting force and based on a mono objective model in the multi-pass operation of the milling process. The objective takes into account the variable energy, the cutting time, the surface roughness and the machining cost. In this paper, the objective function is formulated as the sum of four normalized objective functions: cutting time, variable cutting energy, machining cost and surface roughness.

2 Objective Functions

2.1 Cutting Time

The quantity of time needed to cut material is presented in a previous work of (Ben Jdidia et al. 2020).

2.2 Cutting Variable Energy

During a milling process, the energy is variable due to the non-linearity of the cutting forces. Energy consumption mathematical formulation is as follow:

$$\begin{aligned} f_2 = E_{machining} &= E_{start\ up} + P_{set\ up} \times t_{set\ up} + P_{tool} \times t_{tool} \\ &+ n_{pass} \times \left(E_{cutting} + E_{air}\right) \end{aligned} \tag{1}$$

As mentioned in the papers of (Ben Hassen et al. 2023) a numerical model of an axis feed and spindle rotation powers are proposed. In order to improve the cutting power prediction, the dynamic behavior of the cutting forces using a dynamic model of the work piece and the tool are considered. The consumed cutting power by the spindle system is deduced from the numerical values of the tangential component of cutting force F_t and the cutting velocity V_c. The tangential cutting force is written using the tool displacements in the instant t and t-τ as presented in (Ben Hassen et al. 2023). The consumed cutting power by the axis feed is deduced from the numerical values of the cutting forces and the axis feed characteristics. The variable axis feed mechanical cutting power is obtained by multiplying the variable cutting torque $T_{cutting}$ by the angular velocity ω. In order to calculate the tangential component of the cutting force F_t, the equation of motion of the machine tool is written and then resolved using the FEM method.

2.3 Surface Quality

The surface quality is presented by the roughness included in our work to minimize it regarding the other objective function. It is formulated in a previous work of (Ben Jdidia et al. 2020).

2.4 Machining Cost

By summing machine, tool and energy costs, the production cost can be calculated and written in a previous work of (Ben Jdidia et al. 2020).

3 Constraints

In our model, three constrains are taken into account as presented in a previous work of (Ben Jdidia et al. 2020): the first one is related to cutting power, the second one is related to the cutting power and the third one is related to the material.

4 Mathematical Formulation

In this investigation, the goal is to achieve the minimum value of the variable cutting energy regarding, at the same time, the cutting time, the machining cost and the surface quality during a multi-pass face milling operation. Thus, the model considers four normalized objective functions (, the minimum time denoted f_1*, the minimum energy denoted,f_2*, the minimum surface roughness denotedf_3* and the minimum cost denoted f_4*). Three decision variables are considered, such as rotational speed Ω, feed per tooth f_z and axial depth of cut a_p. The obtained analytical formulation is described as follows:

$$\begin{cases} \min(F) = \dfrac{f_1}{f_1^*} + \dfrac{f_2}{f_2^*} + \dfrac{f_3}{f_3^*} + \dfrac{f_4}{f_4^*}. \\ s.c: \begin{cases} g_1 \leq f_{\max} \\ g_2 \leq P_{\max} \\ g_3 \leq \tau_{\max} \end{cases} \end{cases} \quad (2)$$

To obtain accurate results of optimization, some limit will be included in our formulation as following:

$$\begin{cases} \Omega_{\min} \leq \Omega \leq \Omega_{\max} \\ f_{z\ \min} \leq f_z \leq f_{z\max} \\ a_{p\ \min} \leq a_p \leq a_{p\ \max}. \end{cases} \quad (3)$$

5 Comparison Between Static and Dynamic Cutting Energy Model

To show the potentialities of involving variable cutting force in strengthening the energy evaluation and the optimization result accuracy, a comparison between variable and static tangential cutting force is performed. To calculate the static model of the cutting force, Alberteli's (Alberteli et al. 2016) model for a face milling operation is used where the cutting forces are modeled without considering the dynamic behavior of the cutting force. From Fig. 1, it is clear that the evaluation of the tangential force with consideration of its dynamic behavior gives more accurate estimation than the others which neglect it. Thus, less accurate energy estimation is obtained (Fig. 2). Indeed the error between dynamic axis feed energy and static axis feed energy is equal to 18.66%. Furthermore, the error between dynamic spindle rotation energy and static spindle rotation energy is equal to 17.76%. This comparison proves the role of considering the variable load caused by the feed and tangential forces in strengthening axis feed and spindle rotation energy consumption estimation and in developing a significant opportunity for obtaining robust energy optimization.

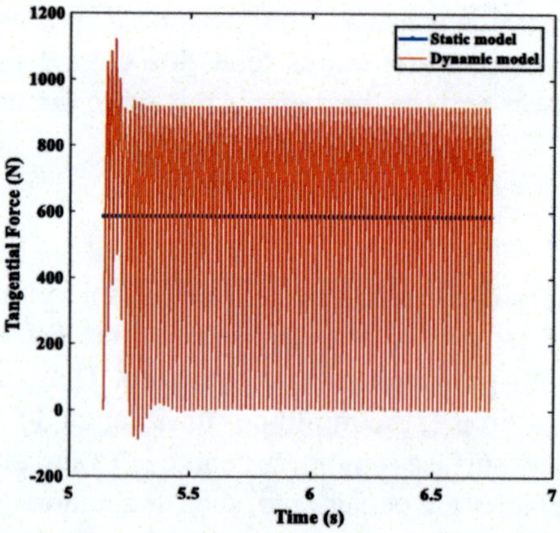

Fig. 1. Comparison between dynamic and static tangential.

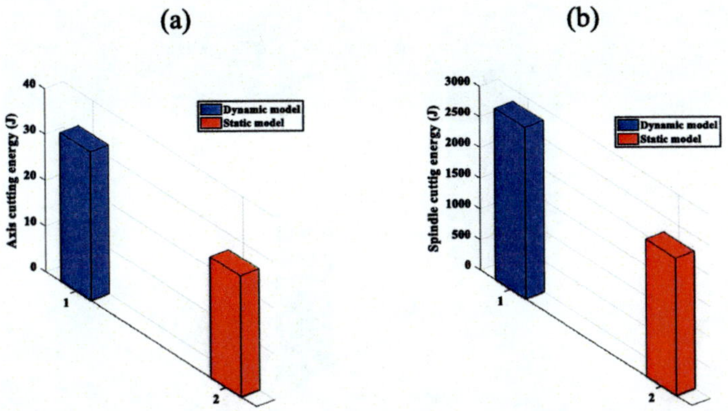

Fig. 2. Comparison between dynamic and static cutting energy for axis feed (a) and spindle rotation system (b)

6 Results and Discussions

As a primarily work, two optimizations problems are resolved to calculate the minimum cutting energy consumed by the cutting system formed by the spindle rotation and the axis feed. The first one is regarding the dynamic behavior of the cutting force and the second one is regarding the static behavior of the cutting force. The parameters used during the simulation are described in Table 1.

Table 1. Simulation parameters

Parameters	Values
Length (mm)	150
Spindle speed(tr/min)	$[3.978 \times 10^2 \ 2.387 \times 10^3]$
Feed per tooth (mm/tooth)	[0.1 0.6]
Axial depth of cut (mm)	[1 4]

Results show that the optimization with the dynamic behavior is better than the optimization with the static one as shown by the following Fig. 3. The error between dynamic optimized energy and static optimized energy is equal to 20.15%. Indeed, these results are explained by the ability of the dynamic behavior to better estimate the quantity of the cutting energy as shown in the above section. So, the importance of considering dynamic behavior of the cutting is needed because it's well known that the better estimation lead to a better optimization.

The dynamic behavior of the cutting energy is considered in our work. As a first step, four mono objective problems are resolved to calculate the minimum cutting time f_1^* (model a), the minimum cutting energy f_2^* (model b), the minimum surface roughness f_3^* (model c), and the minimum cutting cost f_4^* (model e). The obtained results are presented in Table 2.

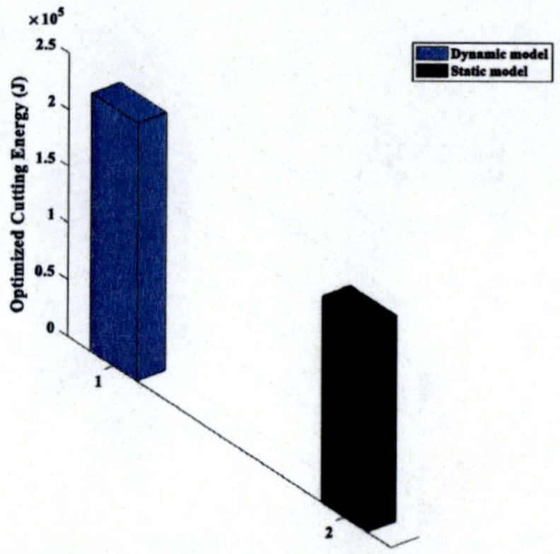

Fig. 3. Comparison between dynamic and cutting energy optimization.

Table 2. Results of each function

Model (a)	Model (b)	Model (c)	Model (e)
$f_1^* = 7.27$	$f_2^* = 3.93$	$f_3^* = 1.82\ 10^5$	$f_4^* = 0.79$

It's clear from the Table 2 that the optimization results of each model conduct to different values of minimum time, minimum energy, minimum cost and minimum roughness. Consequently, a global model to optimize at the same time these four objective functions is needed to ensure a balance between them during face milling operation performed with multi-pass. The normalized mono-objective model is resolved using PSO algorithm. Results show a balance between each function (minimum cutting time, minimum cutting energy, minimum roughness and minimum machining cost) as shown by the next Table 3. One can conclude that the proposed global model is efficient to optimize at the same time the four objective functions during face milling operation performed with multi-pass.

Table 3. Global model optimization results

Min(F)	f_1	f_2	f_3	f_4
4.31	7.16	3.77	$1.74\ 10^5$	0.75

To show the importance of the global proposed model during face milling operation performed with multi-pass, a second comparison is performed with a recent work elaborated by (Ben Jdidia et al. 2020) for the case of face milling operation performed with a single pass. The comparison results are presented in the next Table 4.

The optimization based on global model for the case during face milling operation performed with multi-pass show that the machine tool consumes a minimum quantity of energy equal to $2.13\ 10^4$ (J) and requires a minimum time equal to 37.20 (s) to

Table 4. Comparison between mono-pass and multi-pass face milling operation

	f_3	$E_{cutting} + E_{air}$	f_1	$t_{cuttig} + t_{air}$
Single-pass	1.75 105	6.25 10^3	450.6	27.57
Multi-pass	1.74 10^5	2.13 10^4	429.6	37.20

remove a thickness of 6 mm. However, the optimization results for face milling operation performed with a single pass show that the machine tool consumes a minimum quantity of energy equal to 6.25 10^3 (J) and needs a minimum quantity of time equal to 27.59 (s) to remove just a thickness equal to 1 mm. Thus, to remove a thickness equal to 6mm the machine tool consumes a quantity of energy and time multiply by 6 comparing with the ones needed to remove just a thickness equal to 1 mm. The obtained results are described in Table 5 which led to conclude that our proposed model during multi-pass case is more accurate and robust than the one proposed in the single case. That's why, the number of pass must be included during the optimization of a multi-pass operation.

Table 5. The error between single and multi pass face milling operation

	$E_{cutting} + E_{air}$	$t_{cutting} + t_{air}$
(Single-pass) × 6	3.75 10^4	165.42
Multi-pass	2.13 10^4	37.20
Error (%)	43.2	77.53

7 Conclusion

This work aimed to elaborate a mono objective optimization for face milling operation performed with multi-pass based on PSO algorithm. The first originality of this work was that the optimization was performed using a dynamic model of the cutting energy. A first comparison elaborated between the results obtained by the global model with ones obtained from each objective function proved that to ameliorate the optimization results, it's important to consider all the objective functions, at the same time, in a global model. A second comparison performed between the result obtained from the single and multi-pass cases showed that the last one ensures the best optimization than the first one. As perspective we propose to compare the PSO algorithm results with GA.

References

Albertelli, P., Keshari, A., Matta, A.: Energy oriented multi cutting parameter optimization in face milling. J. Clean. Prod. **137**, 1602–1618 (2016)

Ben Hassen, D., Ben Jdidia, A., Hentati, T., Abbes, M.S., Haddar, M.: A novel intelligent reasoning method to estimate the cutting system energy consumption for a sustainable manufacturing. J. Chin. Inst. Eng. **46**(1), 74–80 (2023)

Ben Jdidia, A., Hentati, T., Bellacicco, A., Khabou, M. T., Rivier, A., Haddar, M.: Optimizing cutting conditions in single pass face milling for minimum cutting energy, time, cost, and surface roughness. In: Chaari, F., Barkallah, M., Bouguecha, A., Zouari, B., Khabou, M. T., Kchaou, M., Haddar, M. (eds.) Advances in Materials, Mechanics and Manufacturing. LNME, pp. 214–222. Springer, Cham (2020). https://doi.org/10.1007/978-3-030-24247-3_24

Department of Energy Statistics, National Bureau of Statistics.,2018. China energy statistical yearbook 2018. Beijing: China Statistics Press

Jang, D.-Y., Jung, J., Seok, J.: Modeling and parameter optimization for cutting energy reduction in MQL milling process. Int. J. Precis. Eng. Manuf.-Green Technol. **3**(1), 5–12 (2016). https://doi.org/10.1007/s40684-016-0001-y

Ben Jdidia, A., Hentati, T., Bellacicco, A., Khabou, M. T., Rivier, A., Haddar, M. : "Optimisation des conditions de coupe en surfaçage en une seule passe pour une énergie de coupe, un temps, un coût et une rugosité de surface minimum". *Dans Advances in Materials, Mechanics and Manufacturing* (pp. 214–222). Springer, Cham 2020

Luan, X., Zhang, S., Li, J., Mendis, G., Zhao, F., Sutherland, J.W. : "Trade-off analysis of tool wear, machining quality and energy efficiency of alloy cast iron milling process". *Procedia Manuf* 26, 383e393 (2018c). https://doi.org/10.1016/j.promfg.2018.07.046

Mushtaq, R.T., Wang, Y., Rehman, M., Khan, A.M., Mia, M.: State-of-the-art and trends in CO2 laser cutting of polymeric materials—a review. Materials **13**(17), 3839 (2020)

Nguyen, T.T., Nguyen, T.A., Trinh, Q.H.: Optimization of milling parameters for energy savings and surface quality. Arab. J. Sci. Eng. **45**(11), 9111–9125 (2020)

Wang, Y.C., Kim, D.W., Katayama, H., Hsueh, W.C.:Optimization of machining economics and energy consumption in face milling operations". Int. J. Adv. Manuf. Technol. **99**, 2093–2100 (2018a) https://doi.org/10.1007/s00170-018-1848-6

Xiao, Y., Jiang, Z., Gu, Q., Yan, W., Wang, R.: A novel approach to CNC machining center processing parameters optimization considering energy-saving and low-cost. J. Manuf. Syst. **59**, 535–548 (2021)

Zarei, O., Fesangharyb, M., Farshia, B., Jalili, S.R., Razfarb, M.R.: Optimization of multi-pass face-milling via harmony search algorithm". *J. Mater. Process.* Technol. **209**, 2386e2392 (2009)

Dynamic Modelling of High-Speed Spindle Supported by Active Magnetic Bearings in Presence of Defects

Abdessalem Jarraya[1,4](✉), Saeed Rubaiee[1,2], Abdullah Salmeen Bin Mahfouz[3], Slim Bouaziz[4], and Mohamed Haddar[4]

[1] Department of Mechanical and Materials Engineering, Faculty of Engineering, University of Jeddah, PO Box 80327, Jeddah 21589, Saudi Arabia
ajarraya@uj.edu.sa

[2] Department of Industrial and Systems Engineering, Faculty of Engineering, University of Jeddah, PO Box 80327, Jeddah 21589, Saudi Arabia

[3] Department of Chemical Engineering, Faculty of Engineering, University of Jeddah, PO Box 80327, Jeddah 21589, Saudi Arabia

[4] Laboratory of Mechanical Modeling and Production (LA2MP), National School of Engineers of Sfax (ENIS), University of Sfax, BP. 1173, 3038 Sfax, Tunisia

Abstract. During the machining process, the instantaneous cutting forces generate vibrations which reduce the effects of the machine's performances. In general, the majorities of cutting force models assume that the cutting tool is geometrically perfect and focus only on the impact of different cutting parameters. Thus, several sources of errors in machining should not be ignored as they can generate vibrations and therefore affect surface quality. Accordingly, this paper will examine the surface topography resulting from the integration of machining errors, such as tool run out, tool tilting and workpiece displacement. The global studied system includes: the milling machine structure that is modeled using the Finite Element Method (FEM), two Active Magnetic Bearings (AMB), supporting the spindle, are modeled by dynamic stiffness and damping coefficients, and finally a predictive cutting force model was also incorporated. Based on this model, a modal analysis of the machine tool is established using the dynamic substructure method. The machined surface profile as well as time and frequency tooltip responses under the influence of the presented errors are also investigated.

Keywords: Peripheral milling · errors modeling · run out · tilting · surface topography · Active Magnetic Bearings

1 Introduction

The complexity of the machine tool structure is a first essential factor that should be investigated and understood in order to capture its influence on dynamic properties. Thus, (Pedrammehr et al. 2012) have applied the Finite Element Method (FEM) to show the dynamic behaviour and modal parameters of a milling machine structure. Besides,

A. Akrout et al. (Eds.): ICAV 2022, ACM 22, pp. 367–377, 2023.
https://doi.org/10.1007/978-3-031-34190-8_39

(Xiangsheng et al. 2012) have used the model synthesis method to be able to analyse a milling machine. A finite element model is then utilized and verified by the Experimental Modal Analysis (EMA). The dynamic response of a Machine-Tool-Part system in an end milling operation is established by (Yangui et al. 2012). All the components of the milling machine are meshed using volume elements. In addition, three cutting force models are presented. The experimental findings showed the influence of cutting models and parameters on the dynamic response of the milling machine. Many works, however, are concentrated on examining the different sources of machining errors and cutting force conditions. Many cutting force prediction models using theoretical and experimental assumptions have been developed by (Liu et al. 2002), they have proposed a dynamic cutting force model in peripheral milling by incorporating the effects of undeformed chip thickness and cutting parameters. Empirical results demonstrated that the finished part is highly affected by the cutting force distribution. Also, (Flocke et al., 2009) have indicated that the cutting parameters, such as cutting speed and feed per tooth had a significant impact on the surface quality and tool life. They have pointed to the fact that both the feed per tooth and feed rate have a strong influence on the surface quality. In fact, to increase the surface quality, it is necessary to decrease the feed rate value. To enable the geometric optimization of milling tools, (Fontaine et al., 2007) has presented a method to study the impact of geometrical parameters, like helix angle, rake angle and tool tip envelope radius.

Using numerical simulations to handle the dynamics containing the time-varying UMP function, (Jin Hao et al. 2022) studied the effects of built-in motor and bearings parameters, rotor static eccentricity, and other factors on the dynamic characteristics of the spindle system. (Shivang Shekhar et al. 2022) present and demonstrate a systematic approach for effectively modeling the tool-tip dynamics in micromachining. Several case studies are presented on ultra-high-speed (UHS) spindle for different microtools, tool blanks and artifacts. (Weixin et al. 2018) propose a new error prediction method to integrate errors of the process system taking in to account the tool rotation error, the machine geometric and tool deformation errors. All these errors generate the movement of tool edge in the workpiece coordinate system. (Yang and Liu, 2015) have studied the surface generation mechanism using cutting process parameters and the different sources of machining errors. They have concluded that the tool deflection has the most significant effect on the surface profile and then, on the work piece displacement and tool run-out. (Hu et al. 2015) have diagnosed the surface topography caused by the tool assembly errors. Based on this work, the authors have noticed that if the cutter assembly errors exist, each tooth tip trajectory will change, leading to the surface topography phenomena that surface rises and falls and cutting marks' size shrinks and expands. Recently, the research of thermal–mechanical behavior in motorized spindle to eliminate the modeling error of thermal and dynamic interaction characteristics was investigated by (Jin Hao et al. 2023). Authors have assumed that interaction characteristics are achieved by the influence of thermal deformation on bearing stiffness and contact load. In this work, a predictive peripheral milling surface profile is investigated by considering different sources of machining errors. The studied model incorporates not only the spindle system, but also the entire milling machine structure which is modelled with the reduced FEM.

The effect of the geometrical tool errors, such as the tool tilting and tool run-out, on the tooth tip responses is presented.

2 The Global Mechanical Model

2.1 Geometrical Machine Tool Model

In this work, a three-axis milling machine is considered. Its whole structure is discredited with plates FE as presented in Fig. 1. The spindle is considered as beam elements.

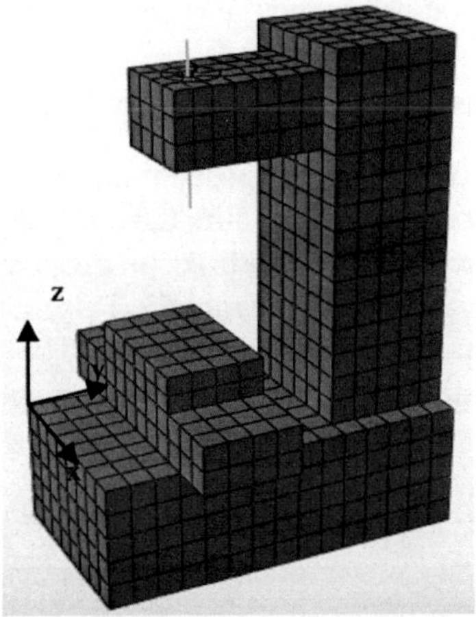

Fig. 1. Machine structure FE model

The lateral tool motion coincides with the X- axis. The motion along the Y-axis is its frontal motion, and finally the third motion (the vertical motion) is along the Z-axis. The FE model presents 959 elements and 974 nodes. Six degrees of freedom at each node are computed. The machine is entirely fixed at its base.

To suspend the spindle shaft, we also use two AMBs. Each one is composed of two pairs of electromagnets modelled by dynamic coefficients.

2.2 Cutting Process Model

The cutting force components F_t, F_r and F_a depend on the instantaneous chip thickness $h(\Phi_j(t))$ and can be calculated from the different values of the rotation angle ϕ_j as follows:

$$
\begin{aligned}
F_t &= K_t a_p h(\Phi_j(t)) \\
F_r &= K_r K_t a_p h(\Phi_j(t)) \\
F_a &= K_a K_t a_p h(\Phi_j(t))
\end{aligned}
\quad (1)
$$

$h(\phi_j(t))$ is compose up of static and dynamic parts: $h(\Phi_j(t)) = h_s(\Phi_j(t)) + h_d(\Phi_j(t))$ where,

$$\begin{cases} h_s = f_z \sin(\Phi_j(t)) \\ h_d = (u_x(t) - u_x(t-\tau)) \sin(\Phi_j(t)) - (u_y(t) - u_y(t-\tau)) \cos(\Phi_j(t)) \end{cases} \quad (2)$$

The rotation angle is expressed as a function of both the number of teeth and time:

$$\Phi_j(t) = \Omega t + j\varphi_p, \quad j = 0, 1, \ldots, Z-1 \quad (3)$$

$u_x(t)$, $u_y(t)$ are the current positions of the tool in x- and y- directions at time t,
$u_x(t-\tau)$, $u_y(t-\tau)$ are the positions of the previous tooth at time t $t-\tau$,
a_p, K_t, K_r and K_a are the axial depth and the specific coefficients of cut, respectively.

2.3 Modal Analysis of the Milling Machine Structure

In this section, a modal analysis of the proposed milling machine is performed to obtain natural frequencies. Due to its complex structure, the dynamic substructure method which aims to reduce the problem size is utilized in this work. Therefore, we can obtain the dynamic characteristics of the machine rapidly. This method divides the entire structure into different simple parts (sub-structures) that will be examined separately. For the kth substructure under consideration, the undamped equation of motion is presented as follows:

$$\left[M^k\right]\left\{\ddot{X}^k\right\} + \left[K^k\right]\left\{X^k\right\} = \left\{F^k\right\} \quad (4)$$

where,

$\left[M^k\right]$ and $\left[K^k\right]$ are mass and stiffness matrices,
$\left\{X^k\right\}$ and $\left\{F^k\right\}$ presents the global displacement and force vectors, respectively.
$\left\{X^k\right\}$ is partitioned into interior DOFs $\left\{X_i^k\right\}$ which can eliminate the exterior DOFs.
$\left\{X_e^k\right\}$ that are in physical contact with the neibour substructure. So, $\left\{X^k\right\}$ is restated as: $\left\{X^k\right\} = \begin{bmatrix} X_e^k \\ X_i^k \end{bmatrix}$.

Reordering the mass and stiffness matrices depending on the DOFs repartition, they are rewritten as follows:

$$\left[M^k\right] = \begin{bmatrix} M_{ee}^k & M_{ei}^k \\ M_{ie}^k & M_{ii}^k \end{bmatrix}, \left[K^k\right] = \begin{bmatrix} K_{ee}^k & K_{ei}^k \\ K_{ie}^k & K_{ii}^k \end{bmatrix} \quad (5)$$

The static modes are determined as follows:

$$\{X_i\} = -[K_{ii}]^{-1}[K_{ie}]\{X_e\} = [\Phi_C]\{X_e\} \quad (6)$$

The normal modes are also calculated using the Eq. 4:

$$\left(\left[K_{ii}^k\right] - \omega_r^2\left[M_{ii}^k\right]\right)\{\varphi_r\} = \{0\} \quad (7)$$

A transformation matrix is then formed using these last modes. It is utilized to formulate the generalized coordinate's vector in the physical domain under the following form:

$$\begin{Bmatrix} X_e \\ X_i \end{Bmatrix} = \begin{bmatrix} I & 0 \\ \Phi_C & \Phi_N \end{bmatrix} \begin{Bmatrix} X_e \\ \Gamma_N \end{Bmatrix} = [P]\{\Gamma(t)\} \tag{8}$$

$\{\Gamma(t)\}$ stands for the state vector in the modal base.

After incorporating this last equation into Eq. 1, we find that:

$$\left[P^k\right]^T\left[M^k\right]\left[P^k\right]\left\{\ddot{\Gamma}^k(t)\right\} + \left[P^k\right]^T\left[K^k\right]\left[P^k\right]\left\{\Gamma^k(t)\right\} = \left\{\Gamma^k(t)\right\}\left\{F^k\right\} \tag{9}$$

Our studied machine model is subdivided into five main substructures: The base, column, Y-axis slide, X-axis slide (worktable), spindle housing and spindle. The same boundary nodes are shared between these different components. Thus, different substructures have the same movement interface based on the interface continuity.

Natural frequencies and modes computed by the finite element model and the substructure method are shown in Table 1.

Table 1. Milling machine Frequencies

Mode N^0	Natural frequencies (Hz)	
	Substructure	FEM
1	56.9	57.1
2	65.4	65.6
3	78.4	78.37
4	91.3	92.1
5	94.5	94.6
6	115.7	113.6
7	117.2	116.1

Based on this table, it should be noted that the results are in good agreement. So, the complex structure of the milling machine does not degrade the computational accuracy.

Figure 2 elucidates the first deformed shape plots. The analysis of these outcomes demonstrates that the most important deformations are seen in both the spindle housing and the column. In fact, this is due to the stiffness of the connections between these parts which are related directly to the spindle. Some deformations observed at the base of the 7th mode should be avoided. In practice, the base is dimensioned using vibration criteria.

3 Sources and Effects of Machining Errors on Surface Topography

3.1 Surface Topography Equation

The surface topography results from the material removal process due to relative motion between the tool and part. Thus, to predict the surface topography, the tooth path equations should be developed. In fact, positions relative not only to the tool axis with respect to the spindle axis (x_{ts} and y_{ts}), but also to the spindle axis regarding the work piece ($y_{s\Omega}$) are given in Eq. 4 by (Arizmendi M. et al., 2010):

$$\begin{cases} x_{ts} = R\cos(\Omega t + \varphi_p - (z\ tg(\beta)/R)) \\ y_{ts} = R\sin(\Omega t + \varphi_p - (z\ tg(\beta)/R)) \\ y_{s\Omega} = f_z \cdot t \end{cases} \tag{10}$$

where, R, β, φ_p, Ω and f_z are the tool radius, helix angle, pitch angle, angular velocity and feed rate respectively.

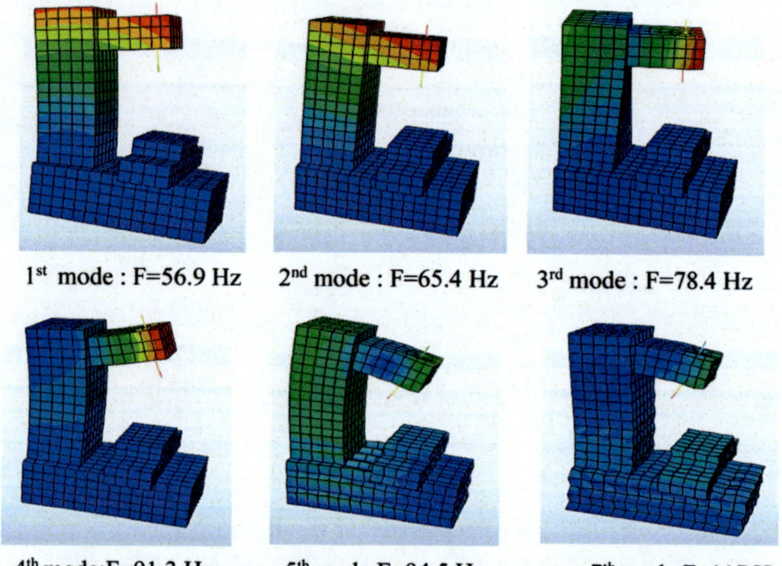

1st mode : F=56.9 Hz 2nd mode : F=65.4 Hz 3rd mode : F=78.4 Hz

4th mode:F=91.3 Hz 5th mode:F=94.5 Hz 7th mode:F=117 Hz

Fig. 2. Milling machine deformed shapes

A time-domain simulation is developed using the Newmark coupled with Newton–Raphson method for the solution of systems. All cutting and AMB parameters are used as input (Table 2). The different sources of machining errors are taken into account in order to generate the surface topography.

Table 2. Cutting parameters

Parameter	Symbol	Value	Unit
Spindle speed	N	20,000	rpm
Feed per tooth	f_z	0.16	mm
Axial depth of cut	a_p	5	mm
Tangential cutting coefficient	K_t	644	N/mm^2
Radial cutting coefficient	K_r	0.38	N/mm^2
Axial cutting coefficient	K_a	0.25	N/mm^2
Teeth number	Z	2	-

Figure 3 presents the simulated surface topography when neglecting the machining error sources. It is clear that sharper undulations occur on the surface. These undulations are dominated by the feed marks.

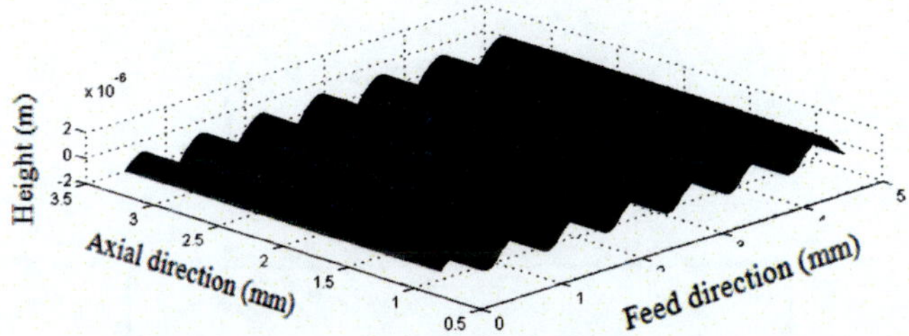

Fig. 3. Surface topography prediction without errors

The impact of the feed f_z and spindle speed Ω on the surface topography is presented in Fig. 4. It can be noted that the surface height increases nonlinearly with the highest feed. In fact, at the smallest feed, successive edge marks on the machined surface will be very close to each other and the high radius cutter will make a mark on the surface. Thus, if the feed rate is increased, successive marks will be separated, and more cutting teeth will leave marks on the surface than others. This result is proved in the work of (Hu et al., 2018). Also, it is noticed that the spindle speed has little effect on the surface height.

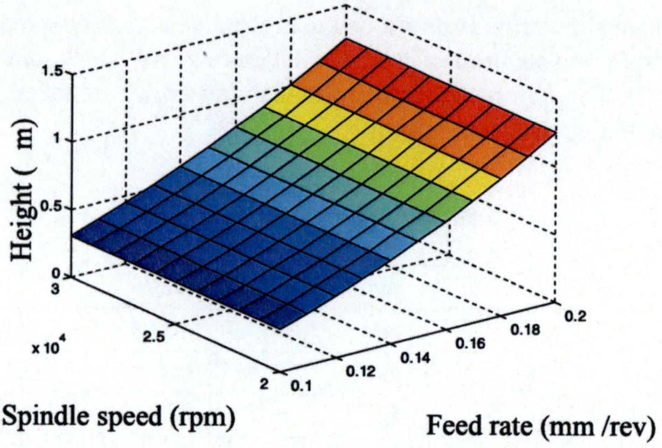

Fig. 4. Effect of feed rate and spindle speed on the machined surface

3.2 Run-Out of the Tool

The run-out error is an imprecision of the rotating tool as it does not rotate in the same line with the rotational axis. It is introduced by two major forms: axial run-out and radial run-out (Fig. 5). The former occurs when the rotating tool present an angle λ. The latter is the result of a decentring tool (distance e). As shown in Fig. 5, the practical tool tip trajectory is larger compared to the ideal one.

The Tool Tip Center (TTC) coordinates in both X- and Y- directions are expressed as follows:

$$\begin{pmatrix} x_{runout,j} \\ y_{runout,j} \end{pmatrix} = e\begin{pmatrix} \cos(\Omega t + \lambda) \\ \sin(\Omega t + \lambda) \end{pmatrix} \quad (11)$$

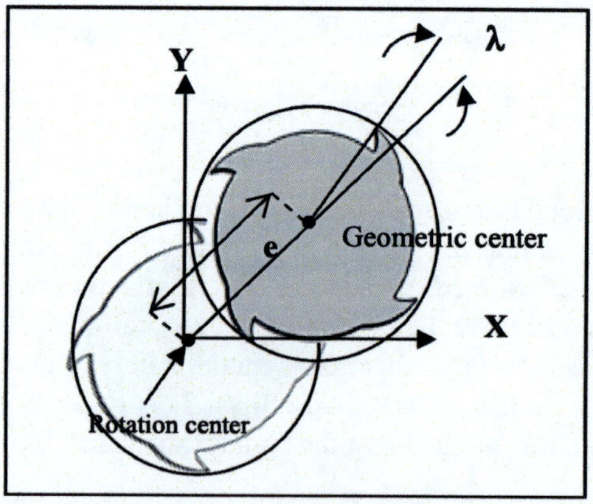

Fig. 5. Tool run-out

3.3 Tool Tilting

It is a geometric error which results from the tool clamping system. It is explained by a declination of the geometric cutting tool axis relative to its rotation axis as presented in Fig. 6. This defect is defined as the declination angle ε between the geometry axis and the rotation axis. Another parameter η represents the angle between (OY) and (OA) axes. This parameter is the projection of the geometrical axis (OZ') onto the (OXY) plane.

The TTC coordinates can be written as:

$$\begin{pmatrix} x_{tilting,j} \\ y_{tilting,j} \end{pmatrix} = \begin{pmatrix} z_j \sin\varepsilon \cdot \sin\eta \\ z_j \sin\varepsilon \cdot \cos\eta \end{pmatrix} \quad (12)$$

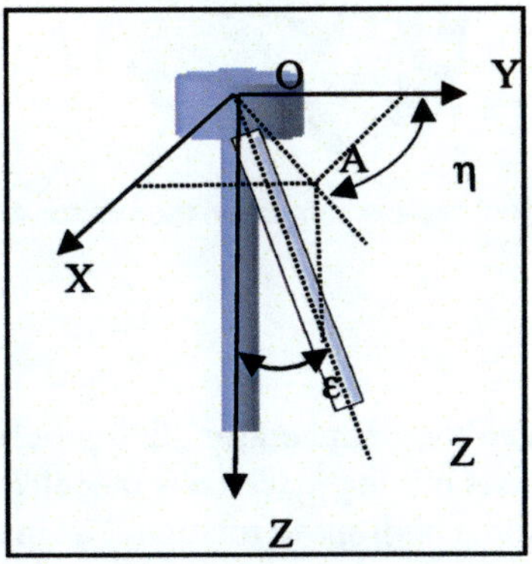

Fig. 6. Tool tilting error

As shown in Sect. 2, the global system equation is written as follow:

$$[M]\{\ddot{Q}\} + [C]\{\dot{Q}\} + [K]\{Q\} = \{F_{c(x,y,z)}(t, \{Q\})\} \quad (13)$$

It involves the mass, stiffness and damping matrices of the entire machine. The stiffness and damping matrices include the instantaneous dynamic coefficients derived from the AMB modeling. This equation is therefore reformulated in a new truncated basis in order to apply the reduction method.

The surface topography in presence of the tool run-out error is illustrated in Fig. 7. The run-out distance is e = 3.5 μm. The run-out angle is deliberately chosen to be zero degree in order to increase the effect of this error on the surface shape. It can be observed that the simulated surface topography presents not only a small run-out deviation on cusp height, but also few subtle changes in cusp patterns resulting from truncation errors. In fact, the chip thickness depends on the cutter vibrations in contact with the part, which are influenced by the cutter run-out value.

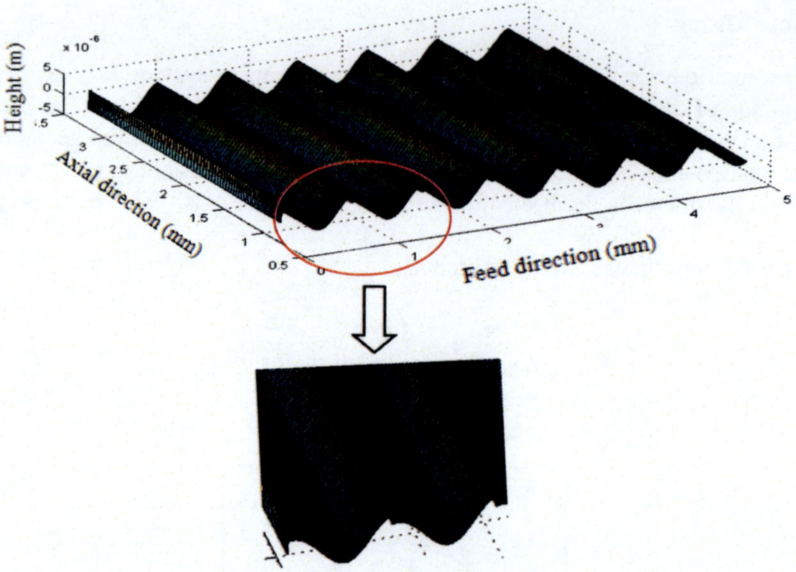

Fig. 7. Tool run-out impact on the surface topography

4 Conclusion

In this paper, we investigated the peripheral milling process with different sources of machining errors. A three-axis milling machine is modeled with the FEM. Two AMBs were also employed for guiding high speed rotation. The modal analysis of the obtained milling structure based on the dynamic substructure method shows that both the spindle housing and column are the most influencial structural parts. In order to predict the machined surface topography, the geometrical tool errors are carried out. Based on this work, we concluded that the surface profile greatly depends on the feed rate, spindle speed, and tool geometry by focusing mainly on its length. Concerning the different tool geometrical impact, it was clear that for the run-out case, the simulated surface topography presents a small deviation in the cusp height and few changes in the cusp pattern resulting from the truncation error. The tool tilting has a great impact on the surface profile compared to both the tool run-out and workpiece displacement. So, the surface topography is affected by various tool errors.

The chip thickness depend on the cutter vibrations at the contact with the part and which are themselves influenced by the cutter run-out value.

Funding. This project was funded by the Deanship of Scientific Research (DSR), University of Jeddah, Jeddah, Saudi Arabia, under grant No. (UJ-20-DR-14). The authors, therefore, gratefully acknowledge DSR technical and financial support.

References

Arizmendi, M., Fernandez, J., Gil, A., Veiga, F.: Identification of tool parallel axis offset through the analysis of the topography of surfaces machined by peripheral milling. Int. J. Mach. Tools Manuf. **50**, 1097–1114 (2010)

Flocke, F., Qwito, F., Arntz, K.: A Study of the influence of cutting parameters on micro milling of Steel with cubic boron nitride (CBN) tools. In: Micromachining and Microfabrication Process TechnologyXIV 7204 (2009)

Fontaine, M., Devillez, A., Dudzinski, D.: Optimisation de la géométrie d'outil en fraisage à partir de la prédiction analytique des efforts de coupe. In: 18ème Congrès Français de Mécanique, Grenoble (2007)

Hu, W., Guan, J., Li, B., Cao, Y., Yang, J.: Influence of tool assembly error on machined surface in peripheral milling process. Procedia CIRP **27**, 137–142 (2015)

Hao, J., et al.: Thermal-mechanical dynamic interaction in high-speed motorized spindle considering nonlinear vibration. Int. J. Mech. Sci. **240**, 107959 (2023)

Hao, J., et al.: Dynamic characteristics analysis of asynchronous motorized spindle considering combined unbalanced magnetic pull and nonlinear bearing restoring force effects. Mech. Syst. Signal Process. **185**, 109807 (2023)

Khabou, M.T., Bouchaala, N., Chaari, F., Fakhfakh, T., Haddar, M.: Study of a spur gear dynamic behavior in transient regime. Mech. Syst. Signal Process. **25**, 3089–3101 (2011)

Liu, X.W., Cheng, K., Webb, D., Luo, X.C.: Prediction of cutting force distribution and its influence on dimensional accuracy in peripheral milling. Int. J. Mach. Tools Manuf. **42**, 791–800 (2002)

Shekhar, Shivang, Bekir Bediz, O., Ozdoganlar, Burak: Tool-tip dynamics in micromachining with arbitrary tool geometries and the effect of spindle speed. Int. J. Mach. Tools Manuf. **185**, 103981 (2023)

Pedrammehr, S., Farrokhi, H., Khani Sheykh Rajab, A., Pakzad, S., Mahboubkhah, M., Ettefagh, M.M., Sadeghi, M.H.: Modal analysis of the milling machine structure through FEM and experimental test. Adv. Mater. Res. **383–390**, 6717–6721 (2012)

Hu, W., Cao, Y., Yang, J., Shang, H., Wang, W.: An error prediction model of NC machining process considering multiple error sources. The Int. J. Adv. Manuf. Technol. **94**(5–8), 1689–1698 (2018)

Gao, X., Zhang, Y., Zhang, H., Wu, Q.: Effects of machine tool configuration on its dynamics based on orthogonal experiment method. Chin. J. Aeronaut. **25**, 285–291 (2012)

Yangui, H., et al.: Influence of cutting and geometrical parameters on the cutting force in milling. Engineering **2**, 751–761 (2012)

Yang, D., Liu, Z.: Surface plastic deformation and surface topography prediction in peripheral milling with variable pitch end mill. Int. J. Mach. Tools Manuf. **91**, 43–53 (2015)

Application of Particle Swarm Optimization to Minimize Active Magnetic Bearing Forces

Salwa Benali[1,2](✉), Anoire Benjdidia[1], Taissir Hentati[1], Slim Bouaziz[1], and Mohamed Haddar[1]

[1] Laboratory of Mechanical Modeling and Production (LA2MP), National Engineering School of Sfax, University of Sfax, Sfax, Tunisia
Salwa.benali96@yahoo.com
[2] National Engineering School of Gabes, University of Gabes, Gabés, Tunisia

Abstract. Active Magnetic bearings (AMB) are widely used in the rotating machine. Its dynamic behavior is impacted by the AMB parameters. That's why, a good selection of the bearing parameters is important to achieve an accurate performance of the rotating machine. This work attempts application of Particle Swarm Optimization (PSO) is applied to select the optimal parameters of bearing leading to minimum bearing forces. Bearing forces are depends on several parameters like rotational speed, control-current, derivative gain and number of windings around the core half. One objective function, three limited parameters and one material constraints are considered in the optimization processes. A rotating composite shaft mounted by two AMB is studied. The rotor motion equation has been established and numerically solved by combining Newmark and Newton-Raphson methods. Results show that bearing parameters affect the bearing force value and the dynamic behavior of the rotor.

1 Introduction

Active magnetic bearings are particularly suitable for high-speed application due to their advantages. In fact, these bearings are capable of supporting rotating shaft without any mechanical contact. They are used in a variety of domains such as machine tools and aerospace industry. It is evident that the parameters of AMBs are a fundamental issue in the controllability of a rotor. In fact, it is difficult to choose the optimal parameters of the AMB controller system and design parameters because the system is subject to different motions during operation. In order to address this problem, several research studies have focused on the optimization of magnetic bearing parameters such as (Bordoloi and Tiwari 2013) in which authors developed a method based on the genetic algorithm to optimize AMB controller parameters. They found that the proportional-factor is the most sensitive parameter. Recently, research developed by (Hamad et al. 2022), in which authors presented and designed an AMB after performing an optimization method by reducing the number of poles and air gap. They showed that the complexities of the AMB control system are inversely proportional to the pole number. More recently, authors proposed in their work (Wu et al. 2023) a multilevel design optimization based

A. Akrout et al. (Eds.): ICAV 2022, ACM 22, pp. 378–388, 2023.
https://doi.org/10.1007/978-3-031-34190-8_40

on the response surface method (RSM) and the PSO algorithm in order to reduce couplings and nonlinearity of hybrid magnetic bearing and improve the bearing capacity per unit area. A fractional order proportional–integral–derivative (FO-PID) controller proposed by (Laldingliana and Biswas 2022) to improve the performance of the AMB system. In their research work PSO was used with three PID control variables and five FO-PID variables where the time specification constraints are considered. In another research work (Bo et al. 2021), PSO is applied to optimize the PID controller parameter for controlling the stability of AMB. They used a combination of online and offline optimization and they showed that this methodology is efficient to control the AMB system. In the study of (Yadav et al. 2021), an optimization based on the multi-objective genetic algorithm is carried out for a rotational speed of 22,000 rpm in order to optimize AMB design parameters. Another multi-objective optimization method is proposed by (Jin et al. 2022) to improve three-pole AMB performance. In their research, the authors applied the Kendall correlation coefficient and Kriging model to improve the optimization process efficiency. They use three parameters and three level optimizations. Results showed the effectiveness and robustness of the proposed optimization method compared with experimental results. Borque Gallego et al. (2021) presents a strategy to improve efficiency of rotating system supported with AMB. This strategy aimed to maximize the bearing and motor parameters by combining the generated forces maximization and loss minimization. Results of efficient optimization showed an increase between 33% and 58%.

From the literature, there are a variety of optimization methods, like GA and PSO algorithms. In the current study, the PSO algorithm is applied to optimize AMB parameters. In the first part, the magnetic forces are presented as an objective function. Thus, this work aims to highlight the impact of AMB parameters on the AMB forces. In order to minimize the objective function, an optimization problem based on the PSO algorithm with a minimization function, one constraint, and three parameters is presented.

2 Particle Swarm Optimization Method

In this research work, the particle swarm optimization (PSO) is used to minimize the bearing forces of a rotating system. The PSO presents a population inspired by fish schooling social behavior or bird flocking social behavior. It is based on optimization technique of stochastic. Compared to other optimization methods, PSO is an easily implemented method, Hyperparametric method, and an efficient optimization algorithm with high global search ability. In this paper, the PSO is applied with one objective function and three parameters.

2.1 Objective Function

The objective function can be presented as one objective function describing AMBs forces. As portrayed in Fig. 1, AMBs with four electromagnets are studied. An electromagnetic force is generated by each pair of electromagnets depending on the control current i_j (j = 1, 2, 3, 4). Magnetic forces of AMBs in the vertical and horizontal directions can be expressed (Bouaziz et al. 2011), respectively, as Eq. (1) and (2):

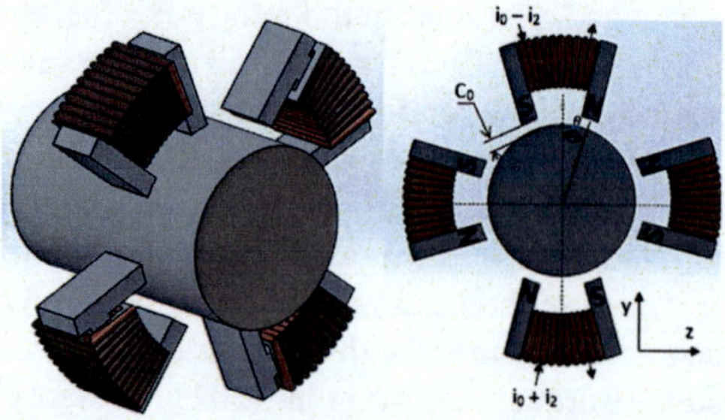

Fig. 1. Cross section of AMBs with four electromagnets.

$$F_y = a\left[\left(\frac{1-\frac{k_p v}{I_0}-\frac{k_d \dot{v}}{I_0}}{1-\frac{v}{C_0}}\right)^2-\left(\frac{1+\frac{k_p v}{I_0}+\frac{k_d \dot{v}}{I_0}}{1+\frac{v}{C_0}}\right)^2\right] \tag{1}$$

$$F_z = a\left[\left(\frac{1-\frac{i_0}{I_0}-\frac{k_p w}{I_0}-\frac{k_d \dot{w}}{I_0}}{1-\frac{w}{C_0}}\right)^2-\left(\frac{1+\frac{i_0}{I_0}+\frac{k_p w}{I_0}+\frac{k_d \dot{w}}{I_0}}{1+\frac{w}{C_0}}\right)^2\right] \tag{2}$$

where v, $\dot{v}$, w and $\dot{w}$ present respectively the displacements and the velocities in y and z directions; K_d present the derivative gain; Kp is the proportional gain; C_0 describe the air gap; I_0 is the control current; i_0 is the bias current and a is the electromagnetic force coefficient written as:

$$a = \frac{\mu_0 A N^2 I_0^2}{4C_0}\cos\theta \tag{3}$$

where A is the electromagnet cross-sectional area, μ_0 present the vacuum permeability; C_0 describe the air gap; N represent the number of windings around the core half; θ is the radial electromagnetic angle and I_0 present the bias current. To estimate AMBs forces, the equation of motion of a composite shaft supported by tow AMBs using the FEM is elaborated (Bouaziz et al. 2011; Chaker 2017), as shown in Eq. (4).

$$[M]\{\ddot{q}\} + [C]\{\dot{q}\} + [K]\{q\} = \{F_{bearings}\} \tag{4}$$

where $\{q\}$, $\{\dot{q}\}$ and $\{\ddot{q}\}$ denotes respectively displacement, velocity and acceleration vectors associated to each node. The second member represents the AMBs force. $[M]$, $[C]$ and $[K]$ are respectively the mass, stiffness and damping matrices of a composite shaft. The mass and stiffness matrices are obtained by assembling the elementary matrices $[M_e]$ and $[K_e]$ given respectively by (Chaker 2017):

$$[M^e]_{8\times 8} = \begin{bmatrix} [M_v] & 0 & 0 & 0 \\ 0 & [M_w] & 0 & 0 \\ 0 & 0 & [M_{\theta_y}] & 0 \\ 0 & 0 & 0 & [M_{\theta_z}] \end{bmatrix} \tag{5}$$

$$[K^e]_{8\times 8} = \begin{bmatrix} [K_v] & 0 & [K_1] & [K_2] \\ 0 & [K_w] & [K_3] & [K_4] \\ [K_1]^T & [K_3]^T & [K_{\theta_y}] & [K_5] \\ [K_2]^T & [K_4]^T & [K_5]^T & [K_{\theta_z}] \end{bmatrix} \tag{6}$$

The elementary mass and stiffness components are given by Eq. (7) and (8):

$$\begin{aligned} [M_v]_{2\times 2} &= \int_{-1}^{1} N_v^T I_m N_v \det(J) d\xi \\ [M_w]_{2\times 2} &= \int_{-1}^{1} N_w^T I_m N_w \det(J) d\xi \\ [M_{\theta_y}]_{2\times 2} &= \int_{-1}^{1} N_{\theta_y}^T I_d N_{\theta_y} \det(J) d\xi \\ [M_{\theta_z}]_{2\times 2} &= \int_{-1}^{1} N_{\theta_z}^T I_d N_{\theta_z} \det(J) d\xi \end{aligned} \tag{7}$$

$$[K_v]_{2\times2} = -\int_{-1}^{1} B_v^T k(\mathrm{A}_{55} + \mathrm{A}_{66}) B_v \det(J) d\xi$$

$$[K_1]_{2\times2} = -\frac{1}{2}\int_{-1}^{1} B_v^T k\mathrm{A}_{16} B_{\theta_y} \det(J) d\xi$$

$$[K_2]_{2\times2} = \int_{-1}^{1} B_v^T k(\mathrm{A}_{55} + \mathrm{A}_{66}) N_{\theta_z} \det(J) d\xi$$

$$[K_w]_{2\times2} = \int_{-1}^{1} B_w^T k(\mathrm{A}_{55} + \mathrm{A}_{66}) B_w \det(J) d\xi$$

$$[K_3]_{2\times2} = \int_{-1}^{1} B_w^T k(\mathrm{A}_{55} + \mathrm{A}_{66}) N_{\theta_y} \det(J) d\xi \quad (8)$$

$$[K_4]_{2\times2} = -\frac{1}{2}\int_{-1}^{1} B_{\theta_y}^T k\mathrm{A}_{16} B_{\theta_z} \det(J) d\xi$$

$$\left[K_{\theta_y}\right]_{2\times2} = \int_{-1}^{1} B_{\theta_y}^T \mathrm{A}_{11} B_{\theta_y} \det(J) d\varepsilon + \int_{-1}^{1} N_{\theta_y}^T k(\mathrm{A}_{55} + \mathrm{A}_{66}) N_{\theta_y} \det(J) d\xi$$

$$[K_5]_{2\times2} = \frac{1}{2}\int_{-1}^{1} B_{\theta_y}^T k\mathrm{A}_{16} N_{\theta_z} \det(J) d\xi - \frac{1}{2}\int_{-1}^{1} B_{\theta_z}^T k\mathrm{A}_{16} N_{\theta_y} \det(J) d\xi$$

$$\left[K_{\theta_z}\right]_{2\times2} = \int_{-1}^{1} B_{\theta_z}^T \mathrm{A}_{11} B_{\theta_z} \det(J) d\xi + \int_{-1}^{1} N_{\theta_z}^T k(\mathrm{A}_{55} + \mathrm{A}_{66}) N_{\theta_z} \det(J) d\xi$$

where N_v, N_w, $N_{\theta y}$ and $N_{\theta z}$ presents the shape functions matrices; B_v, B_w, $B_{\theta y}$ and $B_{\theta z}$ are the deformation matrices and J is the Jacobian matrix; $A_{11}, A_{16}, A_{55}, A_{66}$ are terms related to elastic behaviors of composite materials (Sino et al. 2008; Chaker 2017); I_m, I_d and I_p present the inertia moments of the composite shaft (Sino et al. 2008; Boukhalfa 2014).

The damping matrix is given by the structural damping. It can be given by Eq. (9) (Thomas and Laville 2007), where η present the pert factor of composite material and Ω is the rotational speed of the system.

$$[C] = \frac{\eta}{\Omega}[K] \quad (9)$$

2.2 Constraints and Limits

The maximum shear stress theory says that the failure of the composite rotor will occur when the maximum shear stress overpasses the shear stress in uniaxial loading. It's, in fact, the only constraint to respect. Thus, the rupture resistance condition of a composite rotor is written as following:

$$\frac{F}{S} \leq \tau_{\max} \tag{10}$$

F represent here AMBs forces and S represent the cross-sectional area of the shaft.

In this research work, the unknowns' parameters for the objective function are k_d, I_0 and N. The bounds are defined according to the literature reviewed (Bordoloi and Tiwari 2013) and in the present work these are listed in Table 1.

Table 1. Upper and Lower parameters limit.

AMBs parameters	Lower limit	Upper limit
K_{d}	0	45
I_0	0	8
N	50	400

3 Results and Discussions

3.1 Parameters Effect

This section led to highlight the effect of AMB properties on electromagnetic force. The model studied is a hollow shaft supported by two identical AMBs. The magnetic force given by Eq. (1) and (2) must be minimized. The equation of motion given by Eq. (4) was numerically solved by coupling Newton-Raphson and Newmark's methods. Material properties and shaft dimension are presented in Table 2.

Table 2. Shaft dimension and material parameters.

Parameters	Value
Outer radius	9 mm
Inner radius	6.4 mm
Length	907 mm
Number of layers	20 layers
Fiber's orientation	[0, 0, 0, 0, 90, 90, 45, −45, 0, 0, 0, 45, −45, 90, 90, 0, 0, 0, 0, 0]
ρ	1500 kg/m^3
E_1	130 GPa
E_2	10 GPa
G_{ij}	7 GPa
ν	0.25
τ_{max}	6.9 GPa

Figures 2, 3, and 4 depict respectively the impact of the control current, number of coils, and derivative gain on the magnetic force of the AMB. It can be found that the AMB force increased significantly with increasing control current. It can be seen also that the control current has a significant effect on the AMB force intensity. It can be found from the 2nd figure that force varies proportionally with the number of coils. Figure 4 shows the AMB forces at different derivative gain values. It can be seen that derived gain does not have a significant effect on force intensity but the variation of k_d causes a phase delay. Hence, it is necessary to choose a good combination between these parameters to better choose a minimum force that respects the rotor shear stress

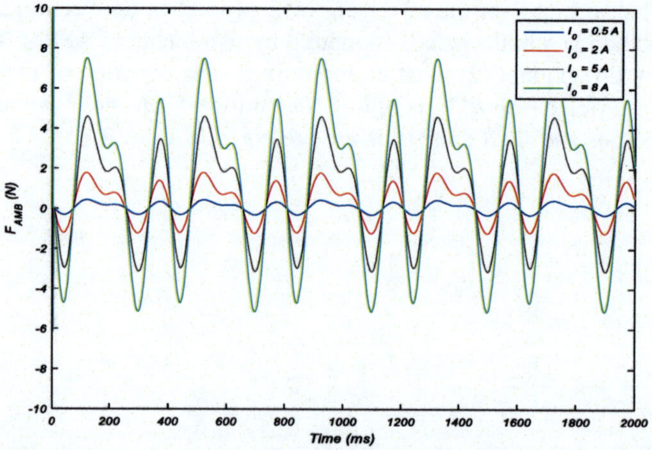

Fig. 2. Control current impact

and makes the shaft more balanced. Thus, the PSO algorithm is applied to achieve this objective.

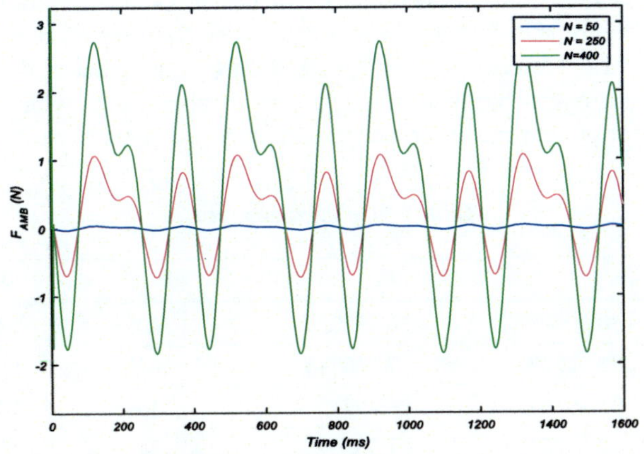

Fig. 3. Coils turn impact

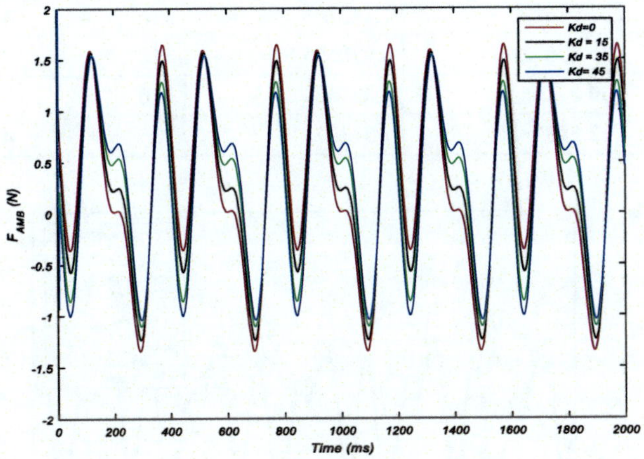

Fig. 4. Derivative gain impact

3.2 Optimum Parameters

The objective of this section is to find the optimum values of decision variables (k_d, I_0 and N) that lead to minimize the AMB's forces. To achieve this goal, the PSO algorithm is used regarding its simplicity and accuracy. PSO algorithm presents a simple method to search an optimal solution in the particle space. It is different from other optimization processes in such a way that only the objective function is needed, also this algorithm taken into account a multi-parameter. An optimization process applied with 100 iterations and 100 particles. The optimization was repeated 10 times with 100 iterations for each one. The results of the 10 tests are listed in the Table 3. The obtained results are very close for the 10 tests, these results are satisfactory. Figure 5 depicts the difference between forces before and after test optimization. It's clear that results of optimization tests are very close, the figure shows that a good selection of the bearing parameters leads to minimize the bearing forces between 29% and 44%. One deduces the importance of bearings parameter optimization in minimizing the bearing forces which will impact the dynamic behavior of the rotor machine supported by such type of bearings.

Table 3. Optimization results.

Tests	K_d	I_0	N	F_{best}
1	30.80505	6.937338	340	4.716
2	35.62503	7.260145	347	5.137
3	35.49391	6.89495	346	4.81
4	37.11883	7.532098	344	5.243
5	36.73808	7.654678	320	4.622
6	33.17804	7.124133	345	4.985
7	38.98643	7.143208	316	4.154
8	35.87622	6.970555	318	4.113
9	39.07166	7.524484	346	5.296
10	32.293	7.51497	335	4.973

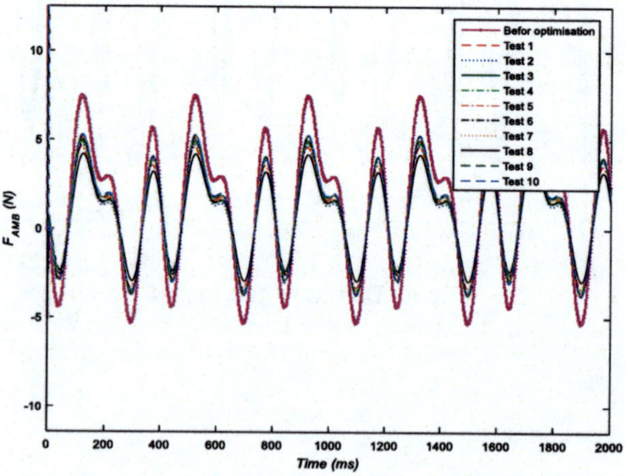

Fig. 5. AMB forces before and after optimization.

Figure 6 shows the effects of the bearing parameter optimization on the shaft vibrations, these results are plotted in the presence of the mass unbalance as sources of excitations. As shown in this figure, an only peak is observed, it corresponds to rotational frequency. It can be observed that optimum parameters lead to decreases of displacement amplitude.

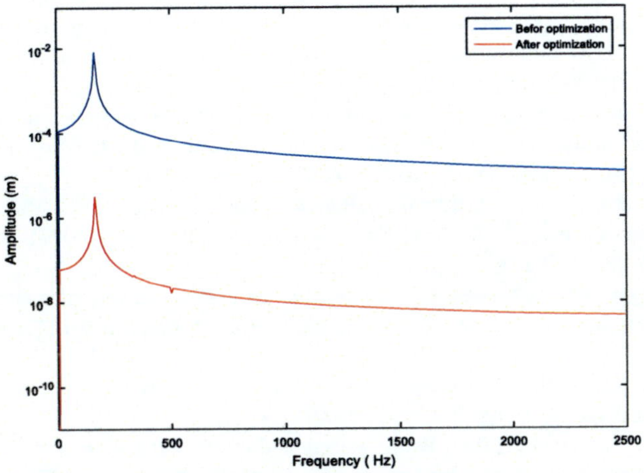

Fig. 6. Mass unbalance response

4 Conclusion

This paper aimed to study the effect of the AMB bearing parameters on the value of the bearing's forces. Results showed that AMB force varies proportionally with the number of coils and control current, but the derivative gain has no significant effect on the AMB forces. After that the PSO algorithm is applied to obtained a good combination between these parameters to minimize bearings forces. Results of optimization showed that a good selection of the bearing parameters leads to minimize the AMB forces with 44% and to decrease the vibration level. As perspective, the results of this study will be used to improve performance and efficiency of composite spindle.

References

Bo, W., Haipeng, G., Hao, L., Wei, Z.: Particle swarm optimization-based fuzzy PID controller for stable control of active magnetic bearing system. J. Phys. Conf. Ser. **1888**, 012022 (2021). https://doi.org/10.1088/1742-6596/1888/1/012022

Bordoloi, D.J., Tiwari, R.: Optimization of controller parameters of Active Magnetic Bearings in rotor-bearing systems. Adv. Vib. Eng. **12**, 319–327 (2013)

Borque Gallego, G., Rossini, L., Achtnich, T., et al.: Efficiency optimization of slotless magnetic-bearing machines. IEEE Trans. Ind. Appl. **57**, 6833–6843 (2021). https://doi.org/10.1109/TIA.2021.3072614

Bouaziz, S., Messaoud, N.B., Mataar, M., Fakhfakh, T., Haddar, M.: A theoretical model for analyzing the dynamic behavior of a misaligned rotor with active magnetic bearings. Mechatronics **21**(6), 899–907 (2011)

Boukhalfa, A.: Dynamic analysis of a spinning functionally graded material shaft by the p-version of the finite element method. Lat. Am. J. Solids Struct. **11**, 2018–2038 (2014). https://doi.org/10.1590/S1679-78252014001100007

Chaker, S.B.A.: Composite shaft rotordynamic analysis using the finite element method (2017)

Hamad, M.N., Khalifa, M.Z., Mohammed, J.A.: Optimization of design parameters for manufacturing a radial active magnetic bearing with 12-poles. Eng. Technol. J. **40**(1), 207–216 (2022). https://doi.org/10.30684/etj.v40i1.2202

Jin, Z., Sun, X., Chen, L., Yang, Z.: Robust multi-objective optimization of a 3-pole active magnetic bearing based on combined curves with climbing algorithm. IEEE Trans. Ind. Electron. **69**, 5491–5501 (2022). https://doi.org/10.1109/TIE.2021.3088380

Laldingliana, J., Biswas, P.K.: Artificial intelligence based fractional order PID control strategy for active magnetic bearing. J. Electr. Eng. Technol. **17**, 3389–3398 (2022). https://doi.org/10.1007/s42835-022-01102-6

Sino, R., Baranger, T.N., Chatelet, E., Jacquet, G.: Dynamic analysis of a rotating composite shaft. Compos. Sci. Technol. **68**, 337–345 (2008). https://doi.org/10.1016/j.compscitech.2007.06.019

Thomas, M., Laville, F.: Simulation des vibrations mécaniques par Matlab. Simulink et Ansys—Presses de l'Université du Québec—9 p. (2007)

Wu, M., Zhu, H., Zhang, H., Zhang, W.: Modeling and multilevel design optimization of an AC–DC three-degree-of-freedom hybrid magnetic bearing. IEEE Trans. Ind. Electron. **70**, 233–242 (2023). https://doi.org/10.1109/TIE.2022.3148744

Yadav, V.K., Kumar, P., Bhushan, G.: Multi-objective optimization in geometric design of active magnetic bearing based on force slew rate, overall volume, and total losses considerations through genetic algorithms. J. Inst. Eng. Ser. C **102**(6), 1473–1487 (2021). https://doi.org/10.1007/s40032-021-00746-z

Estimation of the Uncertainties Effect in the Acoustic Performance of Locally Reactive Materials

Hanen Hannachi[1,2](✉), Mohamed Taktak[1,2], Hassen Trabelsi[1], and Mohamed Haddar[1]

[1] Laboratory of Mechanics, Modelling and Production (LA2MP), National School of Engineers of Sfax, University of Sfax, B.P. W3038, Sfax, Tunisia
hanenhannachi672@gmail.com, mohamed.taktak@fss.usf.tn, mohamed.haddar@enis.rnu.tn

[2] Department of Physics, Faculty of Sciences of Sfax, University of Sfax, B.P. 1171, 3000 Sfax, Tunisia

Abstract. In the industrial sector today, uncertainties pose a significant problem that impacts the performance of any material. Many studies are carried out to reduce the error's effects. Uncertainty analysis is an important and necessary stage for conception under uncertainty conditions. The assessment of uncertainties must be taken into account in these situations. Therefore, to assess the impact of these uncertainties, numerical approaches are needed. The Monte Carlo simulation, which is seen as a tool to the study of the propagation of inputs uncertainties in theoretical models, is one of these techniques. it was conceived as a stochastic probabilistic method for solving difficult problems. The proposed study suggests the use of the Monte Carlo simulation to determine the influence of each input error on the calculation of the acoustic absorption coefficient of locally reactive materials. The 95% confidence intervals of the outputs along with the corresponding errors obtained by using the Monte Carlo method are estimated and discussed.

Keywords: Locally Reactive Material · Physical parameters · Acoustic absorption coefficient · Monte Carlo Method

1 Introduction

To reduce the noise radiated by systems such as fans, aircraft turbojet engines, compressors, and automobile exhaust silencers, several types of research have been carried out in recent years to identify the sources of noise, and several works are devoted to study the attenuation and absorption of acoustic waves. The most often used method consists on coating the internal surfaces of the guides with absorbing materials. One of these absorbing materials is the locally reactive materials.

The materials with localized reaction are often made of perforated plates associated with a honeycomb structure where each cell behaves like a resonator. They are used in the aeronautical industry in turbojet engines where thermal constraints do not allow the use of porous materials. These materials are represented by physical models estimating

A. Akrout et al. (Eds.): ICAV 2022, ACM 22, pp. 389–397, 2023.
https://doi.org/10.1007/978-3-031-34190-8_41

the local boundary condition in the form of a complex impedance Z at a wall [1, 2]. This local representation is a good approximation for modeling the presence of absorbing materials on the walls. This kind of absorbent is dependent on the frequency of the acoustic waves.

Many semi-empirical theoretical models have been developed allowing the determination of the acoustic characteristics of these materials, such as Beraneck and Ver [3], Melling [4], Elnady [5] and Ingard [6]. These models were used in recent researches like Othmani et al. [7], which allows the computation of the acoustic impedance of these materials and then deduce their absorption coefficient. These direct models use as inputs the intrinsic acoustic parameters of the material such as plate thickness, plate perforation diameter, cavity length and plate perforation rate. They are used as input to predict the acoustic propagation and radiation in and from duct systems. Experimental and theoretical studies are performed to investigate the acoustic performance of these materials. There is a distinction between theoretical and experiential outcomes. This study aims to find out the source of this difference. For this purpose, an uncertainty analysis based on the Monte Carlo method to study the propagation of uncertainties using the Elnady model [3] has been developed.

The outlines of this paper are as follows: in Sect. 2, a theoretical recall on the calculation of the acoustic parameters of locally reactive materials. Section 3 presents the details of the introduction of uncertainties using the Monte Carlo method. Finally, numerical results are presented and discussed in Sect. 4.

2 Computation of the Acoustic Absorption Coefficient

The sound absorption coefficient is calculated from the surface acoustic impedance of a locally reactive material as follows Lee and Kwon [8]:

$$\alpha = \frac{4\mathrm{Re}(Z)}{[1 + \mathrm{Re}(Z)]^2 + [\mathrm{Im}(Z)]^2} \tag{1}$$

with Z is the normalized surface acoustic impedance of the of locally reactive material (Fig. 1).

The global impedance of a system composed of a perforated plate and a cavity can be composed of two distinct impedances: the impedance of the cavity Z_{cav} and the impedance of the plate Z_p.

$$Z = Z_{CAV} + Z_P \tag{2}$$

The cavity impedance is known analytically.

$$Z_{cav} = -j\cot(kl_{cav}) \tag{3}$$

l_{cav} The Length of the cavity.

The used acoustic impedance model is the one presented by Elnady and Boden [3] for the perforated plate which is expressed as follows:

$$Z = \mathrm{Re}\left\{\frac{ik}{\sigma C_D}\left[\frac{t}{F\left(k_s'/2\right)}\right]\right\} + i\mathrm{Im}\left\{\frac{ik}{\sigma C_D}\left[\frac{t}{F\left(k_s'/2\right)} + \frac{\delta_{im}}{F(k_s d/2)}\right]\right\} \tag{4}$$

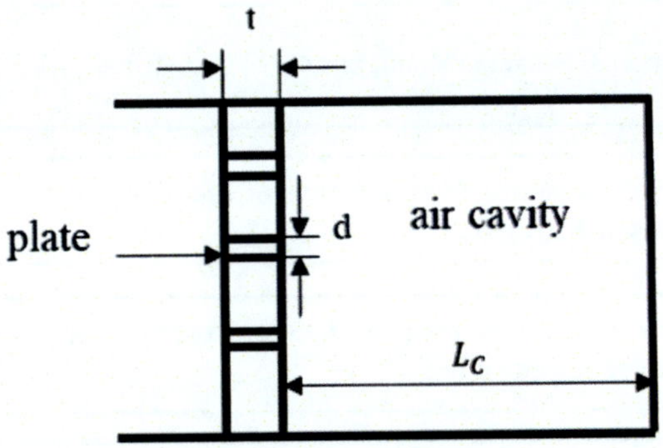

Fig. 1. Studied material.

Elnady and Boden [3] proposed an empirical model, based on the Melling model, but which adapts the length corrections and removes the Fok's functions.

It replaces the corrections $\delta_{im} = 8d/3\pi$ and $\delta_{re} = d$ used by Ingard [6], by a correction resulting from its measurements without flow:

$$\delta_{re} = 0.2d + 200d^2 + 16000d^3 \text{ and } \delta_{im} = 0.2856d \quad (5)$$

with C_d is the discharge coefficient, d is the pore diameter, t is the plate thickness, σ is the plate porosity.

3 Uncertainty Analysis: The Monte Carlo Method

The Monte-Carlo method is a very efficient numerical method for the propagation of distributions. This method can be used to evaluate the uncertainty of analytical processes as presented in Bouazizi, et al. [9], Taktak et al. [10], (Trabelsi et al. [11], (Gallardo et al. [12], (Hwang and Ta [13] and Öztürk and Kahraman [14].

The computational steps of the Monte Carlo simulation used in this analysis are grouped in the diagram shown in Fig. 2. The first step is to identify the probability distribution for each input. Then, a set of random inputs x is selected according to the chosen distribution. After that, the algorithm is started for each value of the input to create a set of outputs. Finally, the outputs are grouped and analyzed.

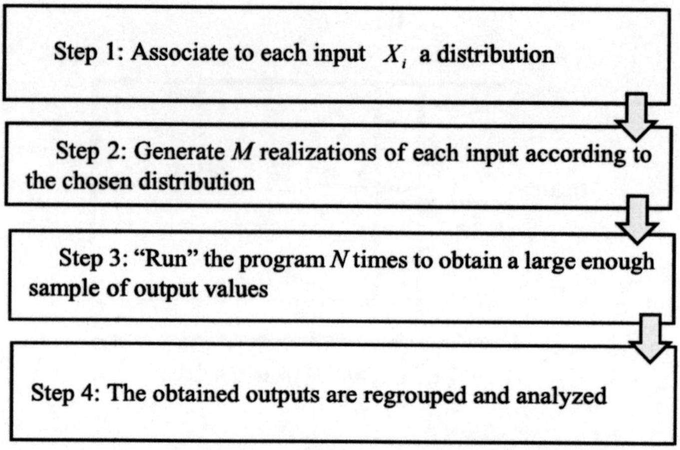

Fig. 2. Uncertainty analysis algorithm using the Monte Carlo method

4 Uncertainty Analysis Results

Uncertainties have been added to the mean value for each parameter of a locally reactive material placed in a duct element in the frequency band [0–4000 Hz]. It is simulated using the Monte-Carlo technique by MATLAB software.

The properties of this locally reactive material studied are shown in Table 1. This table presents the mean value used for each parameter.

Table 1. Proprieties of the studied locally reactive material.

Parameters	Value
The plate thickness t (mm)	1
The pore diameter d (mm)	1
The plate porosity σ	0.025
The Length of the cavity l_{cav} (mm)	16

In this part, the influence of the errors of parameters of the plate-air structure on the sound absorption coefficient is studied.

The plate-air structure has a very strong absorption over a restricted frequency range but is independent of the incidence of the wave. Perforated plate liners are a widespread technology to attenuate noise production. The behavior of these liners being local can be described by a surface impedance, spatially constant.

Once the Gaussian distribution associated with the input quantities is defined, a ±5% variation is added to the mean value of each input keeping the other inputs fixed. After a random selection of a set of each input (plate thickness, plate hole diameter, cavity length, and plate perforation rate), simulations are run to get the distribution of the output quantity (absorption coefficient). 10,000 iterations are performed, a sufficient number to ensure convergence of the results (the 95% confidence intervals bounded by the minimum and maximum of the *N* values of the absorption coefficient of the locally reactive material and the corresponding mean value). Figures 3, 4, 5 and 6 show the results of the uncertainty analysis.

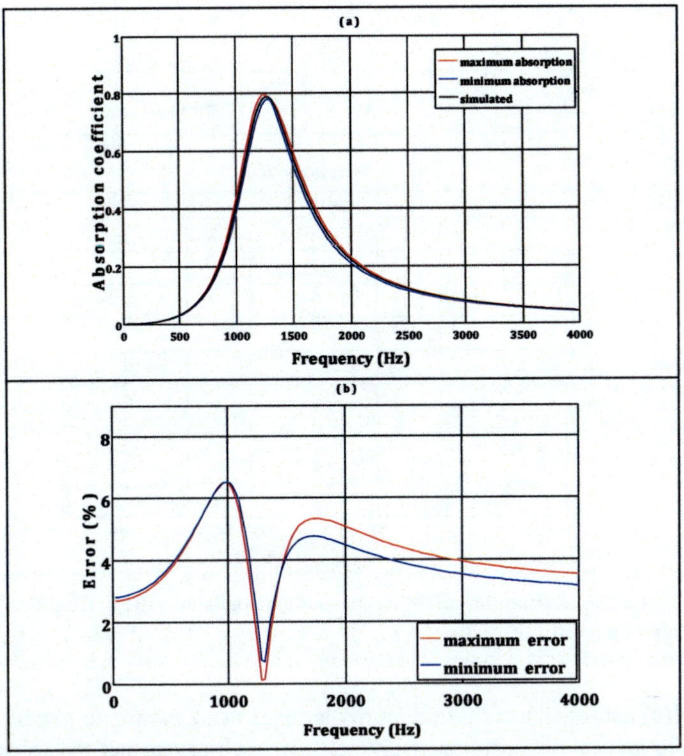

Fig. 3. Plate thickness uncertainties effect on the acoustic behaviors: 95% Confidence interval (a) and the corresponding errors (b).

The results of the uncertainty analysis using plate thickness are shown in Fig. 3. This figure shows that this parameter has a negligible influence on the absorption of the locally reactive material. Indeed, the thickness of the confidence interval is very small (Fig. 3(a)), indicating that the effect of this parameter is negligible. This result is validated by the results presented in Fig. 3(b) which illustrates the variation of the maximum and minimum errors of the plate thickness variation. Thus, a variation of ±5% of the nominal value of the plate thickness generates a minimum error equal to 6% and a maximum error equal to 6%.

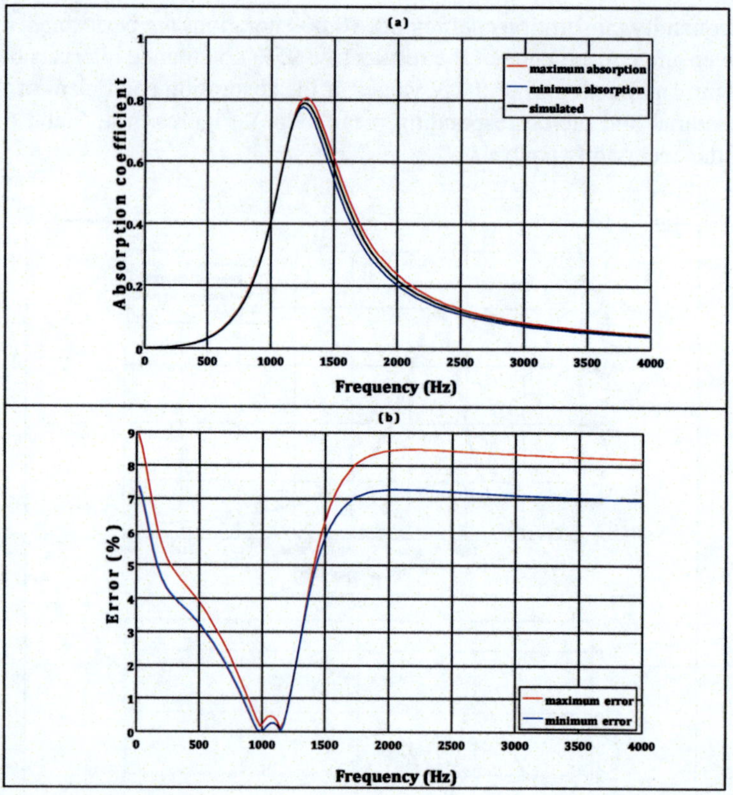

Fig. 4. Pore diameter uncertainties effect on the acoustic behaviors: 95% Confidence interval (a) and the corresponding error (b).

Figure 4(a) demonstrates that the pore diameter has a moderate impact on the estimation of the acoustic absorption coefficient. It is observed that the thickness of the confidence interval is medium.

The variation of the maximum and minimum errors due to the pore diameter errors is illustrated in Fig. 4(b). It shows that the variation of ±5% of the nominal value of pore diameter affects the results with a minimum error equal to 7% the and a maximum error equal to 8.5% at high frequency in the frequency band between [2000–4000 Hz].

a variation of 5% of the nominal value of the plate porosity has a more important impact on the absorption coefficient. Indeed, the thickness of the confidence interval is significant (Fig. 5(a)) indicating that the effect of this parameter is important.

Figure 5(b) shows that a variation of ±5% of the nominal value of the viscous and thermal length results in a minimum error equal to 14% at a frequency equal to 1000 Hz, then decreases to 8% in the band [1500–4000 Hz], and a maximum error equal to 19% at a frequency equal to 1000Hz then decreases to 8% in the band [1500–4000 Hz].

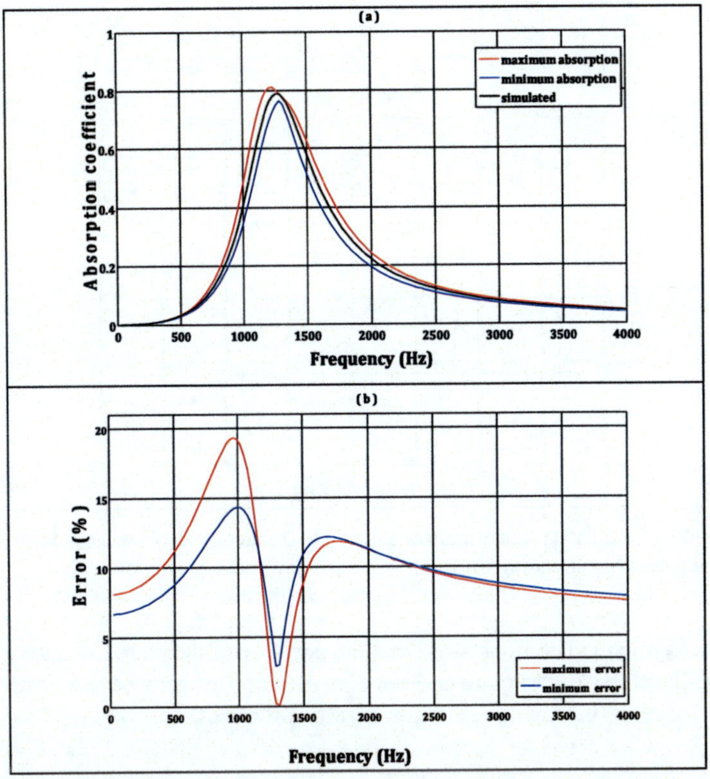

Fig. 5. Plate porosity uncertainties effect on the acoustic behaviors: 95% Confidence interval (a) and the corresponding error (b).

Figure 6 shows that the effect of cavity length on sound absorption is negligible, as shown in Fig. 6(a) which shows that the thickness of the confidence interval is very small.

This result is confirmed by the results presented in Fig. 6(b) which shows the variation of the maximum and minimum errors due to the normal distribution of the cavity length variation. Thus, a variation of ±5% of the nominal value of the cavity length gives a minimum error equal to 2% (at low frequency) and decreases to 0.3% at 4000 Hz and a maximum error equal to 1.8% and decreases to 0.3 Hz at 4000H.

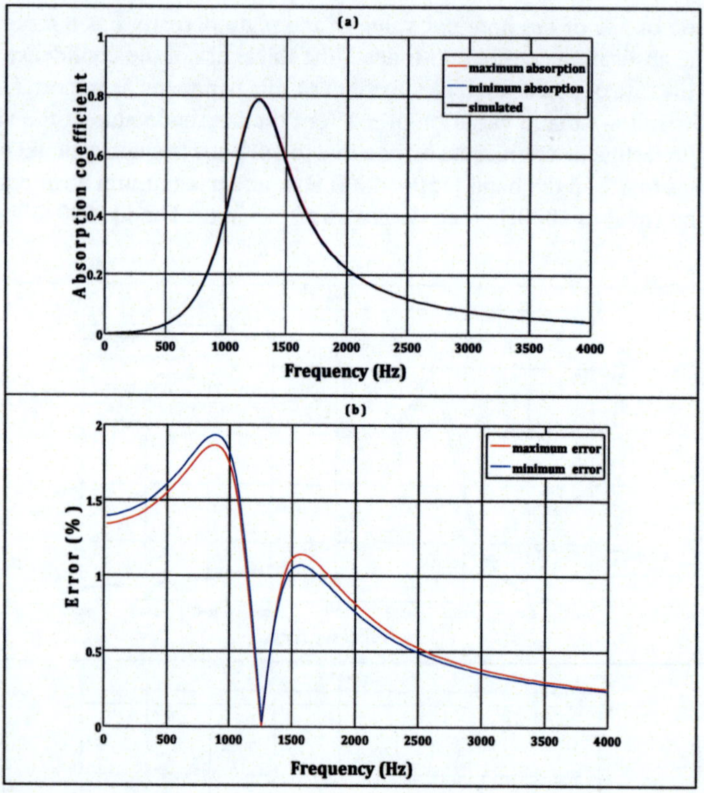

Fig. 6. Length of the cavity uncertainties effect on the acoustic behaviors: 95% Confidence interval (a) and the corresponding error (b).

The results obtained demonstrated that the porosity of the plate has a very important impact. The thickness of the plate and the diameter of the pores have a little significant effect. The length of the cavity has a non-significant effect.

5 Conclusion

To demonstrate the impact of mistakes in the characteristics of locally reactive materials on sound absorption, an uncertainty analysis is provided. The influential material factors that must be taken into account while designing these absorbers were identified by the uncertainty analysis.

It's intriguing to estimate uncertainty using the Monte-Carlo approach. For calculating 95% confidence intervals for each input parameter, it is the best choice.

In fact, this study demonstrates that the plate's porosity has the greatest impact. The diameter of the pores and the plate thickness have minimal impact.

The local material's ability to absorb sound is little impacted by variations in the cavity length.

References

1. De L'épine, Y.B.: Identification de l'impédance d'un traitement en présence d'un écoulement. Ph.D thesis, Université de Technologie de Compiègne (2017)
2. Delattre, G.: Impédance de paroi des matériaux à réaction localisée. Ph.D thesis, Université Pierre et Marie Curie - Paris VI (2009)
3. Beranek, L., Ver, I.L.: Noise and Vibration Control Engineering: Principles and Applications (1992)
4. Melling, T.H.: The acoustic impedance of perforates at medium and high sound pressure levels. J. Sound Vib. **65**(1), 1–65 (1973)
5. Elnady, T., Boden, H.: On semi-empirical liner impedance modeling with grazing flow. In: Proceedings of 9th AIAA/CEAS, South Carolina, USA (2003)
6. Ingard, U.: On the theory and design of acoustic resonators. J. Acoust. Soc. Am. **25**, 1037–1061 (1953)
7. Othmani, C., Hentati, T., Taktak, M., Elnady, T., Fakhfakh, T., Haddar, M.: Effect of liner characteristics on the acoustic performance of duct systems. Arch. Acoust. **40**(1), 117–127 (2015)
8. Lee, D.H., Kwon, Y.P.: Estimation of the absorption performance of multiple layer perforated panel systems by transfer matrix method. J. Sound Vib. **278**, 847–860 (2004)
9. Bouazizi, L., Trabelsi, H., Othmani, C., Taktak, M., Chaabane, M., Haddar, M.: Uncertainty and sensitivity analysis of porous materials acoustic behavior. Appl. Acoust. **144**, 64–70 (2019)
10. Taktak, M., Ville, J.-M., Gabard, G., Haddar, M., Foucart, F.: A 3D two port scattering matrix based method for educing liner impedance: simulation and error evaluation. Adv. Acoust. Vibr. 1–17 (2009)
11. Trabelsi, H., Abid, M., Taktak, M., Fakhfakh, T., Haddar, M.: Effect of the aerodynamic force modeling on the tonal noise prediction model for axial fan: sensitivity and Uncertainty analysis. Appl. Acoust. **117**, 61–65 (2017)
12. Gallardo, S., Querol, A., Ortiz, J., Ródenas, J., Verdú, V.G.: Uncertainty analysis in environmental radioactivity measurements using the Monte Carlo code MCNP5. Radiat. Phys. Chem. **116**(2), 14–218 (2015)
13. Hwang, Y.L., Ta, T.N.: Uncertainty analysis of CNC machine tools based on Monte Carlo method. In: Applied Mechanics and Materials, vol. 900, pp. 9–13 (2020)
14. Öztürk, S., Kahraman, M.F.: Modeling and optimization of machining parameters during grinding of flat glass using response surface methodology and probabilistic uncertainty analysis based on Monte Carlo simulation. Measurement **145**, 274–291 (2019)

A Multi-objective Model Case Study for the Sustainable Flow-Shop Scheduling Issue

Hager Triki[1] (✉), Hanen BenAmmar[2], and Yasmine Tchaicha[1]

[1] University of Sfax, Laboratory of Mechanic, Modeling and Production (LA2MP), Engineering School of Sfax, Tunisia, Higher Institute of Industrial Management of Sfax, Sfax, Tunisia
Hager_Triki@yahoo.fr

[2] Mechanics, Modeling and Production Research Laboratory, National Engineering School of Sfax (ENIS), Sfax University, Route de Sokra B.P.1173, 3038 Sfax, Tunisia
hanen.benammar@enis.tn

Abstract. Sustainable scheduling of distributed production has gained increasing attention as sustainable manufacturing and economic globalization have grown. Nevertheless, in most real industrial cases, the majority of sustainability frameworks cannot engage three industrial sustainability criteria holistically. Despite the fact that there are numerous other environmental and social variables, the majority of the study on sustainable scheduling focuses on environmental issues like electricity usage or carbon emissions. This study addresses a real industrial SFSP (Sustainable Flow-Shop Scheduling Problem) that exists within the Med light firm in this context. This company manufactures a variety of lighting products. A sustainable production parameters are included in this industrial SFSP model. With the aim of reducing makespan, ineffective social impact (ISI), and overall energy consumption, a novel mathematical model is formulated of this SFSP in this company. To determine the set of Pareto-optimal solutions, the problem is solved using the augmented epsilon-constraint approach. Investigating the impact of variation in the three industrial sustainability criteria in this industry is the aim of this study. The set of Pareto-optimal solutions provides the decision maker with many compromise job sequences, allowing him to choose the best option based on his preferences between three sustainability criteria.

Keywords: flow-shop scheduling Augmented epsilon-constraint · makespan · ineffective social impact · overall energy consumption

1 Introduction

Product, process, and system—the three key components of manufacturing—should minimize waste, maximize resource efficiency, lessen adverse environmental effects, provide operational safety, and enhance individual health while maintaining and/or improving the quality of the product and the process to achieve sustainable production (Jawahir et al. 2013). In fact, sustainable manufacturing is the process of interdependently taking into account economic, social, and environmental factors (Huang and Badurdeen 2018). According to data, the industrial sector generates more than 38% of all

A. Akrout et al. (Eds.): ICAV 2022, ACM 22, pp. 398–406, 2023.
https://doi.org/10.1007/978-3-031-34190-8_42

carbon emissions and uses almost half of the energy consumed globally (Yin et al 2017, Zheng et al. 2015). The majority of research on sustainable scheduling is concerned with different environmental factors (Duan et al 2021; Akbar and Irohara 2021). However, there are few studies involving the three sustainability pillars (Lu et al 2020; Abedini et al 2019). Many businesses typically prioritize economic criteria while ignoring social concerns. With regard to the social component, the harmful effects of manufacturing industries, such as overtime, noise pollution and high-risk industrial operations, are also an important social problem (Garcia-Nieto et al 2012, Lu et al 2019). The Sustainable Flow-Shop Scheduling Problem (SFSP) is a broad-ranging and significant problem in the green manufacturing processes. This problem is characterized by the scheduling of n jobs on m machines while adhering to clearly defined limitations and achieving certain sustainable aspects. This study analyzes a SFSP process at Med Light Company. It is a manufacturer of lighting equipment with a focus on lighting (https://medlight.com.tn). A wide variety of lighting products are designed, produced, and marketed as part of its business process for the domestic and international markets.

As a result, this company is being used as a case study to investigate at the Sustainable Flow-Shop Scheduling Problem (SFSP), which entails lowering the energy consumption (Em), ineffective social impact (ISI), and makespan (Cmax) of the current machines all at once.

The remainder of this essay is organized as follows: The mathematical model suggested to address the industrial challenge is described in Sect. 2. The chosen resolution strategy is demonstrated in Sect. 3. The computation results are shown in Sect. 4. Finally, Sect. 5 presents final observations.

2 Mathematical Formulation

The SFSP includes allocating n jobs across m linearly organized machines. This SFSP type scheduling model aims to minimize Cmax, ISI and Em by the current machines in the production line while adhering to the restrictions of a conventional SFSP. The mathematical model is an extension of Triki et al. (2021) flow shop model. The following is a technical description of this model:

2.1 Objective Functions

The sustainability in this SFSP must adhere to three aspects (Wen et al. 2020). As a result, a SFSP is addressed while taking makespan, ineffective social impact (ISI), and overall energy usage into account.

- *Economic criterion*

The makespan in shop scheduling issues typically serves as a representation of the economic criterion (Elkington 2010).

From some extent, Makespan can indicate the economic value of an organization. Therefore, reducing makespan is the economic goal of SFSP. Below is a definition of it:

$$\mathrm{F1} = \mathrm{Cmax} = \max(FT_{ij}) \tag{1}$$

– *Ineffective social impact criterion*

Health, happiness, the working environment, and social capital in businesses make up the social criterion (Elkington 2010).

The social criterion in this research work primarily pertains to the workplace environment, which measures the level of safety and comfort of businesses. These workplace environmental concerns, like noise pollution, are typically high-risk.

The efficiency of the workforce and the company's reputation are negatively impacted by production and overwork.

Generally, the social aspect is related to job processing times (Amrina and Vilsi 2015). A penalty coefficient, or ineffective social impact(ISI), is posed to this part into this social aspect.

This penalty coefficient relates to how quickly an operation is processed. The penalty coefficient increases with processing time of the job.

$$\mathrm{F2} = \mathrm{ISI} = \sum_{j=1}^{m}\sum_{i=1}^{n}\sum_{R=1}^{n} T_{ij}W_{ij}F_{iR} \tag{2}$$

– *Environment criterion*

Massive energy use contributes to environmental issues like global warming. The environment criterion in this study is to reduce Em which this objective is determined using the formula below:

$$\mathrm{F3} = E_m = \sum_{j=1}^{m} C_j'\left[P_j^{idle}(1 - K_j) + P_j^{load}K_j\right] \tag{3}$$

where,

$$K_j = \frac{\sum_{i=1}^{n} T_{ij}}{C_j'}, \text{ with K1} = 1 \text{ and } C_j' = \max_{1\le i\le n}\{Y_{ij} + T_{ij}\} - \min_{1\le i\le n}\{Y_{ij}\} \tag{4}$$

2.2 Indexes, Parameters and Variables

The Table 1 exhibites the indexes and parameters of the proposed mathematical model.

Table 1. Indexes, parameters and tables

Indexes	Signification
N	number of jobs
J	the machine, $j = 1..m$
i,k	the position of the job, $i, k = 1..n$
R	job position index
Parameters	Signification
$C_j^{'}$	the completion time of all jobs of the machine j
T_{ij}	processing time of job i on the machine j
K_j	Utilization rate of machine j
P_j^{idle}	the idle power of the machine j during the idle state
P_j^{load}	the loading power of the machine j during the loading state
w_{ij}	coefficient of labor force i employment on machine j
Variable	Signification
Y_{ij}	Start date of job i on machine j
X_{ik}	Boolean variable, equal to 1 if job i is performed after job k and 0 otherwise. (job i enters after job k)
FT_{ij}	downtime of job i on machine j
F	equal to 1 if job i occupies position P, 0 otherwise

2.3 Constraints

$$\sum\nolimits_{i \in J} \sum_{K \in J|\{i\}} X_{ik} = n - 1 \; \forall i \in J \tag{5}$$

$$\sum_{k \in J|\{i\}} X_{ik} \leq 1 \; \forall i \in J \tag{6}$$

$$Y_{11} = 0 \text{ for } i = j = 1 \tag{7}$$

$$Y_{k1} \geq \sum_{k \in J|\{i\}} X_{ki}(Y_{i1} + T_{i1}) \forall i \in J|\{n\}, \forall k \in J|\{1\} \tag{8}$$

$$Y_{kj} \geq \sum_{k \in J/\{i\}} X_{ki} \max\left[\left(Y_{ij} + T_{ij}\right); \left(Y_{k,j-1} + T_{k,j-1}\right)\right] \forall j \in M/\{1\} \, and \, \forall i \in J/\{n\} \tag{9}$$

$$C_{\max} \geq Y_{im} + T_{im} \forall i \in \{1..n\} \tag{10}$$

$$X_{ik} \in [0, 1] \forall i \in \{1..n\} \, and \, \forall k \in J|\{i\} \tag{11}$$

According to the limitations (5), jobs *i* and *k* are part of a chain of n jobs. Constraints (6) set the decision variables' range. The start date of the first task completed on machine 1 is equal to zero, according to constraint (7). It indicates that machine 1 is never left idle. The start date of job *k* on the first machine is determined by constraint number eight (8). Because of the constraints (9), job *k*'s start date in machine *j* must be later than the later of the two dates:

* date1: the total of the job's start date, which is carried out by machine *j* before machine *k*, and this machine's processing time,

*date2: the total of job *k*'s start date on machine *j-1* (prior) with this machine's processing time.

The upper bound of the total of the start time of job *i* for the final machine m with its processing duration on this machine is determined by constraints (10). This binding provides the makespan. As indicated in Eq. (11), X_{ik} is a binary variable.

3 Resolution Method

The MOOP (Multi-Objective Optimization Problem) has been solved using a variety of methods that have been described in the literature (Fan et al 2016,Yan et al. 2014). One of the well-known techniques for treating the MOOP and locating Pareto-optimal fronts is the Ɛ-constraint method (Mavrotas 2009). In this paper, a set of Pareto-optimal solutions is produced using the augmented epsilon constraint. This section presents the augmented epsilon-constraint used after first describing the epsilon-constraint approach.

3.1 Description of the Epsilon-Constraint Method

In order to identify Pareto-optimal fronts for MOOP, Chankong and Haimes (1983) have developed the epsilon-constraint technique. One of the objective functions is selected as the primary objective function to be optimized in this approach, and the other objective functions are considered as constraints parts of the model by giving each of them some permitted levels. The set of Pareto-optimal solutions is formed by changing the value of *i*. The reformulated issue is written as follows:

Minimize $F_1(x)$.

$$\text{Subject to } F_i(x) \leq \varepsilon_i \; \forall i = 2..N;$$

$$x \in X$$

where M is the number of objective functions

3.2 Description of the Augmented Epsilon-Constraint Method

In comparison to the weighing approach, the epsilon-constraint method has a number of advantages. Nevertheless, Mavrotas (2009) suggested three points before its execution that require extra care:

(1) Determining the range of different objective functions inside the useful set.
(2) Confirmation of the viability of the Pareto solutions presented.

(3) The more drawn-out response time for answers to problems with multiple objectives.

To avoid these three issues, the epsilon constraint augmented method is employed. By adding surplus variables or positive slack, this approach converts the inequality of the objective function constraints into equality constraints. As a result, the new issue is:

$$\text{Minimize F} = \left(F_1(x) + \delta \times \left(\frac{f_1}{c_1} + \frac{f_2}{c_2} + \ldots \frac{f_N}{c_N}\right)\right)$$

$$\text{Subject to } F_i(x) + f_i = \varepsilon_i \forall i = 2..M;$$

$f_i \in \mathbb{R}^+$, $x \in X$, M is the quantity of objective functions.

δ: A small number typically falls between 10–3 and 10–6. The payoff table is used to determine the value of the c_i, which represents the ith objective function's range.

Calculated through individual optimization are the maximum and minimum values of each objective function. Therefore, by deducting the maximum and minimum values of each objective function, the range of each is determined.

4 The Application of the Augmented Epsilon-Constraint Method

To solve the proposed SFSP, the mathematical formulation is implemented with LINGO software 64–19 on a PC within Windows 10 Professional.

Within the Med Light Company, the SFSP is utilized. The industrial instance of SFSP handles 7 jobs, which are carried out on four machines. Tables 2, 3 and 4 include the machine's necessary experimental data. The processing times for each job on each machine are listed in Table 2. The power information for the devices in an idle and active condition is contained in Table 3. Table 4 lists the ineffective social impact for jobs.

Table 2. Processing Times of jobs (in min)

	Job1	Job2	Job3	Job4	Job5	Job6	Job7
Machine1	25.972	51.944	12.819	76.916	128.194	33.33	20.51
Machine2	80.27	160.5	35.972	240.83	401.388	99.05	61.06
Machine3	128.55	511.77	127.361	509.583	1276.11	110.138	67.77
Machine4	52.361	103.33	21.25	157.083	258.33	67.167	40.889

Table 3. Power for each machine in Watt

Power type (in watt)	Machine 1	Machine 2	Machine 3	Machine 4
pidle	3800	3900	3800	3900
pload	4180	5850	4560	6240

Table 4. Ineffective social impact

	Job1	Job2	Job3	Job4	Job5	Job6	Job7
Machine1	0.2	0.3	0.1	0.6	0.4	0.3	0.4
Machine2	0.4	0.2	0.3	0.3	0.2	0.1	0.2
Machine3	0.3	0.1	0.1	0.2	0.1	0.4	0.3
Machine4	0.5	0.6	0.5	0.7	0.8	0.8	0.7

5 Results and Discussion

The outcomes of the chosen methodology are displayed in Fig. 1 and Table 5 below. The derived non-dominated solutions are typically found in the first front (the Pareto front in Fig. 1). Four non-dominated solutions (job sequences) are, in fact, found.

Additionally, our approach solves the industrial problem with a fairly reasonable computation time of just a few seconds (1 min).

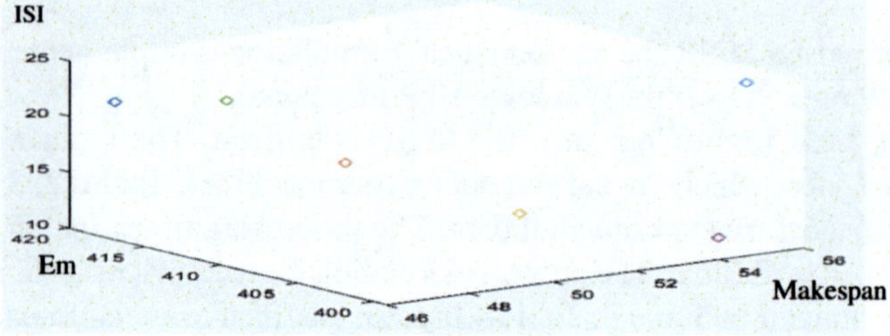

Fig. 1. The Pareto front

Table 5. The non-dominated solutions

Non dominated solutions (NS)	Job sequence	Makespan (in hour)	Em (KWh)	ISI (in hour)
NS1	1234567	47,12	420	20,7
NS2	1453276	49,4	411,1	17
NS3	3256721	50,3	402	15
NS4	4172635	54	399	12

The goal of this study is minimize three criteria: makespan, Em, and SI existing in the chosen company. The four obtained non-dominated solution by the Fig. 1 present the same evolution, in Table 2, of the three aspects of sustainability. Indeed, when the makespan,increases (47,12 to 54), the Em decreases(420 to 399). This proves that these two aspects are confluent. However, the ISI decreases (20.7 to 12) because this aspect is related to the speed with which an operation is processed.The decision-maker has indeed more possibilities to elaborate a plan which achieves the right compromise between the three opposing objectives. He can choose the best compromise between Makespan and Em taking into account the social aspect that is ISI.

6 Conclusion

The enterprise Med Light is the setting for this paper's industrial case study of a scheduling issue of the sustainable flow-shop variety. While optimizing the three pillars of sustainability within production process, the proposed work is described as a multi-objective optimization problem. The created multi-objective SFSP aims to decrease simultaneous the makespan (C_{max}), ineffective social impact (ISI) and overall energy consumption of the line's current machines.

In comparison to the genuine solution that was executed, the proposed method offers a variety of trade off job sequences solution that lower energy consumption and makespan time taking into account the social labor penalty (ISI). The performance of the company lines is thus enhanced by the suggested method epsilon constraint.

By finding a trade-off between three sustainability pillars, this model can be extended by adding many aspects concerning:

- the inverse logistic in this flow shop type
- the set up time of the machine

Taking into account the configuration of the industrial challenge would be a relevant extension.

References

Chankong, V., Haimes, Y.: Multi-objective Decision Making Theory and Methodology. Elsevier, New York (1983)

Jawahir, I.S., Badurdeen, F., Rouch, K.E.: Innovation in sustainable manufacturing education. In: Proceedings of 11th Global Conference on Sustainable Manufacturing. Berlin, Germany, 23–25 September, pp. 9–16 (2013). ISBN 978-3-7933-2609-5

Huang, A.H., Badurdeen, F.: Metrics-based approach to evaluate sustainable manufacturing performance at the production line and plant levels. J Cleaner Produc. **192**, 462–476 (2018)

Yin, L.J., Li, X.Y., Gao, L., Lu, C., Zhang, Z.: A novel mathematical model and multi-objective method for the low-carbon flexible job shop scheduling problem. Sustain. Comput.-Inform Syst. **13**, 15–30 (2017)

Zheng, H.Y., Wang, L.: Reduction of carbon emissions and project makespan by a Pareto-based estimation of distribution algorithm. Interna J. Prod. Econ. **164**, 421–432 (2015)

Duan, J.G., Zhang, Q.L., Zhou, Y., Wang, Y.: Sustainable scheduling optimization of mixed-line production for large marine power components. J. Cleaner Prod. **280**, 124461 (2021)

Akbar, M., Irohara, T.: Scheduling for sustainable manufacturing: a review. J. Cleaner Prod. (2018). https://doi.org/10.1016/j.jclepro.2018.09.100

Lu, C., Gao, L., Gong, W., Hu, C., Yan, X., Li, X.: Sustainable scheduling of distributed permutation flow-shop with non-identical factory using a knowledge-based multi-objective memetic optimization algorithm. Swarm Evol. Comput. (2020). https://doi.org/10.1016/j.swevo.2020.100803

Abedini, A., Li, W., Badurdeen, F., Jawahir, I.S.: Sustainable production through balancing trade-offs among three metrics in flow shop scheduling. In: Procedia CIRP, vol.80, pp. 209–214 (2019)

Garcia-Nieto, J., Alba, E., Olivera, A.C.: Swarm intelligence for traffic light scheduling: application to real urban areas. Eng. Appl. Artif. Intell. **25**, 274–283 (2012)

Lu, C., Gao, L., Pan, Q.K., Li, X.Y., Zheng, J.: A multi-objective cellular grey wolf optimizer for hybrid flowshop scheduling problem considering noise pollution. Appl. Soft Comput. **75**, 728–749 (2019)

Elkington, J.: Cannibals with forks : the triple bottom line of 21st century business. Environ Qual. Manage **8**, 37–51 (2010)

Amrina, E., Vilsi, A.L.: Key performance indicators for sustainable manufacturing evaluation in cement industry. In: Selige, G., Yusof, N.M, (eds.) 12th Global Conference on Sustainable Manufacturing - Emerging Potentials, pp. 19–23 (2015)

Wen, X., Li, X., Gao, L., Wang, K., Li, H.: Modified honey bees mating optimization algorithm for multi-objective uncertain integrated process planning and scheduling problem. Int. J Adv. Robot. Syst. **17**, 1729881420925236 (2020)

Triki, H., BenYahia, W., Masmoudi, F.: A case study of a bi-objective model for flow-shop scheduling problem. In: International Conference Design and Modeling of Mechanical Systems CMSM 2021: Design and Modeling of Mechanical Systems – V, pp. 655–663 (2021)

Wen, X., Li, X., Gao, L., Wang, K., Li, H.: Modified honey bees mating optimization algorithm for multi-objective uncertain integrated process planning and scheduling problem. Int. J. Adv. Robot. Syst. **17**, 1729881420925236 (2020)

Mavrotas, G.: Effective implementation of the E-constraint method in multi-objective mathematical programming problems. Appl. Math. Comput. **213**(2), 455–465 (2009). https://doi.org/10.1016/j.amc.2009.03.037

Fan, Z., et al.: An improved epsilon constraint handling method embedded in MOEA/D for constrained multi-objective optimization problems. In: IEEE Symposium Series on Computational Intelligence (SSCI) (2016)

Yang, Z., Cai, X., Fan, Z.: Epsilon constrained method for constrained multiobjective optimization problems: some preliminary results. GECCO Comp 2014. In: Proceedings of the Companion Publication of the 2014 Annual Conference on Genetic and Evolutionary Computation July 2014, pp. 1181–1186 (2014). https://doi.org/10.1145/2598394.2610012

A Phase Field Numerical Modelling of Quasi-brittle Material Fracture Applied to Low Velocity Impact

Mariem Saidane[1](✉), Sana Koubaa[1], Zoubeir Bouaziz[1], and Radhi Abdelmoula[2]

[1] Laboratory of Applied Fluids Mechanics of Process Engineering and Environment, ENIS, University of Sfax, Sfax, Tunisia
{mariem.saidane,sana.kouba,zoubeir.bouaziz}@enis.tn

[2] University of Paris 13, Laboratory of Process and Materials Sciences, 93430 Villetaneuse, France
radhi.abdelmoula@univ-paris13.fr

Abstract. Cracking is the primary damage and failure mechanism for most brittle materials. The development of numerical simulation tools of failure phenomena has become necessary to evaluate the risk of initiation and propagation of defects in structures. Phase field approaches are one of the most promising ways to analyze the crack behavior of such materials. In this study, we introduce a numerical phase field technique for modeling the damage of a quasi-brittle material applied to a low velocity impact. The allowed models are framed by applying variational principles. Two numerical examples are used to predict bifurcation mechanism. Results prove the influence of the impact speed on brittle solids. A great agreement with literature is depicted.

Keywords: phase field model · quasi-brittle material · variational principles · failure

1 Introduction

The failure of quasi-brittle materials such as ceramics or concrete can be represented schematically by the sequence of nucleation and coalescence stages of microcracks [1]. Nucleation-coalescence processes can occur in the original undamaged structure in the most stressed regions, which then lead to the formation of macroscopic cracks. The modeling of this failure process is a particularly important issue when considering the strength of concrete structures, and especially when predicting the permeability of damaged structures [2].

The development of numerical simulation tools of failure phenomena has become necessary to anticipate material behavior [3, 4].

In recent years, fracture variational methods has become an effective and popular method for predicting damage and failure of engineering materials. In the method proposed by Francfort and Marigo [5], Griffith's brittle fracture is modified as a global

A. Akrout et al. (Eds.): ICAV 2022, ACM 22, pp. 407–414, 2023.
https://doi.org/10.1007/978-3-031-34190-8_43

minimization problem of the total energy to fracture a solid [6]. The damage theory describes the evolution of the phenomena between the virgin state and the initiation of the mesoscopic crack. This evolution is often accompanied by permanent deformation. The first continuous modeling of the damage was proposed by Kachanov [7] concerning the creep of metals under one-dimensional stress.

A considerable number of recent contributions have advanced our knowledge in this field. However, many open issues still stand in the way of a definitive answer including the lack of consideration for micro-inertia when modeling dynamic fracture. Parrinello's work [8] and Kamensky et al.'s [9] considered micro-inertia factor that was incorporated into the models. Dynamic cracks have not been considered when calculating the durability of a brittle material.

Cross-cutting research reveals the physical ramifications of damage evolution within the phase field framework [10]. This leads to the creation of a damaged phase-field model based on Hamilton's principle.

This paper is organized as follows: Sect. 2 addresses the gradient damage modelling. Section 3 is devoted to detail the numerical implementation. Section 4 presents two numerical examples to demonstrate the efficiency of the proposed model. And finally, conclusions are presented in Sect. 5.

2 Gradient Damage Modelling

In this section, an overview of the gradient damage models is presented. The bulk energy density depends on the gradient of damage. The presence of the damage gradient $\nabla\alpha$ approximates the non-locality of constitutive model. In doing so, it is equivalent to replacing the discontinuity surface of the displacement field by small areas in which the displacement gradient is large (damaged areas). In fact, the bulk energy density can be introduced as the sum of the stored elastic energy Ψ, the local part of the dissipated energy w(α) and its non-local part involving $\nabla\alpha$ (Eq. 1). According to Table 1, the scalar damage field is assumed to be bounded in [0, 1], where α equals to zero is denoting the undamaged material and α equals to one is referring to the completely damaged material.

$$W\left(\varepsilon^{e}, \alpha, \nabla\alpha\right) = \Psi\left(\alpha, \varepsilon^{e}\right) + w(\alpha) + \frac{1}{2} w_1 l^2 \nabla\alpha.\nabla\alpha \quad (1)$$

The elastic energy is described as:

$$\Psi\left(\alpha, \varepsilon^{e}\right) = \frac{1}{2} A(\alpha)\varepsilon(u) : \varepsilon(u) \quad (2)$$

The damage model can be given by:

$$A(\alpha) = \left(1 - \alpha^2\right)A_0, \quad w(\alpha) = w_1\alpha \quad (3)$$

where: $A_0 is$ the stiffness tensor of the undamaged material.

The local and non local parts of the energy density lost due to damage in (1), (2), and (3) can be considered as equivalent functional densities for the surface of Griffith cracks in damage gradient models. The crack is then represented by a gradient-damage

Table 1. Constitutive Model Parameters

Scalar damage field	α
Elastic strain tensor	ε^e
Gradient damage field	$\nabla\alpha$
State of the volume element	$(\varepsilon^e, \alpha, \nabla\alpha)$
The bulk energy density of the material	$W(\varepsilon^e, \alpha, \nabla\alpha)$
The stiffness tensor of the material	$A(\alpha)$
The linearized strain of displacement u	$\varepsilon(u)$
Damage constitutive laws	$\alpha \to A(\alpha)$ $\alpha \to w(\alpha)$
Material toughness	$\sigma_c = \sqrt{w_1 E}$
Yield strength	$w_1 = \frac{3G_c}{8l}$
Internal length	l
Effective fracture toughness	G_c
The thickness coordinate centered at the crack	x

model, where α (the damage profile) can be regarded as a gradient-damage crack, for small value of l. The damage profile is given by:

$$\alpha(x) = \begin{cases} \left(1 - \frac{|x|}{\sqrt{2}l}\right)^2 & if\ |x| < \sqrt{2}l \\ 0 & else \end{cases} \tag{4}$$

3 Numerical Implementation

3.1 Low Velocity Impact in Compact Tension

Firstly, compact tension tests are considered in concrete specimens (200 × 200 mm) with two different loading rates: Case 1: 0.304 m/s; Case 2: 1.375 m/s. As shown in Fig. 1 displaced loads were applied to the right edge of the notch, while its left edge remained fixed [14]. The material parameters are set in Table 2. The velocity constraints imposed are as follows, for $t_0 = 100\ \mu s$.

$$v(t) = \begin{cases} v_0 \frac{t}{t_0} & if\ t \leq t_0 \\ v_0 & else \end{cases} \tag{5}$$

3.2 Dynamic Fractured Plate

In this example, a fractured rectangular plate dynamically loaded in tension is considered. As shown in Fig. 2, the design contains an initial horizontal crack whose length is 50 mm

Table 2. Material and geometrical Parameters [13]

Young's modulus	$E_0 = 3.6 \times 10^4$ MPa
Poisson's ratio	$\nu_0 = 0.18$
Mass density	$\rho = 2400$ kg/m^3
Failure strength	$f_t = 3.80$ MPa
Fracture energy	$G_f = 65$ J/m^2
Length scale	$b = 2.5$ mm

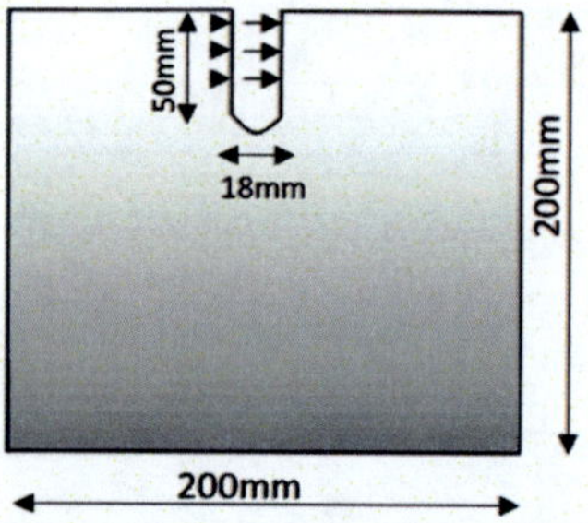

Fig. 1. Geometry and boundary conditions of compact tension specimen.

and width is 0.5 mm. A uniform traction $\sigma_0 = 1.0$ MPa is applied to the top and bottom edges of the specimen and held constant throughout whole process. Three distinguished cases of structure mesh are compared, for three element size, $h_1 = 0.2$ mm, $h_2 = 0.1$ mm and $h_3 = 0.08$ mm. The material parameters in the simulations are set as depicted in Table 3.

Table 3. Material and geometrical Parameters [13]

Young's modulus	$E_0 = 3.2 \times 10^4$ MPa
Poisson's ratio	$\nu_0 = 0.2$
Mass density	$\rho = 2450$ kg/m^3
Failure strength	$f_t = 12$ MPa
Fracture energy	$G_f = 3$ J/m^2
Micro mass density	$k = 3 \times 10^{-16}$ kg $\cdot$ mm^{-1}
Length scale	$b = 0.5$ mm

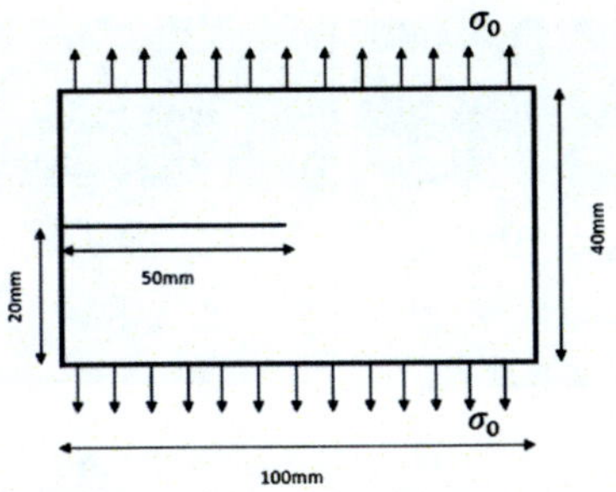

Fig. 2. Geometry and boundary conditions of the example of the dynamic fractured plate

4 Results and Discussion

4.1 Low Velocity Impact in Compact Tension

As shown in Fig. 3, the failure patterns depend on the loading rates. When the loading rate is low, the crack is perpendicular to the loading direction. While, when the loading rate increases, the crack presents an incline angle and this matches well with the experimental and the dynamic fracture phenomena in concrete results by Ožbolt et al. [14].

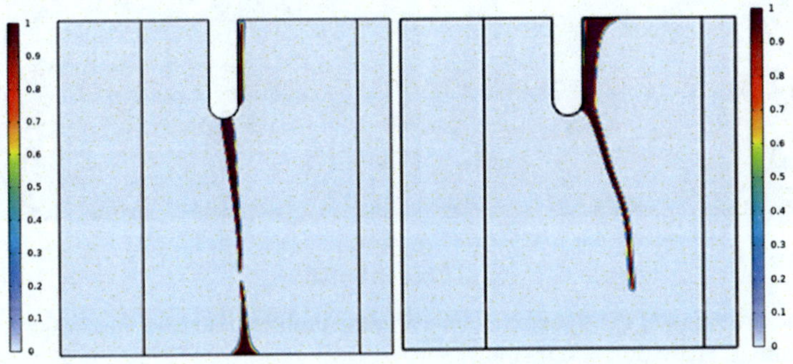

(a) Case1: 0.304m/s (b) Case2 :1.375m/

Fig. 3. Numerical final damage profiles in compact tension: (a) case 1: 0.304 m/s; (b) case 2: 1.375 m/s

4.2 Dynamic Fractured Plate

Figure 4 shows the cracks resulted from the three considered cases: $h_1 = 0.2$ mm, $h_2 = 0.1$ mm and $h_3 = 0.08$ mm.

As shown, branches into two smaller cracks at a specific point can be revealed. This matches well with the results of phase-field simulations reported by Borden et al. [11] and Nguyen and Wu [12]. In addition, this model demonstrates the quantity of elastic energy stored in the structure and the fracture energy.

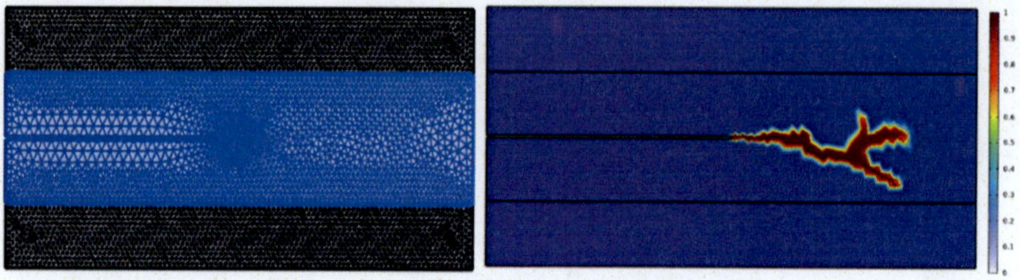

(a) Case 1: h=0.2mm

(b) Case 2: h=0.1mm

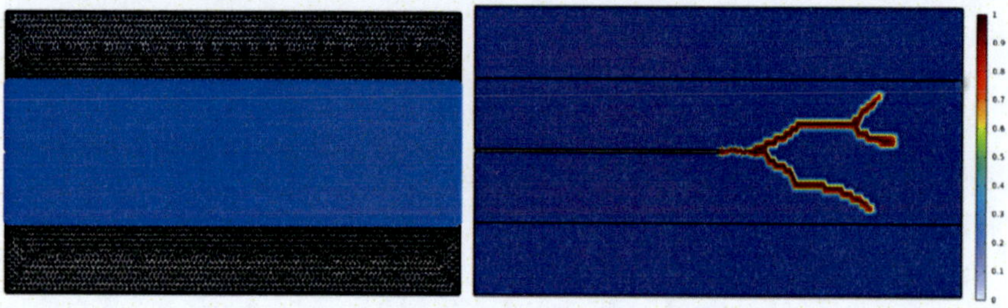

(c) Case 3: h=0.08

Fig. 4. Crack patterns of the dynamic fractured plate problem: (a) case 1: h = 0.2 mm; (b) case 2: h = 0.1 mm; (c) case 3: h = 0.08 mm

A clear comparison between the energy evolutions can be depicted in Fig. 5. The obtained curves correlate well with the experimental study of Borden et al. [11] and numerical simulation of Lu Hai et al. [13]. It is important to notice that this finding is more correlated with cases 2 and 3 rather than case 1. Indeed, according to Lu Hai et al. [13], case 1 does not satisfy the condition $h \leq b/5$. As such, Case 1 needs a mesh size of $h = b/5$ to properly resolve the damage field. This was confirmed by similar crack characteristics and damage fields across different cases.

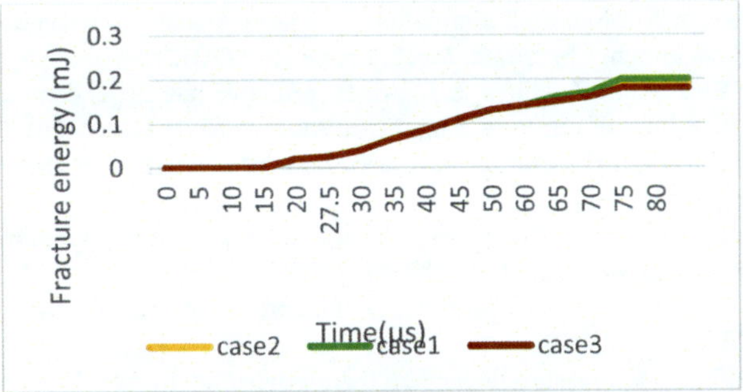

(a) Fracture energy

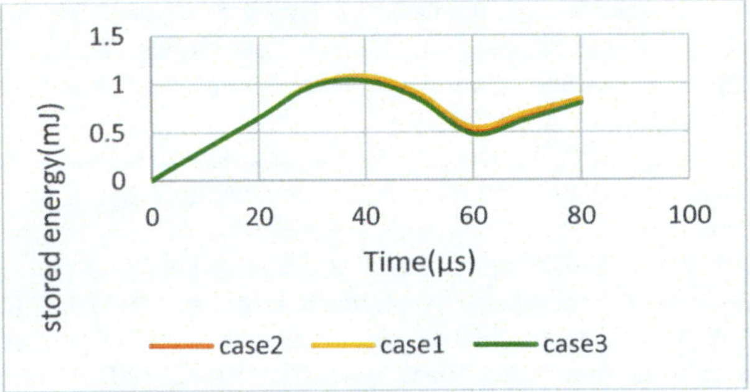

(b) Stored energy

Fig. 5. Comparisons of the energy evolution results in three cases: (a) Stored elastic energy; (b) Fracture energy.

5 Conclusion

In this work, a phase field numerical modelling of quasi-brittle material fracture is developed where the structure is applied to low velocity impact. The effect of loading rates on the crack propagation and mesh refining were studied in compact tension and dynamic fractured plate, respectively. Results prove that the failure patterns is depending on the loading rates. In addition, it is shown that failure branches into two smaller cracks at a specific point. The energy evolution resulting for different mesh size is useful to provide a geometrical criterion in FE refinement definition. Further track of work will be devoted to study dynamic indentation process while considering non local Griffith damage approach.

References

1. Ravi-Chandar, K., Yang, B.: On the role of microcracks in the dynamic fracture of brittle materials. J. Mech. Phys. Solids **45**, 535–563 (1997)
2. Shockey, D.A., Curran, D.R., Seaman, L., Rosenberg, J.T., Petersen, C.F.: Fragmentation of rock under dynamic loads. Int. J. Rock Mech. Min. Sci. Geomech. Abstr. **11**, 303–317 (1974)

3. Ožbolt, J., Sharma, A.: Numerical simulation of dynamic fracture of concrete through uniaxial tension and L-specimen. Eng. Fract. Mech. **85**, 88–102 (2012)
4. Pereira, L.F., Weerheijm, J., Sluys, L.J.: A new effective rate dependent damage model for dynamic tensile failure of concrete. Eng. Fract. Mech. **176**, 281–299 (2017)
5. Bischoff, P.H., Perry, S.H.: Compressive behaviour of concrete at high strain rates. Mater. Struct. **24**, 425–450 (1991)
6. Francfort, G.A., Marigo, J.J.: Revisiting brittle fracture as an energy minimization problem. J. Mech. Phys. Solids **46**, 1319–1342 (1998)
7. Kachanov, L.M.: Time of rupture process under creep conditions. Isv. Akad. Nauk. SSR. Otd Tekh. Nauk **23**, 26–31 (1958)
8. Parrinello, A.F.: A rate-pressure-dependent thermodynamically-consistent phase field model for the description of failure patterns in dynamic brittle fracture. Ph.D. dissertation, Oxford University (2017)
9. Kamensky, D., Moutsanidis, G., Bazilevs, Y.: Hyperbolic phase field modeling of brittle fracture: Part I—Theory and simulations. J. Mech. Phys. Solids **121**, 81–98 (2018)
10. Wu, J.Y.: A unified phase-field theory for the mechanics of damage and quasi-brittle failure. J. Mech. Phys. Solids **103**, 72–99 (2017)
11. Borden, M.J., Verhoosel, C.V., Scott, M.A., Hughes, T.J.R., Landis, C.M.: A phase-field description of dynamic brittle fracture. Comput. Methods Appl. Mech. Eng. **217–220**, 77–95 (2012)
12. Nguyen, V.P., Wu, J.Y.: Modeling dynamic fracture of solids with a phase-field regularized cohesive zone model. Comput. Meth. Appl. Mech. Eng. **340**, 1000–1022 (2018)
13. Hai, L., Wu, J.-Y., Li, J.: A phase-field damage model with micro inertia effect for the dynamic fracture of quasi-brittle solids. Eng. Fract. Mech. **225**, 106821 (2020)
14. Ožbolt, J., Bošnjak, J., Sola, E.: Dynamic fracture of concrete compact tension specimen: experimental and numerical study. Int. J. Solids Struct. **50**, 4270–4278 (2013)

Morphological Analysis of Brake Lining Material for Railway Type Application

Mouna Baklouti[1,2](✉), Anne-Lise Cristol[3], Yannick Desplanques[3], and Riadh Elleuch[1]

[1] National School of Engineers of Sfax (ENIS), LASEM, University of Sfax, 3038 Sfax, Tunisia
mouna.bakloutikharrat@gmail.com, riadh.elleuch@gnet.tn
[2] Faculty of Sciences of Gafsa, University of Gafsa, 2112 Gafsa, Tunisia
[3] UMR 9013-LaMcube-Laboratoire de Mécanique Multiphysique Multiéchelle, University Lille, CNRS, Centrale Lille, 59000 Lille, France
anne-lise.cristol@ec-lille.fr,
yannick.desplanques@centralelille.fr

Abstract. The origin of the frictional system's vibration as in the case of screeching during braking is necessarily related to the microstructure of frictional materials. The friction mechanisms of those later are reputed to be complex to understand, due to the diversity of scales and couplings involved. In fact, brake lining rubbed surfaces are highly heterogeneous, with multiple load-bearing localizations which bear the contact and determine the performance in braking. The characterization of the morphology of the interface constitutes the key to understanding the friction-induced phenomena. In this context, the analysis of the morphological properties of rubbed surfaces after braking tests was carried out by complementary techniques such as scanning electron microscopy and optical profilometry, focusing on the various scales involved. Correlations between the results are established to better describe the load-bearing localizations. Results show that the lower part of the pad is more rubbed and contains more compacted third body plates. This is attributed to the impact of the thermal localization migration from the outer perimeter to the inner perimeter of the disc's friction track during braking. This study allowed us to highlight the importance of a multiscale approach in brake lining rubbed surfaces characterization.

Keywords: braking · third body · heterogeneous morphology · optical profilometry · scanning electron microscopy

1 Introduction

Brake lining materials must present several physical, mechanical, and thermal properties to confront the mechanical stress-heat solicitations during braking. All these properties are the result of combining several ingredients which interact together in the order to respond to these requirements. The multitude of ingredients makes a combination of properties but also a very complex microstructure that impacts the tribological mechanisms activated at the interface and then the behavior of the material in braking (Baklouti

A. Akrout et al. (Eds.): ICAV 2022, ACM 22, pp. 415–424, 2023.
https://doi.org/10.1007/978-3-031-34190-8_44

et al. 2015; Mutlu et al. 2006; Cho et al. 2005). As, for example, the case of vibrational behavior which seems to be correlated in time to the activated tribological mechanisms (Davin et al. 2019). The tribological mechanism describes the way that a third body takes place in the contact's interface between the pad and the disc during braking. There is a process of accumulation and compaction of the powder in privileged trapping sites related to the behavior of the different ingredients and their ability to attract the third body (Eriksson and Jacobson 2000; Baklouti et al. 2013). The latter is arranged in flat plates adhering to the surface of the composite material and constituting the load-bearing areas of contact. Eriksson et al. clearly showed that secondary plateaus are relatively persistent and are the seat of speed accommodation and energy dissipation (Eriksson et al. 2001). However, the powder that remains non-adherent circulates in the contact and provides the internal flow necessary for the renewal of the third body. This movement is governed by a complex dynamic of formation, fragmentation, destruction and reconstitution of flat plates of the third body called the tribological circuit giving rise to a transient interface (Desplanques et al. 2006). Furthermore, thermal phenomena developed during braking and consisting of hot band and spots migration on the friction track surface of the disc can also have an impact on the manner of third body arrangement (Cristol-Bulthé et al. 2007). The role of the interface on the braking behavior has been shown also by other authors in particular the importance of the amount and the arrangement of the third-body layer (Massi et al. 2008; Lee and Jang 2018). Massi et al. demonstrate that the topography of the contact surface after the braking phase with and without squeal caused by vibrations were found to be completely different in the two cases (Massi et al. 2008).

Moreover, the phenomena involved in braking are multi-scale, from the macro-scale of the rubbing surfaces (the geometry of the pad gives a friction surface of 1100 mm^2 which constitutes 1/9 of the rubbed surface of the disc) to the micro-scale of the ingredients forming the material and the developed flat plates.

It is in this context that this study was conducted. The idea is to find a criterion for scanning the majority of rubbed surface in different localizations for morphological characterization in order to pursue and understand the distribution of the third body and its arrangement in relation to the tribological circuit and thermal phenomena.

2 Materials and Methods

2.1 Pair of Studied Materials

The pair of studied materials is an organic composite material constructed for railways type application rubbing against a lamellar graphite cast iron disk. The pad is made of particles and fibers (mineral, metallic and organic) of different natures and shapes distributed in a phenolic resin matrix. The detailed volume distribution is given in Table 1. The choice of the tested material is due to its countenance of an important percentage of fibers and particles which contribute to load bearing during braking especially those of metallic and mineral nature by forming a support for the development of the secondary plates (Baklouti et al. 2013, Baklouti et al. 2017).

Table 1. Composition of the pad's material.

Ingredient	Vol (%)
Fibers	32,5
Particles	35,5
Phenolic resin	32

2.2 Experimental Protocol for Braking Tests and Characterization Techniques

The braking performed on the pad are a succession of run-in type test with a final stop-braking which simulates that of a Z-TER train running at an initial speed of 80 km/h and aim to produce a rubbed surface with a low gradient of properties. A visual inspection of the surface of the pad after rubbing allows us to judge a preliminary homogeneity of the contact between the disk and the pad with an entire distribution over the surface of the third body powder. The maximum temperatures reached are recorded at 2 mm depth under the surfaces of the disc and the pad on the average radius of friction. For these low dissipated energy tests, the mass temperature of the pad reaches a maximum of 75 °C and that of the disc 105 °C.

After braking tests, the objective is to study the morphology of the rubbed surface. Three zones are investigated, namely zones 1, 2 and 3, which are differentiated by their position in relation to the input and output of the contact indicated by the sliding orientation (Fig. 1). These zones can retain the third body in different quantities or arrangements due to the impact of the tribological circuit. In investigations, different localisations at the top and the bottom of each zone are also considered in relation respectively to the outer and inner radius of disc and pad sliding tracks in the objective to know if localized thermal phenomena such as hot bands and hot spots can also impact the third body distribution. In order to achieve this objective, complementary techniques were used to study the pad's surface. The techniques are scanning electron microscopy (SEM) and optical profilometry.

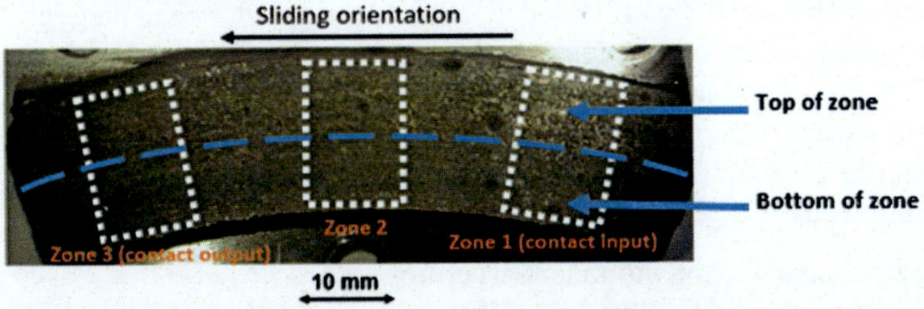

Fig. 1. Pad on a geometry of a sector (mean friction R: 100 mm, friction track width: 17 mm) and characterized zones bounded by white dotted frames on the pad surface

3 Results and Discussion

3.1 Macroscale and Mesoscale Morphological Characterization

After the pad has been subjected to braking, a first approach to characterize the pad surface by interferometric profilometry consists of making measurements by scanning the entirety of zones 1, 2 and 3 defined previously. These measurements allow having a general description of the rubbed surface and verifying the state of contact between the disk and the pad by considering the input of the contact and its output. Each stitching measurement corresponds to an area of about 140 mm^2 (Fig. 2).

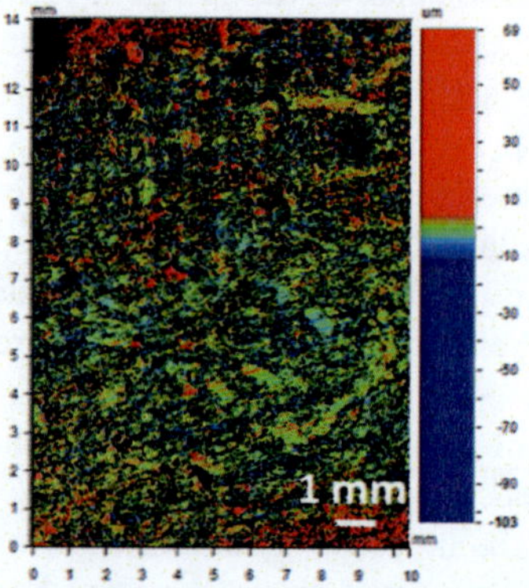

Fig. 2. Optical profilometry measurements obtained by stitching, zone 2 as an example

As a criterion for analyzing the obtained topographies, the Abbott curve was used. It is the bearing length curve that describes the percentage of material crossed by a cutting level. This curve allows us to qualitatively understand the wear that will occur in service (Ech et al. 2007). In this study, the idea is to use this tool with the aim to identify if there are morphological differences between several zones in the pad surface and between the bottom and the top radius of the pad. This investigation plans to link with the role of the tribological circuit and the thermal localizations, that the surface undergoes in braking, on the third body distribution and arrangement in blocks of powder or flat plates. This study is based on previous works where the authors suggest that the size of the plates and their thicknesses tend to be greater at the contact output and at the bottom of the pad (Cristol-Bulthé et al. 2007).

An Abbott curve is generally divided into three parts (Fig. 3):

1. Part I corresponds to the prominent asperities.
2. Part II represents the section that will provide the surface function. The lower the slope, the lower the wear rate.
3. Part III represents the section of depressed cavities which are not worn.

The Abbott curves for the three studied zones are practically superposed (Fig. 3). At this macroscopic scale, we can note that no difference is generated by these curves. Therefore, the quantitative distribution of the third body on the entire surface of the

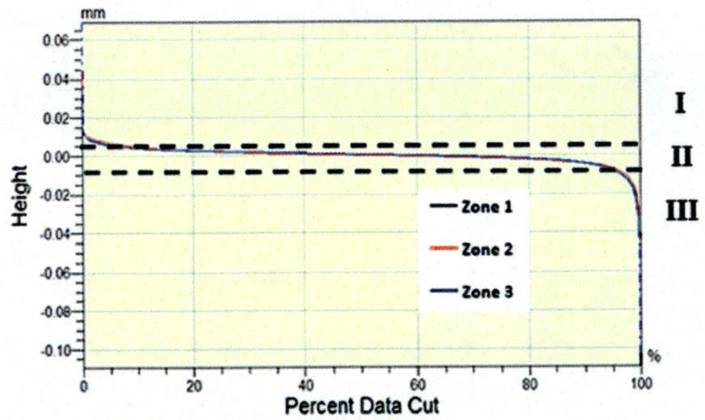

Fig. 3. Abbott curves of zones 1, 2 and 3.

pad proves to be homogeneous with equivalent percentages of asperities, cavities and bearing capacity for the three zones.

By decreasing the size of the treated zones, an analysis of three subdivisions of about 40 mm^2 tacked from the same area is made (here zone 3 is considered) (Fig. 4). Similar to the 3 zones in Fig. 3, the Abbott curves of the three subdivisions of the same zone remain almost superposed. This indicates that, for the considered scale, the percentage of material per cut level is equivalent also for the high, middle and low subdivisions, so the quantities and thickness of the third body seem to be similar and no clear difference is reported.

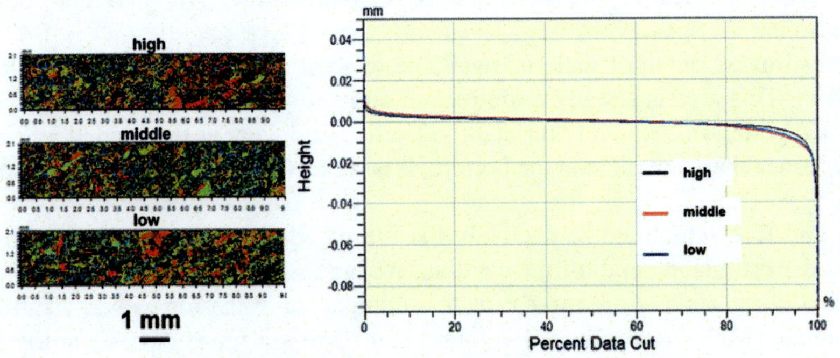

Fig. 4. Abbott curves of three subdivisions of zone 3

By reducing the scale of analysis, distinct Abbott curves for even smaller localizations with an approximate size of 1 mm^2 chosen at the top and bottom of zones are seen (Fig. 5). A slightly lower slope of the low localization will probably be an indication of a smaller wear rate giving a more stable load-bearing capacity. Parameters significations and values associated with these curves, namely Mr1, 100-Mr2, R_k, R_{pk} and R_{vk} are given in Table 2.

The prominent pics are normally those of third-body powder remaining friable and non-compacted. Its proportion Mr1 is lower at the bottom than at the top. This result leads to the assumption that the lower part of the pad is more rubbed and contains

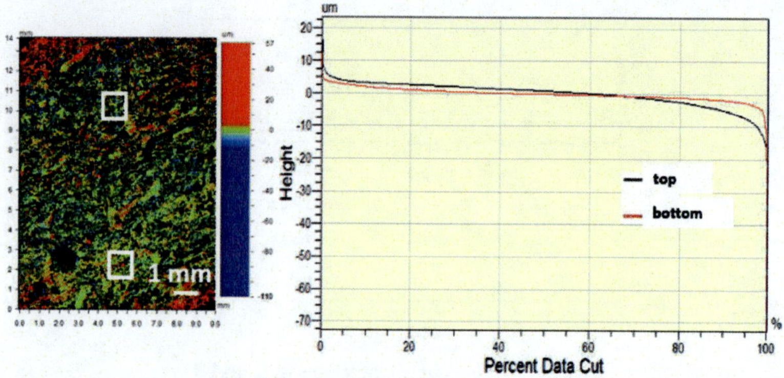

Fig. 5. Abbott curves of the two localizations at the top and bottom of zones (zone 3 as an example)

more compacted third body plates, supported by fibers and particles, and fewer pics of a friable third body. This can be attributed to the impact of the thermal localization migration from the outer to the inner perimeter of the disc's friction track during braking, in contact respectively with the top and the bottom of the pad (Kumar et al. 2011). A migration, accompanied by an increase in thermal intensity, leads to increased wear and thickness of plates on the bottom (Cristol-Bulthé et al. 2007).

100-Mr2 which represents the proportion of cavities is higher at the top than at the bottom. More cavities mean more depressed sites uncovered of the third body. This can be explained by the more limited formation and growth of plates at the top and subsequently, there are more cavities. This result is in logical agreement with the previous one for the same reasons.

The estimated depth of material available to wear and load-bearing R_k is lower at the bottom. This area is already more rubbed and compacted as mentioned. Moreover, the formed plates are relatively resistant to friction and wear as mentioned by Eriksson et al. and since there are more at the bottom, less wear results in this area (Eriksson et al. 2001).

R_{pk} and R_{vk}, which are respectively the estimated altitude of the most prominent pics and the estimated depth of the cavities, are higher at the top. The variation of R_{pk} is justified because there are more pics less compacted at the top as was identified by the parameter Mr1. Similarly, it was found that there are fewer cavities at the bottom (100-Mr2), the surface of the pad in this area is more pressed and more covered with third body, resulting in a lower depth of cavities R_{vk} compared to those at the top.

All these results support the observations realized in previous work regarding the distribution of the third body on the pad surface (Cristol-Bulthé et al. 2007) and thus offer validation of this surface morphological characterization tool. A second conclusion is that from a mesoscale characterization of the brake lining surfaces, more relevant information about the arrangement of the third body can be obtained.

3.2 Mesoscopic and Microscopic Characterization of Localized Details

Local measurements are performed to characterize more locally the rubbed surface, and more precisely the load-bearing plates. The same plate is illustrated by both optical

Table 2. Abbott curve's parameters corresponding to the top and bottom localizations in diverse zones.

Parameter	Signification	Values for 1 mm^2 localizations in the top and bottom of zones	
		Top	Bottom
Mr1 (%)	Proportion of prominent pics	13,33 (2,42)	7,30 (2,31)
100-Mr2 (%)	Proportion of cavities	19,49 (2,69)	12,27 (1,76)
R_k (nm)	Estimated depth of material available to wear and load-bearing	6240 (354)	3393 (232)
R_{pk} (nm)	Estimated altitude of the most prominent pics	4625 (1346)	1858 (451)
R_{vk} (nm)	Estimated depth of the cavities	9541 (2302)	4617 (881)

The values in parentheses are the standard deviation.

profilometry and SEM. It is distinguished by the green color in the profilometry's image (Fig. 6b) and, with SEM illustrations, it appears covered with the third body (Fig. 6c). The sliding direction is distinguished by different colored and parallel traces on the bearing plate (Fig. 6b).

As authors describe, the primary plates are formed by flattened ingredients which in turn will serve to develop the secondary plates by being preferential sites for the attachment of the third body (Eriksson and Jacobson. 2000; Mosleh et al. 2004). The ingredient forming the base of this plate is distinguished by a white color thanks to uncovered grooves with the third compacted body. This latter is also driven in a direction parallel to sliding and comes to cling in front and behind the primary plate to develop blocks of powder that will compact as the rubbing proceeds and gradually increase the size of load-bearing localizations (Fig. 6c, d and e).

The microscopic scale of characterization is adequate for the exploration of bearing plates. It shows the way of development of the secondary plates around the primary plates and the preferential sites of their compaction.

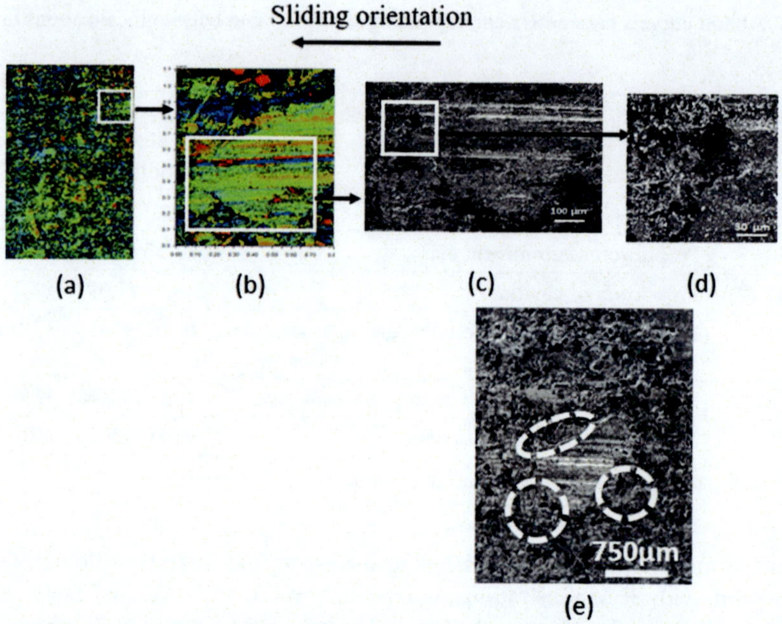

Fig. 6. A bearing plate by profilometry and SEM, (a) localization of the plate in a zone, (b) plate in profilometry, (c) plate in SEM, (d) and (e) preferential sites of third body powder accumulation

4 Conclusion

In this study, a brake lining material surface was characterized. The interest was on morphological properties offering indications about the behavior under braking and wear, especially since it is the distribution of compacted third body that manages the bearing sites in contact between disc and pad and thus influences the expected performance.

The use of the bearing length curve (Abbott curve) has made it possible to evaluate the evolution of a pad's friction surface after braking. The parameters determined from this curve make it possible to quantify this evolution according to the localization and scale. This method seems able to give different informations on the evolution of the studied surface: cavities, pics, quantities of material available for wear… In this case of solicitation, results show that the lower part of the pad, in contact with the inner radius of the disc track, is more rubbed and contains a more compacted third body forming secondary plates bearing the contact. This can be attributed to the impact of thermal localizations which migrate from the outer to the inner perimeter of the disc friction track with accentuated intensity leading to more wear.

On a macroscopic scale of exploitation, the results ensure a control of the homogeneity of the surface and give indications on the relevant zones to be characterized microscopically.

On a mesoscopic scale, the results show different evolutions specific to the characterized zone with relevant indications on the state of development of the load-bearing capacity and the compaction of the third body. This will allow a better understanding of the distribution mechanisms of the third body that will support the contact and manage the frictional behavior.

The microscopic scale of characterization is more appropriate for the description of load-bearing plates. It shows the way of development of the secondary plates around the primary ones and the preferential sites of their compaction.

References

Baklouti, M., Cristol, A.-L., Desplanques, Y., Elleuch, R.: Impact of the glass fiber addition on tribological behavior and breaking performances of organic matrix composites for brake lining. Wear **330–331**, 507–514 (2015). https://doi.org/10.1016/j.wear.2014.12.015

Mutlu, I., Eldogan, O., Findik, F.: Tribological properties of some phenolic composites suggested for automotive brakes. Tribol. Int. **39**, 317–325 (2006). https://doi.org/10.1016/j.triboint.2005.02.002

Cho, M.H., Kim, S.J., Kim, D., Ho, J.: Effects of ingredients on tribological characteristics of a brake lining: an experimental case study. Wear **258**, 1682–1687 (2005). https://doi.org/10.1016/j.wear.2004.11.021

Davin, E., Cristol, A.-L., Brunel, J.-F., Desplanques, Y.: Wear mechanisms alteration at braking interface through atmosphere Modification. Wear **426–427**, 1094–1101 (2019). https://doi.org/10.1016/j.wear.2019.01.057

Eriksson, M., Jacobson, S.: Tribological surfaces of organic brake pads. Tribol. Int. **33**, 817–827 (2000). https://doi.org/10.1016/s0301-679x(00)00127-4

Baklouti, M., Elleuch, R., Cristol, A.-L., Najjar, D., Desplanques, Y.: Relationships between the heterogeneous microstructure, mechanical properties and braking behaviour of an organic brake lining material. Proc. IMechE Part D: J. Automobile Eng. **227**, 549–560 (2013). https://doi.org/10.1177/0954407012461751

Eriksson, M., Lord, J., Jacobson, S.: Wear and contact conditions of brake pads: dynamical in situ studies of pad on glass. Wear **249**(3–4), 272–278 (2001). https://doi.org/10.1016/s00431648(01)00573-7

Desplanques, Y., Degallaix, G., Roussette, O., Francois, M., Bulthe, A.-L., Sabatier, L.: A reduced-scale test for pad-disc contact tribological analysis in railway braking. In: Barton, D.C., Fieldhouse, J.D. (eds.) Advances in Vehicle Braking Technology, pp. 218–230. I Mech E Publications (2006)

Massi, F., Berthier, Y., Baillet, L.: Contact surface topography and system dynamics of brake squeal. Wear **265**, 1784–1792 (2008). https://doi.org/10.1016/j.wear.2008.04.049

Lee, S., Jang, H.: Effect of plateau distribution on friction instability of brake friction materials. Wear **400–401**, 1–9 (2018). https://doi.org/10.1016/j.wear.2017.12.015

Baklouti, M., Cristol, A.-L., Elleuch, R., Desplanques, Y.: Brass in brake linings: key considerations for its replacement. Proc. IMechE Part J.: J. Eng. Tribol. **231**, 461–468 (2017). https://doi.org/10.1177/1350650115588966

Ech, M., Morel, S., Yotte, S., Breysse, D., Pouteau, B.: Étudier l'évolution de la macrotexture de chaussée par une méthode fine. 25ème rencontres de l'AUGC, 23–25 mai 2007, Bordeaux (2007). https://www.iut.u-bordeaux.fr/gc/augc07/index/pdf/AMC/Ech.pdf

Kumar, M., Boidin, X., Desplanques, Y., Bijwe, J.: Influence of various metallic fillers in friction materials on hot-spot appearance during stop braking. Wear **270**(5–6), 371–381 (2011). https://doi.org/10.1016/j.wear.2010.11.009

Cristol-Bulthé, A.-L., Desplanques, Y., Degallaix, G.: Coupling between friction physical mechanisms and transient thermal phenomena involved in pad–disc contact during railway braking. Wear **263**(7–12), 1230–1242 (2007). https://doi.org/10.1016/j.wear.2006.12.052

Mosleh, M., Blau, P., Dumitrescu, D.: Characteristics and morphology of wear particles from laboratory testing of disk brake materials. Wear **256**(11–12), 1128–1134 (2004). https://doi.org/10.1016/j.wear.2003.07.007

Enhanced Disc Brake Thermo-Kinetics for Better Wear Test Reproducibility

Sellami Amira[1,2](✉), Guesmi Mohamed Hedi[1], and Elleuch Riadh[2]

[1] Higher Institute of Transport and Logistics of Sousse, University of Sousse, Sousse, Tunisia
amira_sellami@hotmail.com

[2] National School of Engineers of Sfax, LASEM, University of Sfax, 3038 Sfax, Tunisia
riadh.elleuch@gnet.tn

Abstract. Brake discs are an important part of the braking system, which are subject to a lot of friction. The principle of a braking system is to convert kinetic energy into heat through friction, in order to allow a vehicle to be stopped by pressing the brake pedal. In fact, one of the key factors of its conception is related to her thermo-mechanical deformation that is especially affected by the thermal factor. Therefore, they are designed to withstand heat, but problems can arise when the temperature gets too high for the brake pads that reduces the effectiveness of the braking system.

In this study, a reflection on the disc thickness effect on the thermo-kinetics of braking systems at higher solicitation is discussed. For that, the disc brake is modeled and simulated and analysis with Finite Elements software. For this purpose, different samples of brake material having the form of pad and disc made from cast iron steel were considered with varied disc thickness. Results show that the 15 mm thick of brake disc provides good thermal kinetic even for the disc and for the brake lining materials.

Thus, grey cast iron with thickness of 15 mm is preferred for disc brake to achieve better thermal performance.

Keywords: Disc brake thickness · Numerical analysis · thermo-kinetics

1 Introduction

With the challenges of brake lining manufacturing industry: in terms of efficiency, reliability, comfort and cost. It is necessary to supplement braking tests with numerical analyzes in order to minimize study times and remedy certain limitations of laboratory tests. Several researchers have demonstrated that brake pressure affects the mechanical, chemical and thermal properties of the disc brake Zhang et al. 2020; Leslie 2004). Tamasho et al. described that brake pressure can induce brake shudder caused by a mechanically deformed disc under the action of the brake (Tamasho et al 2000; Fieldhouse and Beveridge 2001). The deflection in the disc is inversely proportional to the disc flexural rigidity that depends of the thickness of the disc. Thus, thick discs cause lower deflection and therefore less thermal impact (Okamura and Yumoto 2006). Previous work has shown that the power dissipated during braking is not sufficiently satisfactory to compensate for energy losses by convection (Baklouti et al. 2015).

A. Akrout et al. (Eds.): ICAV 2022, ACM 22, pp. 425–432, 2023.
https://doi.org/10.1007/978-3-031-34190-8_45

The ultimate objective of this research work is to simulate the thermal gradients produced by the sliding contact on the pin-on-disc tribometer. The frictional heat is generated in the contact between the stationary pin and the rotating disc then it is partitioned between the two bodies. A transient heat transfer analysis studies the heat partitioning for the pin-on-disc configuration.

The heat flux specified on the surface of the disc is compared to the analytic data. The prediction of temperature distribution during the braking process is obtained by using experimental tests and theoretical calculations. Among the reliable outcomes of the experimental investigations is the validation of the numerical solutions.

For the sake of overcoming the design flaws of the tribometer without affecting the mechanical properties of the disc, this study focuses on the study of the modification of the thickness of the disc. The numerical study allows verifying the disc thermal kinetics and the experimental study is dialed in order to determine the optimum thickness, which guarantees good friction properties.

2 Materials and Methods

2.1 Friction Materials

The friction material is a composite material with a phenolic matrix reinforced with fibers and particles of different natures, which interact to better respond to braking stress (Manoj et al. 2023). The friction surface is characterized by scanning electron microscopy (SEM) observations in order to study its morphology (Fig. 1). The density of the designed friction materials is determined using Archimedes principle, according to ISO standard (Yigrem et al. 2022). Thermal properties were measured using hot-disk methods (Table 1).

Table 1. Thermo physics properties of brake lining materials

	Brake lining materials
λ (W.m^{-1}.K^{-1})	0,056
Cp (J Kg^{-1} K^{-1})	36,34
ρ (kg m^{-3})	2211

2.2 Numerical Modeling

The numerical model used in our study is a two-dimensional axisymmetric transient thermal finite element model developed using Abaqus software (Fig. 2) which was developed by Makni (Makni et al. 2022).

Materials properties (Table 2) introduced in the model was calculated properties of the real geometries from Eq. 1 and 2.

$$\lambda_{num} = \lambda_{real}\frac{Sd}{Sp} \tag{1}$$

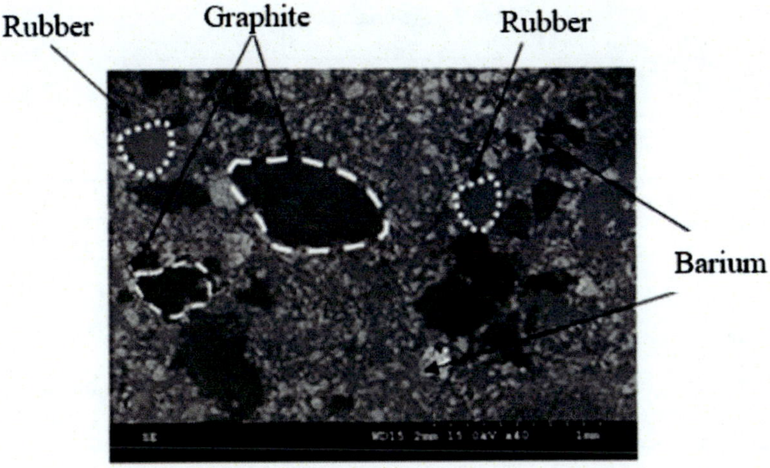

Fig. 1. MEB of brake lining materials

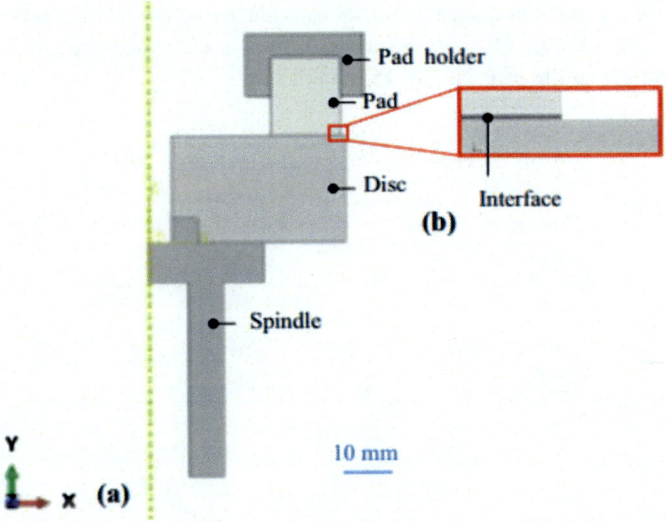

Fig. 2. 2D axisymmetric thermal models for a pin-on-disc tribometer, b) interface

$$(\rho\ Cp)_{num} = (\rho\ Cp)_{real}\ \frac{Sd}{Sp} \tag{2}$$

Different boundary conditions were undertaken to properly estimate temperature distribution:

- At the end of spindle the temperature is fixed at the outside room temperature 25 °C.
- The emissivity at the level of pin holder is 0.55.

Table 2. Materials properties

	Pin		Pin holder	
	Real	model	Real	model
λ ($W.m^{-1}.K^{-1}$)	1.041	0.056	42	9.23
Cp ($J\ Kg^{-1}\ K^{-1}$)	673	36.35	480	105.5
ρ ($kg\ m^{-3}$)	2211	2211	7790	7790

- The heat is generally dissipated by convection.

The heat flux generated by the friction between the pin and disc was introduced in the interface that represent the third body generated by friction between disc and pin (Laraqi et al. 2009). The heat flux density was calculated from the coefficient of friction taken for different times of the friction test (Table 3). In fact, the frictional heat is dissipated by conduction towards the different components of the tribometer and by convection from the free rotational surface. Thus, to increase the thermal kinetics of the disk, it is essential to minimize the cooling free surface, in particular the lateral surface of the disc, by reducing its thickness. To achieve our goal, four models are proposed by modifying only the thickness of the disc 22, 20, 15 and 10 mm.

Table 3. Heat flux density (Baklouti et al. 2016).

Time (s)	Friction coefficient	heat flux density (10^{10} W/m^3)
10	0,262	1.036
100	0,333	1.316
1000	0,233	1.922
1500	0,276	1.089

2.3 Experimental Set-Up

The wear test was conducted on a pin-on-disc tribometer. The interest of the study is based on the failures of the results found by Baklouti et al. (2014) in the friction test in the case of severe braking stress. For this, the experimental parameters are chosen as follows reproducing severe braking solicitation (Table 4). The experimental protocol is illustrated in Fig. 3. For the first cycle, the pin was in contact with the disc until the disc reach 350 °C then the load was removed and the disc was cooled to 300 °C. For the second cycle, the load was applied again until the disc temperature reached 350 °C and the disc was again cooled to 300 °C.

The dimension of the pin is 14 mm * 16 mm and the disc in lamellar graphite cast iron with a diameter of 40 mm.

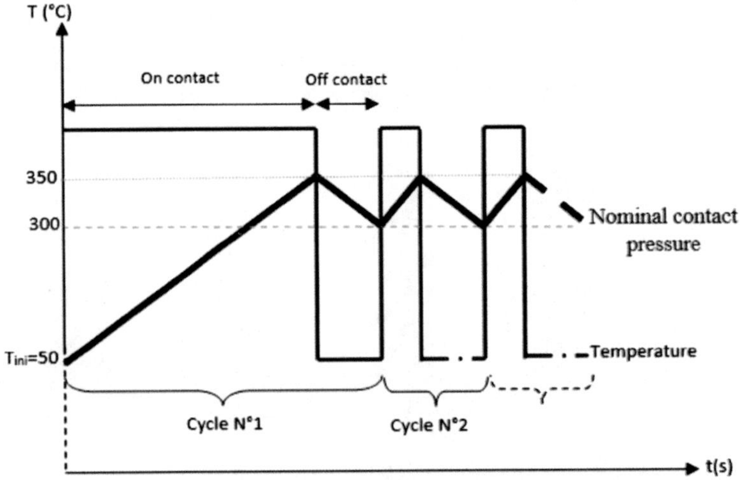

Fig. 3. Experimental protocol

Table 4. Experimental parameters

Parameters	Values
Temperature (°C)	300–350
Contact pressure (MPa)	1,8
Sliding speed (m/s)	9
Number of cycles	30

3 Results and Discussions

3.1 Numerical Results

The results of the numerical simulation (Fig. 4) show that for thickness equal to 10 mm the kinetics of temperature rise is 48 °C/min, on the other hand for thickness equal to 22 mm the disc thermal kinetics is equal to 27 °C/min. Therefore, the increase in disc thermal kinetics is inversely proportional with the thickness of the disc.

The increasing the thickness of the disc, induce a decrease of the heat dissipation of the disc. In fact, it affects its mechanical properties, thus inducing many defects such as the presence of a strong thermal gradient, deformation and deflection of the disc. Therefore, there will be a generation of vibrations and the phenomenon of squealing. Although the increase in the thickness of the disk has induced a reduction in the kinetics of the temperature rise. In order to validate the choice to raise the thickness of the disc, it would have been necessary to check the impact of this change on the behavior of the lining and especially its behavior with respect to the coefficient of friction (stability and loss of efficiency) which is conditioned by the rapid rise in temperature on the disc side.

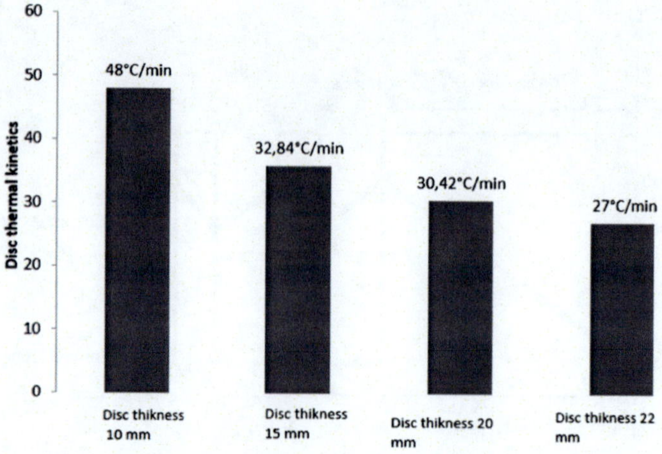

Fig. 4. Disc thermal kinetics in function of disc thickness

3.2 Friction Test Results

The disc temperature evolutions for the different disc thickness are shown in Fig. 5. The rapid increase of the temperature recorded during the friction test is very significant for disc thickness 10 mm. This maximum temperature value resulting from the reduction in thickness and therefore in the convective exchange surface. Figure 6 presents the friction coefficient evolution for the different configuration. It shows that the minimum friction coefficient increases from 0.3 to 0.32 than 0.35 to 0.37 respectively for disc thickness 22 mm, 20 mm, 15 mm to 10 mm. Withal for the maximum friction coefficient, the values for the different discs are approximately similar. It is obvious that for the disc thickness 10 mm the friction range between minimum and maximum friction is smaller so the coefficient of friction is more stable from one cycle to another. Otherwise, for disc thickness 10 mm, the friction coefficient evolution is marked by a reduction in the coefficient of friction, which is called "Fading" (Zhang et al. 2019). This loss of efficiency excludes the choice of the 10 mm thick disc. For this reason and given its friction response, which is stable during cycling, the 15 mm thick disc is the best choice for friction tests carried out on the tribometer.

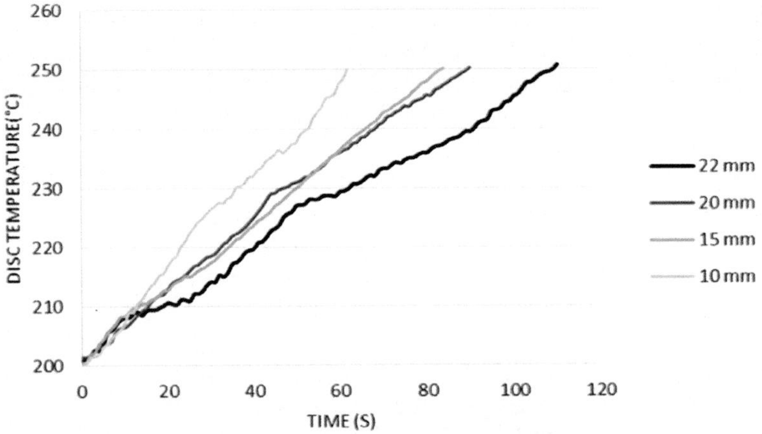

Fig. 5. Disc temperature evolution

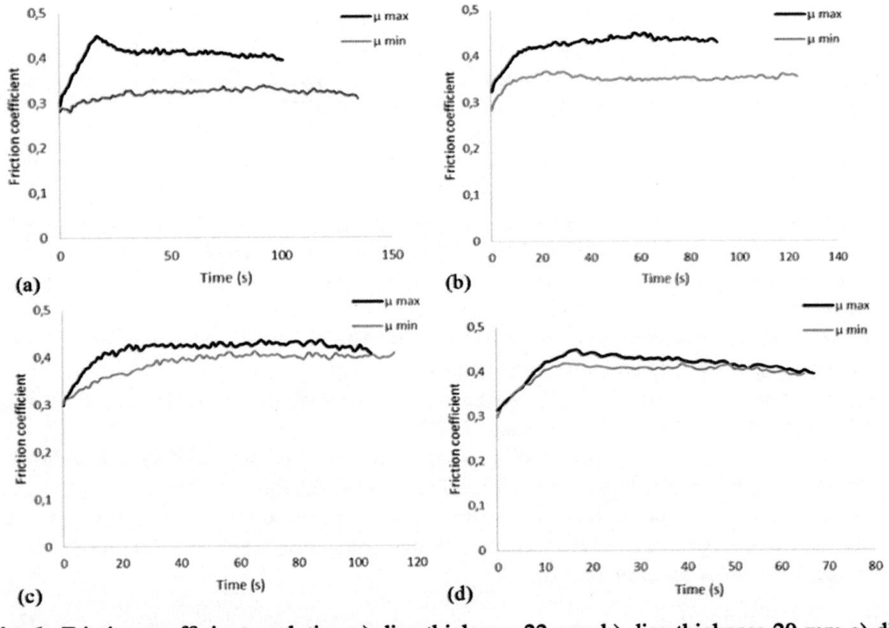

Fig. 6. Friction coefficient evolution a) disc thickness 22 mm b) disc thickness 20 mm c) disc thickness 15 mm d) disc thickness 10 mm

4 Conclusion

The main interest of our study is to study the effect of the modification of disc thickness on both thermal kinetics and friction stability. The numerical model simulating the thermal behavior of friction contact shows that more it is thin more the power dissipated by friction is sufficient to compensate for the losses by convection. The experimental study shows that for disc thickness equal to 10 mm, friction material behavior presents a loss of efficiency due to the rapid increase in temperature. Moreover, the good repeatability of friction coefficient from one cycle to another with disc thickness equal to 15 mm that made it possible to properly consider this change in disc thickness.

Acknowledgements. The authors are grateful for every contribution that led to the realization of this modest research work and make it possible.

References

Zhang, P., et al.: Effect of carbon fiber on the braking performance of copper-based brake pad under continuous high-energy braking conditions. Wear **458–459** (2020)

Leslie, A.C.: Mathematical Model of Brake Caliper to Determine Brake Torque Variation Associated with Disc Thickness Variation (DTV) Input. SAE 2004-01-2777 4 (2004)

Tamasho, T., Doi, K., Hamabe, T., Koshimizu, N., Suzuki, S.: Technique for reducing brake drag torque in the non-braking mode. JSAE **21**(1), 67–72 (2000)

Fieldhouse, J.D., Beveridge, C.: Experimental Observations of Hot Judder. SAE 2001-01-3135 (2001)

Okamura, T., Yumoto, H.: Fundamental study on thermal behavior of brake discs SAE transactions. J. Passenger Car: Mech. Syst. J. **115**, Section 6, 1731–1746 (2006)

Baklouti, M., Cristol, A., Elleuch, R., Desplanques, Y.: Brass in brake linings: key considerations for its replacement. Proc. Inst. Mech. Eng. Part J: J. Eng. Tribol. **231**(4), 461–468 (2017)

Makni, F., Cristol, A., Elleuch, R., Desplanques, Y.: Organic brake friction composite materials: impact of mixing duration on microstructure, properties and tribological and wear behavior. Polymers **14**(9), 1692 (2022). https://doi.org/10.3390/polym14091692

Chen, A., Kienhöfer, F.: The failure prediction of a brake disc due to nonthermal or mechanical stresses. Eng. Failure Anal. **124** (2021)

Zhang, P., et al.: Fade behaviour of copper-based brake pad during cyclic emergency braking at high speed and overload condition. Wear **428–429**, 10–23 (2019)

Manoj, E., et al.: Investigation on the mechanical and tribological properties of silicon in an autmotive brake pad. Mater. Today Proc. (2023)

Yigrem, M., Fatoba, O., Tensay, S.: Tribological and mechanical properties of banana peel hybrid composite for brake-pad application. Mater. Today: Proc. **62**, Part 6 (2022)

Laraqi, N., Alilat, N., Garcia de Maria, J.M., Baïri, A.: Temperature and division of heat in a pin-on-disc frictional device—exact analytical solution. Wear **266**(7–8), 765–770 (2009). https://doi.org/10.1016/j.wear.2008.08.016

Towards the Development of a Numerical Model for the Simulation of Thermal Behavior of Disc Brake Systems

Sellami Amira[1,2](✉), Zerai Kawther[2], and Elleuch Riadh[1]

[1] National School of Engineers of Sfax, LASEM, University of Sfax, 3038 Sfax, Tunisia
amira_sellami@hotmail.com, riadh.elleuch@gnet.tn

[2] Higher Institute of Transport and Logistics of Sousse, University of Sousse, Sousse, Tunisia

Abstract. In a disc braking system, the contact of a rotating disc with a stationary brake lining converts mechanical energy into thermal energy. This energy induces the heating of the disc and the brake lining that can reach high temperatures in certain cases of severe friction. This complexity makes necessary the numerical study in order to be able to optimize the design of the friction system and to a better understanding of her thermal behavior.

In this study, 3D modeling of the thermal behavior of the pin-disc assembly was performed on the calculation code Ansys v15 with the consideration of the loading and boundary conditions around the disc and the pad. To achieve our goal, an experimental investigation was dialed of on a specific pin-on-disc tribometer to determine boundary conditions using an inverse resolution algorithm. The creation and validation of the numerical simulation make it possible to analyze the evolution and distribution of temperatures at the level of the contact zones.

Thus, the analyses made on the thermal behavior of this model show that these types of technological solutions are the keys of the improvement in the brake systems design.

Keywords: 3D numerical modelisation · braking system · thermal behavior · Code Ansys

1 Introduction

The problem of braking requires the study of different physical phenomena, at different scales, taking into account their couplings that are considered as scientific challenges. The complexity of small-scale experimentation, such as monitoring temperature, pressure and reproducing real braking conditions, leads to the use of numerical modeling to predict the thermal behavior of brake systems (Jafari and Akyüz 2022). In some studies, it has been shown that the finite element method is adopted as one of the most appropriate for the investigation of a thermal problem during braking (Pranta et al. 2022). Several researchers (Talati and Jalalifar 2008; Yevtushenko and Grzes 2011) have studied finite element modeling of axisymmetric problems. The concept of the third body generated at the interface between the disc and the pad and responsible for generating the heat flow

A. Akrout et al. (Eds.): ICAV 2022, ACM 22, pp. 433–439, 2023.
https://doi.org/10.1007/978-3-031-34190-8_46

during the braking action was taken into account (Berthier and Godet 1996; Majcherczak et al. 2005). The multi-physics problem of thermal generation and dissipation at sliding contacts, with consideration of the third body is considered in this study. This type of problem, despite having been the subject of some work in recent decades (Elhilali et al. 2021; Maria Vinsiya et al. 2022), remains poorly controlled from a thermal point of view.

Therefore, our study is based on an iterative numerical resolution of an inverse problem, in order to determine the input parameters of the numerical model. The validation of the proposed model is conditioned by the good agreement between the experimental evolution of the temperature and the numerical results.

2 Numerical Model

A three-dimensional transient thermal finite element model was developed using ANSYS software. The numerical model illustrated in Fig. 1 shows the geometry studied in numerical model, composed of disc, spindle and interface. The third body is considered an interface layer has a cylindrical shape 14 mm in diameter and 10 μ in thickness (Baklouti et al. 2015). It rubs against a gray cast iron disc with a diameter of 80 mm and a thickness of 22 mm. The average friction radius is 32 mm. In order to simplify the numerical study:

- Kinetic energy is converted into thermal energy by friction at the sliding interface, without any other loss of energy during braking.
- Perfect contact between disc and spindle.
- The radiation of the disc is neglected due to the rapidly braking action.
- Dissipation of the heat is limited to convection throw free surfaces of the tribometer (Fig. 2).

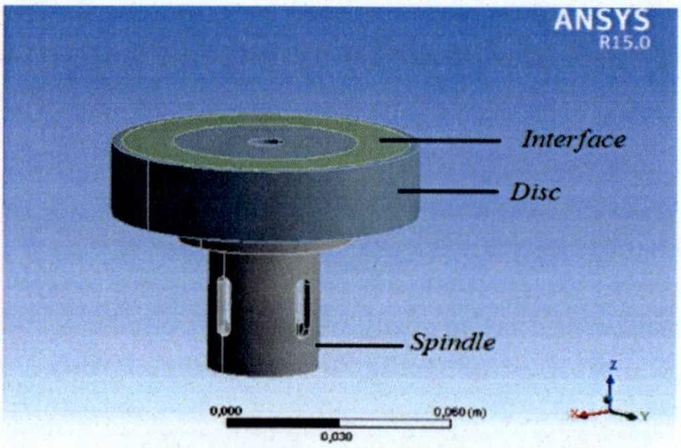

Fig. 1. 3D numerical model

The numerical model adopted consists of disc, spindle and the third body represented on the entire friction surface. Recognizing that the modeling is interested in the evolution of the temperature, the thermo physical properties of the various components were introduced into the model (Table 1). Third body properties are evaluated by (Day 1990) and it is modeled by a 10 μm thick interface. In this study, for meshing, the elements used is 3-node linear elements adopted to the heat transfer study. To ensure more efficient thermal results, the mesh was finely refined in a depth of 2 mm under the rubbed disc surface and at the level of the interface.

Table 1. Thermo-physical properties of the disc, spindle (Baklouti et al. 2015) and third body (Day 1990).

	Disc (grey cast iron)	Spindle (iron)	Third body
λ ($W.m^{-1}.K^{-1}$)	12	42	0.007
Cp ($J\ Kg^{-1}\ K^{-1}$)	500	480	1000
ρ ($kg\ m^{-3}$)	7293	7790	0.1

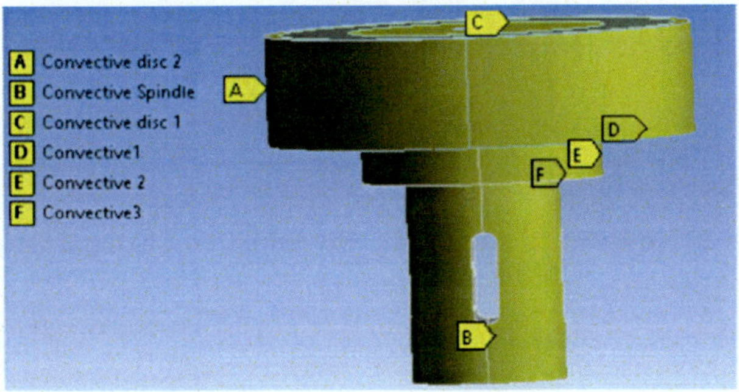

Fig. 2. Boundary conditions

3 Experimental Model

3.1 Wear Tests

Wear tests were carried out on a pin on disc type tribometer (Baklouti et al. 2015). A continuous friction test was performed under severe friction condition: Pressure of 1.8 MPa and a sliding speed of 9 m/s. Three wear tests were run in order to readjust all the convective exchanges, on the faces exposed to ambient air of different rotating solids: disc and spindle. Figure 3 presents the experimental setup of the wear test.

3.2 Heat Flux Density

Figure 4 represents the evolution of the heat flux density for three tests which is determined by (Eq. 1) from the means friction coefficients, pressure and sliding speed.

$$\varphi = \frac{\Phi}{S} = \mu PV \quad (1)$$

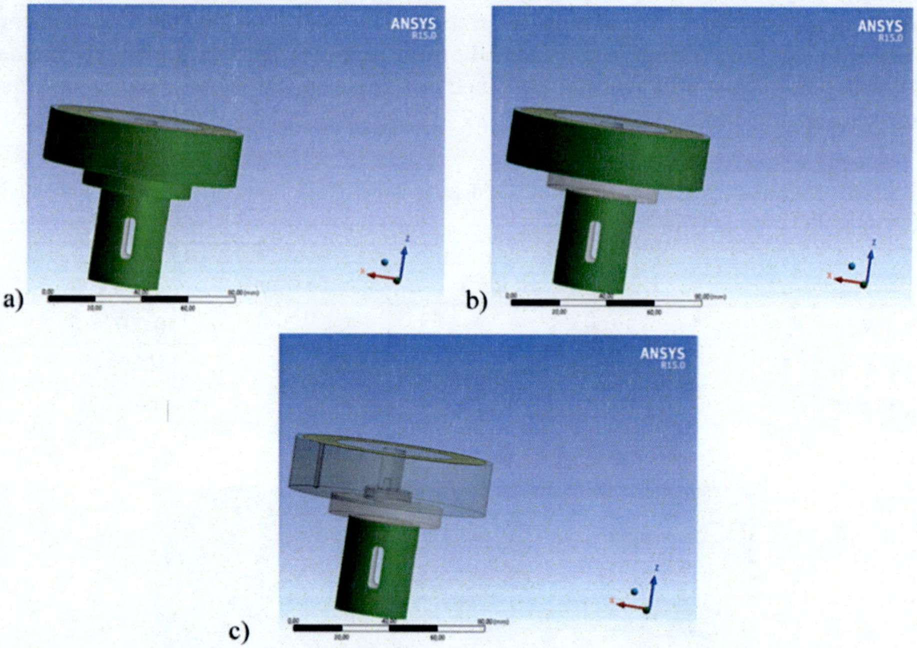

Fig. 3. Experimental setup of the wear test a) test 1 b) test 2 c) test 3

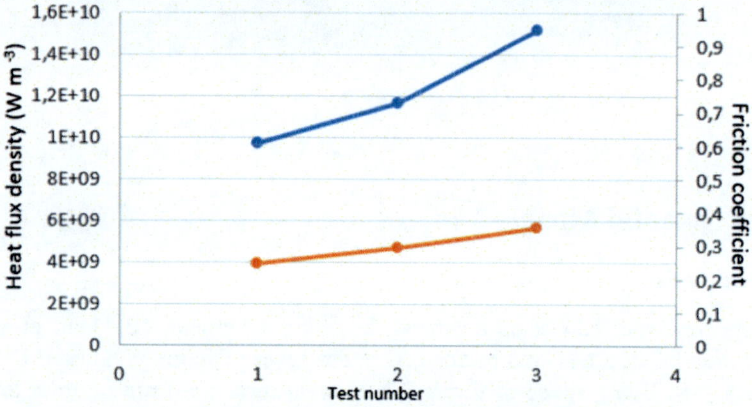

Fig. 4. Friction coefficient and density heat flux for three wear tests

4 Results and Discussions

With the inverse identification method (Fig. 5) the numerical temperature results which were in good agreement with experimental results (Fig. 5a) led to the determination of the convective exchange for free surfaces used for the test 1 (Table 2).

Table 2. Convective heat exchange Coefficients

Convective heat exchange coefficients	Convective disc 1	Convective1	Convective 2	Convective 3
Values (W m^{-2} K^{-1})	140	20	90	110

Subsequently, for test 2, it suffices to add a convection coefficient to the spindle (convective spindle) which is equal to 600 $Wm^{-2}K^{-1}$. This value is high due to the presence of four radial and parallel holes at this level of the tribometer. This value is confirmed since the good concordance of numerical and friction test results (Fig. 6b). For test 3, in the same way, the good agreement between both numerical and experimental results (Fig. 6c) confirm the convective heat exchange coefficient (convective disc 2) added on the lateral surface of the disc which is equal to 40 $Wm^{-2}K^{-1}$.

The values obtained in terms of temperature appear to correspond to the experimental measurements. It has also been shown that the numerical flux partition coefficient is consistent with the estimated value of material effects and test conditions. All these findings validate the numerical model.

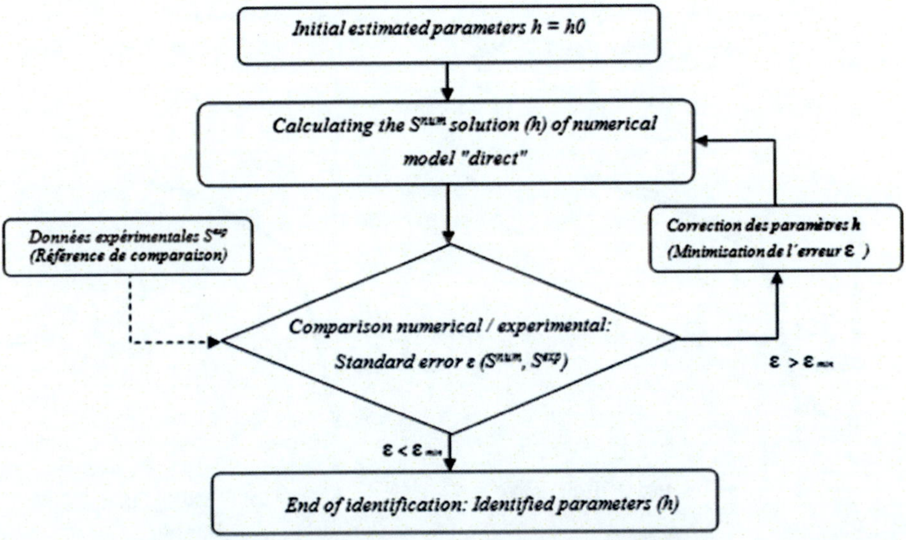

Fig. 5. Principle of the inverse identification method

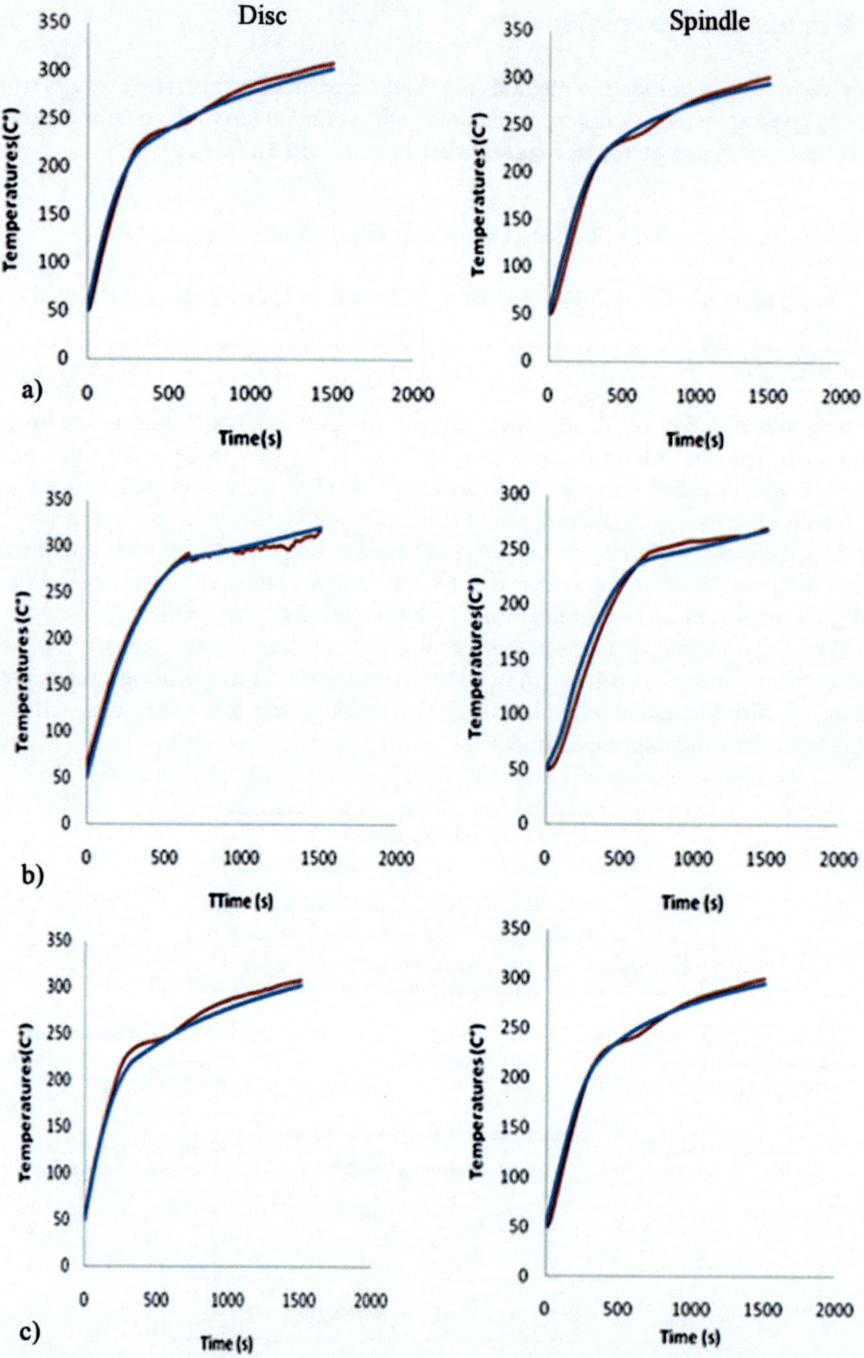

Fig. 6. Comparison of numerical (in red) and experimental (in blue) temperature evolutions for test a) test 1 b) test 2 c) test 3

5 Conclusions

During braking, the brake discs undergo several types of macroscopic damage that are mainly of thermal origin. The objective of our study is to model the thermal behavior of brake discs and this through the development of a 3D model using the ANSYS software. The results show that the proposed numerical method makes it possible to correctly predict the evolutions of the temperatures at the level of the disc, lining and pin. The three-dimensional simulation takes into account the third body generated by friction.

References

Jafari, R., Akyüz, R.: Optimization and thermal analysis of radial ventilated brake disc to enhance the cooling performance. Case Stud. Therm. Eng. **30**, 101731 (2022)

Pranta, M.H., Rabbi, M.S., Banik, S.C., Hafez, M.G., Chu, Y.-M.: A computational study on structural and thermal behavior of modified disk brake rotors. Alexandria Eng. J. **61**(3), 1882–1890 (2022)

Talati, F., Jalalifar, S.: Investigation of heat transfer phenomena in a ventilated disk brake rotor with straight radial rounded vanes. J. Appl. Sci. **8**, 3583–3592 (2008)

Yevtushenko, A., Grzes, P.: Finite element analysis of heat partition in a pad/disc brake system. Numer. Heat Trans. Part A Appl. **59**, 521–542 (2011)

Berthier, Y., Godet, M.: Third body approach. Tribology **32**, 21–30 (1996)

Majcherczak, D., Dufrénoy, P., Nait-Abdelaziz, M.: Third body influence on thermal friction contact problems: application to braking. ASME J. Tribol. **127**(1), 89–95 (2005)

Elhilali, F., Fihri-Fassi, H., Ourihi, R.: Towards the development of an optimized numerical model of the brake system pad with natural material. Mater. Today Proc. **45**, Part 6, 5419–5425 (2021)

Vinsiya, A.M., et al.: Comparative structural and frictional analyses on various lightweight materials for aircraft disc brake. Mater. Today Proc. **59**, Part 3, A22–A35 (2022)

Day, A.J.: Brake interface temperature prediction. In: Second Brakes Workshop University of Bradford (1990)

Baklouti, M., Cristol, A., Elleuch, R., Desplanques, Y.: Brass in brake linings: key considerations for its replacement. Proc. Inst. Mech. Eng. Part J: J. Eng. Tribol. **231**, 461–468 (2015). https://doi.org/10.1177/1350650115588966

Investigation on the Effect of Mesh Phasing on the Vibration Response of a Damaged Planetary Gear Transmission

Ayoub Mbarek[1,3](✉), Ahmed Hammami[3], Alfonso Fernández Del Rincón[2], Fakher Chaari[3], Fernando Viadero Rueda[2], and Mohamed Haddar[3]

[1] ENIB - Ecole Nationale d'Ingénieurs de Bizerte, Menzel Abderrahmen, 7021 Bizerte, Tunisia
ayoubmbarekenit@gmail.com

[2] Department of Structural and Mechanical Engineering, Faculty of Industrial and Telecommunications Engineering, University of Cantabria, Avda de los Castros s/n, 39005 Santander, Spain
{alfonso.fernandez,fernando.viadero}@unican.es

[3] Laboratory of Mechanics, Modelling and Production (LA2MP), National School of Engineers of Sfax, BP1173, 3038 Sfax, Tunisia
mohamed.haddar@enis.rnu.tn

Abstract. For any mechanism, and in the case of planetary gear (PG), the vibration response always has a good concept that can give information about the system in any case. It allows to clarify properly the crack defect characteristics of PG train.

In fact, a lumped-parameter-model of PG system is used to study the impact of the internal excitations due to the cracked sun. In this model, Mesh Stiffness Function (MSF) with crack damage are computed using an analytical approach and then they are adopted to drive the model. In addition, the effect of geometry and design parameters are highlighted through the study of the mesh phasing effect in the presence of failure crack. Three kinds of PG trains, namely In-Phase (IP), Sequentially Phased (SQ) and Counter-Phased (CP) gear meshes are discussed in simulation analyses.

The obtained results shows the validity and the ability of the developed dynamic model to solve the problem related to the complex structure. It shows also the importance of such computationally techniques that are more efficient predictive models for the design purposes of PG.

Keywords: Two stages planetary gear · crack defect · mesh phasing · vibration signal

1 Introduction

PG system requirement has increased during the last decades regarding as far as view their unique characteristic is concerned, which makes the research on these systems more interesting. PGs are quite complex systems and include several important parameters which should be considered in design, modelling or diagnostic purposes.

A. Akrout et al. (Eds.): ICAV 2022, ACM 22, pp. 440–448, 2023.
https://doi.org/10.1007/978-3-031-34190-8_47

In this section, a brief overview of PG system is presented. These systems can generally be modelled by a lumped parameter model [1], a finite element model [2], a hybrid model which combines the two-precedent models [3] or a phenomenological model [4]. Thus, these models can be exploited in several conditions depending on the application type.

However, as any other rotating system, PGs are subjected to external excitation due to the variations in the speed and/or load of functions [5], or any internal excitation that are related principally to the MSF. These functions can be modelled using several modelling approaches cited by [6–11, 12–18]. These approaches are divided into four categories. First, the Finite Element Methods (FEM) that were used to evaluate MSF [6, 7].

For any mechanism, and in the case of PG, the vibration response always has a good concept that can give information about the system in any case. It allows to clarify properly, as listed above, the crack defect characteristics of PG train.

In fact, a lumped-parameter-model of PG system is used to study the impact of the internal excitations due to the cracked sun. In this model, MSF with crack damage are computed using an analytical approach and then they are adopted to drive the model. In addition, the effect of geometry and design parameters are highlighted through the study of the mesh phasing effect in the presence of failure crack. Three kinds of PG trains, namely (IP), (SQ) and (CP) gear meshes are discussed in simulation analyses.

The main contributions are summarized as follows:

1- A rigid body PG model, based on a lumped mass modelling approach is built. It takes account of the influencing functions such as stiffness, damping and external excitation.
2- The system is studied and analysed in a steady state condition and some parameters are considered to show the importance of PG configuration and its impact on cracked tooth features.

2 Numerical Model

Figure 1 presents the sketch of numerical model which is developed using the lumped mass method. The main components of each stage are a sun (S),a carrier (C), a ring (R) and three planets (P1, P2, P3) which are characterized by three DOF and defined by the masses (mij), Inertia (Iij) and stiffness (kijk) where i = c, r, s, p1, p2, p3; j = r, t in direction k = u, v, w, ϕ, Ψ, θ. As presented in the previous section, the meshing stiffness associated to the sun-planets (S-P) and ring-planets (R-P) functions are Krpr1, Krpr2, Krpr3 and Kspr1, Kspr2, Kspr3 for the reaction gear set and by the stiffness Krpt1, Krpt2, Krpt3 and Kspt1, Kspt2, Kspt3 for the test gear set. The shafts connecting the two gears set are modelled by an axial stiffness ksa, kca; flexural stiffness ksf, kcf and torsional stiffness kst, kct, respectively. It is a back-to-back PG system. It is proposed for the reasons of higher load and compactness. In fact, this concept of PG test bench consists of injecting an output power in the shaft connecting the motor to the input. The injected power should have the same speed as the motor. So, another PG called reaction gear is mounted. The suns are connected through a common shaft and the carriers are connected through a hollow shaft. The reaction PG set, and the test PG set have the same transmission ratio. The test ring is fixed, and the reaction ring is free, allowing the introduction of an external load. All the values of the parameters inserted in the model are listed in Table 1.

More details about this model are listed in [7–10].

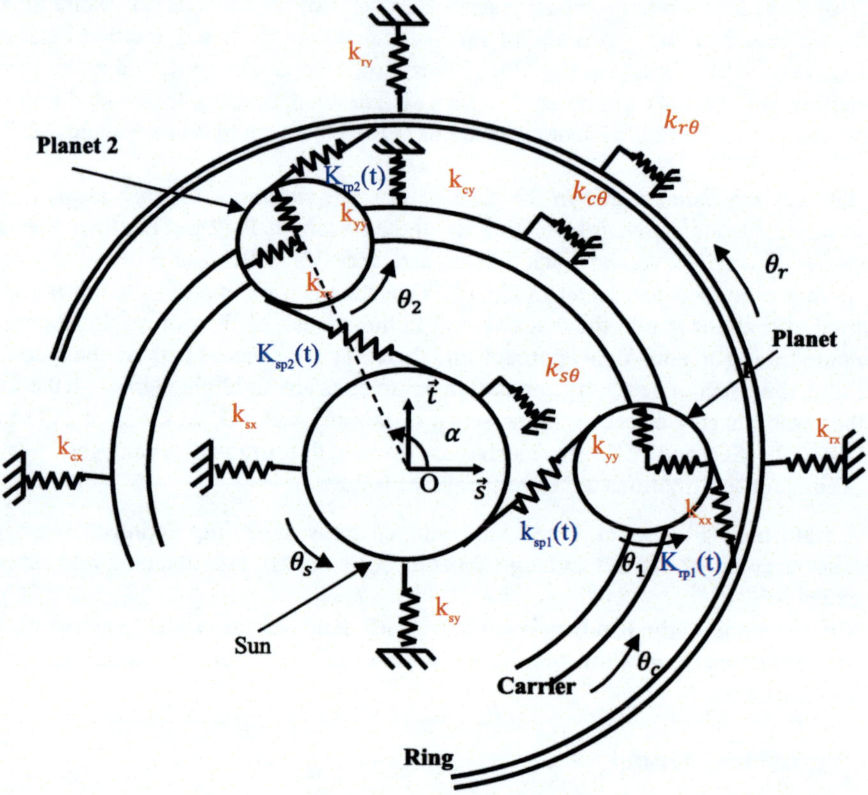

Fig. 1. PG numerical model

3 Results

Substituting all parameters described previously such as MSF in normal condition, i.e. without defect and in presence of a crack defect in the sun, into the numerical model, the vibration response of each gear component is extracted. The results are presented on the stationary ring components where all the vibration paths can be met. Under fixed speed condition (1500 rpm) and fixed external torque (100 N-m), the vibration response signals are plotted as shown in Fig. 2. The time and spectrum domain were chosen to present signals.

The vibration response is simulated on the torsional component of the test ring gear due to its sensitivity to the nonlinear factor and to the excitations.

Table 1. Gear parameter.

Component	Sun	Planet (3)	Ring	Carrier
Test and reaction gear: Axial bearing Stiffness [N/m]	$K_{sx} = K_{srz} = 1.5 \times 10^8$ $K_{sz} = 3 \times 10^8$	$K_{px} = K_{pry} = 1.1 \times 10^8$ $K_{pz} = 3 \times 10^8$	$K_{rx} = K_{ry} = 8 \times 10^8$ $K_{rz} = 1 \times 10^9$	$K_{cx} = K_{cz} = 1 \times 10^8$ $K_{cz} = 5 \times 10^8$
Reaction gear: Torsional bearing stiffness [Nm/rad]	$K_{sr\phi} = K_{s\Psi} = 6 \times 10^9$	$K_{pr\phi} = K_{pr\Psi} = 6 \times 10^9$	$K_{rr\phi} = K_{rr\Psi} = 1.5 \times 10^9$	$K_{cr\phi} = K_{cr\Psi} = 6 \times 10^9$
Test gear: Torsional stiffness [Nm/rad]	$K_{st\phi} = K_{st\Psi} = 5 \times 10^9$	$K_{pt\phi} = K_{pt\Psi} = 6 \times 10^9$	$K_{rt\phi} = K_{rr\Psi} = 5 \times 10^9$ $K_{rt\theta} = 8 \times 10^6$	$K_{ct\phi} = K_{ct\Psi} = 5 \times 10^9$

The time response in Fig. 2b shows the presence of peaks in the case of defect compared to the same signal without defect. This behaviour is normal because an extra excitation is added, and it is principally related to the defect. Also, the vibration amplitude of the acceleration signal is increased.

The geometry of the PG on its dynamic vibration response is already previously studied by Inalpolat and Kahraman, [10]. Cases of PG were under consideration to study the relationship between the geometry and the generated vibrations. They found that certain geometries produce similar spectral structures. According to these studies and based on the planet's position as well as the mesh phasing, PGs can be classified into five groups as indicated by Peng et al. [11].

The IP, SQ, and CP configurations are the design parameters.

In this work, we focus on the influence of mesh phasing with respect to an equal space between the planets, i.e., two scenarios are studied; the first is an equally spaced planets' gear and IP mesh phase and the second is a spaced planet gear and an SQ mesh phased.

The trend of the ring planets and the sun planets meshing functions is displayed in Fig. 3 and 4. It is well observable that the meshing functions are well in phase for the SQ configuration where the red markers are well aligned. However, for Fig. 3 which presents the IP configurations, a phasing between the three functions is presented.

In gear system, the main source of the vibrations is the internal excitations that are principally due to the MSF and precisely to the gear mesh forces. In case of PGs, and depending on the number of planets, the Dynamic Mesh Force (DMF) are presented between the sun-planets components and ring- planets components. So, these functions have a great potential to inform about the PG vibratory response.

To have further insight about the potential of the numerical model, the system is analysed in both phasing configurations (IP and SQ).

Considering a PG system with no geometrical errors, and since the planet gears are equally distributed, the Gear Mesh Frequency (GMF) magnitudes are the same across all planet components. To better visualize and comprehend the evolution of GMF and avoid its presentation with the input speed, the spectrum of acceleration are plotted in frequency domain. Figure 5 displays the spectrum of acceleration on the ring for both categories. We notice that the GMF are moved from the SQ to the IP configuration, this fact is to the modification of the number of teeth and the kinematic of the system.

When the cracked tooth sun enters in contact with another tooth, the vibration amplitudes of PG system increase. Also, the spectrum response of both configuration (IP or SQ) is the same. Only the sidebands related to the faulty sun are detected because the force of the rotation of carrier is neglected. Besides, only GMF is presented in healthy case.

The effects of mesh phasing are quantified using a dynamic factor (DF) for each gear mesh.

Where the DMF is the Dynamic Mesh Force, and the Static Mesh Force is the quasi-static mesh force. This factor changes in time because it is related to the DMF. It is typically used as a critical parameter for design engineering. To distinguish the difference between the dynamic response of different phasing conditions, the dynamic

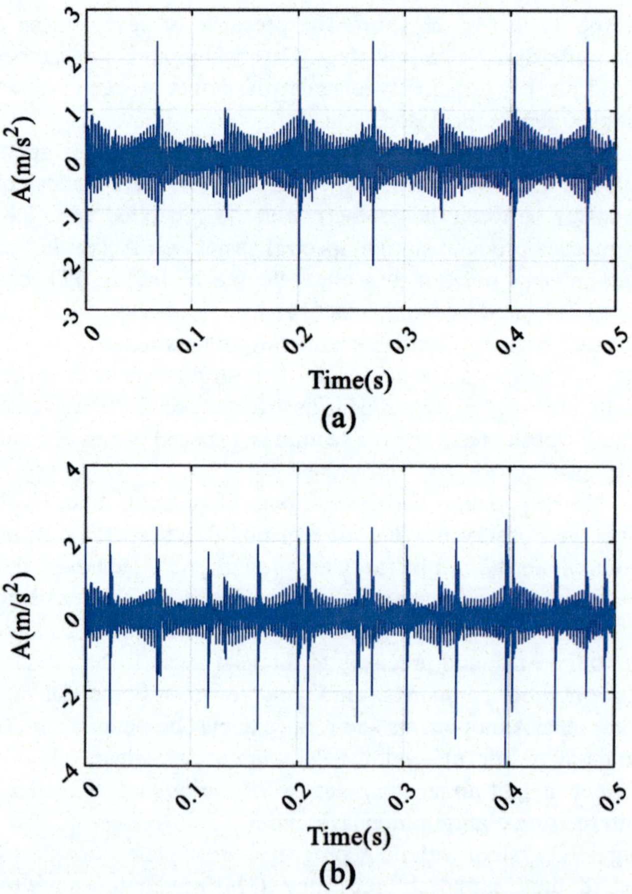

Fig. 2. Vibration waveform (a) Normal case (b) Crack case

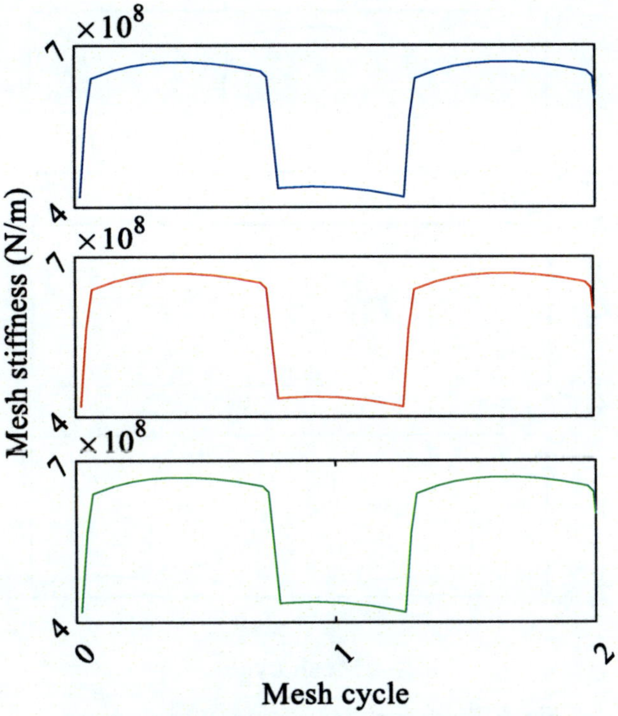

Fig. 3. Ring-planets MSF trend in SQ configuration

factors of the PG system in IP and SQ configurations are illustrated in Fig. 6. In fact, In the studied system, the planetary gears stages are mounted back-to-back. The design parameter adopted which is the SQ configuration shows an interesting result in diagnosis process comparing to the IP configuration.

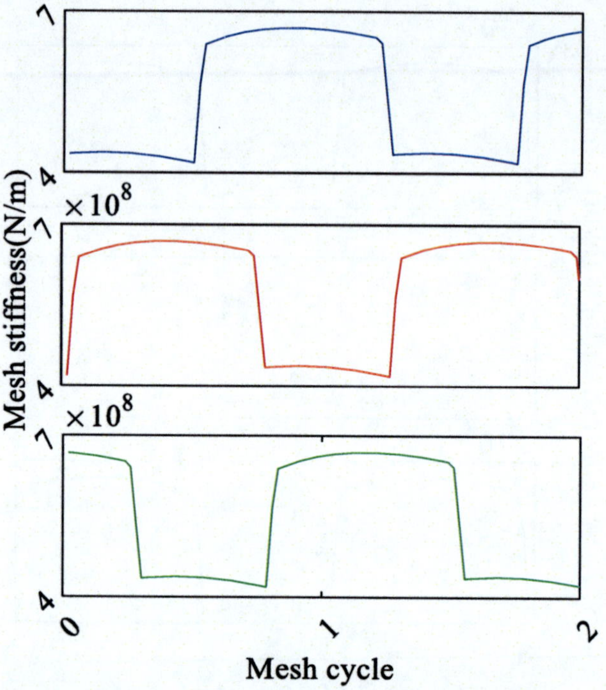

Fig. 4. Ring-planets MSF trend in IP configuration

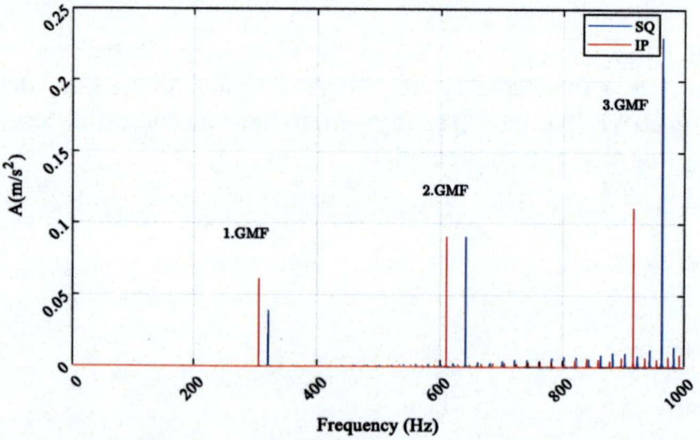

Fig. 5. Spectrum of acceleration simulated on the test ring.

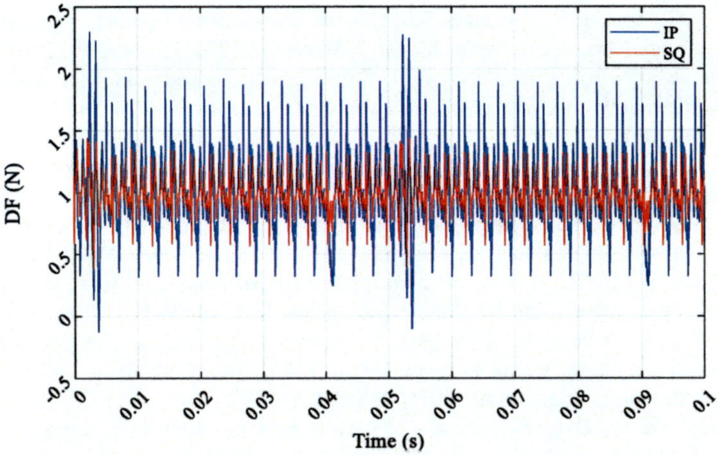

Fig. 6. Dynamic factor between sun-planets for IP and SQ mesh configuration

4 Conclusion

Using lumped mass model of PG transmission with a power recirculation energy that considers the gear components as a rigid body and each element is characterised by 3 DOF, and by using the MSF, which are the responsible functions of the internal excitations and which reflect the presence of cracked tooth on the behaviour of the system, the dynamic properties behaviour of the system under cracked sun gear are investigated numerically and experimentally using a test rig, these properties are contributed to validate the proposed model and could be a good indicators for cracked tooth features. Some useful conclusions are listed below:

- The cracked defects introduce a variation on the vibration waveform. The time domain analysis shows many peaks that will occur. In the frequency domain, the frequency of the cracked sun and its multiple frequencies are presented of the first GMF and its harmonic. Since the defect feature of each gear component in the PG is different, the frequency characteristic of the defect and the source of the fault can be determined.
- When the cracked tooth sun enters in contact with another tooth, the vibration signals of PG system as well as the gear mesh configuration (IP or SQ) have the same sidebands features. Only the sidebands related to the faulty sun are detected.

Acknowledgements. The authors would like to acknowledge Project DPI2017-85390-P funded by the Spanish Ministry of Economy, Industry, and Competitiveness for supporting this research.

References

1. Liang, X., Zuo, M.J., Hoseini, M.R.: Vibration signal modeling of a planetary gear set for tooth crack detection. Eng. Fail. Anal. **48**, 185–200 (2015)

2. Parker, R.G., Agashe, V., Vijayakar, S.M.: Dynamic response of a planetary gear system using a finite element/contact mechanics model. J. Mech. Des. **122**(3), 304–310 (2000)
3. Abousleiman, V., Velex, P.: A hybrid 3D finite element/lumped parameter model for quasi-static and dynamic analyses of planetary/epicyclic gear sets. Mech. Mach. Theory **41**(6), 725–748 (2006)
4. Zhang, M., et al.: An improved phenomenological model of vibrations for planetary gearboxes. J. Sound Vib. **496**, 115919 (2021)
5. Chaari, F., Abbes, M.S., Rueda, F.V., del Rincon, A.F., Haddar, M.: Analysis of planetary gear transmission in non-stationary operations. Front. Mech. Eng. **8**(1), 88–94 (2013)
6. Lin, T., Ou, H., Li, R.: A finite element method for 3D static and dynamic contact/impact analysis of gear drives. Comput. Meth. Appl. Mech. Eng. **196**(9–12), 1716–1728 (2007)
7. Mbarek, A., Hammami, A., del Rincon, F.A., et al.: Mesh force analysis of planetary gear system with defect. In: Walha, L., et al. (eds.) CMSM 2021. LNME, pp. 680–686. Springer, Cham (2023). https://doi.org/10.1007/978-3-031-14615-2_76
8. Mbarek, A., et al.: Dynamic behavior of back to back planetary gear in presence of pitting defects. In: Fakhfakh, T., Karra, C., Bouaziz, S., Chaari, F., Haddar, M. (eds.) ICAV 2018. ACM, vol. 13, pp. 16–22. Springer, Cham (2019). https://doi.org/10.1007/978-3-319-94616-0_2
9. Mbarek, A., Hammami, A., Del Rincón, A.F., Chaari, F., Rueda, F.V., Haddar, M.: Early damage detection in planetary gear transmission in down-time regime. In: Hammami, A., Heyns, P.S., Schmidt, S., Chaari, F., Abbes, M.S., Haddar, M. (eds.) MOSCOSSEE 2021. ACM, vol. 20, pp. 31–37. Springer, Cham (2022). https://doi.org/10.1007/978-3-030-85584-0_4
10. Inalpolat, M.: A theoretical and experimental investigation of modulation sidebands of planetary gear sets. Dissertation, the Ohio State University (2009)
11. Peng, D., Smith, W.A., Randall, R.B., Peng, Z.: Use of mesh phasing to locate faulty planet gears. Mech. Syst. Signal Process. **116**, 12–24 (2019)

Exponentially Weighted Moving Average Control Chart for Fault Detection of the Spur Gear Transmission System

Rasheed Majeed[1,3](✉), Maroua Haddar[1,2], Fakher Chaari[1], and Mohamed Haddar[1]

[1] Laboratory of Mechanics, Modeling, and Production (LA2MP), National School of Engineers of Sfax, BP 1173–3038, Sfax, Tunisia
{rasheed-majeed-ali.al-jorani,maroua.haddar, fakher.chaari}@enis.tn, mohamed.haddar@enis.rnu.tn

[2] National School of Engineers of Sousse, University of Sousse, 4003 Sousse, Tunisia

[3] Ministry of Construction, Housing, Municipalities and Public Works, Diyala Governorate, Diyala IQ32001, Iraq

Abstract. A tooth fracture is one of the most common failures in the gear systems used in industrial sectors and manufacturing. In order to prevent fracture defects, great research interest has emerged in fault detection in gear systems. Statistical process control charts have been recently used for fault detection based on experimental data. Despite the advantages of simulation data for dynamic models recognized in gear systems, this approach has yet to be used to train SPC charts. This paper proposes to apply an exponentially weighted moving average chart (EWMA) to monitor and detect fracture deterioration generated by the 5% reduction in gear mesh stiffness in the dynamic model of the spur gear system. Next, white Gaussian noise is added at different levels to raw signals, and the appropriate noise level is chosen based on the highest signal-to-noise ratio (SNR). Then, time-domain features of both root mean square (RMS) and kurtosis are extracted as statistical control indicators. After that, EWMA charts are constructed using the statistical features of the gear signal under healthy conditions. Finally, EWMA charts are being tested based on faulty status features. Results show that the EWMA by RMS is more effective than the EWMA by kurtosis in detecting fracture faults.

Keywords: spur gear system · slight fracture fault · signal-to-noise ratio (SNR) · statistical features · exponentially weighted moving average chart (EWMA)

1 Introduction

The gearbox system is a crucial component of current mechanical equipment. It is significant in transmitting power and motion in various applications, including power generation, manufacturing systems, and other modern mechanical systems (Shi et al., 2021). Gearboxes are often subjected to failure due to harsh operating conditions and heavy loads. Gear failure can result in unexpected system breakdowns, expensive maintenance costs, production loss, and risks to human life (Mohammed et al., 2022). Therefore,

A. Akrout et al. (Eds.): ICAV 2022, ACM 22, pp. 449–461, 2023.
https://doi.org/10.1007/978-3-031-34190-8_48

research to develop gear system fault detecting and diagnosis methods has become a necessary and effective topic for researchers (Chen et al., 2019). The use of dynamic models for monitoring the state of gear systems and diagnosing their faults has received significant interest since it is a viable alternative for obtaining data from expensive experiments or field measurements (Mohammed et al., 2015). Several dynamic models with vibration analysis methods have been used in the literature for condition monitoring and fault detection of various gear systems (Chen and Shao 2011) (Jiang et al. 2020).

One of the techniques used for fault detection is statistical process control (SPC), which was used to diagnose faults in industrial processes (Harrou et al., 2016). The primary goal of the SPC technique is to monitor the variations in the process over time. In recent years, different types of SPC have been used to detect faults in rotating

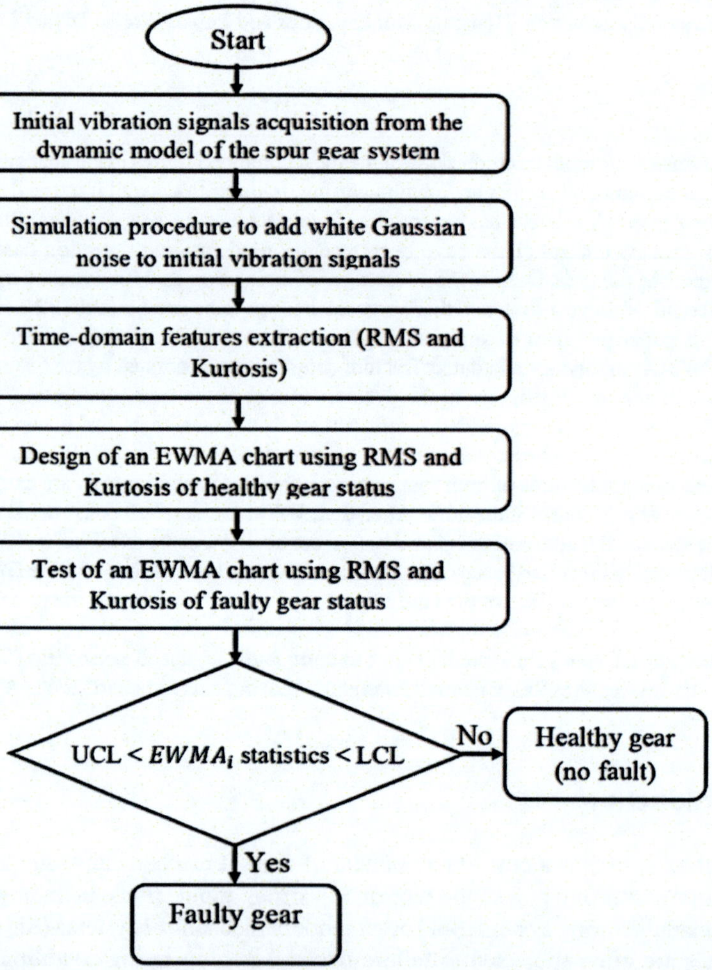

Fig. 1. The flowchart for the proposed methodology to detect fracture faults by apply EWMA control chart

machines, such as gears and bearings. For example, (Lal and Kane 2019) applied the exponentially weighted moving average chart (EWMA) using the time series model to detect spur gear tooth fracture at three levels, i.e., 37.5, 75, and 100%. In contrast, using the oil analysis method, (Maraş et al., 2021) used one of the Shewhart charts, the X-S bar chart, to detect tooth wear in the spur gear system. Along the same lines, (Jawad and Jaber 2021) presented the cumulative sum chart (CUSUM) for detecting bearing failures using statistical features of time vibration signals. Based on their findings, the researchers concluded that statistical control charts are reliable for finding faults in rotating machines. In addition, it is easy to implement, does not necessitate sophisticated computations, and can visually spot faults and unexpected shifts simultaneously (Maraş et al., 2021).

What distinguished the proposed methodology in this study, which was carried out by Lal and Kane (2019), was applying one of the SPC charts to detect a slight tooth fracture. From a practical point of view, detecting gear tooth fracture with large levels of more than 40% is not a benefit; the tooth may break due to heavy loads during operation. Therefore, this paper proposes the detection of gear tooth fracture caused by simulating a gear mesh stiffness loss of 5%, which is difficult to simulate in experimental tests. The EWMA control scheme was designed and tested based on the vibration simulation data obtained from a dynamic model of the spur gear system. Two statistical indicators of time domain analysis, RMS and kurtosis, were adopted to design and test the EWMA charts. The performance of the EWMA control charts was compared with the sensitivity improvement of the statistical features to detect gear fracture deterioration. Figure 1 shows a flowchart of the proposed methodology, which is explained in the following sections of the paper.

2 Mathematical Modeling of Spur Gear System

The proposed EWMA control chart is evaluated for fault detection using initial simulated vibration signals from a dynamic model of a spur gear system with 8 degrees of freedom (DOF) developed by (Chaari et al., 2012),a as shown in Fig. 2. It is made up of two blocks. A pinion has 20 teeth, and a wheel has 40 teeth. Each block has four DOF, two translations k_{xi}, k_{yi} and two rotations $k_{\theta i}$, $k_{\theta j}$ of the pinion and wheel. The two gear bodies are assumed to be rigid spur discs, and the shafts are presented with torsional stiffness. The bearings support the shafts, each represented by two linear springs. A linear spring moving along the motion line of the meshing teeth yields a time-varying gear mesh stiffness K(t). The description of the displacement on the path of action is expressed by

$$\delta(t) = (x_1 - x_2)\sin(\alpha) + (y_1 - y_2)\cos(\alpha) + \theta_{12} rb_{12} + \theta_{21} rb_{21} \tag{1}$$

x_i and y_i are the translations of block i (i = 1, 2). θ_{ij} is the angular displacement of the wheel j in block i (i, j = 1, 2). $rb12$ and $rb21$ are respectively the base radius of the pinion and the wheel and α is the pressure angle.

The equation of motion of the system is acquired by applying Lagrangian formulation as follows by:

$$M\ddot{q} + C\dot{q} + K(t)q = F \tag{2}$$

F is the external applied torques vector given by: $F = \{0, 0, C_m, 0, 0, 0, 0, C_r\}^T$.

q is a vector of the degree of freedom of the system defined as $q = \{x_1, y_1, x_2, y_2, \theta_{11}, \theta_{12}, \theta_{21}, \theta_{22}\}^T$.
M is a diagonal mass matrix of the system given by: $M = diag(m_1, m_1, I_m, I_1, m_2, m_2, I_2, I_r)$.
m_1 and m_2 are the mass of the block 1 and block 2,respectively. I_m, I_1, I_2, I_r are the inertia moment of the motor, the pinion, the wheel, and the receiver, respectively.

The implicit Newmark algorithm is applied to calculate the dynamic system response, and the mesh stiffness *K(t)* is inserted into the motion Eq. (2) to simulate a local defect. It is represented by a slight fracture of the gear tooth caused by a 5% loss in the gear mesh stiffness, as shown in Fig. 3.

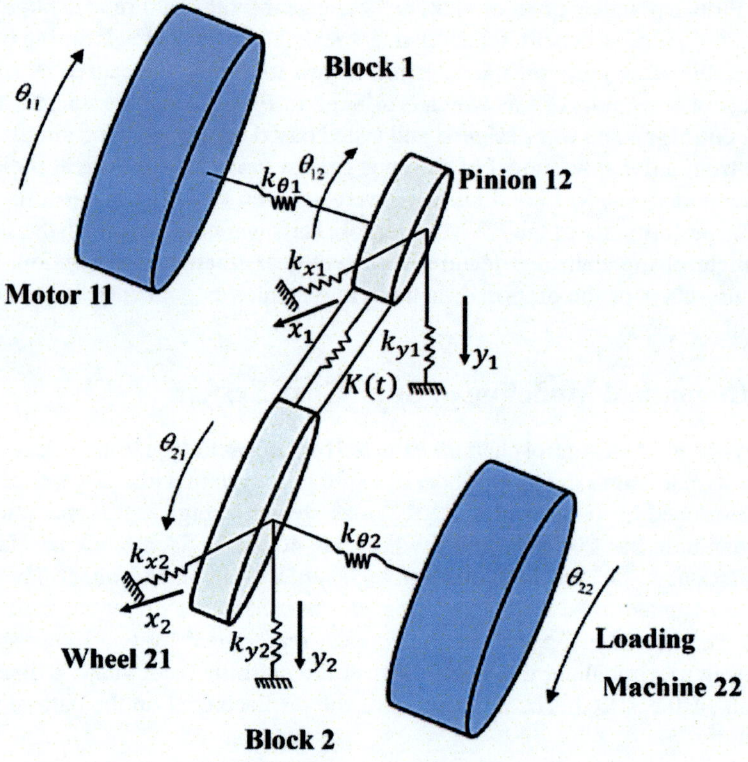

Fig. 2. Dynamic model of a one-stage spur gear system

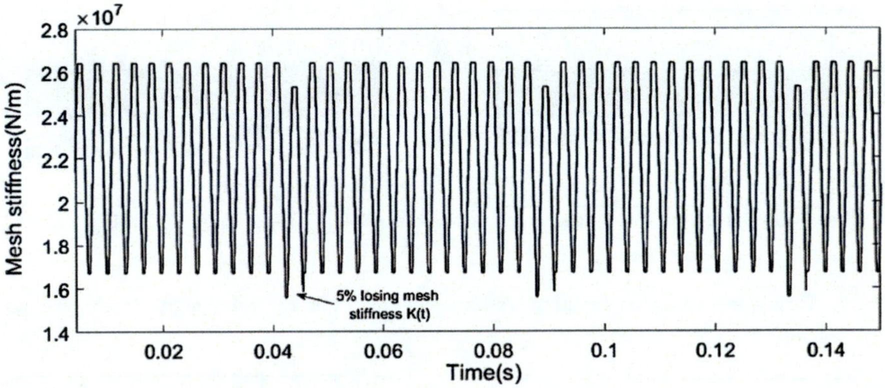

Fig. 3. Gear mesh stiffness function with a reduction of 5% to a simulation of the local defect

3 Simulation of White Gaussian Noise

This section explains the simulation procedure of adding noise to the initial vibration signals of the spur gear system under healthy and faulty conditions. White Gaussian noise (WGN) was added with different noise levels to simulate the operating conditions of the gear system in real life. The study considered examples of commonly used noise levels, with values of 5 dB, 10 dB, and 15 dB used by (Zhao et al., 2016) (Zhang et al., 2015). The signal-to-noise ratio (SNR) values were computed using an equation provided by (Li et al., 2017).

$$SNR = 10 \times log_{10}(\frac{P_x}{P_\varepsilon}) \tag{3}$$

where $p_x = \sum_{n=1}^{N}[x(n)]/N$, is the average power of the signal and, $P_\varepsilon = \sum_{n=1}^{N}[\varepsilon(n)]/N$, is the average power of the noise.

A higher SNR indicates that the fault signal power is higher than the background noise power, which will make it easier to detect. Therefore, the suitable noise level was chosen to be equal (15 dB) for all signals since it gave the highest SNR ratio, as shown in Table 1. Figures 4 and 5 show the gear system's initial and noisy vibration signals under healthy and faulty conditions.

Table 1. Signal-to-noise ratio (SNR) of gear signals under healthy and faulty conditions

Noise levels	5 dB	10 dB	15 dB
SNR of healthy gear signal	11.5489	23.0304	34.5648
SNR of faulty gear signal	11.4833	23.0718	34.5330

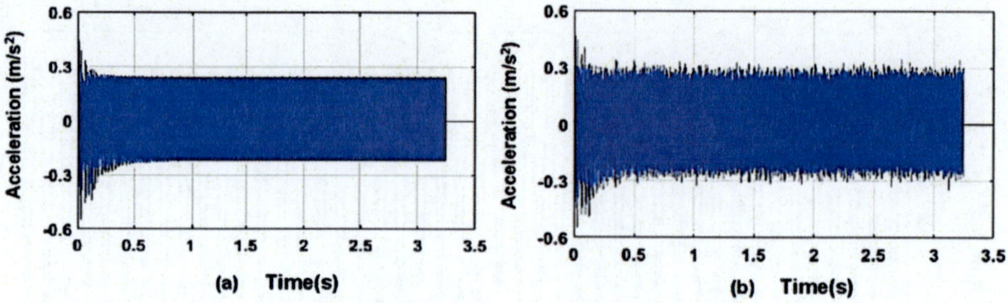

Fig. 4. Simulation vibration signal of healthy gear condition: (a) initial signal (b) noisy signal

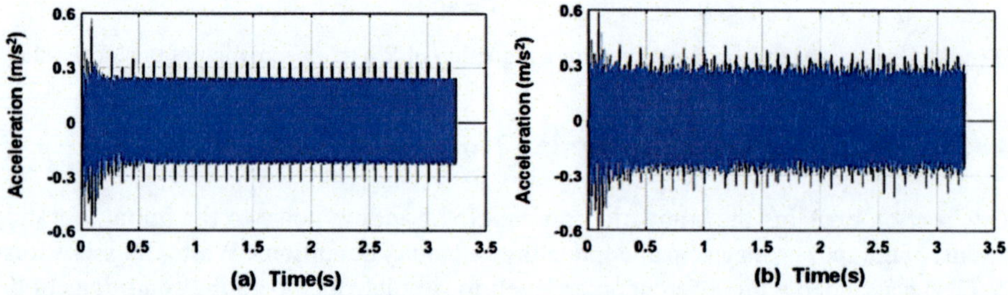

Fig. 5. Simulation vibration signal of faulty gear condition: (a) initial signal (b) noisy signal

4 Time-Domain Analysis

The time-domain vibration signal features are often used to monitor a gear's health condition. Root mean square (RMS) and kurtosis were widely used statistical indicators applied in time domain analysis to assess the health of gear systems (Mohammed et al., 2015).

In short-time signals, fault feature extraction is more accurate than long-time signals. Therefore, the noisy signals shown in Figs. 4b and 5b were divided into 35 segments as short-time signals. Each short-time signal represents the period between two fault-related pulses at 0.09 s, as shown in Fig. 6. The short-time signal of the faulty gear in Fig. 6b seems similar to the healthy gear signal in Fig. 6a. Since no periodic pulse appearance can indicate the fault.

As the next step, the RMS and kurtosis statistical features are extracted as fault-sensitive features from all short-time signals (35 signals) under healthy and faulty conditions. Then, statistical features are used to design and test the exponentially weighted moving average (EWMA) control chart for crack fault detection, which we will explain in Sect. 5.

4.1 Root Mean Square (RMS) Feature

The root mean square (RMS) is a statistic used to assess the severity of a vibrational signature. The detection of defects in rotating machinery is greatly helped by this feature. Using the root-mean-square (RMS) method is the simplest way to measure flaws in the temporal domain, and it is often sufficiently sensitive to detect fault onset (Yang et al., 2021).

$$\text{Root mean square(RMS)} = \sqrt{\frac{1}{N}\sum\nolimits_{n=1}^{N}[x(n) - \overline{x}]^2} \tag{4}$$

where $x(n)$ is the i-th member of points in the dataset or samples x, N is the number of data points in dataset samples x, and $\overline{x} = \frac{1}{N}\sum_{n=1}^{N} x(n)$.

4.2 Kurtosis Feature

Kurtosis has many applications in the monitoring and maintenance of mechanical systems, including gear systems including gear systems to detect tooth cracks (Yang et al., 2021).

$$Kurtosis = \frac{\frac{1}{N}\sum_{n=1}^{N}[x(n) - \overline{x}]^4}{\{\frac{1}{N}\sum_{n=1}^{N}[x(n) - \overline{x}]^2\}^2} \tag{5}$$

where x_n is the n- member of points in the dataset or samples, N is the number of sampling points and $\overline{x}$ is mean of samples $\overline{x} = \frac{1}{N}\sum_{n=1}^{N} x(n)$.

5 Exponentially Weighted Moving Average (EWMA) Chart

The exponentially weighted moving average (EWMA) chart is one of the statistical process control charts (SPCC). The EWMA chart provides the desired quality control conditions in a product's production process, detects product defects, and improves the process. Statistical control charts monitor the process samples by scheduling the quality indicators of a product, such as dimensions, temperatures, and vibration amplitude (Selvamuthu and Das 2018). EWMA effectively monitors small shifts in the process since it calculates all previous and current data in the entire process history (Hynek et al., 2014). EWMA statistics are calculated as explained in Eq. (6).

$$\begin{array}{ll} Z_{t_i} = \lambda x_{t_i} + (1 - \lambda) z_{t_i - 1} & \text{if } 1 < \text{t} \\ Z_0 = \mu_0 & \text{if t} = 0 \end{array} \tag{6}$$

where the Z_t represents the output for the observation statistics values of EWMA, and x_t is the value corresponding to the observation in the monitoring process in real-time. The Z_t statistic value is initially dependent on the average of the normal data (healthy condition) to be equal to μ_0. The λ is the smoothing constant, its value lies between (0 −1) (Montgomery 2009).

The EWMA chart consists of three limits which include both the central limit (CL), which is the average of the process, and the upper and lower control limits plotted above and below the chart, which are calculated as in Eqs. (7)–(9). The center limit represents the target value of the quality characteristic. The area delimited by the upper and lower

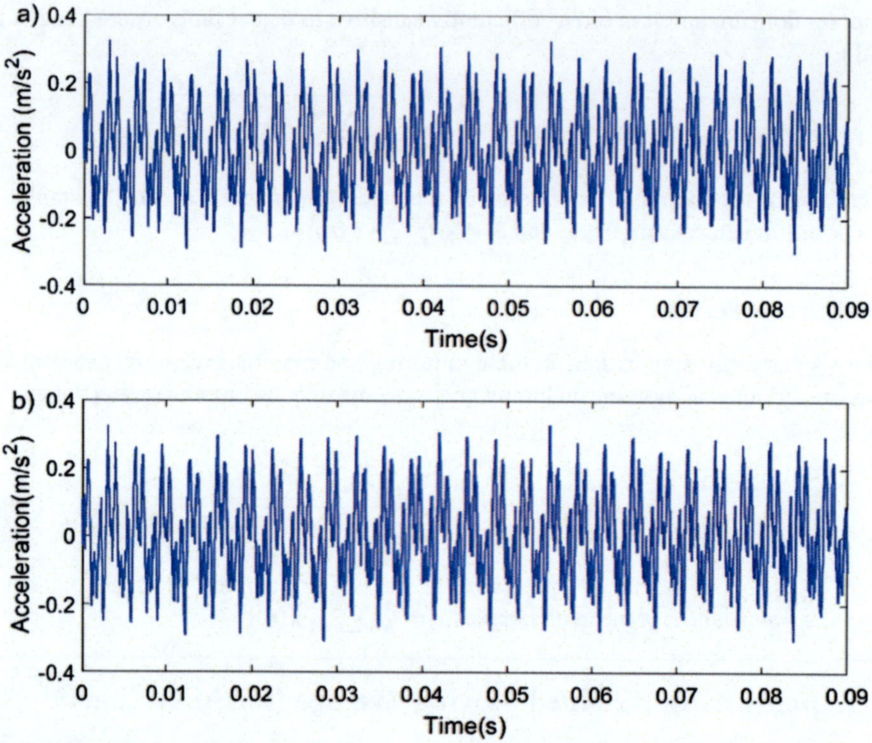

Fig. 6. Short-time signals for every two cycles of the pinion rotation: (a) healthy gear signal, (b) faulty gear signal

control limits indicates the permitted controlled area (3 sigma area). The obtained values are recorded on the chart as the process continues.

$$\textit{Upper control limit (UCL)} = \mu_0 + L\sigma\sqrt{\frac{\lambda}{2-\lambda}[1-(1-\lambda)^{2i}]} \tag{7}$$

$$\textit{Central limit (CL)} = \mu_0 \tag{8}$$

$$\textit{Lower control limit (LCL)} = \mu_0 - L\sigma\sqrt{\frac{\lambda}{2-\lambda}[1-(1-\lambda)^{2i}]} \tag{9}$$

where L is the design parameter affecting the sensitivity of the control chart and σ is the standard deviation.

The control charts use the measurement values collected from the product or the statistical value of the process being watched. It is determined that the process is under control if the value in question falls within the values that constitute the control region. If the value lies outside the control region, it is regarded as evidence that the process is out of control and a cause for concern. However, it is widely known that when control charts are used, two types of errors occur: Type-I and Type-II (Yang et al. 2018). In this study, we classify Type-I errors as those in which the gearbox is healthy (without fault), but its monitored data falls outside the control area. In type II errors, the gearbox is faulty, but controlled data does not fall outside the control area.

6 Results and Discussion

This section discusses the results of EWMA charts for gear fault detection. The evaluation of results is divided into two phases: the design of the EWMA chart and the testing of the EWMA chart.

6.1 Results of the EWMA Chart Design Stage

The EWMA charts were designed using root mean square (RMS) and kurtosis statistics for healthy gear system conditions. The control area limits (UCL, CL, and LCL) in the

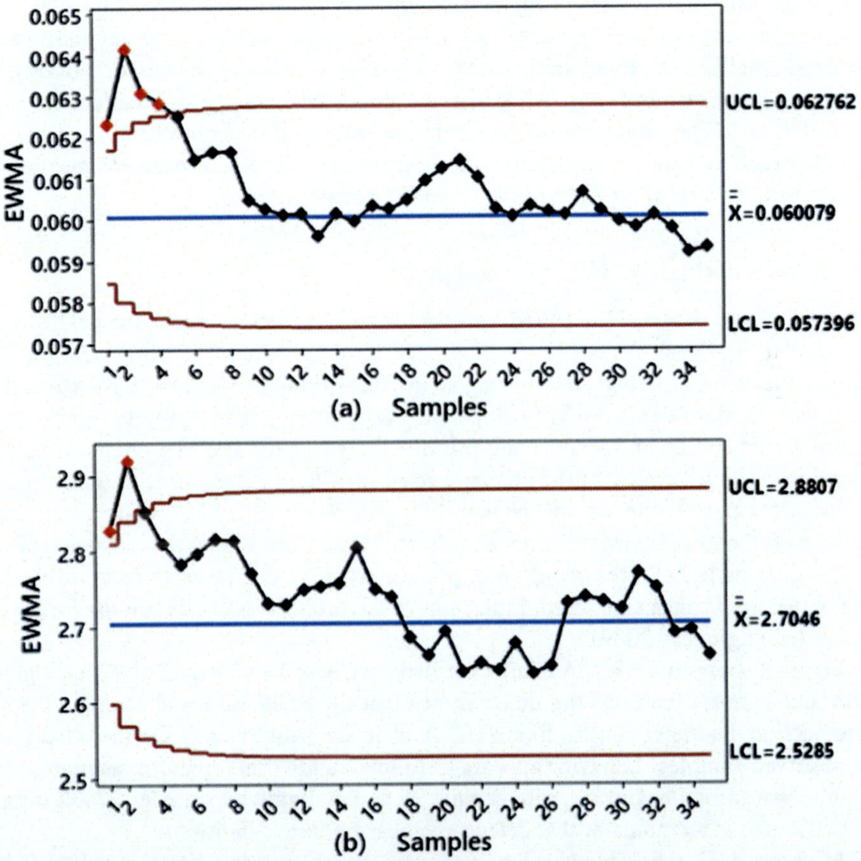

Fig. 7. Design EWMA control chart of the spur gear system under healthy conditions based on:(a) Root Mean Square (RMS) features (b) Kurtosis features

EWMA chart are calculated using Eqs. (7)–(9). The design parameters are determined for L = 3 and L = 0.2, as used in (Lal and Kane 2019). Moreover, the small value of $\lambda = 0.2$ is chosen because its small values increase the sensitivity of the EWMA scheme to detect small shifts. In addition, it provides a good balance of data in observing historical and current observation data. On the other hand, L = 3 is chosen, making about 99% of the observed data fall within the monitoring region (region 3 Sigma) (Selvamuthu and Das 2018).

Figure 7 shows the design of two EWMA charts under healthy gear conditions using RMS, and kurtosis statistical features, to achieve whether the proposed charts yield better results for monitoring the situation of the gear system. In Fig. 7a EWMA chart designed based on RMS shows a behavior of the observed statistics through non-random patterns appearance, represented by more than one point outside the upper control limits, specifically at samples1, 2, 3, and 4, although the normal for the gear system. In contrast, the exact behavior of the observed statistics is repeated in the EWMA designed using Kurtosis features where two first points are out of control, as shown in Fig. 7b.

As a result, the performance of both charts points to the occurrence of a type I error, often known as a false positive. This error suggests that an error happened even though it did not occur. This type I error can be explained due to an increase in the vibration amplitude to simulate the onset of rotational velocity acceleration during the first 0.36 s, as shown in Fig. 4. After that, the shift starts moving in the direction of the controlled variables to become stable within the bounds of the control region. Then, the shift changes to a constant amplitude since the motor's rotational speed has reached the rated speed. As a result, it becomes the operating system stable.

6.2 Results of the EWMA Chart Test Stage

After the design phase of the EWMA charts for the gear system under the healthy state is complete, the gear system is tested for fracture detection. The RMS and kurtosis extracted features of faulty vibration signal are transferred to Minitab 19 software, then mean and standard deviation are computed. The next step is to apply the same upper control limits (UCL) and lower control limits (LCL) of the EWMA charts in healthy conditions. The reliability of the charts in respect of fault detection is evaluated based on the SPC rules. These are specialized rules connected with SPC charts and utilized to make predictions regarding abnormal shifts. These rules reflect the existence of any non-random patterns in the monitoring process, such as one point or more exceeding the control limits, as a sign alarm indication of the problem occurring in the process or system (Montgomery 2009).

Figure 8 shows the EWMA chart's performance based on the defective gear signal's RMS and kurtosis features. Figure 8a shows that the RMS values at samples 1 and 2 were outside the upper control limits(UCL), as in the healthy case. On the other hand, the observed samples' behavior appears to follow an abnormal pattern, as represented by an upward trend beginning with sample 25 and ending with samples 27–32 outside UCL. Thus, it is an indicator that detects the gear fracture condition.

In contrast, Fig. 8b shows the behavior of the observation kurtosis values in the EWMA chart, revealing an abnormal pattern through clustering to one side far from the central limit (CL). In addition, more than one point exists near the upper control limits.

However, no exit of the observed samples was observed outside the control limits (the control region), although the signal represents the condition of the defective gear. In this case, the EWMA performance suffers from a type II error (the inability to detect the gear fracture as it occurs).

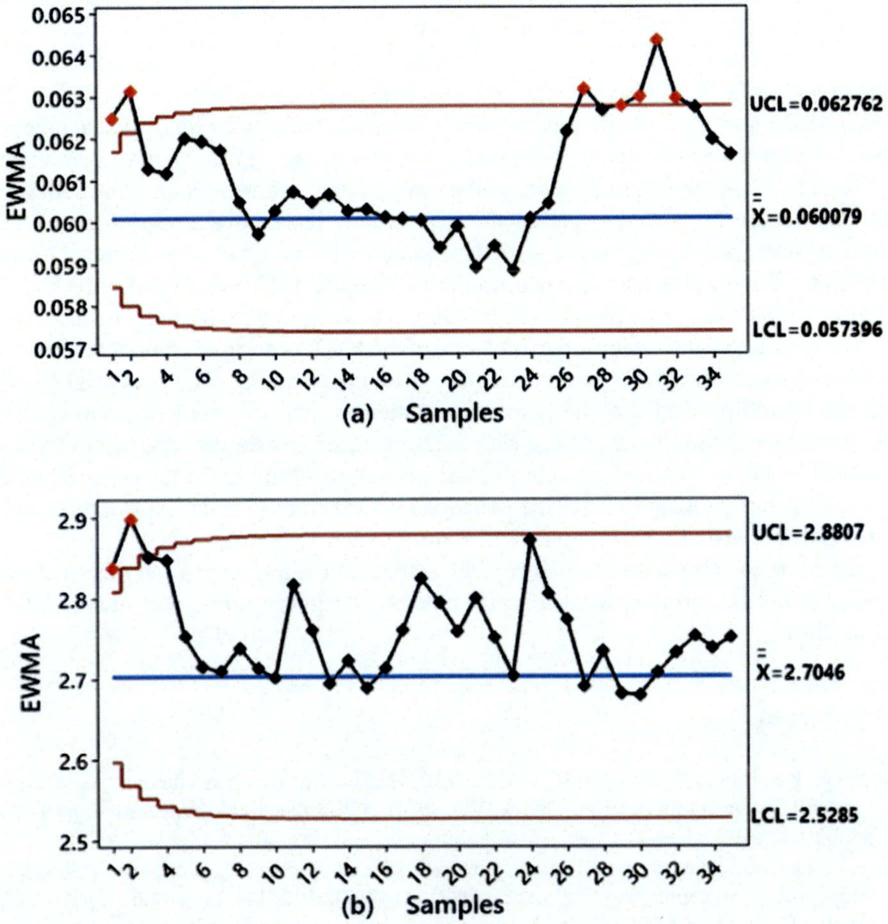

Fig. 8. Testing EWMA control charts of the spur gear system under faulty conditions based on:(a) Root Mean Square (RMS) features (b) Kurtosis features

According to these results, the EWMA chart established using RMS statistical features is an effective and successful tool for detecting minor fractures. Compared to the results presented by Lal and Kane 2019, using the EWMA chart for detecting tooth fracture at a level of 37.5, 75, and 100% From a practical point of view, detecting gear tooth fracture at these levels is not beneficial; the tooth may break due to heavy loads during operation.

7 Conclusions

The main aim of this study was to investigate the exponentially weighted moving average (EWMA) chart to detect a slight fracture resulting from a 5% mesh stiffness reduction. The EWMA control charts are introduced using statistical indicators extracted from time domain signal analysis, specifically, the root mean square (RMS) and kurtosis, as a univariate to design the EWMA control chart.

Raw simulated vibration signals are acquired from a dynamic model of the spur gear system with 8 DOF. Different levels of Gaussian white noise (5 dB, 10 dB, and 15 dB) were added to the original vibration signals to simulate their conditions in actual practice. The noise level was chosen at 15 dB since it provides the highest signal-to-noise ratio (SNR). Then, the statistical features of the time domain vibration signal for both RMS and kurtosis are calculated as fault-sensitive features. The control limits of the EWMA chart are designed based on the statistical features of the gear system under healthy conditions. Escorted to the same control limits designed for healthy conditions,EWMA scheme is tested based on the statistical indicators of the system in faulty conditions.

The results showed that the performance of the EMWA designed with RMS features responded more effectively to gear fracture detection than the EWMA designed with kurtosis. In addition, an abnormal ascending pattern is seen as a result of the monitoring statistic values gradually increasing with the occurrence of gear defects, particularly for the RMS-EWMA chart, which is higher than the control limits. On the other hand, the onset of fracture propagation and the variations in the behavior of the controlled samples were observed visually on the statistical control charts simultaneously.

Based on the above results, the EWMA statistical control chart is an important tool that can be used to monitor and detect different faults in gear systems and other industrial applications.

References

Chaari, F., Bartelmus, W., Zimroz, R., Fakhfakh, T., Haddar, M.: Gearbox vibration signal amplitude and frequency modulation. Shock. Vib. **19**(4), 635–652 (2012). https://doi.org/10.3233/SAV-2011-0656

Chen, Z., Shao, Y.: Dynamic simulation of spur gear with tooth root crack propagating along tooth width and crack depth. Eng. Fail. Anal. **18**(8), 2149–2164 (2011). https://doi.org/10.1016/J.ENGFAILANAL.2011.07.006

Chen, Z., Zhai, W., Wang, K.: Vibration feature evolution of locomotive with tooth root crack propagation of gear transmission system. Mech. Syst. Signal Process. **115**, 29–44 (2019). https://doi.org/10.1016/j.ymssp.2018.05.038

Montgomery, D.C.: Introduction to Statistical Quality Control, 6th edn. John Wiley & Sons Inc, New York, NY, USA (2009). 978-0-470-16992-6

Harrou, F., Ramahaleomiarantsoa, J.F., Nounou, M.N., Nounou, H.N.: A data-based technique for monitoring of wound rotor induction machines: a simulation study. Eng. Sci. Technol., An Int. J. **19**(3), 1424–1435 (2016). https://doi.org/10.1016/J.JESTCH.2016.04.008

Hynek, M., Smetanová, D., Stejskal, D., Zvárová, J.: Exponentially weighted moving average chart as a suitable tool for nuchal translucency quality review. Prenat. Diagn. **34**(4), 367–376 (2014). https://doi.org/10.1002/pd.431410

Jawad, S.M., Jaber, A.A.: Rolling bearing fault detection based on vibration signal analysis and cumulative sum control chart. FME Trans. **49**(3), 684–695 (2021). https://doi.org/10.5937/fme2103684M

Jiang, H., Liu, F., Wang, M.: Dynanic simulation of helical gears with crack propagating in to gear body. IOP Conf. Ser.: Earth Environ. **440**, 022108 (2020). https://doi.org/10.1088/1755-1315/440/2/022108

Li, C., Zhang, W., Peng, G., Liu, S.: Bearing fault diagnosis using fully-connected winner-take-all autoencoder. IEEE Access **6**(92), 6103–6115 (2017). https://doi.org/10.1109/ACCESS.2017.2717492

Maraş, S., Arslan, H., Birgören, B.: Detection of gear wear and faults in spur gear systems using statistical parameters and univariate statistical process control charts. Arab. J. Sci. Eng. **46**(12), 12221–12234 (2021). https://doi.org/10.1007/s13369-021-05930-y

Mohammed, O.D., Rantatalo, M., Aidanpää, J.O.: Dynamic modelling of a one-stage spur gear system and vibration-based tooth crack detection analysis. Mech. Syst. Signal Process. **54**, 293–305 (2015). https://doi.org/10.1016/j.ymssp.2014.09.001

Mohammed, S.A., Ghazaly, N.M., Abdo, J.: Fault diagnosis of crack on gearbox using vibration-based approaches. Symmetry **14**(2), 417 (2022). https://doi.org/10.3390/sym14020417

Lal, H., Kane, P.V.: Gearbox fault detection using exponentially weighted moving average control charts. In: Badodkar, D.N., Dwarakanath, T.A. (eds.) Machines, Mechanism and Robotics. LNME, pp. 39–47. Springer, Singapore (2019). https://doi.org/10.1007/978-981-10-8597-0_4

Selvamuthu, D., Das, D.: Introduction to statistical methods, design of experiments and statistical quality control. In: Introduction to Statistical Methods, Design of Experiments and Statistical Quality Control (2018). https://doi.org/10.1007/978-981-13-1736-1

Shi, J.-F., Gou, X.-F., Zhu, L.-Y.: Generation mechanism and evolution of five-state meshing behavior of a spur gear system considering gear-tooth time-varying contact characteristics. Nonlinear Dyn. **106**(3), 2035–2060 (2021). https://doi.org/10.1007/s11071-021-06891-5

Yang, H.H., Huang, M.L., Lai, C.M., Jin, J.R.: An approach combining data mining and control charts-based model for fault detection in wind turbines. Renew. Energy **115**, 808–816 (2018). https://doi.org/10.1016/j.renene.2017.09.003

Yang, L., Wang, L., Yu, W., Shao, Y.: Investigation of tooth crack opening state on time varying meshing stiffness and dynamic response of spur gear pair. Eng. Failure Anal. **121**, 105181 (2021). https://doi.org/10.1016/j.engfailanal.2020.105181

Zhang, Y., Tang, B., Liu, Z., Chen, R.: An adaptive demodulation approach for bearing fault detection based on adaptive wavelet filtering and spectral subtraction. Meas. Sci. Technol. **27**(2), 25001 (2015). https://doi.org/10.1088/0957-0233/27/2/025001

Zhao, M., Lin, J., Miao, Y., Xu, X.: Detection and recovery of fault impulses via improved harmonic product spectrum and its application in defect size estimation of train bearings. Measurement **91**, 421–439 (2016). https://doi.org/10.1016/j.measurement.2016.05.068

Intelligent Diagnosis of Gear Transmission Systems of Robots Based on a Digital Model

Anis Frej[1,2(✉)], Fakher Chaari[1], Xavier Chiementin[2], Fabrice Bolaers[2], and Mohamed Haddar[1]

[1] Laboratory of Mechanics, Modeling and Production (LA2MP), National School of Engineers of Sfax, BP1173, 3038 Sfax, Tunisia
anis.fredj@enis.tn, mohamed.haddar@enis.rnu.tn

[2] Institute of Thermics, Mechanics and Material (ITHEMM), University of Reims, Moulin de la Housse, 51687 Reims Cedex 2, France
{xavier.chiementin,fabrice.bolaers}@univ-reims.fr

Abstract. In a highly competitive industrial context, the maintenance field has relied in recent years on various tools such as artificial intelligence algorithms (AI algorithms), in order to achieve its main objective, which is the continuous availability of machines. Previous applications of these algorithms in machine monitoring were mainly relying on historical operating datasets, which restricted the reliability and accuracy of diagnosis. This is mainly because the collected databases that will be exploited in AI algorithms are often difficult to obtain, especially for complex machines like robots. For this reason, this chapter proposes a new approach for monitoring of gear systems inside robots using AI algorithms, but which avoids the requirements of experimental data and replaces them with datasets generated by a numerical model. In order to realize this approach, a robot model able to simulate the joints vibration behavior without and with the presence of gears defects, is developed. Then, from the numerical simulations, fault indicators well known in the literature of gear systems diagnosis, are extracted. Finally, three common classifiers, SVM (multiple kernel support vector machine), DT (decision trees) and KNN (k-nearest neighbor algorithm) are driven by the most relevant simulated features and subsequently validated by experimental datasets.

Keywords: Robot monitoring · Predictive maintenance · Machine learning · Robot dynamic modelling · Gears defect

1 Introduction

For several years, manufacturers have been tending to the massive use of robots in order to improve their productivity. This comes from the ability of robots to work continuously without interruption, unless due to imperative reasons, such as uncontrolled failures. For this reason, these latter can be considered a real anxiety for managers, given the enormous financial losses that may be incurred due to the unscheduled shutdowns that can occur. One of the sources of failures in robots are the articulations or more precisely the gear transmission systems, since they represent the most exposed components to mechanical

A. Akrout et al. (Eds.): ICAV 2022, ACM 22, pp. 462–473, 2023.
https://doi.org/10.1007/978-3-031-34190-8_49

loads and solicitations. Therefore, the implementation of a monitoring strategy of these elements can be considered as essential and unavoidable because it will allow to provide quick interventions at the appropriate time and under the best conditions.

In this context, the efforts deployed by researchers and engineers over the last decades to develop and improve the AI algorithms, have led to their intensive use in the maintenance field as well as to the emergence of many monitoring systems. Indeed, the effectiveness of these tools in robot diagnosis has been proven in many previous studies. Pan et al. [1], developed a methodology to detect the backlash fault that can occur in industrial robot joints. This study is designed based on an ANN (Artificial Neural Network) algorithm, which has been used to classify the relevant features extracted from the vibration signals. On the other hand, Datta et al. [2], also used the ANN algorithm as a classifier driven by characteristics of torque signals applied by the joint actuators in order to detect failures in industrial robots. In addition, Ikbal et al. [3], proposed a strategy that relies on the analysis of the articulation vibration state of an industrial welding robot, to predict the maintenance period with a RBNN (Radial-Based artificial Neural Network) algorithm. Whereas Hsu et al. [4], presented an intelligent system to diagnose the health status of six-axis robots based on a multi-class SVM algorithm to classify faults from the temporal and frequency features of signals acquired from encoders and current sensors of the motors, as well as an external accelerometer. Similarly, Long et al. [5], implemented a methodology to analyze defects in the mechanical transmissions of a multi-articulated robot, based on the SVM algorithm that has been driven by a set of data collected by an attitude sensor installed on the last robot joint. Elangovan et al. [6], also employed the SVM algorithm to obtain a system that could detect autonomously defects of the bio-inspired reconfigurable robot through data collected by an inertial measurement unit. In these previous works, the signals used by the AI algorithms are usually robot operation histories and experimental data, which limits the credibility of the diagnostics for cases with a narrow database, due to the difficulty of collecting histories of all the operating states or rather all the failures that may occur in the robot. In addition, having a rich database can be expensive, due to the specific sensors and equipment that may be needed to acquire the data or even the acquisition conditions in some cases.

Therefore, to address this problem, in this chapter a new strategy is proposed for the surveillance of gear systems in robots based on a dynamic model and AI algorithms. This work is structured as follows: the process of the diagnostic method is presented in Sect. 2. In Sect. 3, the dynamic model as well as the experimental configuration and the obtained results with the new approach are described.

2 Methodology of the Proposed Monitoring Approach

In this work, the vibration analysis technique has been chosen due to its effectiveness in diagnosis, the simplicity of its implementation, and the consensus on its reliability in the field of monitoring, especially of gear systems. In addition, the proposed approach in this chapter for the monitoring of these components in robots has been based on supervised or automatic algorithms that consist of the design of a model able to extract links between the data and their classes. These artificial intelligence methods usually proceed in three phases: (1) processing and extraction of relevant features from the acquired data, (2) learning, and (3) classification.

2.1 Processing and Extraction Phase

In the first phase, a study based on the calculation of statistical indicators of the vibration signals has been carried out, in order to determine the healthy and defective conditions as well as to describe the fault trend. The choice of the statistical parameters analysis can be explained by the high efficiency of this approach in the detection of an eventual failure. The choice of statistical condition indicators is not done in a standard way and is closely related to the objective of the diagnosis. In fact, some indicators are more sensitive to the presence of a fault and allow its detection in the best time. While others are more sensitive to variations of operating conditions than to the presence of defects. In the present study, only the five temporal indicators: RMS, Kurtosis, Crest factor, STD and Peak, have been considered (see Table 1), given their wide use in the diagnosis of vibration signals in rotating machines and their better response to constraints [7].

Table 1. Extracted statistical features

Indicator	Expression
Root Mean Square (RMS)	$x_{rms} = \sqrt{\frac{\sum_{i=1}^{N} x_i^2}{N}}$
Crest factor (CF)	$x_{Fc} = \frac{x_{max}}{x_{rms}}$
Kurtosis (Ku)	$x_{Ku} = \frac{1}{N.\sigma^4} \sum_{i=1}^{N} (x_i - \overline{x})^4$
Standard deviation (STD)	$x_\sigma = \sqrt{\frac{1}{N-1} \sum_{i=1}^{N} (x_i - \overline{x})^2}$
Peak	$x_{max} = \max(\lvert x_i \rvert)$

where:

- x_i is the amplitude of the sampled signal and N is the number of samples.
- $\overline{x}$ is the average value of the signal.

The increase of the number of employed statistical indicators can contribute to improve the performance of the classification algorithms. However, such parameters may contain unnecessary information that can increase the process and diagnostic delays as well as falsify the results. Thus, in order to solve these problems, it is essential to proceed to the second part of this first phase, which consists of reducing the number of these parameters and selecting those that are most relevant and most sensitive to faults. In the literature, the extraction of the most pertinent indicators can be performed using two types of approaches. Firstly, the projection techniques such as PCA (Principal Component Analysis) [8] and Kernel-PCA [9], which allow the projection of all the indicators into a reduced-dimensional space and to create linear or non-linear combinations, in order to maximize the variance between them. Secondly, the selection or optimization methods that aim to identify the most informative indicators among the available parameters [10]. In this work, the SBS (Sequential Backward Selection) method, which belongs to this last category, has been used due to its ability to discern the data corresponding to the different conditions, preserving them as much as possible and without losing the physical signification of each feature. Therefore, among the previously selected five statistical indicators, only the three relevant ones will be retained by the SBS, in order to represent each vibratory signal by a point in a three-dimensional space. SBS is an iterative selection approach that relies on a selection criterion J to remove the least pertinent features. The computation of this criterion consists of calculating two types of dispersion: inter-class, D_{inter}, and intra-class, D_{intra}, from the set of indicators through the following equation [11]:

$$J = D_{inter}/D_{intra} \tag{1}$$

$$D_{intra} = \sum_{i=1}^{M}\sum_{j=1}^{n}(x_{ij} - g_i)\cdot(x_{ij} - g_i)^T \qquad D_{inter} = \sum_{i=1}^{M}(g_i - g)\cdot(g_i - g)^T$$

where:

- x_{ij} is j^{th} that contains the indicators of the ith class.
- g_i and g are respectively the center of gravity of the ith class and that of the set of classes.

2.2 Learning and Classification Phases

In the second phase of the operation process of an artificial intelligence algorithm, which is the training stage, the most relevant characteristics retained by the selection method (SBS) will be used to form the classifier that will be used in the third phase of classification. This operation is carried out by means of a database, of which the classes are known in advance and a label will be assigned to each of them. On this level, the numerical model interferes in order to avoid the disadvantage of the classical approach in this step which consists in the dependence of a huge experimental database representing all the damages (historical data) (see Fig. 1-(a)). Thus, as proposed by Farhat et al. [12], the role of the numerical model in the proposed monitoring strategy will be the training of AI algorithms by simulated signals representing the different operating states, in order to optimize time and cost (see Fig. 1-(b)).

3 Implementation of the Monitoring Approach

3.1 Experimental Configuration

In this work, the robot FANUC LR Mate 200iC (see Fig. 2-(a)) that exists at the University of Reims Champagne-Ardenne has been studied, and more precisely the gear defects that can occur in its fourth joint has been chosen to be studied. Thus, experimental tests

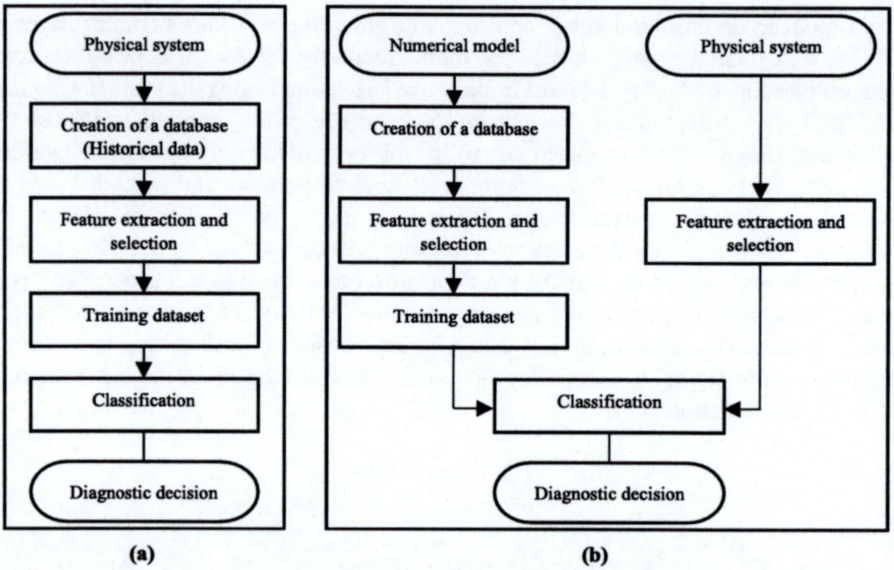

Fig. 1. The operation of an artificial intelligence algorithm based on (a) experimental data (b) data generated by a numerical model

that consist in the measurement of vibrations existing in this joint have been conducted with two triaxial piezoelectric accelerometers which have been placed on the structure of the joint (see Fig. 2-(b)). Finally, the acquisition of the signals of these sensors has been made through a data collector of the type OROS36 with a sampling frequency fixed at 51.2 kHz. Besides, the data has been recorded during a continuous rotation cycle performed by the articulation between two different positions.

The transmission chain of this joint is composed of a motor, which drives a single stage spur gear reducer, connected at its output to another special reducer called Harmonic Drive. In addition, in this chapter, the two localized faults: tooth crack and total tooth rupture, which can be classified as the most common gear defects, have been chosen to be considered. For this reason, in order to carry out the experimental tests, these two types of defects have been artificially created in the pinion of the first reducer of this joint. For this purpose, two gears with the same mechanical dimensions have been manufactured (see Fig. 3), the first one with a removal of part of a tooth while the second one with a complete extraction of a tooth.

3.2 Robot Dynamic Modelling

This section focuses on the presentation of the FANUC robot numerical model that has been used to generate the data needed to verify the new monitoring approach for the robot's gear transmission systems. In order to obtain the dynamic model of the FANUC robot, in other words to recover its equations of motion, the Euler Lagrange formalism has been used. This methodology consists in the determination of the total kinetic and potential energies of the different robot components. Knowing that, in this model, the

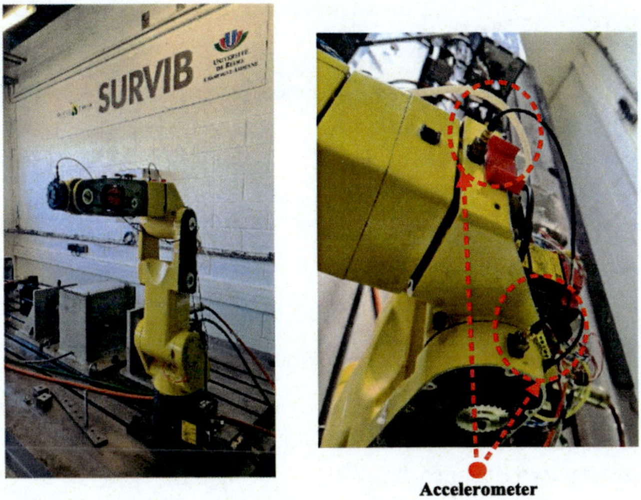

(a) (b)

Fig. 2. (a) The Fanuc robot, (b) The fourth articulation of the robot

(a) (b) (c)

Fig. 3. The experimental pinions in the: (a) healthy case, (b) case with a tooth crack defect, and (c) case with a tooth rupture defect

bodies and joints have been considered as rigid and only the flexibility at the fourth articulation has been taken into account. Moreover, in this model, a rotational DOF (degree of freedom) "q_i" is assigned for each joint of the robot. But, in contrast to the other joints, the flexible one has been represented by three nodes N_{m4}, N_{g4} and N_{l4} (see Fig. 4), which have four additional translational DOFs x_{m4}, y_{m4}, x_{g4} and y_{g4}, as well as, two additional rotational DOFs q_{m4} and q_{g4}.

Thus, as shown in Fig. 4, the elasticity of this fourth joint is modeled by a new approach that relies mainly on four springs. A linear one "k_{m4}", which represents the stiffness of the meshing between the pinion and the wheel of the first simple stage reducer, and a torsion spring "k_4", which models the elasticity of the connection between the fourth actuated body and the output shaft of the first reducer. Besides, the gears of

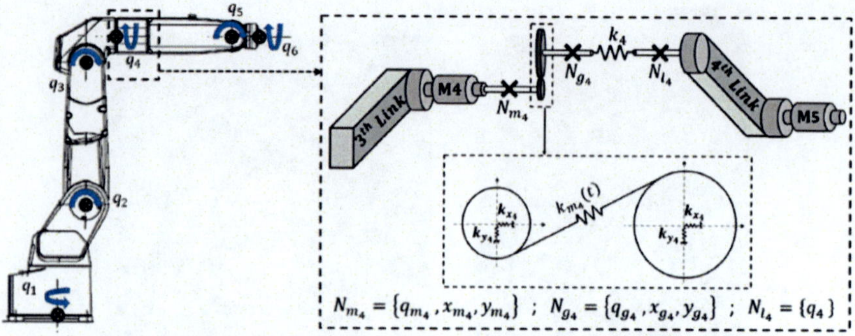

Fig. 4. Representative scheme of the robot's fourth flexible joint

the single stage reducer are considered as rigid and the flexibility of their transmission axes is modelized by two linear springs "k_{x4}" and "k_{y4}" for each shaft.

With this new approach, the mechanical effects related to gear systems, such as the excitation of the internal meshing between the teeth of the gears, and the different types of defects that can occur in these systems, have been incorporated into the robot dynamic model, and therefore the behavior with and without gear defects has been simulated. Referring to the research of Chaari et al. [13], in the healthy case, the meshing stiffness "k_{m4}" can be modeled by a square signal with an amplitude varying between a minimum and a maximum value according to the number of pairs of teeth in contact, as illustrated in the Fig. 5. Furthermore, the tooth crack failure may be modeled by a periodic reduction of the amplitude of this stiffness at each mesh with the damaged tooth.

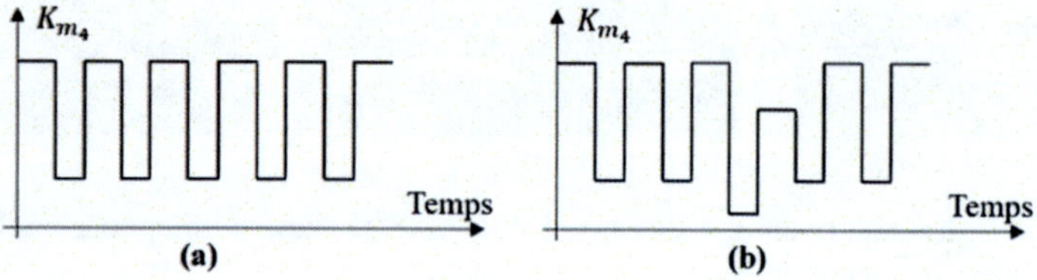

Fig. 5. The mesh stiffness in the case of: (a) Healthy teeth, (b) Tooth crack defect with a 30% stiffness reduction rate.

As mentioned previously, according to the Euler Lagrange formalism, the global robot equation of motion has been obtained in the following form:

$$[\mathcal{M}] \cdot \ddot{\Theta} + [D] \cdot \dot{\Theta} + [C] + [G] = \Gamma \tag{2}$$

where:

- $[\mathcal{M}]$ and $[D]$ are respectively the inertia matrix and the global dissipation constant of the robot.
- $[C]$ and $[G]$ are respectively the vector containing the terms that represent the centrifugal and Coriolis forces and the vector of the gravitational and elastic forces.

– Θ and Γ are respectively the vector of generalized coordinates which is defined by the DOFs of the robot and the vector of torques applied by the servomotors.

Then, the displacement "Θ", velocity "$\dot{\Theta}$", and acceleration "$\ddot{\Theta}$" are determined by solving the above equation of motion using MATLAB's ode45 function, which is based on an explicit fourth and fifth order Runge-Kutta formula.

3.3 Results and Discussion

The aim of this section is to investigate the performance of the classification approach based on a numerical model presented in Sect. 2, for the diagnosis of gear failures which can occur in the articulations of a robot. Thus, in this investigation a dataset that contains the vibration signals acquired with three different severities of gear defects (healthy, crack and rupture), has been developed. Knowing that each experimental test has been repeated three times under the same conditions. Moreover, this database is characterized by a mixing of the signals acquired by the two accelerometers in the two directions "x" and "y". In addition, in order to have a rich database, each signal has been divided into two parts, which allowed us to finally obtain 72 vibration signals with 150000 samples each one. Through the model presented in the previous subsection, the same experimental operating modes have been simulated and a numerical database has been constructed containing precisely the accelerations of the fourth joint of the FANUC robot along the "x" and "y" directions in the following cases: healthy, with a cracked tooth and with a broken tooth. Considering that the rupture tooth defect has been numerically implemented as a cracked tooth defect with a reduction rate of 100% of the mesh stiffness.

Therefore, in the first phase, the SBS has been applied to the indicators calculated from the experimental signals, since they include many random phenomena, unlike the digital signals which can be regarded as perfect. From this selection step, the RMS, Kurtosis and Crest factor have been selected as the most relevant indicators among the five already chosen at the beginning, given the fact that having the highest values of the selection criteria (see Table 2), and therefore it is the features that provide the best dispersion between the different states.

Table 2. The quality criterion calculated by the SBS

Indicator	Quality criterion "J"
RMS	0.1257
Peak	0.1077
STD	0.1207
Ku	0.1208
CF	0.1306

Finally, before proceeding with the test of the model-based approach, a verification of the conventional monitoring method has been carried out in order to confirm the quality of the acquired experimental data and to compare then its results with those of the new diagnostic approach. For this purpose, with the use of the MATLAB toolbox "Classification Learner", 75% of the experimental database has been randomly selected in each class, in order to train several classifiers, which have then been tested with the rest of the data (75%). In this context, the obtained results are shown that the three best classifiers are: SVM, KNN and Decision Trees (DT). For evaluating the performance of these classifiers, the confusion matrices (see Table 3) in which each row represents a real class and each column corresponds to an estimated class, as well as the classification accuracy, have been determined. Regarding the calculation of the classification precision, a counting of the labelled data such as true positive, true negative, false positive, and false negative, has been done.

Table 3. The confusion matrix for the conventional surveillance approach based on experimental dataset with the classifier (a) SVM, (b) Decision Trees, (c) KNN

(a), Accuracy = 94.4%				
		Predicted Class		
		Healthy	*Crack*	*Rupture*
True Class	*Healthy*	6	0	0
	Crack	1	5	0
	Rupture	0	0	6

(b), Accuracy = 94.4%				
		Predicted Class		
		Healthy	*Crack*	*Rupture*
True Class	*Healthy*	5	1	0
	Crack	0	6	0
	Rupture	0	0	6

(c), Accuracy = 94.4%				
		Predicted Class		
		Healthy	*Crack*	*Rupture*
True Class	*Healthy*	6	0	0
	Crack	1	5	0
	Rupture	0	0	6

In addition, in order to demonstrate the ability and efficiency of the model-based method in the diagnosis of teeth crack and rupture defects that may occur in the joint, the three previous classifiers have been trained in this case with strictly numerical data. As for the test phase, this time the experimental data set that was adopted previously for training in the conventional approach (75% of the experimental data), is used. Thus, the results of this investigation are shown by the confusion matrices and the accuracies of the three classifiers in the following table:

The results obtained with the conventional approach have demonstrated the reliability of the diagnosis with an accuracy of 94.4% for the three classifiers (see Table 3). On the other hand, the results achieved with the use of the simulated data by the model (see Table 4), reveal a degradation of the precision rates of the different classifiers, namely an accuracy of 40.7% for SVM, 66.7% for KNN and finally 70.4% for DT. This precision

Table 4. The confusion matrix for the novel monitoring approach based on the robot dynamic model with the classifier (a) SVM, (b) Decision Trees, (c) KNN

(a), Accuracy = 40.7%

		Predicted Class		
		Healthy	*Crack*	*Rupture*
True Class	*Healthy*	1	17	0
	Crack	6	12	0
	Rupture	9	0	9

(b), Accuracy = 70.4%

		Predicted Class		
		Healthy	*Crack*	*Rupture*
True Class	*Healthy*	18	0	0
	Crack	11	7	0
	Rupture	0	5	13

(c), Accuracy = 66.7%

		Predicted Class		
		Healthy	*Crack*	*Rupture*
True Class	*Healthy*	18	0	0
	Crack	18	0	0
	Rupture	0	0	18

decay can be explained by the lack of recalibration of the model and the simplifications taken into account during the modelling, as well as by the various sources of non-linearity that have not been taken into account. However, it does not prevent the ability of the model to simulate data close to reality and to detect the difference between the different conditions. Besides, thanks to the confusion matrices (see Table 3 and Table 4), it can be seen that the high error percentages in the classification with numerical data comes from the confusion between the healthy and the crack defect cases. This confusion can be explained by the convergence between the results of the two situations (healthy and with a crack defect), which has been observed also when strictly experimental data have been employed. Thus, these results have shown that the new proposed approach reliably diagnosed gear defects in the joints without any previous knowledge of the experimental data.

4 Conclusion

In this chapter, the phases necessary to have an intelligent methodology able to detect the gear system failures that can occur in the robot joints, are described. Firstly, the operation process of this methodology is presented, noting that it is based on AI algorithms that are driven by numerical data generated by a dynamic model of the robot, unlike existing approaches. Secondly, this chapter introduces the dynamic model that has been adopted to generate the numerical data set in different joint gear system conditions (with and without defects), as well as the experimental configuration that has been used to create an experimental database to test the classification algorithms. Finally, some results of a physical robot joint monitoring, through a conventional and a model-based method, have been exposed. These results have proven the effectiveness of the proposed model-based approach in the detection of different types of gear defects without any prior knowledge of the experimental data.

In a future work, and in order to reduce the accuracy decrease observed in the obtained results by the model-based approach, many improvements will be performed such as model recalibration. The purpose of these enhancements can minimize the differences between the relevant characteristics of the measured and simulated signals, and thus have a perfect correlation between the two modes. In addition, this surveillance technique will be used to monitor other components inside the joints such as bearings.

References

1. Pan, M.-C., Van Brussel, H., Sas, P.: Intelligent joint fault diagnosis of industrial robots. Mech. Syst. Signal Process. **12**(4), 571–588 (1998). https://doi.org/10.1006/mssp.1997.0124
2. Datta, A., Mavroidis, C., Krishnasamy, J., Hosek, M.: Neural network based fault diagnostics of industrial robots using wavelt multi-resolution analysis. In: 2007 American Control Conference, pp. 1858–1863 (2007). https://doi.org/10.1109/ACC.2007.4283012
3. Eski, I., Erkaya, S., Savas, S., Yildirim, S.: Fault detection on robot manipulators using artificial neural networks. Robot. Comput.-Integr. Manuf. **27**(1), 115–123 (2011). https://doi.org/10.1016/j.rcim.2010.06.017
4. Hsu, H.-K., Ting, H.-Y., Huang, M.-B.,Huang, H.-P.: Intelligent fault detection, diagnosis and health evaluation for industrial robots. Mechanics **27**(1), Art. no 1 (2021). https://doi.org/10.5755/j02.mech.24401
5. Long, J., Mou, J., Zhang, L., Zhang, S., Li, C.: Attitude data-based deep hybrid learning architecture for intelligent fault diagnosis of multi-joint industrial robots. J. Manuf. Syst. **61**, 736–745 (2021). https://doi.org/10.1016/j.jmsy.2020.08.010
6. Elangovan, K., Krishnasamy Tamilselvam, Y., Mohan, R.E., Iwase, M., Takuma, N., Wood, K.L.: Fault diagnosis of a reconfigurable crawling–rolling robot based on support vector machines. Appl. Sci. **7**(10), Art. no 10 (2017). https://doi.org/10.3390/app7101025
7. Mohammed, S.A., Ghazaly, N.M., Abdo, J.: Fault diagnosis of crack on gearbox using vibration-based approaches. Symmetry **14**(2), Art. no 2 (2022). https://doi.org/10.3390/sym14020417
8. Vidal, R., Ma, Y., Sastry, S.S.: Principal component analysis. In: Vidal, R., Ma, Y., Sastry, S.S. (eds.) Generalized Principal Component Analysis, pp. 25–62. Springer, New York (2016).https://doi.org/10.1007/978-0-387-87811-9_2
9. Schölkopf, B., Smola, A., Müller, K.-R.: Kernel principal component analysis. In: Gerstner, W., Germond, A., Hasler, M., Nicoud, J.D. (eds.) Artificial Neural Networks - ICANN 1997, pp. 583–588. Springer, Heidelberg (1997). https://doi.org/10.1007/BFb0020217
10. Janecek, A., Gansterer, W., Demel, M., Ecker, G.: On the relationship between feature selection and classification accuracy. In: Proceedings of the Workshop on New Challenges for Feature Selection in Data Mining and Knowledge Discovery at ECML/PKDD 2008, pp. 90–105 (2008)
11. Frini, M., Soualhi, A., El Badaoui, M.: Gear faults diagnosis based on the geometric indicators of electrical signals in three-phase induction motors. Mech. Mach. Theory **138**, 1–15 (2019). https://doi.org/10.1016/j.mechmachtheory.2019.03.030

12. Habib Farhat, M.: SURveillance VIBratoire pour l'industrie 4.0: Classification par jumeau numérique de machines tournantes (SURVIB 4.0). These de doctorat, Université de Reims Champagne-Ardenne (2021)
13. Chaari, F., Fakhfakh, T., Haddar, M.: Simulation numérique du comportement dynamique d'une transmission par engrenages en présence de défauts de dentures. Mech. Ind. **6**(6), 625–633 (2005). https://doi.org/10.1051/meca:2006008

Author Index

A. Akrout et al. (Eds.): ICAV 2022, ACM 22, pp. 475–476, 2023.
https://doi.org/10.1007/978-3-031-34190-8